Quantum Fields and Solids-1983
(Sanibel Island, Florida)

Sanibel Island, Florida
April 11 – 15 , 1983

AIP Conference Proceedings
Series Editor: Hugh C. Wolfe
Number 103

Quantum Fluids and Solids-1983
(Sanibel Island, Florida)

Edited by
E. D. Adams and G. G. Ihas
University of Florida

American Institute of Physics
New York   1983

L.C. Catalog Card No. 83-72440
ISBN 0-88318-202-5
DOE CONF- 830456

Symposium on Quantum Fluids and Solids

Sanibel Island, Florida

April 11-15, 1983

Sponsored by

Division of Materials Research
National Science Foundation

and

Division of Sponsored Research
University of Florida

Additional support by

Department of Physics
University of Florida

Organizing Committee

E.D. Adams	N.S. Sullivan
G.G. Ihas	D.B. Tanner
P. Kumar	S.B. Trickey

INDUSTRIAL SPONSORS

Air Products & Chemicals, Inc., Allentown, Pennsylvania

American Magnetics, Inc., Oak Ridge, Tennessee

EG&G Princeton Applied Research, Princeton, New Jersey

Harra Technical Sales, Clearwater, Florida

Lake Shore Cryotronics, Inc., Westerville, Ohio

Linseis, Princeton Junction, New Jersey

Oxford Instruments, Inc., Burlington, Massachusetts

Scientific Instruments, Inc., West Palm Beach, Florida

S.H.E. Corporation, San Diego, California

ADVISORY COMMITTEE

G. Ahlers, Santa Barbara

V. Ambegaokar, Cornell

A.J. Berlinksy, British Columbia

J. Clarke, Berkeley

S. Doniach, Sanford

G. Frossati, Leiden

A. Goldman, Minnesota

R. Guyer, Massachusetts

H. Hall, Manchester

A. Leggett, Sussex

K. Maki, Los Angeles-USC

Y. Masuda, Nagoya

H. Meyer, Duke

P. Nozières, Grenoble

F. Pobell, Jülich

R. Richardson, Cornell

J. Serene, Yale

I. Silvera, Harvard

H. Smith, Copenhagen

E. Varoquaux, Orsay

D. Vollhardt, Munich

Preface

These proceedings are a collection of papers presented at the International Symposium on Quantum Fluids and Solids held at Sanibel Island, Florida, April 11-15, 1983. In addition to the invited papers, 30 papers were presented in poster sessions. Abstracts of the poster papers have been included at the discretion of the authors.

This conference was organized to satisfy requests from colleagues for "another Sanibel Conference" on Quantum Fluids and Solids. (There had been two previous conferences in this field at Sanibel, in 1975 and 1977.) The timing and subject matter of the conference were selected bearing in mind the various other conferences in the field, such as the 1982 Gordon Conference and LT-16 and LT-17. The enthusiastic support by the quantum fluids and solids community confirmed the need for the conference and justified the effort put forth in organizing it. Suggestions for the program were provided by an Advisory Committee, whose members are listed at the beginning of this volume.

The conference would not have been possible without financial support from several sources. For their role in arranging this support, the organizers wish to thank Dr. Donald Liebenberg of the National Science Foundation, Dr. Gene Hemp of the Division of Sponsored Research, University of Florida, Dr. Charles Hooper, Chairman Department of Physics, University of Florida, and several industrial sponsors (listed at the beginning of this volume).

Thanks are extended to a number of people who contributed to the success of the conference. Other members of the organizing committee were Neil Sullivan, David Tanner, and Samuel Trickey. The experience, advice, and effort of Sam Trickey were particularly valuable and Neil Sullivan provided a useful "European connection". We are especially grateful to Sheri Hill, conference secretary, for her able performance of numerous tasks in connection with all phases of the conference. Also we thank Diana Tonnessen for assistance with several of these duties. The leg work during the conference was performed by the "Gator Gofers": Bobby Berg, Brad Engel, Ying Lin, Lie Liu, Simon Phillpot, Vijay Samalam, Greg Spencer, Deepak Srivastava, Ralph Rosenbaum, and Kurt Uhlig. The many thankless tasks which they performed were invaluable and appreciated.

There are two major omissions from this volume: the banquet speeches by Professor Per Löwdin on the history of Sanibel Symposia and by Professor John Wheatley on the milestones of low temperature physics in the last thirty years, a reminder that proceedings are a poor substitute for active conference participation.

We thank the speakers for their cooperation in preparing the photo-ready manuscripts and the participants and others who willingly refereed the manuscripts. The many new results presented and the rapid publication possible with the format chosen for the proceedings should result in a current account of some of the most active areas of research in 1983.

E. Dwight Adams
Gary G. Ihas

Gainesville, Florida
June 20, 1983

Table of Contents

Part I. Quantum Solids

PRECISION MEASUREMENTS OF SOLID ^{3}He MAGNETIC SUSCEPTIBILITY

Z. Olejniczak, W. P. Kirk, A. A. V. Gibson, P. Kobiela
and A. Czermak
Department of Physics, Texas A&M University,
College Station, Texas 77843-4242

ABSTRACT

High-precision magnetic susceptibility measurements on solid ^{3}He samples ranging in molar volumes from 21.0 to 24.0 mL/mole have been made from $\approx$16 mK to 510 mK in a low magnetic field (17.1 mT). New values of the Curie-Weiss temperature θ have been determined and are found to be nearly a factor of two smaller than the values from all previous susceptibility measurements. These new values of θ, however, are in good agreement with P(T,H) measurements which leads to satisfactory thermodynamic consistency between these two types of experiments. Description of a double-sample-cell measuring technique and the use of Fourier-Transform pulsed NMR spectroscopy are given.

INTRODUCTION

In recent years, interest in the magnetic properties of solid ^{3}He has focused mainly on the very low temperature range near the phase transition. These very low temperature studies have provided much new information. In addition, this work has also made it clear that understanding the magnetic behavior at much higher temperatures, where the high-temperature series expansions of the spin-exchange Hamiltonians can be utilized, is of great importance for a consistent theoretical picture.

Over a period of more than two decades, a large number of thermodynamic and NMR (nuclear magnetic resonance) experiments have been performed on solid ^{3}He at temperatures much greater than the transition temperature, Tc. From these measurements a wide range of important parameters related to magnetic effects have been determined and studied with much interest. By 1978, however, it was evident from the review work of Guyer[1] that substantial inconsistencies existed in the high temperature data base and that a consistent theoretical picture was not possible.

Specifically, high-field pressure measurements P(T,H) and magnetic susceptibility measurements χ(T) displayed quite different results when, for example, the Curie-Weiss temperature θ was used to compare the experiments. In the case of the P(T,H) experiments, the molar volume derivatives dθ/dV were measured and the θ values determined indirectly by an integration procedure. Whereas, in the χ(T) experiments the values of θ were measured directly from the temperature dependent behavior of χ(T).

Not many P(T,H) experiments have been reported, but the recent high precision P(T,H) measurements by Van Degrift, Pipes, McQueeny, and Bowers at NBS[2] substantially corroborate the results of the early P(T,H) experiment by Kirk and Adams at Florida[3]. On the other hand, there have been many $\chi(T)$ measurements made on solid ^{3}He[4-12]. All of the $\chi(T)$ experiments tend to fall into one of two categories. In one category, the measurements[4-8] have been made at high temperatures ($\approx$10 mK to 500 mK) where the deviation from Curie law behavior is too small to obtain reliable values of θ without doing high precision measurements. It should be noted that these early high temperature $\chi(T)$ measurements were done primarily for the purpose of establishing the sign of the exchange interaction or the type of magnetic ordering the solid would undergo at the phase transition. The other category of measurements[9-12] have all been made at very low temperatures near Tc ($\approx$1 mK to 15 mK). Unfortunately these measurements encounter significant higher order deviations from the Curie-Weiss law. As a result, the data can be easily misinterpreted when determining the value of the Curie-Weiss θ because the measurements were not carried to high enough temperatures for the high-T series expansions to apply. In summary, the value of θ as determined from $\chi(T)$ measurements have not been accurately determined.

In this paper, we will report on a high-precision magnetic susceptibility measurement of solid ^{3}He that was done at high temperatures ($\approx$16 mK to 510 mK) where the data can be compared reliably with the high-temperature series expansions of the spin exchange Hamiltonians. In addition several molar volume samples (ranging from 21.0 mL/mole to 24.0 mL/mole) were studied. A small magnetic field (17.1 mT) was used so that the zero-field limit of the magnetic susceptibility would apply. Briefly we find the Curie-Weiss temperatures to be substantially smaller than those determined by any of the previous $\chi(T)$ measurements. Our new values, however, are in very good agreement with the values of θ determined from P(T,H) experiments.

TECHNIQUE

In order to obtain more reliable susceptibility measurements, we used a sample chamber with two identical cells. Each cell was filled independently with ^{3}He. The cells were connected thermally by a bundle of fine copper wires through epoxy seals. The wires were tightly packed to provide a large surface area for thermal contact with the ^{3}He. One cell was used to hold a high density solid ^{3}He sample, which closely followed Curie law behavior because of the very small exchange energy. In effect this high density sample served as an NMR thermometer to measure the susceptibility of a lower density solid ^{3}He sample, which was formed in the other cell. This method helped to significantly overcome problems associated with achieving thermal equilibrium between thermometer and sample, with possible thermal gradients, and with minimizing thermometer calibration errors.

A pulsed NMR technique was used to measure the magnetic susceptibility $\chi(T)$ of solid ^{3}He based on the fact that in a low magnetic field $\chi(T)$ is proportional to the temperature dependent dc magnetization M_0. The M_0, as produced by the nuclear spin system in a constant magnetic field H_0, was tipped a small angle away from equilibrium by a second time dependent magnetic field H_1 (generated by external radio-frequency pulse) that was perpendicular to H_0. The precession of M_0 about H_0 induced an emf, as a free induction decay (FID), in a suitably placed coil. The initial free precession amplitude is proportional to M_0 as it existed just prior to the application of the rf pulse. M_0 cannot be measured directly following the rf pulse because of the finite recovery time of the coil and receiver system. Consequently, a Fourier transform of M(t) is done and then the area under the spectrum is calculated. The area is proportional to M_0 and hence to $\chi(T)$. This technique allowed us to measure the temperature dependence of $\chi(T)$ rather than its absolute value.

The NMR resonance coils surrounded each cell and were connected in series. This meant that the FIDs from the two cells were measured simultaneously. Any fluctuations of the measurement system coming from thermal effects or from the spectrometer, such as drift of the spectrometer gain, etc., were therefore added equally to each signal. Hence, the final results, as extracted from the data, were much less dependent on these types of problems. The applications of a small magnetic field gradient allowed the signals from the two cells to be distinguished after a Fourier transform into the frequency domain.

Using a high-temperature series-expansion of a model independent spin Hamiltonian[13], we can express the inverse magnetic susceptibility in the zero-field limit as a function of temperature to second order as follows,

$$\frac{1}{\chi} = \frac{1}{C}\left(T - \theta + \frac{B}{T}\right) \tag{1}$$

where C is the Curie constant, θ the Curie-Weiss temperature, and B a second order coefficient. In the case of the double sample cell, this expression can be applied to the samples in each cell to give two equations. The temperature T can be eliminated between these equations and the signals normalized at a reference temperature. This leads to the following expression, in which terms higher than second order have been neglected,

$$\frac{\chi_H}{\chi_L} = \frac{C_H}{C_L}\left[1 - \Delta\theta\left(\frac{\chi_H}{C_H}\right) + A\left(\frac{\chi_H}{C_H}\right)^2\right] \tag{2}$$

$$\text{where } \Delta\theta = \theta_L - \theta_H \text{ and } A = B_L + \theta_H^2.$$

The subscripts H and L correspond to signals from high and low pressure samples respectively. The B_H coefficient has been neglected. χ_H and χ_L represent the quantities measured in the experiment. Constants C_H and C_L were determined at a high temperature reference point (≈ 510 mK), where the susceptibilities of both samples follow Curie law. Specifically, $C_H = T_{ref} \cdot \chi_H(ref)$, and $C_L = T_{ref} \cdot \chi_L(ref)$. All the experimental data was fitted to expressions of the form given by equation (2). When the dependent

4

variable is expressed as the ratio of the two measured quantities, χ_H/χ_L, the advantage of the double sample cell becomes apparent; namely, any dependences due to drift of the spectrometer gain, etc., and other types of fluctuations will cancel out to first order. The uncertainty in the independent parameter χ_H is small compared to the range over which this quantity changes and can be neglected in the data analysis. Coefficients $\Delta\theta$ and A are related to the interesting quantities θ and B. The unknown parameter θ_H can be estimated with the aid of the strong molar volume dependence of the exchange energy.

APPARATUS

The lower portion of the cryostat used in the experiment is shown in Fig 1a. The figure illustrates some of the major components such as the dilution refrigerator mixing chamber, the interface platform with pressure gauges, the double sample cell, the NMR magnet, and a nuclear demagnetization stage that was not used in the present experiment. The ^{3}He sample chamber (located at the center of the superconducting NMR solenoid) was thermally linked to a CMN thermometer, then to an LCMN thermometer, and finally to a Superconductive Fixed Point Device, NBS SRM767 (Fig 1b). Both the CMN and LCMN thermometers relied on an AC-mutual inductance bridge with a SQUID detector for determining the temperature dependent susceptibilities. Temperature control of the system was achieved by using a carbon resistor thermometer and a heater, both located inside the mixer. Another resistance thermometer was located on the interface platform close to the pressure gauges. All thermometers were calibrated against the NBS SRM767 using the superconducting transition temperatures of Cd, Zn, Al, and In. The detection of these superconducting transitions was done with a stable mutual-inductance AC bridge. Two pressure capacitive-strain gauges were used to measure the pressures of the ^{3}He samples separately. Capacitances were measured by an AC-bridge technique with a reference capacitor held at liquid nitrogen temperature. The pressure gauges were calibrated against a Texas Instruments Fused Quartz Precision Pressure Gauge.

A detailed diagram of the sample chamber is shown in Fig. 1c. The two cylindrical cells are shown with a copper wire bundle ($\approx$3700 wires, 25 μm dia) linking them thermally. Each cell has an effective volume (for the ^{3}He) equal to $\approx 8\times10^{-3}$ cm^3 and a surface area (of the copper wires) equal to about 13 cm^2. Each sample was formed at the appropriate molar volume by the blocked capillary technique and annealed for about 0.5 hours near melting. The impurity level of the sample gas was 27 ppm of ^{4}He. The samples were cooled to the minimum temperature and nearly all measurements were taken on warm-up.

SPECTROMETER

A pulsed NMR spectrometer was used to transmit an rf pulse to the sample and receive the resulting response. The response was

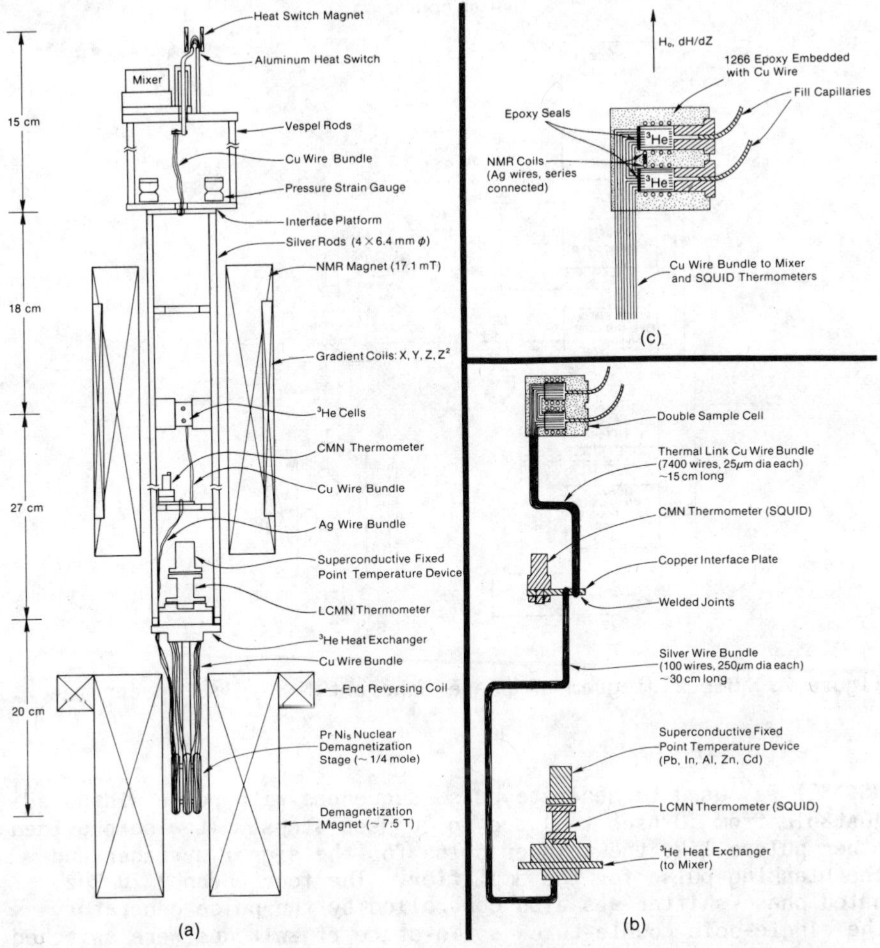

Figure 1. Schematic layout of (a) cryostat (lower portion only), (b) thermal linkage between sample chamber and auxiliary thermometers, and (c) sample chamber.

then recorded and analyzed by the data processing system. To allow operation over a wide frequency range with a minimum of retuning, a heterodyne spectrometer[14,15] was built. All the phase shifting, most of the gating, amplification, and phase detection were performed at an intermediate frequency (IF) of 20 MHz. A block diagram of the spectrometer, including the data processing system and coupling circuits to the sample, is shown in Fig. 2. A frequency synthesizer (General Radio model 1061) was used as a very stable and precise frequency source in the range 400 kHz - 160 MHz. Its internal frequency standard was also used to control the pulse generator as well as provide the IF frequency for the spectrometer. The pulse generator (Interface Technology model

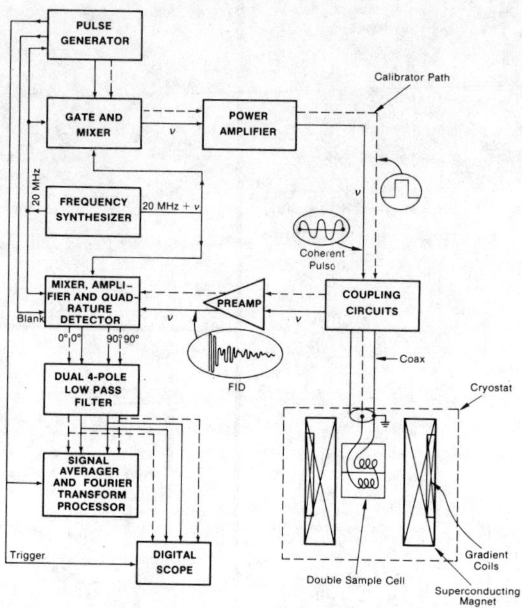

Figure 2. Block diagram of phase coherent NMR pulse spectrometer.

RS648E) was used to generate pulse sequences with pulse widths adjustable from 50 nsec to 1 sec in 50 nsec steps. It also provided other pulses like the trigger pulse for the signal averager and the blanking pulse for the amplifier. The four channel 20 MHz gated phase shifter was also controlled by the pulse generator. The single-pole double-throw solid-state rf switches were switched in the direction appropriate for the desired rf phase. There was an additional pulse stretching circuit in each switch driver to fine adjust the length of the pulse by as much as 50 nsec. The 20 MHz pulse was then mixed in a single-sideband mixer with 20 MHz + ν to give a pulse of chosen frequency ν (= 550 kHz in this experiment). The 20 MHz + ν frequency from the synthesizer was applied to the transmitter mixer only when pulses were generated in order to reduce rf leakage.

The power amplifier (ENI model 320L) had a bandwidth of 250 kHz- 110 MHz and a 50 dB gain. A 90° pulse corresponded to 100 μsec. The H_1 amplitude produced a field at the sample equal to 0.8 Gauss. The actual pulse length used in the experiment was 8 μsec, corresponding to a tipping angle of 7°. This was done to reduce heating effects in the sample, as well as to minimize eddy current heating in the surrounding copper. In the range of small tipping angles, the amplitude of the FID was proportional to the

magnitude of applied pulse and this enabled more effective calibration of the spectrometer.

The input of the preamplifier was overload protected from the
transmitter pulse by a duplexer circuit that was specially designed to form a high-impedance parallel resonant circuit during pulse
transmission and a low-impedance series resonant circuit during
signal reception. The duplexer on/off impedance ratio was equal
to 4×10^3. The electrical parameters of the sample coil were the
following: self resonance frequency, 536 kHz; resonant impedance,
12 kΩ; quality factor, 4; and a recovery time of 2 μsec. When the
sample was at the lowest temperatures, the noise performance of
spectrometer was solely determined by the preamplifier. To minimize this problem, a monolithic preamplifier (Optical Electronics
model AH0013) was used. It had the following characteristics:
input impedence 10^{11} Ω, gain 30 dB, and a measured effective noise
temperature of 7 K for a 12 kΩ source impedance.

The unit shown in Fig. 2 containing mixer, amplifier, and
phase detector was used to mix the signal from the preamplifier,
at the operating frequency ν with a 20 MHz + ν frequency from the
synthesizer. This mixing was done only during reception of signals to prevent IF amplifier overload. The lower 20 MHz sideband
was then amplified by the IF amplifier and phase detected by two
double-balanced mixers in quadrature. The amplitudes of the signals differed by a factor of 30 between low and high temperatures;
therefore the spectrometer gain was adjusted by using a precision
step attenuator. Since the accuracy of this attenuator had a direct effect on the results of the experiment, it was calibrated
against an Ailtech 3230 precision waveguide-beyond-cutoff attenuator with a ± 0.05 dB uncertainty. The $0°$ and $90°$ phase signals
from the phase detector were passed through two matched 24 dB per
octave phase-flat filters (Rockland model 442) to prevent alias-
ing. Cutoff frequency was equal to 48 kHz. The signals were then
digitized at a sampling rate of 20 kHz and 1000 points of each
signal were recorded. The time delay between the trigger pulse
and the actual start of data recording was carefully chosen to
eliminate the need of frequency-dependent phase correction of the
absorption spectrum.

The internal calibrator circuit was a unique feature of the
spectrometer.[16] It was used to compensate for temperature dependent effects of the sample circuit and for spectrometer drift. A
calibrator pulse was generated in the same transmitter channel as
the NMR pulse and it also followed the same path as the FID in the
receiver-part of spectrometer (dashed line in Fig. 2) The calibrator pulse was recorded and processed the same way as the NMR
signal; except, only the magnitude was calculated since it did not
have the same phase as the NMR signal. Switching between "NMR
Mode" and "Calibrator Mode" did not affect the phase coherence of
the spectrometer.

The amplitude of the signal at the output of the spectrometer
can be expressed as follows: $E_{out} \approx AE_T r^{-1} G(a,g) M_o$ where M_o is
the dc magnetization, $G(a,g)$ a geometrical factor (a is radius, g
the length of cylindrical coil), A the overall amplifier and phase

detector gain, E_T the amplitude of the transmitter pulse, and r
the loss resistance of the sample coil. Fluctuations of the
parameters A, E_T, and r were compensated for by dividing the NMR
signal by the calibrator signal. At low temperatures (<4 K),
changes in the parameter G due to thermal expansion can be neglected. Application of the calibrator circuit improved the spectrometer performance by one order of magnitude.

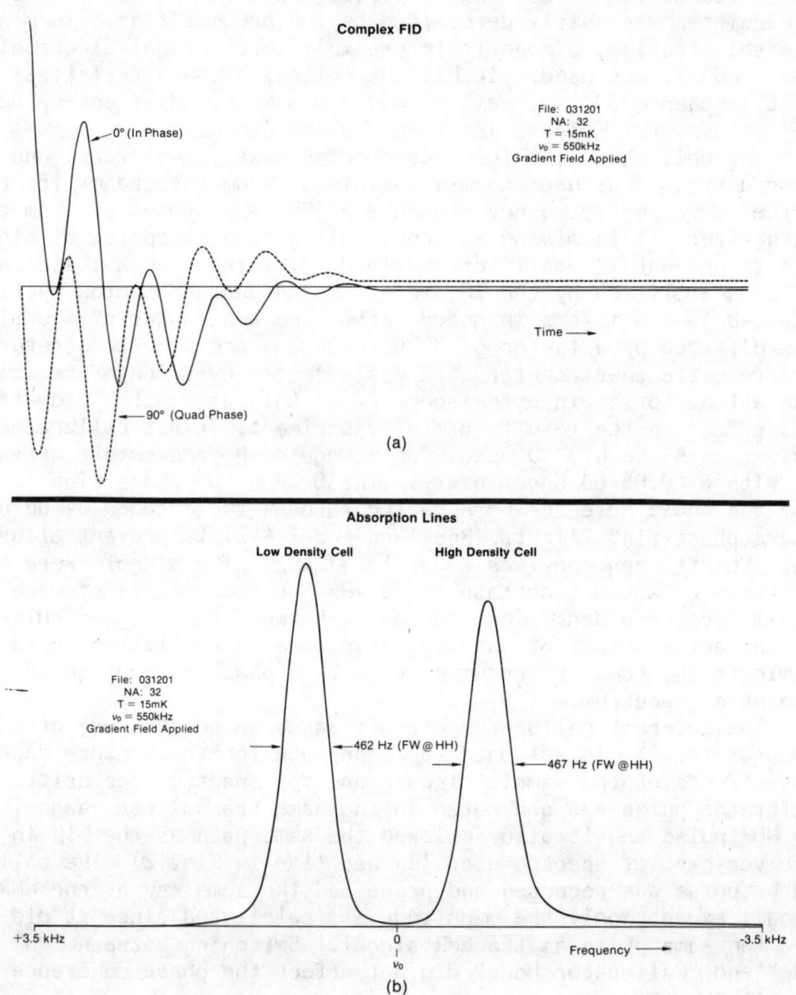

Figure 3. (a) Free induction decay (FID) from double cell showing
In-phase and Quad-phase components. (b) Fourier transform of the
FID in (a). Dispersion lines not shown.

DATA RECORDING AND PROCESSING

After forming the sample and cooling down, an "on resonance" signal was used to adjust the magnetic field H_0 and the phase of the quadrature detector. The magnetic field gradient was then applied (0.7 Gauss/cm). At each temperature interval, thermal equilibrium was determined by monitoring the CMN and LCMN thermometers. At the lowest temperatures, it took about 6 hours to establish thermal equilibrium. Each data point was taken twice, with a calibrator signal measured at the start, between, and at the end of each measuring sequence. The number of acquisitions was increased with increasing temperature from 32 to 16,000 to keep the statistical error approximately constant over the entire temperature range. At the reference temperature (T_{ref} = 509.7 mK), data acquisitions were done in sixteen segments of 1000 acquisitions with a calibrator signal measured after each block. A typical FID measured at 15 mK is shown in Fig. 3a. Both the in-phase and out-of-phase components were recorded in two quadrature channels. Because of the large number of acquistions, the S/N (signal to noise) ratio of the high temperature data looks similar to the data shown in Fig. 3a.

Recorded time-domain data were ortho-normalized so as to compensate for possible gain and phase errors arising from the phase detector.[17] If A_i and B_i are corresponding data points in the in-phase and out-of-phase channels, the algorithm replaces A_i by $\gamma(A_i - \alpha B_i)$ and leaves B_i unchanged. Coefficients α and γ were determined by applying a strong external signal as part of a tune-up procedure before each run. Typical values in our case were α = 0.026, γ = 0.996. The data was then digitally filtered by using a Gaussian multiplication algorithm to improve the S/N ratio. The data were then zero filled by adding to the FID an equal number of zeros. This was done twice, giving in effect 4000 data points in each channel. The zero filling algorithm improves resolution.[18] Finally a Fourier transform was performed using a FFT algorithm. The resulting real part of the spectrum is presented in Fig. 3b. The accompanying dispersion spectrum is not shown, although it was recorded.

A small asymmetry of the spectrum can be noted. This corresponds to a zero-order phase error equal to 0.2°. All zero-order (frequency-independent) phase corrections were done manually with a ±0.1° accuracy. This type of correction contributed only ±0.05% error to the final values of the areas under the absorption peaks. First order (frequency-dependent) phase corrections were not needed because the delay between the trigger pulse and the start of data recording was properly chosen. The spectrum was then baseline corrected using 128 points on each wing of the spectrum so that an average baseline offset value could be calculated and then subtracted from the spectrum. The areas under both peaks were then found by integrating over ±2 kHz wide windows centered on the spectrometer frequency ν. The resulting integral values were divided by the appropriate integral value of the calibrator signal which then yielded the final values for χ_H and χ_L.

RESULTS AND DISCUSSION

In Fig. 4 (a) and (b), we show susceptibility data of two ^{3}He samples (24.06 and 21.03 mL/mole) plotted as χT vs. $1/T$ where the temperature has been determined from the CMN thermometer. From these plots it is clear that, with the use of a CMN temperature scale, there are systematic errors in both the low and high temperature limits. Most of the curvature and deviations displayed by the data can be attributed to NMR spectrometer fluctuations, calibration uncertainties, SQUID instrument drift, and thermal

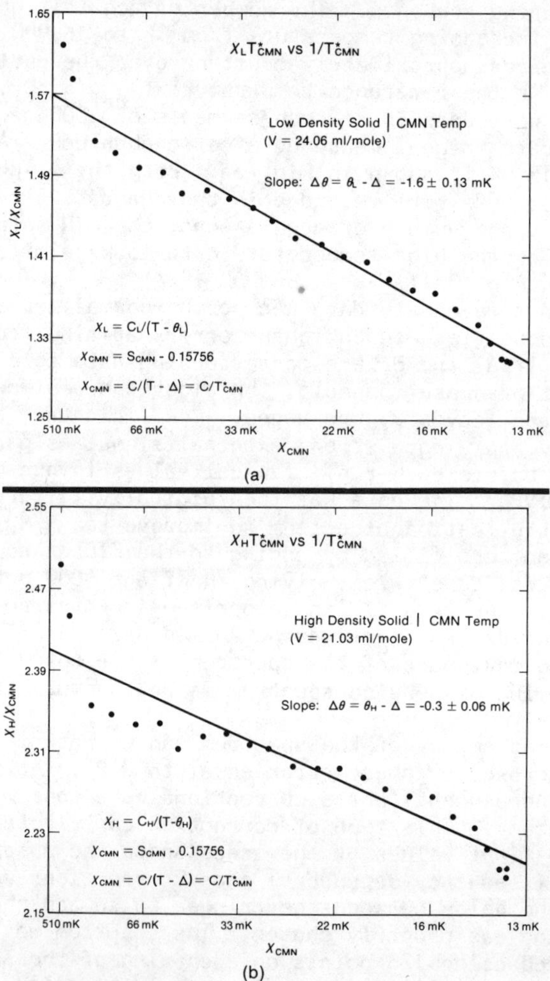

Figure 4. χT vs. T^{-1} plots: (a) low density solid ^{3}He and (b) high density solid ^{3}He plotted respectively as χ_L/χ_{CMN} and χ_H/χ_{CMN} versus inverse temperature, χ_{CMN}, from the CMN thermometer. CMN thermometer electronic background = 0.15756.

gradients between the sample and the CMN thermometer. The straight lines drawn through the data were determined by a numerical least squares fit to the data. Using the slope and intercept values from these curves, the $\Delta\theta$ values for the high and low density solids were determined as shown. The errors associated with these $\Delta\theta$ values represent the accumulated error from T_{ref}, χ_H(ref) or χ_L(ref), and the standard error of the intercept and slope values of the fitted linear curves.

In the case of the CMN thermometer, the calibration was done at high temperatures (0.5 K to 1.2 K). The best straight line fit of the CMN susceptibility was extrapolated to lower temperatures assuming it obeyed a Curie law with a correction parameter Δ. Temperatures were determined after subtracting the electronic background value corresponding to the CMN signal at high temperatures. This method cannot be justified in general, so one does not expect good accuracy for the temperature scale. This procedure, as is well known, is operationally useful for determining temperatures. Nevertheless, the slope values, $\Delta\theta$, of the best straight line fits obtained with the CMN temperature scale are quite consistent with the data calculated by using the double-sample-cell technique described earlier. Only, in this case, the standard deviations are much larger. Furthermore, it is possible using the data in Fig. 4(b) to set an upper limit for θ_H, the Curie-Weiss temperature of the high density solid (used as the reference solid in all the double-sample-cell measurements). Specifically, -0.3 mK $< \theta_H < 0$, assuming that Δ is a positive number such at $0 < \Delta < 0.3$ mK. This last condition represents a range of values for Δ that is not untypical of the values many workers have found applicable for CMN thermometers.

In Fig. 5(a) we illustrate the advantage of the double-sample-cell measuring technique by showing a χT vs. $1/T$ plot of the data (for the same samples as discussed above) when the high density sample was used as a thermometer. It is obvious from the plot that this technique improves the susceptibility measurement considerably. There are fewer systematic errors and the standard deviation has been decreased by a factor of two. In general, we find that, for all data taken this way, the scatter is never greater than 1/3%. Again, the straight line drawn through the data was determined by a least squares fit. The slope of this curve provides a value for the difference between the Curie-Weiss temperatures of the low density and high density samples, $\Delta\theta = \theta_L - \theta_H$.

We have measured, using the double-sample-cell technique, the susceptibility of several density samples while holding the high density sample constant at 21.03 mL/mole. The results are summarized in Fig. 5(b) where the data has been plotted as $1/\chi T$ vs. $1/T$. This plot is equivalent to the previous χT plot in so much as the first order coefficient is concerned. We believe the $1/\chi T$ plot is a better way to plot the data because there is less correlation between the first and second order coefficients. The various molar volume samples were measured in the following order (in mL/mole): 24, 22, 23, 21. The role of the cells was then reversed;

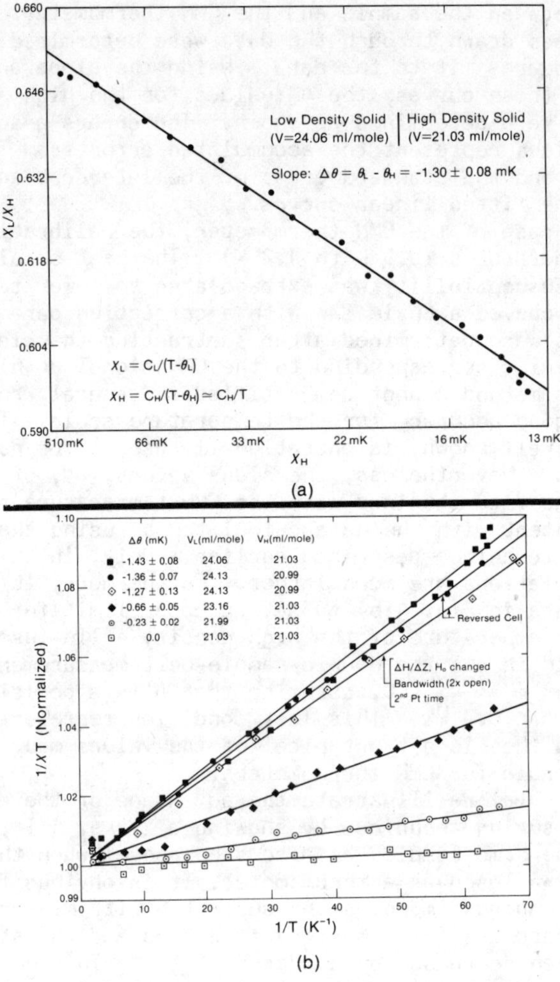

Figure 5. (a) χT vs. T^{-1} plot using double-sample-cell technique. χ_L is susceptibility of low density solid. χ_H is susceptibility of high density solid and also serves as a thermometer with $\chi_H \propto T^{-1}$. (b) Solid ^{3}He susceptibility measurements of several samples, at various molar volumes, plotted as $1/\chi T$ vs. T^{-1} using double-cell technique with a high density reference sample of $V \approx 21$ mL/mole.

that is a high density sample was formed in the cell with the previous low density sample and vice versa. Again samples were formed at 24 and 21 mL/mole and measurements were made to check for systematic errors and for general reproducibility of the experiment. In addition some other critical parameters were changed

for purposes of fine tuning and checking out systematics and reproducibility. Namely, the main polarizing field was shifted by
0.2%, the magnetic field gradient increased by 10%, the bandwidth
of the receiver circuit was doubled, and the timing of the spectrometer readjusted to determine the sensitivity of the data to
any error associated with the second point on the FID (2nd point
error causes spectrum baseline curvature effects). As shown in
Fig. 5(b), the result of all these types of changes is not significant and all the data is found to be consistent within 8% accuracy.

The straight lines shown drawn through the data were determined by a least squares fit of the data to an expression of the
form given by equation 2. From the slopes of these curves, the
values for $\Delta\theta = \theta_L - \theta_H$ were determined. The errors associated
with these values are the result of the accumulated error from

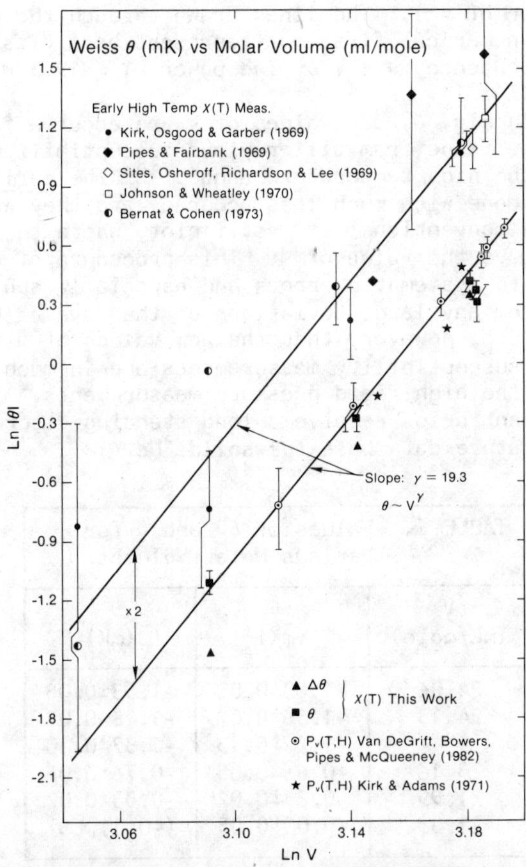

Figure 6. Comparison of the Curie-Weiss temperature θ, at various
molar volumes V, from this experiment with P(T,H) experiments and
previous χ(T) experiments. Values are plotted as $\ell n|\theta|$ vs. ℓnV.

14

T_{ref} ($\approx$1% error), $\chi(ref)$ ($\approx$3%), and the standard error of the slope ($\approx$1%) and intercept ($\approx$0.1%).

We have analyzed the data for the second order coefficient, and find that with the present accuracy only an upper limit of the B coefficient (Eq. 1) can be determined; namely, $B = 0 \pm 1$ mK^2 at a molar volume of 24.13 mL/mole.

The comparison of our results with previous data is shown in Fig. 6. To determine the absolute values of θ, it was necessary to estimate the value of θ_H. This estimate was done by plotting $\ln|\Delta\theta|$ versus $\ln V$, where V is the molar volume. By calculating the best straight line fit and extrapolating the curve to the molar volume 21.03 mL/mole, a value of $\theta_H = -0.1$ mK was determined. This value of θ_H was then added to each of the $\Delta\theta$ values and from this we obtained the final θ values for each sample. These values of θ are shown plotted (as closed squares) in Fig. 6 and are listed in Table I. The error limits are exactly the same as those associated with the $\Delta\theta$ values (no attempt was made to estimate the error of θ_H). The lines drawn through the points in Fig. 6 are not numerical fits, but represent best fits by eye for a power law dependence, $\theta \propto V^\gamma$. The power, $\gamma = 19.3$ was determined graphically.

As shown in Fig. 6, our values of θ are about a factor of two smaller than the value from all previous susceptibility experiments done in the high temperature range.[4-8] The early $\chi(T)$ measurements were done with much less accuracy and they were all analyzed using the conventional χ^{-1} vs. T plot, where the intercept on the T-axis gave the value of θ. This procedure of determining θ is sensitive to systematic errors and especially subject to errors arising from any large deviations of the data at higher temperatures. We see, however, that the new values of θ from the high-precision susceptibility measurements are in much better agreement with the high-field pressure measurements.[2,3] Presumably, this agreement helps resolve a long standing discrepancy in the high-temperature data base for solid ^{3}He.

TABLE I.	Values of $\Delta\theta$ and θ for Various Molar Volumes.	
V (mL/mole)	$\Delta\theta$ (mK)	θ (mK)
24.06	-1.43±0.08	-1.53±0.08
24.13	-1.36±0.07	-1.46±0.07
"	-1.27±0.13	-1.37±0.13
23.16	-0.66±0.05	-0.76±0.05
21.99	-0.23±0.02	-0.33±0.02
21.03	-0.00±0.05	(-0.1) EST

ACKNOWLEDGEMENTS

We are grateful to Dr. Michael Twerdochlib for contruction of the double sample cell. Other colleagues who have provided useful advice and assistance from which we have greatly profited include, Drs. Horst Armbruster, Jeevak Parpia, and Robert Arnold. This work was supported by NSF - Low Temperature Physics grant DMR 82-05902.

REFERENCES

1. R. A. Guyer, J. Low Temp Phys. $\underline{30}$, 1 (1978); R. A. Guyer, Phys. Rev. $\underline{A9}$, 1452 (1974).
2. C. T. Van Degrift, W. J. Bowers, Jr., P. B. Pipes, and D. F. McQueeney, Phys. Rev. Lett. $\underline{49}$, 149 (1982).
3. W. P. Kirk and E. D. Adams, Phys. Rev. Lett. $\underline{27}$, 392 (1971).
4. W. P. Kirk, E. B. Osgood, and M. Garber, Phys. Rev. Lett. $\underline{23}$, 833 (1969).
5. P. B. Pipes and W. M. Fairbank, Phys. Rev. Lett. $\underline{23}$, 530 (1969); Phys. Rev. A4, 1590 (1971).
6. J. R. Sites, D. D. Osheroff, R. C. Richardson, and D. M. Lee, Phys. Rev. Lett. $\underline{23}$, 836 (1969).
7. R. T. Johnson and J. C. Wheatley, Phys. Rev. $\underline{A1}$, 1836 (1970).
8. T. R. Bernat and H. D. Cohen, Phys. Rev. $\underline{A7}$, 1709 (1973).
9. T. C. Prewitt and J. M. Goodkind, Phys. Rev. Lett $\underline{39}$, 1283 (1977).
10. D. M. Bakalyar, C. V. Britton, E. D. Adams, and Y. C. Hwang, Phys. Rev. Lett. $\underline{64A}$, 208 (1977).
11. Y. Morii, K. Ichikawa, T. Hata, C. Kanamori, H. Okamoto, T. Kodama and T. Shigi, in Physics of Ultralow Temperatures, edited by T. Sugawara et al. (Physical Society of Japan, Tokyo, Japan, 1978) p. 196.
12. T. Hata, S. Yamasaki, Y. Tanaka, and T. Shigi, Physica (Utrecht) $\underline{107B}$, 201 (1981).
13. M. Roger, J. H. Hetherington, and J. M. Delrieu, Rev. Mod. Phys. $\underline{55}$, 1 (1983).
14. J. D. Ellet, Jr., M. G. Gibby, U. Haeberlen, L. M. Huber, M. Mehring, A. Pines, and J. Waugh, Adv. Magn. Reson. $\underline{5}$, 117 (1971).
15. A. A. V. Gibson, J. R. Owers-Bradley, I. D. Calder, J. B. Ketterson, and W. P. Halperin, Rev. Sci. Instrum. $\underline{52}$, 1509 (1981).
16. A. A. V. Gibson and W. P. Kirk, to be published.
17. S. I. Parks and R. B. Johannsen, J. Magn. Reson. $\underline{22}$, 265 (1976).
18. J. C. Lindon and A. G. Ferrige, Prog. NMR Spectr. $\underline{14}$, 27 (1980).

EXPERIMENTAL CONSTRAINTS ON THE PARAMETERS DESCRIBING
UNORDERED bcc ^{3}He

Craig T. Van Degrift
Temperature and Pressure Measurements and Standards Division
National Bureau of Standards, Washington, D.C. 20234

ABSTRACT

A wide variety of experimental results on the unordered phase of bcc ^{3}He are reviewed in light of recent high precision P(T,H,V) measurements made at NBS. Specific formulas are given for the volume dependence of the elastic constants, Debye temperature, exchange parameters J_t and K_p, and the Zeeman-exchange spectral density function. Some topics for further research are identified.

INTRODUCTION

In the 30-year history of the study of solid ^{3}He, progress toward a coherent physical description of its properties has been surprisingly slow. This is not an indication of a lack of effort, but rather reflects the extent to which solid ^{3}He research has enhanced our understanding of many-body physics and tested our ability to make measurements. Studies of the unordered phase prior to 1971 led researchers to believe that solid ^{3}He might be adequately described as a Heisenberg nearest neighbor antiferromagnet. With the realization that this is not the case, more complicated models appeared.

This paper is not a historical discussion of solid ^{3}He, but rather a guide to the most recent and reliable measurements of parameters describing the unordered phase. The reader must consult the cited papers to gain an understanding of the evolution of solid ^{3}He research. Particularly valuable in that regard are a 1971 review by Guyer, Richardson, and Zane[1] of the study of excitations using NMR, a 1972 review by Trickey, Kirk and Adams[2] of thermodynamic, elastic, and magnetic properties, and a 1983 review by Roger, Hetherington, and Delrieu[3] of magnetic effects emphasizing the multiple-exchange theories and the behavior in the ordered phase.

The phase diagram for the magnetically unordered part of the bcc phase of solid ^{3}He is shown in Figure 1. It is bounded at low pressures and high temperatures by the melting curve which, if measured with high precision, can provide temperature and pressure reference points[4,5]. At high pressures there is a transition to an hcp structure[6] while at temperatures below about 1 mK a magnetically ordered phase exists[7].

The physics of the low temperature magnetically ordered bcc phase and of the lower 150 mK of the unordered bcc phase are both thought to be governed by multiple atom exchange[3]. Although there have been many interesting experiments performed in the ordered phase, the theory can only be calculated in the mean field approximation at such temperatures. Thus, the constraints imposed by those experiments on the parameters of the theory are of limited quantitative value. On the other hand, experiments at temperatures above 25 mK in the unordered phase may be compared with exact calculations

0094-243X/83/1030016-16$3.00 American Institute of Physics

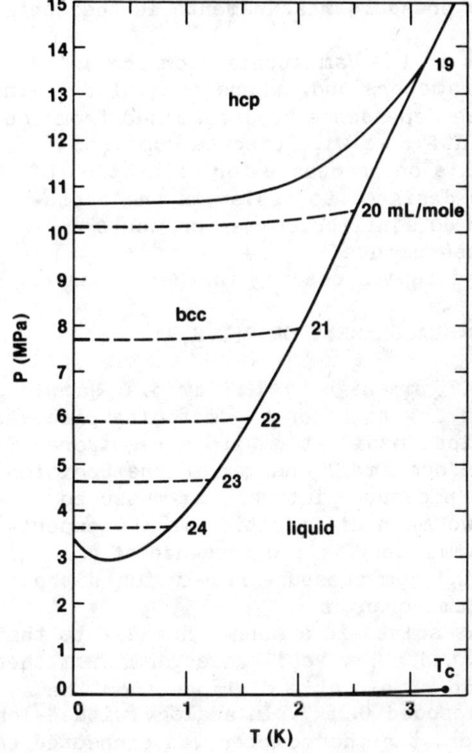

Fig. 1. bcc phase diagram.

of the initial terms of the high temperature expansion of the theory. Unfortunately, at higher temperatures, the size of exchange effects diminishes and experiments require greater sensitivity. Furthermore, in specific heat and pressure experiments, exchange effects are mixed with phonon effects above 150 mK and vacancy effects above 300 mK.

Measurements of the isochoric pressure of solid ^{3}He have been pioneered by Adams and coworkers at Florida as a probe of exchange, phonon, and vacancy excitations. Particularly valuable have been the measurements of Straty and Adams[8] of the basic PVT data throughout the bcc phase and into the hcp phase, Panczyk and Adams[9] which gave the exchange pressure in zero magnetic field, and that of Kirk and Adams[10] which first showed that the Heisenberg nearest neighbor model did not work in solid ^{3}He. The results of Straty and Adams and those of Panczyk and Adams fit in well with evidence from other experiments and the theoretical expectations of the time and were published in full detail. The Kirk and Adams experiment, on the other hand contradicted both existing susceptibility measurements, and the prevailing theoretical expectation. Unfortunately, some recent theoretical models fail to consider these results.[3]

The NBS experiment, inspired by P. B. Pipes of Dartmouth, was designed to measure P(T,H,V) with sufficient thoroughness that its results would not be easily discounted even in the presence of contrary experimental results or adverse theoretical prejudices. Our initial publication[11] verified the Kirk and Adams result (as interpreted by Guyer[12]) and strongly suggested that a large number of nuclear susceptibility measurements had been misinterpreted. Recent susceptibility measurements at Texas A&M and reported in these proceedings[13] now dramatically confirm this.

The primary goal of this paper, however, is to examine all ways in which the best available data constrain the parameters describing unordered solid ^{3}He. P(T,H,V) measurements, however, affect many of these and we include a few additional details regarding our experiment and, where appropriate, show how our data restrict possible values of parameters. P(T,H,V) measurements alone, however, constrain only the volume derivatives of the usual theoretical parameters and are best interpreted using plausible functional forms for the volume dependence of the parameters. It is valuable, if not

18

essential, to carefully analyze pressure measurements in the context
of all related measurements.

We will consider each term in the Hamiltonian for bcc solid
^{3}He, describing the relevant parameters and, where possible, giving
simple formulas for their volume dependence as determined from equilibrium measurements. Subsequently, we will discuss some non-
equilibrium results with emphasis on processes for relaxation of the
Zeeman energy. That section is designed to assist in the cross-
fertilization between those who do equilibrium measurements and
those who do NMR relaxation measurements.

In closing, we list related topics needing further research.

PROGRESS ON THE NBS MEASUREMENT OF P(T,H,V)

In the NBS experiment, a 0.75 mm high by 12.7 mm dia. sample
chamber was constructed by sintering together two matching pieces of
coin silver. A pulsed tunnel diode oscillator with a reentrant LC
resonator was then used to transform small changes in the position
of a sample chamber wall (i.e., pressure) into easily measured
frequency changes. There are two main difficulties in the experiment -- performing accurate thermometry in the presence of a large
magnetic field and attaining a 0.1 ppm pressure resolution without
allowing significant sample volume changes.

The thermometry problem was solved in a manner similar to that
used by Kirk and Adams. We located a low-resistance germanium thermometer (400 Ω @ 30 mK) in a lead shield at a distance from the
sample where the field never exceeded 0.06 T. In analogy with 4-lead
resistance measurement techniques, the thermometer was connected to
the sample by a different thermal path than the refrigerator. Heat
flowing from the sample chamber and pressure sensor to the refrigerator could not cause a significant temperature difference between
the sample and thermometer. As an added precaution, magneto-thermal
resistance measurements were made along the relevant thermal circuit.

The volume-constancy problem was solved by simply making the
wall especially thick (4 mm) and the capacitive gap of the reentrant
resonator very small (5 μm). An adequate pressure sensitivity of
0.3 Pa (corresponding to 3×10^{-12} cm wall displacement) was retained
while the error resulting from volume changes was only 0.32% of the
measured pressure changes.

Corrections for the temperature, magnetic field and voltage dependences of the oscillator and for zero shifts, which often occurred
during liquid helium transfers, greatly complicated the data analysis. Fortunately, their effect on our final uncertainties is small
and comparable to the temperature, magnetic field, and molar volume
errors. The standard deviation of our fits in the low temperature
region was 0.6 Pa with systematic errors in the temperature and magnetic field dependence of sample pressure being typically 1 to 2 Pa.

Further details of the apparatus were given in our initial
publication[11] and full details of the apparatus, corrections, and
analysis will be published with our final results[14]. In appropriate
sections of this review, we present the current state of our analysis and results. After our final calibrations, specific values will
be slightly changed and uncertainties reduced.

CONSTRAINTS FROM EQUILIBRIUM MEASUREMENTS

Phonon Parameters: In spite of the large zero point motion in solid ^{3}He, its phonon thermodynamics appears similar to that of ordinary solids. The Debye model works very well for solid ^{3}He at lower temperatures and, as is common in solids, a temperature dependent Debye temperature, $\theta_D(T,V)$, can be used to represent deviations from it at higher temperatures. The volume dependence of the zero-temperature limit of the Debye temperature, $\theta_D(0,V)$, can be represented within experimental error by a constant Gruneisen parameter, $\gamma_\theta = d\ln\theta_D / d\ln V$. γ_θ and the value of θ_D at 24 mL/mole, θ_{Do}, can be independently determined by pressure, specific heat, and sound velocity measurements.

Figure 2 illustrates the extent to which the specific heat data of Greywall[15] and our own pressure data restrict the possible values of θ_{Do} and γ_θ.

Pairs of values for θ_{Do} and γ_θ which lie within the smaller ellipse fit the Greywall data within the statistical error. The elliptical contour gives the locus of points with a standard deviation greater than the best fit value by a factor of $1+1/\sqrt{n}$ where n is the number of data points. To estimate the effect of systematic errors, we have also drawn error bars which show the direction and extent to which the center of the ellipse might be displaced as a result of an 0.5% temperature scale error (horizontal bars) and a 1% background correction error (tilted bars).

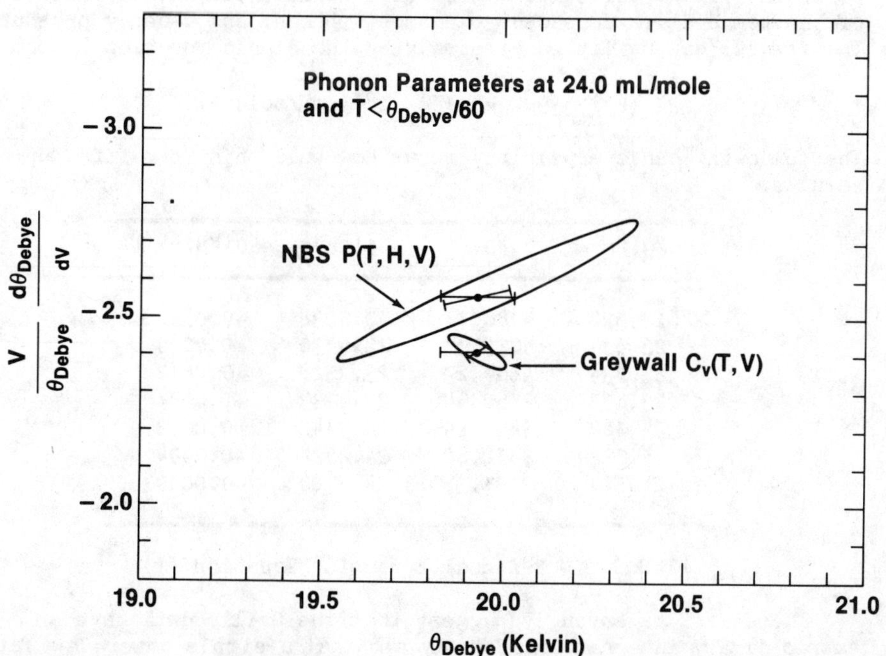

Fig. 2. Constraints on θ_D and its logarithmic derivative, γ_θ.

The larger ellipse shows the constraints on θ_{Do} and γ_θ which come from our pressure data. Our dominant systematic errors are also in the temperature scale and background correction. It is clear that the specific heat data of Greywall provide much tighter constraints on θ_{Do} and γ_θ than do the pressure data. This is easily understood because the pressure is a volume derivative of the free energy and is therefore more sensitive to exchange and vacancy effects (which have a relatively high volume dependence) than to phonon effects.

Using this figure, we have chosen the values $\theta_{Do} = 19.84$ K and $\gamma_\theta = -2.46$ to represent the zero temperature limit of the phonon effects. At higher temperatures, we use the temperature dependence $\theta_D(T,V)/\theta_D(0,V)$ given by de Wette and Werthamer[16] appropriately converted to apply to pressure rather than specific heat.

The analysis described above illustrates how easily deceived one may be by ordinary methods of quoting parameter errors which ignore correlations. The values $\theta_{Do} = 20.4$ K and $\gamma_\theta = -2.40$ might be thought to be consistent with the pressure measurements when in fact they are not. Occasionally, when applying such procedures to other problems, the ellipses should actually be larger than the size dictated by the statistical error alone and may, in fact, be distorted by non-linearities in the equations.

Elastic Parameters: Our measurements are able to relate the melting pressures and Grilly molar volumes[17], $V_G(P_M)$, to the "bare" pressure, P_o, at low temperatures that would exist in the absence of the temperature dependent exchange, phonon, and vacancy pressures. The results can be fitted precisely by the simple functional form

$$V(P_o) = 24 \ (P_o/3635.7 \ \text{kPa})^{-1/5.597} \ \text{mL/mole} \ . \tag{1}$$

The following table explicitly shows how well this form fits our results.

$V_G(P_M)$	P_o	$V(P_o)$	$V_G(P_M)-V(P_o)$
22.522	5186.77	22.5238	−0.0018
22.614	5073.27	22.6130	+0.0010
23.134	4467.29	23.1328	+0.0012
23.834	3780.34	23.8333	+0.0007
24.163	3499.59	24.1642	−0.0012
24.222	3452.80	24.2224	−0.0004
24.371	3337.36	24.3700	+0.0010

TABLE I. Residual Error for Equation (1).

These results strongly suggest that the Grilly data have an extraordinary internal consistency and that a simple power law fits the data with an rms deviation of only 0.0012 mL/mole.

In figure 3, we give formulas and curves showing the molar volume dependence of the elastic constants of bcc solid ^{3}He. Also shown are the results of sound velocity measurements of Greywall[18] (solid circles with error bars) on oriented crystals at 21.7 and 24.4 mL/mole. The upper curve is the bulk modulus derived from Eq. (1) with the estimated absolute error above 22.5 mL/mole being approximately given by the line width. Also shown, for comparison, are the bulk modulus points of Straty and Adams[8] (open squares). The middle curve is simply a fit through the sound velocity measurements of C_{44}. The bottom curve shows what $C_{11}-C_{12}$ must be to obtain mutual consistency between Greywall's specific heat measurements, our bulk modulus, and C_{44}. The discrepancy at 21.7 mL/mole is not unreasonable considering the great difficulty in determining $C_{11}-C_{12}$ from the sound measurements.

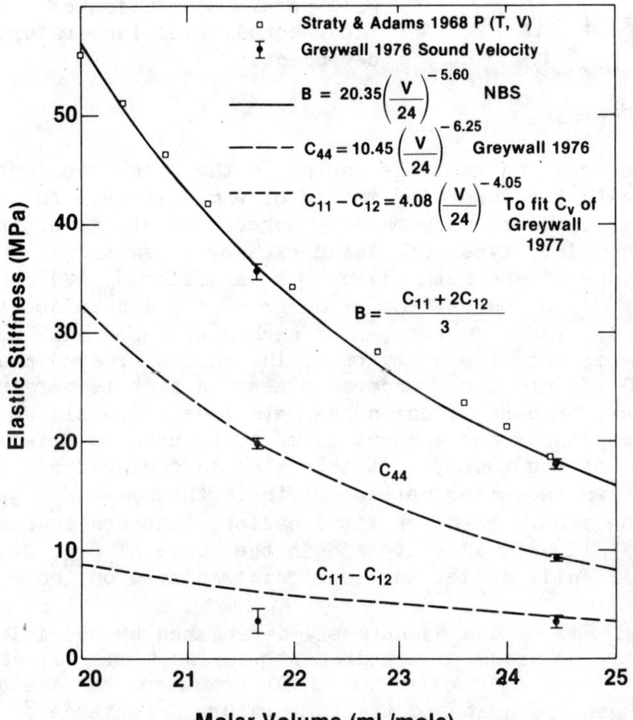

The specific heat calculated from the appropriate angular averages of the elastic constants using the formulas given in Fig. 3 will not have a perfectly constant logarithmic derivative. The difference from constancy, however, is small and does not contradict the specific heat data.

Fig. 3. Elastic Parameters

Vacancy Parameters: Vacancies in bcc solid ^{3}He are non-localized[1,19] with their mobility expressed by a bandwidth to their density of states. Most information about vacancies comes from X-ray lattice parameter measurements at constant volume[20] and NMR relaxation at high temperatures[21]. The basic trends are shown in figure 4.

Thermodynamic measurements, especially at higher molar volumes, cannot yet be reliably used to obtain information on the vacancies[22].

The large bandwidth-to-activation energy ratio at larger molar

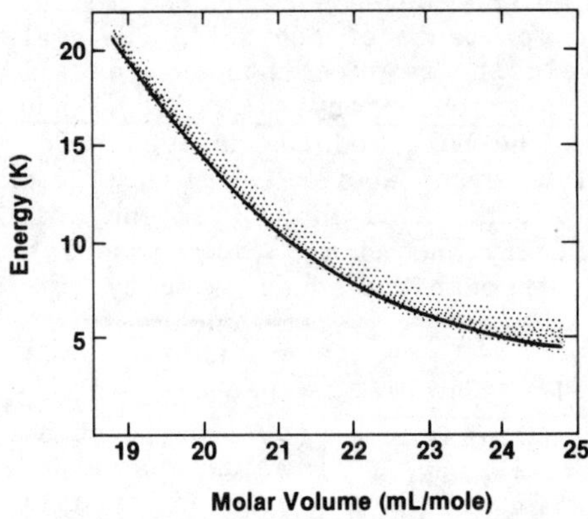

volumes and considerable uncertainty regarding the temperature dependent corrections to the Debye temperature greatly complicate any analysis. We plan to attempt an analysis based on simultaneous fitting of both specific heat data and pressure data with the hope that the different volume dependences of vacancy and phonon effects will permit separation of their complicated temperature dependences.

Fig. 4. Vacancy Energies

Exchange Parameters: The exchange energy is the spin-dependent energy associated with the tunneling motion of whole atoms. An extensive review has been recently made by Roger, Hetherington, and Delrieu[3] who consider four types of atomic exchange: nearest neighbor pair exchange of strength given by a parameter $J_{nn}(V)$, triple exchange, $J_t(V)$, planar 4-atom exchange, $K_p(V)$ and folded 4-atom exchange, $K_f(V)$. These four types of exchange, especially J_t and K_p, are felt to control the behavior of the entire ordered phase and up to about 150 mK into the unordered phase. A high temperature expansion of this exchange Hamiltonian has been made which allows experiments at temperatures above about 25 mK to be used to determine the parameters of the theory. At this time in our analysis we are only using the two parameter version of their theory -- J_{nn} and K_f are assumed to be zero. When our final calibrations are complete we will present a 4-dimensional ellipsoid in the space of J_{nn}, J_t, K_p and K_f which will fully depict the constraints placed on those parameters by our data.

In principle, J_t and K_p can be uniquely determined by specific heat, susceptibility, or pressure measurements as functions of temperature, magnetic field, and molar volume. In practice, the field dependence of the specific heat and the temperature dependence of the susceptibility are dominated by a large exchange-independent contribution coming from the Zeeman part of the Hamiltonian. The pressure, however, being the volume derivative of the free energy, is insensitive to that contribution and provides sufficient information to clearly determine both J_t and K_p. Furthermore, the exchange effects have a much greater volume dependence than the phonon effects which compete with them at higher temperatures. As a result, the temperature dependence of the pressure is dominated by exchange up to a higher temperature than that of the specific heat.

There is, however, a difficulty in the analysis of pressure data. Since only a volume derivative of the exchange parameters was determined from 22.5 to 24.4 mL/mole, a functional form for the volume dependence must be assumed to allow integration with respect to volume. A constant logarithmic derivative with respect to volume is conventionally assumed for each parameter as we did in the analysis of our first results[11]. Recently, however, Avilov and Iordanski[23] have proposed a more complicated form which still has only two adjustable parameters. Specifically,

$$J_t = J_{to} (V/24)^{\gamma_\theta /2-2}\; e^{A_{Jt}[(V/24)^{\gamma_\theta +2/3} -1]} \tag{2}$$

for J_t and a similar form for K_p. Furthermore they claim that the volume dependence parameters for J_t and K_p will be very nearly equal (i.e. $A_{Jt} \simeq A_{Kp}$).

We have used this volume dependence with $\gamma_\theta = -2.46$ and the high temperature expansion for the pressure as a function of J_t and K_p to fit our data. The resulting parameter values are

$$\begin{aligned}
J_{to} &= -0.1221 \pm 0.0061 \text{ mK} & A_{Jt} &= -12.00 \\
K_{po} &= -0.3301 \pm 0.0066 \text{ mK} & A_{Kp} &= -11.81
\end{aligned} \tag{3}$$

Related parameters $e_2 = 12J_{xx}^2$, $e_3 = 12J_{xxx}^3$, $\theta_w = 4J_{xzz}$, $B = \theta_w^2 - 12J_{xxzz}^2$, etc. can be calculated from Eqs. (3) using formulas in ref. 3.

At the present stage of our analysis, $A_{Jt} = A_{Kp}$ seems also to represent the data within experimental error. If this actually turns out to be true and if the two parameter theory adequately represents solid ^{3}He, there should be a remarkable universality in the volume dependence of all exchange effects at all temperatures.

A powerful test of the two parameter theory seems possible using our pressure data. There are two nearly equal dominant terms in the pressure, one varying as $1/T$ and the other varying as H^2/T^2. These can be used to fix J_t and K_p while allowing the three weak higher order terms varying as $1/T^2$, H^2/T^3, and H^4/T^4 to be fitted independently. Alternatively, the theory can then be used to calculate the magnitude of these terms from J_t and K_p. Preliminary tests show that the theory correctly predicts these higher order coefficients within 30% which is comparable with the experimental error. We will repeat this test after our final calibrations when the result may be taken more seriously.

Recent measurements of susceptibility at Osaka City University[24] and Texas A&M[13] appear to have resolved the conflict[11] between pressure and susceptibility determinations of the Weiss temperature, θ_w. The Texas A&M data have particularly small errors because of the use of a differencing method to gain immunity to the large Zeeman contribution mentioned above. In Fig. 5, these results are plotted together with the θ_w calculated from our J_t and K_p (solid line). Although it was necessary to use our extrapolated value for the reference θ_w at 21 mL/mole in plotting the Texas A&M data, the agreement is, nevertheless, quite impressive. The older susceptibility measurements (see ref. 11 for a list) were all clustered around the dashed line which shows the θ_w that would be deduced from our zero field pressure data using the Heisenberg nearest-neighbor model. Our $P(T,H,V)$ measurements are in agreement with the earlier results of Kirk and Adams[10] as analyzed by Guyer[12] (solid squares in Fig. 5). Furthermore, the zero field results agree

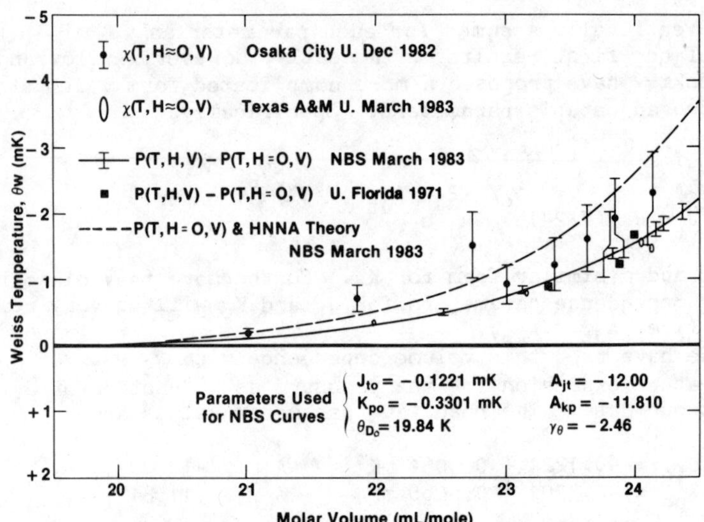

Fig. 5. Volume dependence of Weiss Temperature.

(within our current error estimates) with the work of Panczyk and Adams.[9] Our J_{xx} values are 2, 4, and 16% lower than theirs at 24, 23, and 22 mL/mole, respectively.

Values for the exchange specific heat obtained by Greywall[15] at 24.454, 23.785, and 23.081 mL/mole agree with those deduced from Eqs. (2) & (3) within 0.8, 1.2, and 16%, respectively. The agreement at the two highest volumes is excellent while that at the lowest is not outside plausible error limits considering the insensitivity of specific heat to exchange and the longer thermal time constants encountered by Greywall at that density.

CONSTRAINTS FROM NON-EQUILIBRIUM MEASUREMENTS

The interpretation of non-equilibrium measurements usually involves more theoretical assumptions than that of equilibrium measurements. Nevertheless, a strong theoretical foundation exists and, in addition to supplementing equilibrium measurements, totally new dynamic information can be obtained. For brevity, we will restrict our discussion to nuclear magnetic relaxation times and direct the reader to the literature for a discussion of thermal conductivity[2,25], and attenuation of sound[26].

Solid ^{3}He may be thought of as consisting of interacting systems (phonons, vacancy waves, Zeeman, atomic exchange, impurity and dipolar), each making specific contributions to the thermodynamic properties. Interactions among excitations within a particular system tend to drive that system toward spatial homogeneity (e.g., thermal conductivity and self-diffusion) while interactions between systems tend to equalize their effective temperatures.

NMR relaxation studies perturb the temperature of the Zeeman system and monitor its return to equilibrium. The Zeeman system is only coupled to the other systems in ^{3}He via the weak but well-defined dipole-dipole interaction. The fluctuating magnetic field of nearby atoms resulting from exchange or vacancy motions can cause energy to flow between those systems and the Zeeman system. It is only the frequency components of these fluctuations at the Larmor frequency or its second harmonic, however, which can do this. Hence the characteristic time for the reciprocal of the temperature of the Zeeman system to relax to that of the vacancy or exchange systems can be written as[1]

$$T_1 = 1/[J(\omega) + 4J(2\omega)] \quad \text{where} \quad J(\omega) = \int_{-\infty}^{\infty} G(\tau)\ e^{-i\omega\tau} d\tau \tag{4}$$

is called the spectral density function and $G(\tau)$ is the correlation function characterizing the exchange or vacancy motion. For a static bcc lattice averaged over all angles (powder assumption), it can be shown rigorously that

$$G(\tau) = M_2/3 - M_4/6\ \tau^2 + \ldots \quad \text{with} \quad M_2 = 22.796 \times 10^{10}/V^2 \ \ \text{sec}^{-2}. \tag{5}$$

Here, V is the molar volume in mL/mole.

M_4 depends on the specific details of the system to which the Zeeman energy is being relaxed via the dipole-dipole interaction.

Zeeman-Exchange Relaxation: Traditionally M_4 for the Zeeman-exchange relaxation is calculated assuming the Heisenberg nearest neighbor exchange Hamiltonian with the result

$$M_4 = 517.76 \times 10^{10} J^2/V^2 \quad \text{sec}^{-4} \tag{6}$$

where J is the single volume-dependent exchange parameter of the Heisenberg nearest-neighbor model expressed in radians/sec.

For the more complicated exchange models of Roger et al., it is reasonable to expect by dimensional analysis that $V^2 M_4$ will be proportional to the square of an effective J calculated from J_{nn}, J_t, K_p, and K_f. In light of the results presented above concerning the constraints imposed by equilibrium data on these parameters, it appears that, to a good approximation, all exchange parameters will have the same volume dependence. Under that universality constraint, the equilibrium pressure data give the values

$$J_t = -0.1257, \quad K_p = -0.3315, \quad \text{and} \quad A_{Jt} = A_{Kp} = -11.82\ . \tag{7}$$

At this point one can examine various two-parameter correlation functions, constrain their zero time values to equal $M_2/3$, and require that the volume dependence of their characteristic frequencies be consistent with the second term in Eq. (5) with

$$M_4(V)/M_4(24) = (V/24)^{-8.46}\ e^{-23.64[(V/24)^{-1.793}-1]} \ . \tag{8}$$

Here we have used Eq. (2) with $A_{Jt} = -11.82$ and $\gamma_\theta = -2.46$ and the analogy to Eq. (6).

$G(\tau)$		$J(\omega)$
$\dfrac{M_2}{3}\, e^{-\frac{1}{2}\omega_T^2\tau^2}$	Gaussian	$\sqrt{2\pi}\,\dfrac{M_2}{3\omega_T}\, e^{-\frac{1}{2}\frac{\omega^2}{\omega_T^2}}$
$\dfrac{M_2/3}{1+\omega_T^2\tau^2}$	Lorentzian	$\pi\,\dfrac{M_2}{3\omega_T}\, e^{-\omega/\omega_T}$
$\dfrac{M_2/3}{(1+\omega_T^2\tau^2)^2}$	Lorentzian^2	$\dfrac{\pi}{2}\,\dfrac{M_2}{3\omega_T}\left(1+\dfrac{\omega}{\omega_T}\right)e^{-\omega/\omega_T}$

TABLE II. Proposed Correlation Functions

Table II displays three candidate correlation functions. These are chosen only because of their simplicity and have, at present, no theoretical justification. The first two forms (identified as Gaussian and Lorentzian) are traditional while the third form (identified as Lorentzian[2]) is a new variation of the second.

Thomlinson, Kelly and Richardson[27] showed that the spectral density function diminishes exponentially as ω/ω_T at high fields in the bcc phase, thereby eliminating the use of the Gaussian form. Furthermore, Giffard et al[28] confirmed that conclusion and determined the ratio of ω_T to J_{xx}, an effective exchange parameter related to the zero field specific heat: $\omega_T/J_{xx} = 2.78 \pm 0.22$ where $J_{xx}^2 = 48\, J_t^2 - 42\, J_t K_p + 51\, K_p^2/4$.

The Lorentzian and Lorentzian[2] functional forms are then completely determined and we may compare them to other measurements of Zeeman-exchange relaxation times, T_{ze}. This is done in figure 6 where the vertical coordinate is the value of T_{ze} normalized to the zero field value given by the Lorentzian[2], and the horizontal coordinate is the Larmor frequency normalized to the characteristic frequency ω_T. The dashed line is derived from the Lorentzian correlation function while the solid curve, which fits the data quite well, is derived from the Lorentzian[2] form. Although the Lorentzian curve could be made to agree with the data at large ω/ω_T if ω_T/J_{xx} were increased to 3.22 (two standard deviations above Gifford et al.), the discrepancy at low ω/ω_T would still be excessive (35%). The Lorentzian[2] curve is seen to pass through the low ω/ω_T data even though no parameters were chosen to force a fit at zero-field. Thus, the Lorentzian[2] form appears to be the better choice of the three we have considered, although yet another functional form could undoubtedly be contrived to improve the agreement at intermediate fields.

At very high values of ω/ω_T there are four sets of data --

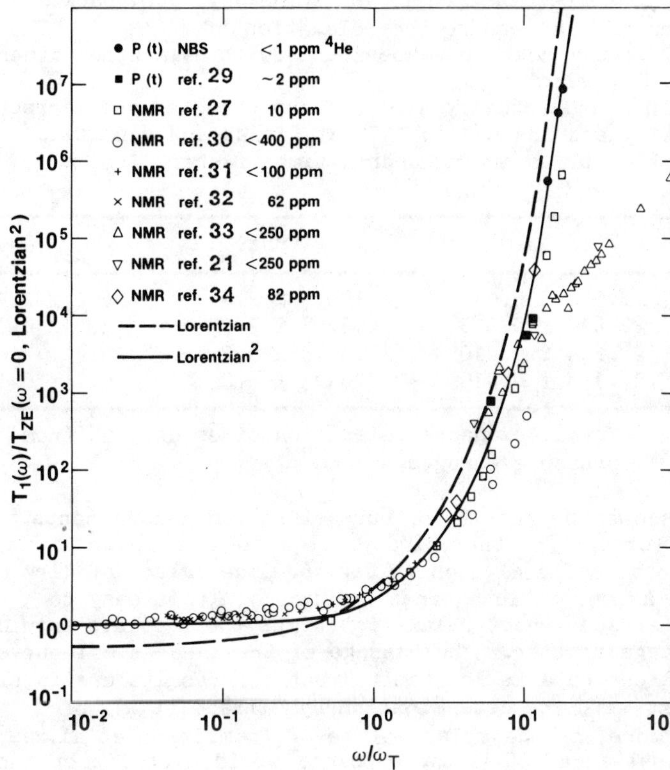

Fig. 6.   Normalized Zeeman-exchange relaxation times

Thomlinson, Kelly, and Richardson[27] (open squares), Sullivan and Chapellier[33] (open triangles), Bernier and Guerrier[34] (open diamonds), and values deduced from our pressure measurements (three solid circles). We will next discuss how Zeeman-exchange relaxation values can be deduced from pressure measurements and later discuss the Sullivan and Chapellier experiment, which strongly deviates from the theoretical curves.

When the partition function is calculated from the Hamiltonian the exchange and Zeeman parts of the Hamiltonian become mixed in the resulting thermodynamic functions because they both depend on the spin variables. If they are not in equilibrium, one must assign a temperature to each system, T_{ex} and T_z, respectively. The resulting high temperature expansion of the pressure will then have terms in $1/T_{ex}$, $1/T_{ex}^2$, H^2/T_z^2, $H^2/T_z^2 T_{ex}$, H^4/T_z^4, etc.

At the lowest molar volumes and highest magnetic fields in our experiments, the Zeeman-exchange relaxation time became extremely long. For example, at 22.614 mL/mole, we made slow measurements lasting 10 days at zero field and obtained the equilibrium pressures. We then increased the field to 8.1 tesla and proceeded to make similar measurements thinking we were measuring the equilibrium pressure. The result, however, was precisely the same as had been obtained in zero field. It is now clear that T_z had been increased to a large value during the raising of the field to 8.1 tesla and had not cooled significantly during the subsequent measurements. Upon lowering the field to 6 tesla, we observed a time dependent pressure which approached the equilibrium value with a time constant of 3.4 days. Furthermore, we could reheat the Zeeman system by raising the sample temperature to 150 mK for a day, cool back down to 35 mK, and once again measure the relaxation of T_z to T_{ex}. T_{ex} always followed the thermometer temperatures with relaxation times too fast to measure reliably.

This phenomenon had actually already been observed and correctly interpreted by Kirk and Adams[29] in 1972 at fields and densities where the relaxation times are a hundred times faster.

T (mK)	V (mL/mole)	f (MHz)	T_{ze} (days)
35 ± 3	22.522 ± 0.010	195.3 ± 3.9	6.2 ± 0.9
32 ± 2	22.614 ± 0.010	196.7 ± 3.9	3.4 ± 0.7
37 ± 3	22.614 ± 0.010	264.5 ± 5.2	> 20
31 ± 1	23.134 ± 0.010	264.3 ± 5.2	0.7 ± 0.11

TABLE III. Zeeman-exchange relaxation times deduced from pressure changes.

Table III shows the results of our relaxation measurements. It is these points (except for the third which is only a lower limit) which appear as the solid circles in Fig. 6. The values of Kirk and Adams are shown as the solid squares in Fig. 6. It is easy to understand that at such short relaxation times their points should have a rather large scatter. The passage of the theoretical curves directly through our data is indicative that our results are in excellent agreement with the determination by Giffard et al. of ω_T/J_{xx}. Furthermore, if the molar volume of Thomlinson et al. were actually 20.51 instead of 20.4, their points would also lie on the Lorentzian[2] curve. This is within their expected error in density determination and, in fact, suggests that the Avilov-Iordanski functional form for the molar volume dependence may actually be significantly better than a simple constant logarithmic derivative form. The latter form would have placed the Thomlinson et al. points a factor of 4 farther from the Lorentzian[2] curve on the other side.

In the earlier discussion of M_2, we explicitly ignored possible anisotropic effects[35]. This assumption appears to have been justified although ^{3}He is expected not to be a powder under the conditions of these experiments. The data in Fig. 6 represent a compilation of results from many different samples and therefore one might expect different orientations.

The results of Sullivan and Chapellier[33], measured up to 80 MHz, clearly contradict this picture. Their results were reproducible during the formation of many samples and covered the same range of molar volumes as our work but at considerably higher temperatures (500-900 mK). Although they quoted a possible ^{4}He impurity concentration of up to 250 ppm, they did not believe that impurity effects were likely to be important.

A comprehensive low field study of impurity effects by Bernier and Deville[36] suggested that ^{4}He impurities can give rise to a temperature independent relaxation mechanism at temperatures near 400 mK which varies as the inverse cube of the impurity concentration. Because of their low field, however, the Zeeman and exchange systems were strongly coupled and it could not be determined whether the impurities were coupling directly to the Zeeman system or indirectly via the exchange system.

Giffard, Stagg, Truscott, and Hatton[28] did a number of experiments at frequencies between 1 MHz and 10 MHz both on bulk samples and on samples containing sintered glass to increase boundary relaxation. They found that the behavior of the samples in sintered glass could be explained by the conventional relaxation processes but that the bulk samples which had impurity concentrations up to 12 ppm, exhibited an unexpected frequency dependence. At 60 mK and from 1 to 5 MHz, the relaxation time exponentially increased with a characteristic frequency 2.5 times greater than ω_T and had a prefactor a million times larger than that for Zeeman-exchange. At higher frequencies, the relaxation time increased less rapidly with frequency.

Very recent measurements at 200 MHz by Bernier and Guerrier[34] on samples containing 82 ppm ^{4}He are also included in Fig. 6. Their Zeeman-exchange relaxation times agree very well with the Lorentzian[2] curve. At temperatures above 200 mK, they confirmed the conventional mechanisms for relaxation of the Zeeman system to the exchange and vacancy systems.

At their lowest temperatures Bernier and Guerrier concluded that the relaxation of the exchange system to the sample chamber walls was slower than the relaxation of the Zeeman system to the exchange system. They offered an extensive discussion regarding the possible role of impurities in pinning dislocations which mediate the exchange-wall relaxation. In our experiment, however, the exchange-wall relaxation times were much shorter than the Zeeman-exchange times. Since our sample chamber is significantly smaller than theirs and our sample purity much higher, a different mechanism is likely to govern our exchange-wall relaxation. Direct diffusion to the walls would be expected to occur more rapidly in our chamber although rough calculations suggest a time which is still about a factor of 10 longer than observed.

It appears that much remains to be understood about low temperature relaxation times, particularly the field dependence of ^{4}He impurity relaxation.

Zeeman-Vacancy Relaxation: Studies of Zeeman-vacancy relaxation provide information on the vacancy number and mobility[1,19-21]. The correlation function is traditionally assumed to be exponential in time with a characteristic frequency ω_V related to the bandwidth of the density of states. Numerous experiments have confirmed the qualitative behavior for the relaxation times which follow from this assumption, but extensive quantitative tests comparable to those available for the Zeeman-exchange mechanism have not yet been made. The work of Sullivan, Deville, and Landesman[21] at 80 MHz seems to indicate that the values deduced for ω_V could be weakly dependent on the magnetic field used -- higher field measurements giving smaller values for ω_V. This might be an indication that a somewhat different correlation function is needed.

SOME AREAS NEEDING FURTHER RESEARCH

In conclusion, it might be useful to list some specific areas where further research would be most useful.

-The high temperature expansion for multiple-exchange processes should be carried to higher order to better utilize low temperature experiments in constraining the possible values of exchange parameters.

-The moments of the Zeeman-exchange spectral density function need to be calculated using the multiple-exchange Hamiltonian. The experimentally observed qualitative difference between the bcc and hcp phases should be calculated.

-The possibility of a universal scaling of exchange phenomena with molar volume should be tested more extensively at all temperatures.

-Additional NMR measurements should be made over the entire range of temperature, density, and magnetic field for up to 20 ppm ^{4}He with particularly careful attention given to density and impurity concentration determinations. Quantitative examination of the spectral density function for vacancy relaxation should be made in both phases.

-Improved calculations for $\theta_D(T)/\theta_D(0)$ are needed.

-Microscopic calculations of atomic and vacancy motion and of elastic constants should be further refined. The ultimate goal of the first-principles calculation of all measured properties does not seem ridiculous given the improvements in computational capacity likely during the next two decades.

ACKNOWLEDGMENTS

The author wishes to thank collaborators P. Bruce Pipes and Walter J. Bowers, Jr., for their contributions to the NBS pressure experiment. Numerous conversations with Robert Guyer, Jack Hetherington, Robert Richardson, and Ralph Simmons were very helpful in clarifying the status of existing measurements and theory. In addition, extensive discussions during the Sanibel conference with Michel Bernier, Maurice Chapellier, and Neil Sullivan were helpful.

REFERENCES

1. R.A. Guyer, R.C. Richardson and L.I. Zane, Rev. Mod. Phys. __43__, 532 (1971).

2. S.B. Trickey, W.P. Kirk and E.D. Adams, Rev. Mod. Phys. __44__, 668 (1972).

3. M. Roger, J.H. Hetherington, and J.M. Delrieu, Rev. Mod. Phys. __55__, 1 (1983).

4. D.S. Greywall, J. Low Temp. Phys. __46__, 451 (1982) and D.S. Greywall, Phys. Rev. __B27__, 2747 (1983).

5. Craig T. Van Degrift, Proc. of 2nd Intern. Conf. on Precision Measurements and Fundamental Constants, NBS, 1981, in press.

6. G.C. Straty and E.D. Adams, Phys. Rev. __150__, 123 (1966).

7. Precise location of the transition line to the ordered phase is difficult. See, for example, T. Shigi, T. Hata, S. Yamasaki, and T. Kodama, these proceedings.

8. G.C. Straty and E.D. Adams, Phys. Rev. 169, 232 (1968).

9. M.F. Panczyk and E.D. Adams, Phys. Rev. 187, 321 (1969) and Phys. Rev. A1, 1356 (1970).

10. W.P. Kirk and E.D. Adams, Phys. Rev. Lett. 27, 392 (1971).

11. C.T. Van Degrift, W.J. Bowers, Jr., P.B. Pipes and D.F. McQueeney, Phys. Rev. Lett. 49, 149 (1982).

12. R.A. Guyer, J. Low Temp. Phys. 30, 1 (1978).

13. Z. Olejniczak, W.P. Kirk, A.A.V. Gibson, P. Kobiela, and A. Czermak, these proceedings.

14. C.T. Van Degrift, W.J. Bowers, Jr. and P.B. Pipes, to be published.

15. D.S. Greywall, Phys. Rev. B15, 2604 (1977) and Phys. Rev. B16, 5129 (1977).

16. F.W. de Wette and N.R. Werthamer, Phys. Rev. 184, 209 (1969).

17. E.R. Grilly, J. Low Temp. Phys. 4, 615 (1971) and E.R. Grilly, J. Low Temp. Phys. 11, 243 (1973).

18. D.S. Greywall, Phys. Rev. B11, 1070 (1975).

19. J.H. Hetherington, Phys. Rev. 176, 231 (1968).

20. S.M. Heald, D.R. Baer and R.O. Simmons, to be published. W.E. Schoknecht and R.O. Simmons, Thermal Expansion - 1971, AIP Conf. Proc. 3, 169 (1971). S.M. Heald, Ph.D. thesis, U. Illinois.

21. N.S. Sullivan, G. Deville, and A. Landesman, Phys. Rev. B11, 1858 (1975).

22. J.H. Hetherington, J. Low Temp. Phys. 32, 173 (1978).

23. V.V. Avilov and S.V. Iordanski, J. Low Temp. Phys. 48, 241 (1982). Also see M. Roger, J. Physique Lett. (to be published).

24. T. Kodama, private communication, Dec. 1982.

25. G. Armstrong and A.S. Greenberg, J. Physique C6, 135 (1978), W.C. Thomlinson, Phys. Rev. 23, 1330 (1969) and B. Bertman, H.A. Fairbank, C.W. White, and M.J. Crooks, Phys. Rev. 142, 74 (1966).

26. J.R. Beamish and J.P. Franck, Phys. Rev. Lett. 47, 1736 (1981).

27. W.C. Thomlinson, J.F. Kelly and R.C. Richardson, Phys. Lett. 38A, 531 (1972). A qualitative difference is demonstrated between the bcc and hcp phases.

28. R.P. Giffard, J.P. Stagg, W.S. Truscott and J. Hatton, J. Low Temp. Phys. 31, 817 (1978).

29. W.P. Kirk and E.D. Adams, Low Temperature Physics - LT 13, K.D. Timmerhaus, W.J. O'Sullivan, and E.F. Hammel, Eds. (Plenum Press, N.Y., 1974), Vol. 2, p. 149.

30. R.C. Richardson, E. Hunt, and H. Meyer, Phys. Rev. 138, A1326 (1965).

31. S. Nishizawa, T. Kobayashi, and Y. Narahara, Phys. Lett. 64A, 103 (1977).

32. M. Devoret, A.S. Greenberg, D. Estève, N.S. Sullivan, and M. Chapellier, J. Low Temp. Phys. 48, 495 (1982).

33. N.S. Sullivan and M. Chapellier, J. Phys. C7, L195 (1974).

34. M. Bernier and G. Guerrier, to be published.

35. G. Deville, J. Physique 37, 781 (1976).

36. M. Bernier and G. Deville, J. Low Temp. Phys. 16, 349 (1974).

ON THE EXCHANGE PARAMETERS OF BCC SOLID ^{3}He

Katsutoshi IWAHASHI and Yoshika MASUDA
Department of Physics, Nagoya University, Nagoya 464
Japan

ABSTRACT

Using the spin Hamiltonian, in which two-spin interactions up to third neighbors and two kinds of large folded and planar cyclic four-spin exchange terms are included, spin wave spectrum of "uudd" and "pseudoferromagnetic (PF)" phase of bcc solid ^{3}He in finite magnetic field has been calculated. The necessary conditions for the exchange parameters to explain the first order phase transition from "uudd" to "PF" at magnetic field of about 0.4T and high temperature experimental results have been obtained within a framework of a three-parameter model. By a calculation using a mean field approximation, the new phase, double spiral spin configuration has been found, and the possibility of the appearance of this phase near T_N are discussed.

INTRODUCTION

During the last two decades, a great deal of experimental and theoretical efforts have been devoted to understanding the magnetism of bcc solid ^{3}He. Due to a discovery of antiferromagnetic resonance of bcc solid ^{3}He, it has been believed that the ordered phase of solid ^{3}He in magnetic field h<0.4T is the so-called "uudd".[1] "uudd" is the state with the ferromagnetic planes perpendicular to a [100] axis arranged in up-up-down-down sequence. At h=0.4T, solid ^{3}He makes the first order transition and in h>0.4T becomes pseudoferromagnetic(PF).[2] It is well known that the magnetic properties of solid ^{3}He can be explained by the Hamiltonian H_0,

$$H_0 = -\sum_{n=1}^{3} (2J_n \sum_{i<j} \vec{S}_i \cdot \vec{S}_j) - \sum_{\alpha=P,F} 4K_\alpha \sum_{i<j<k<\ell} (\vec{S}_i \cdot \vec{S}_j \vec{S}_k \cdot \vec{S}_\ell$$

$$+\vec{S}_j \cdot \vec{S}_k \vec{S}_\ell \cdot \vec{S}_i - \vec{S}_i \cdot \vec{S}_k \vec{S}_j \cdot \vec{S}_\ell). \tag{1}$$

The second term is a four-spin cyclic permutation operator acting around planar (P) and folded (F) rings of nearest neighbors. In the case of K_P, one of the diagonals corresponds to second neighbors and other to third neighbors.

In the case of K_F, both diagonals are second neighbors.

Roger, Hetherington and Delrieu[3] showed that the magnetic properties of solid ^{3}He, except the determination of magnetic field strength at which the first order phase transition occurs, could be explained by a two parameter (j_t, K_P) model, where j_t is the triple spin exchange. However, precise determination of exchange parameters is very difficult. The high temperature expansion coefficients of free energy determined from experimental results are not accurate. On the other hand, the transition temperature, T_N, entropy change at T_N, δS, and mean velocity of spin wave, v_s, deduced from a temperature dependence of melting pressure,[4] seem to be obtained with high accuracy. No theory yet exists to determine T_N and δS accurately. The value of v_s calculated by usual spin wave theory changes appreciably by taking into account the interactions between spin waves.[5]

Since the investigation by Hetherington and Willard[6], various spin configurations such as ferro, PF, uudd, $SSQA\perp$ and $SCAF\perp$ have been proposed,[7] by means of a three-parameter (j_t, K_P, K_F) model. $SSQA\perp$ represents the simple square antiferromagnetic phase, in which the state with the ferromagnetic planes perpendicular to the [110] axis arranged in udud sequence and the spins at corner-site is perpendicular to that at body-center site. $SCAF\perp$ represents a simple cubic antiferromagnetic phase, in which the spin at body-center site is perpendicular to that at corner-site. However, from calculations of spin wave spectra of the uudd and PF phases, we found that there exists the domain in the exchange parameter space, in which both spin wave frequencies of uudd and PF become imaginary. In this region, a new phase other than the above-mentioned, becomes stable.[8]

In the second chapter, we look for the phase diagram in the exchange parameter space by a rather systematic way. In the third chapter, we discuss the first order phase transition at h=0.4 T by taking into account the existence of the new phase, and show the appropriate region for real solid ^{3}He in the three-exchange parameter space. In the fourth chapter we discuss the possibility of an appearance of the new phase at finite temperature.

DOUBLE SPIRAL SPIN CONFIGURATION

In this chapter we deal with the problem by a mean field approximation (MFA). If four spin exchanges K_P and K_F are absent, the ground state of the system described by Eq.(1) apparently has a simple form as

$$<\vec{S}_i> = \text{Re}[\vec{\sigma}\exp(i\vec{k}\cdot\vec{R}_i)], \tag{2}$$

34

where $\vec{R}_i$ is the position vector of the i-th site. In this case, using the Fourier transform of $<\vec{S}_i>$, the energy of the system is represented as,

$$E = \sum_{\vec{q}} -J(\vec{q})\vec{S}_{\vec{q}}\cdot\vec{S}_{-\vec{q}}. \tag{3}$$

At $\vec{q}=\vec{k}$, $J(\vec{q})$ becomes maximum. But when four spin exchanges exist, there is possibility that the ground state has more complicated form. Therefore we examine the energy of the state which has a more generalized form than Eq. (2) such as,

$$<\vec{S}_i> = Re[\vec{\sigma}\,exp(i\vec{k}\cdot\vec{R}_i)+\vec{\sigma}'exp(i\vec{k}'\cdot\vec{R}_i)], \tag{4}$$

with the restriction that all $|<\vec{S}_i>|$ are equal. This restriction yields the condition $\vec{k}'=\vec{G}/2-\vec{k}$, where $\vec{G}$ is a reciprocal lattice vector of the bcc lattice, and $<\vec{S}_i>$ is represented as

$$<\vec{S}_i> = S[\vec{u}\,cos(\vec{k}\cdot\vec{R}_i)+\vec{v}\,sin(\vec{k}\cdot\vec{R}_i)],$$

$$\text{for site } \vec{G}\cdot\vec{R}_i/2 = 2n\pi \tag{5}$$

$$<\vec{S}_j> = S[\vec{u}'cos(\vec{k}\cdot\vec{R}_j)+\vec{v}'sin(\vec{k}\cdot\vec{R}_j)],$$

$$\text{for site } \vec{G}\cdot\vec{R}_j/2 = (2n+1)\pi \tag{5'}$$

where $\vec{u}^2=\vec{v}^2=\vec{u}'^2=\vec{v}'^2=1$, $\vec{u}\cdot\vec{v}=\vec{u}'\cdot\vec{v}'=0$ and n is integer. That is, the generalization of spin configuration of Eq. (2) to Eq.(4) with the restiction that all $|<\vec{S}_i>|$ being equal, mean that the spins of one site, at which $\vec{G}\cdot\vec{R}_i/2$ equal to $2n\pi$ are situated on the plane composed by two vectors $\vec{u}$ and $\vec{v}$, and form spiral or helicoidal structure with wave vector $\vec{k}$. The spins of another site, at which $\vec{G}\cdot\vec{R}_j/2$ is equal to $(2n+1)\pi$, are located on the other plane composed by two vectors $\vec{u}'$ and $\vec{v}'$. Thus w being the angle between two planes, that is $cos(w)=(\vec{u}\times\vec{v})\cdot(\vec{u}'\times\vec{v}')$, $<\vec{S}_j>$ of the site $\vec{G}\cdot\vec{R}_i/2=(2n+1)\pi$, Eq.(5'), can be represented in the following way,

$$<\vec{S}_j> = S\{\vec{u}\cdot cos(\vec{k}\cdot\vec{R}_j+\delta)+[\vec{v}\cdot cos(w)+\vec{u}\times\vec{v}\cdot sin(w)]sin(\vec{k}\cdot\vec{R}_j+\delta)\} \tag{5''}$$

$\vec{G}/2$ can be restricted to be equal to $\pi(2,0,0)/a$, case(I) or $\pi(1,1,0)/a$, case(II), without loss in generality, where a is the lattice constant of conventional cubic unit cell. δ is the phase difference. The exchange energy per spin is easily calculated by making use

of Eqs. (1), (5) and (5') in each case. We look for $\vec{k}$, w
and δ for which the exchange energy becomes minimum.
By computer calculation we found that the spin configuration at which exchange energy has the lowest value, appears in case(I) alone. In this case the exchange
energy is given as,

$$E = S^2\{-4J_1[1+\cos(w)]\cos(\delta)\gamma_{1\vec{k}}-6J_2\gamma_{2\vec{k}}-12J_3\gamma_{3\vec{k}}$$

$$+12S^2\sin^2(w)(K_p\gamma_{p\vec{k}}+K_F\gamma_{3\vec{k}})-6S^2(K_p+K_F\gamma_{2\vec{k}})$$

$$\times[1+\cos(w)]^2\cos(2\delta)\}. \tag{6}$$

In this case(I), if one site is corner site of the bcc
lattice, another site becomes body-center site, and $\gamma_{n\vec{k}}$
are given by Eq. (7):

$$\gamma_{1\vec{k}} = \cos(k_xa/2)\cos(k_ya/2)\cos(k_za/2),$$

$$\gamma_{2\vec{k}} = [\cos(k_xa)+\cos(k_ya)+\cos(k_za)]/3,$$

$$\gamma_{3\vec{k}} = [\cos(k_xa)\cos(k_ya)+\cos(k_ya)\cos(k_za)$$

$$+\cos(k_za)\cos(k_xa)]/3, \tag{7}$$

$$\gamma_{p\vec{k}} = \cos(k_xa)\cos(k_ya)\cos(k_za).$$

Further, when the problem is restricted within the three-
parameter model, only the state in which w is equal to
zero, has a minimum energy for the given set of exchange
parameters. In this case, case(I) with w=0, the spins of
both sites are situated on the same plane, but the spins
of the body-center site have different phase from that of
the corner site. The validity of Eq. (6) can be easily
verified, because the following identity holds for the
planar or folded cyclic four-spin exchange:

$$<\vec{S}_i>\cdot<\vec{S}_j><\vec{S}_k>\cdot<\vec{S}_\ell>+<\vec{S}_j>\cdot<\vec{S}_k><\vec{S}_\ell>\cdot<\vec{S}_i>-<\vec{S}_i>\cdot<\vec{S}_k><\vec{S}_j>\cdot<\vec{S}_\ell>$$

$$= S^4\cos[\vec{k}\cdot(\vec{R}_i-\vec{R}_j+\vec{R}_k-\vec{R}_\ell)-2\delta],$$

and

$$\vec{R}_i-\vec{R}_j+\vec{R}_k-\vec{R}_\ell = \begin{cases} 0 & \text{for any quartet of the planar type} \\ (\pm a,0,0),\ (0,\pm a,0)\ \text{or}\ (0,0,\pm a) \\ \qquad \text{for the folded type.} \end{cases}$$

Now the state is characterized by wave vector $\vec{k}$ and phase difference δ. We introduce the notation $(\vec{k}a, \delta)$ to represent the state, and name the state in which $\vec{k}a$ and δ are not equal to $(\pi/2) \times$ integer, as a double spiral spin configuration, which abbreviates as DS hereafter. If the exchange integrals are selected appropriately, following eight states become stable in the three-parameter model:

$$(0,0,0,0) \equiv \text{ferro}, \quad (0,0,0,\delta) \equiv \text{PF}, \quad (\pi,0,0,\pi/2) \equiv \text{uudd},$$

$$(\pi,\pi,\pi,0) \equiv \text{SCAF}\perp, \quad (\pi,\pi,0,\pi/2) \equiv \text{SSQA}\perp, \quad (x,0,0,\delta) \equiv \text{DS}(100),$$

$$(x,x,0,\delta) \equiv \text{DS}(110) \quad \text{and} \quad (x,x,x,\delta) \equiv \text{DS}(111).$$

The calculated results of energies of states are listed in Table I, where $4K_\alpha S^2$ was replaced by K_α for simplicity.[9]

Table I Exchange energy

state	energy
ferro	$2S^2[-4J_1-3J_2-6J_3-3(K_P+K_F)]$
PF	$2S^2\{-3J_1-6J_2+3(K_P+K_F)+2J_1^{\,2}/[3(K_P+K_F)]\}$
uudd	$2S^2(-J_2+2J_3+3K_P+K_F)$
DS(100)	$2S^2[-J_2+2J_3+3K_P+K_F+(\sqrt{(3K_P+K_F)(J_2+4J_3-K_F)}-\lvert J_1\rvert)^2/K_F]$
SSQA$\perp$	$2S^2(J_2+2J_3+3K_P-K_F)$
DS(110)	$2S^2\{J_2+2J_3+3K_P-K_F+[J_2+J_1\cos(\delta)+K_F\cos(2\delta)] \times [(J_2-2K_F-K_F\cos(2\delta)]/2J_3\}$
SCAF$\perp$	$2S^2(3J_2-6J_3-3K_P+3K_F)$
DS(111)	$2S^2\{-3\cos(x)[J_2+2J_3\cos(x)]-3[K_P+K_F\cos(x)]+J_1^{\,2}[1+\cos(x)]^3/(12[K_P+K_F\cos(x)])\}$

x and δ are given as follows,

$$\cos(\delta) = -J_1/[3(K_P+K_F)], \quad \text{for PF,}$$

$$\begin{cases} \cos^2\left(\tfrac{x}{2}\right)=-\dfrac{1}{2K_F}\left\{\,|J_1|-\sqrt{(3K_P+K_F)(J_2+4J_3-K_F)}\,\right\}\sqrt{\dfrac{3K_P+K_F}{J_2+4J_3-K_F}}\ , \\[4ex] \cos(\delta)=\mathrm{sign}(J_1)\sqrt{\dfrac{J_2+4J_3-K_F}{3K_P+K_F}}\ \cos\left(\tfrac{x}{2}\right), \qquad \text{for DS(100)} \end{cases}$$

$$\begin{cases} K_F^2\cos(3\delta)-2\,[\,3K_PJ_3-K_F(J_1+J_2+J_3)\,]\cos(2\delta)+(J_1^2+K_F^2)\cos(\delta) \\[2ex] \qquad\qquad +J_1(J_2+K_F)=0, \\[2ex] \cos(x)=1-\dfrac{1}{2J_3}[\,J_2+J_1\cos(\delta)+K_F\cos(2\delta)\,], \qquad \text{for DS(110)} \end{cases}$$

$$\begin{cases} J_2-K_F+4J_3\cos(x)-\dfrac{J_1^2\,[1+\cos(x)]^2}{12\,[K_P+K_F\cos(x)]}\left\{1-\dfrac{K_F[1+\cos(x)]}{3\,[K_P+K_F\cos(x)]}\right\}=0, \\[4ex] \cos(\delta)=-\dfrac{J_1\cos^3\left(\tfrac{x}{2}\right)}{3\,[K_P+K_F\cos(x)]}. \qquad\qquad \text{for DS(111)} \qquad (8) \end{cases}$$

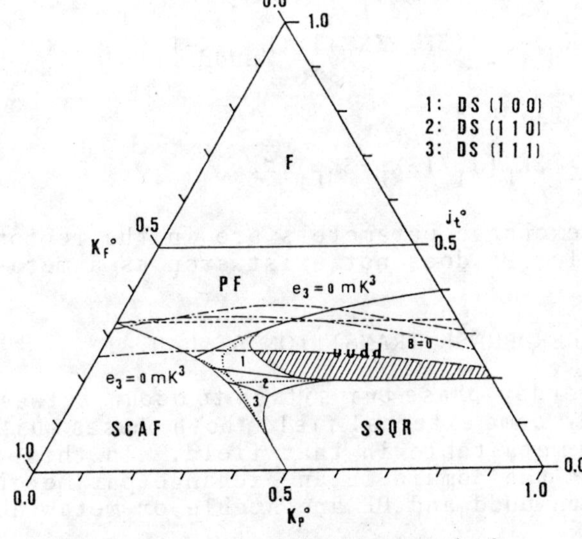

Fig.1 Phase boundaries at h=0.

The phase boundaries are shown in Fig. 1 as functions of relative ratios of exchange parameters:

$$j_t^0=j_t/(j_t+K_P+K_F),$$

$$K_P^0=K_P/(j_t+K_P+K_F)$$

and

$$K_F^0=K_F/(j_t+K_P+K_F).$$

The lines on which the high temperature expansion coefficients e_3 or B becomes zero are also included. In the region between two lines showing $e_3=0$,

e_3 becomes negative, and in an upper part above B=0 line B becomes positive. Therefore exchange parameters must be in the shaded region in Fig.1.

The important fact deduced from Fig. 1 is that there exists the domain of DS(100) being stable in between two domains in which "uudd" and "PF" being stable respectively. The energy of DS(100), $E_{DS(100)}$ is given by,

$$E_{DS(100)} = E_{PF} + [-|J_1| \cdot \sqrt{\frac{3K_P + K_F}{3(K_P + K_F)} + \sqrt{3(K_P + K_F)(J_2 + 4J_3 - K_F)}}]^2$$

$$\times 2S^2/K_F,$$

where E_{PF} is the energy of PF. Therefore it is evident from Table I and the above equation that DS(100) is more stable than uudd or PF, whenever x and δ exist. The condition for the existence of x and δ is obtained from Eq. (8) as follows,

$$J_1[J_1/(3K_P + K_F) - J_1/(3K_P + 3K_F)] \leq (E_{uudd} - E_{PF})$$
$$\leq -2K_F[J_1/(3K_P + 3K_F)]^2,$$

where E_{uudd} is the energy of uudd. On the other hand, one of the necessary conditions for spin wave frequency of uudd being real is given as,

$$J_1[J_1/(3K_P + K_F) - J_1/(3K_P + 3K_F)] \geq (E_{uudd} - E_{PF}),$$

and that for PF is given as,

$$(E_{uudd} - E_{PF}) \geq -2K_F[J_1/(3K_P + 3K_F)]^2.$$

That is, when the exchange parameters are in the region of uudd being stable, PF does not exist even as a metastable state.

FIRST ORDER PHASE TRANSITION AT h=0.4T

For the first order phase transition to occur between the two phases under some external field, both phases must be stable or at least metastable in that field. In this chapter we look for the domain in an exchange parameter space, in which both uudd and PF are stable or metastable at about h=0.4T.

We define "the state being metastable" as

$$E\{\vec{S}_i^0\} \leq E\{\vec{S}_i^0 + \delta\vec{S}_i\}, \tag{9}$$

for any small $\delta\vec{S}_i$, where E is the energy of the phase $\{\vec{S}^0_i\}$ calculated by using the MFA. At 0 K, this means that all spin wave frequencies of the phase $\{\vec{S}^0_i\}$ are real.

We now ask for the stability of PF. The magnetic field is applied in parallel to the x-axis, and S_x' and S_y' are some small quantities. Then $\vec{S}^0_i + \delta\vec{S}_i$ is represented as follows,

$$\vec{S}^0_i+\delta\vec{S}_i=\{S-[\frac{1}{2S}(S^2_{ix'}+S^2_{iy'})-\frac{1}{2}]\}\cdot\begin{pmatrix}\sin\theta\\0\\\cos\theta\end{pmatrix}+S_{ix'}\cdot\begin{pmatrix}\cos\theta\\0\\-\sin\theta\end{pmatrix}$$

$$+S_{jy'}\begin{pmatrix}0\\1\\0\end{pmatrix}\qquad\text{for corner site,}\qquad(10)$$

$$\vec{S}^0_j+\delta\vec{S}_j=\{S-[\frac{1}{2S}(S^2_{jx'}+S^2_{jy'})-\frac{1}{2}]\}\begin{pmatrix}\sin\theta\\0\\-\cos\theta\end{pmatrix}+S_{jx'}\cdot\begin{pmatrix}\cos\theta\\0\\\sin\theta\end{pmatrix}$$

$$+S_{iy'}\begin{pmatrix}0\\-1\\0\end{pmatrix}\qquad\text{for body-center site.}\qquad(10')$$

The above definition of $S_{ix'}$ and $S_{iy'}$ assures the commutation relation that $[S_{ix'},S_{iy'}]=i\vec{S}_i\cdot\vec{S}^0_i/|\vec{S}^0_i|$. The constant $-1/2$ in the parenthesis [ ] of Eqs. (10) and (10') is requested quantum mechanically. θ will be defined later, as a function of the exchange parameters and external field strength h. Inserting Eqs. (10) and (10') into Hamiltonian H, $H=H_0-h\sum_i S_{ix}$, the energy of PF in magnetic field is expressed as follows,

$$E_{PF}(h)=E^M_{PF}(h)+S^{-1}(\frac{\partial E^M_{PF}(h)}{\partial\theta})N^{-1}\sum_{(i)}S_{ix}+E^{(2)}_{PF}-SD,\qquad(11)$$

$$E_{PF}^{M}(h) = 2S^2[4J_1\cos(2\theta) - 3J_2 - 6J_3 - 3(K_P + K_F)\cos(4\theta)$$

$$-\frac{h}{2S}\sin(\theta)] \tag{11'}$$

$$NE_{PF}^{(2)} = D\sum_i (S_{ix}^2 + S_{iy}^2) + \frac{1}{4}A\sum_{(nn)} S_{ix}\cdot S_{jx} + \frac{1}{3}B\sum_{(nnn)} S_{ix}\cdot S_{kx}$$

$$+\frac{1}{6}C\sum_{(3-rd)} S_{ix}\cdot S_{kx} + \frac{1}{4}a\sum_{(nn)} S_{iy}\cdot S_{jy}$$

$$+\frac{1}{3}b\sum_{(nnn)} S_{iy}\cdot S_{ky} + \frac{1}{6}c\sum_{(3-rd)} S_{iy}\cdot S_{ky}, \tag{11''}$$

$$A = 8[-J_1\cos(2\theta) + 3(K_P + K_F)\cos(4\theta)], \quad a = 8[J_1 - 3(K_P + K_F)\cos(2\theta)],$$

$$B = 6[-J_2 + (K_P + 2K_F)\cos(4\theta)], \quad\quad b = 6(-J_2 + K_P + 2K_F),$$

$$C = 12[-J_3 + \tfrac{1}{2}K_P\cos(4\theta)], \quad\quad c = 12(-J_3 + \tfrac{1}{2}K_P),$$

$$D = -a - b - c, \tag{11'''}$$

where primes in $S_{ix'}$ and $S_{iy'}$ were omitted and $4K_\alpha S^2$ was replaced by K_α.

θ is determined by the requirement $\dfrac{\partial E_{PF}^{M}(h)}{\partial \theta} = 0$:

$$-16J_1\sin(\theta) + 48(K_P + K_F)\cos(2\theta)\sin(\theta) - \frac{h}{2S} = 0. \tag{12}$$

The phrase "PF being stable" means that the quadratic polynominal with recpect to S_{ix} and S_{iy}, $NE_{PF}^{(2)}$ is positive definite. By taking Fourier transform

$$\vec{S}_i = \frac{1}{\sqrt{N}}\sum_{\vec{k}} \exp(i\vec{k}\cdot\vec{R}_i)\vec{S}_{\vec{k}},$$

$NE_{PF}^{(2)}$ is transformed into the following simple form,

$$NE_{PF}^{(2)} = \sum_{\vec{k}} (D + A\gamma_{1\vec{k}} + B\gamma_{2\vec{k}} + C\gamma_{3\vec{k}}) S_{\vec{k}x} \cdot S_{\vec{k}x}^{*}$$

$$+ (D + a\gamma_{1\vec{k}} + b\gamma_{2\vec{k}} + c\gamma_{3\vec{k}}) S_{\vec{k}y} \cdot S_{\vec{k}y}^{*}$$

$$\equiv \sum_{\vec{k}} D_x S_{\vec{k}x} \cdot S_{\vec{k}x}^{*} + D_y S_{\vec{k}y} S_{\vec{k}y}^{*}, \tag{13}$$

where number of $\vec{k}$ is N and $\vec{k}$ is some vector in the first Brillouin zone of the bcc lattice, not the simple cubic lattice, and $\gamma_{1\vec{k}}$ and so on are given by Eq. (7). We define h_{c2} for a given set of exchange parameters as the smallest field strength for which D_x and D_y are positive for any $\vec{k}$. In Fig.2. several h_{c2}'s are shown in units of $(-\Theta/3)$, where Θ is the paramagnetic Curie temperature.

In phrase of usual spin wave theory of an antiferromagnet, PF must have two spin wave modes. This is easily shown as the following way from Eq.(13). In Eq.(13), for $\vec{k}$ being outside of the first Brillouin zone of the simple cubic lattice, we define $\vec{k}'$ is to be in the first Brillouin zone of the simple cubic lattice, by the relation of $\vec{k}'=\vec{k}-\vec{G}$. $\vec{G}$ is the appropriate reciprocal lattice vector of the simple cubic lattice. Then putting $S_{\vec{k}}=S_{\vec{k}}'$, Eq.(13) is reexpressed as follows:

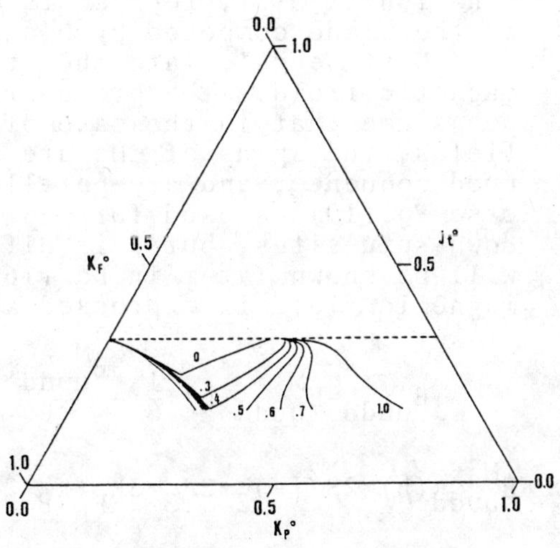

Fig.2 h_{c2} in units of $(-\Theta/3)$.

$$NE_{PF}^{(2)} = \sum_{\vec{k}} F_{1\vec{k}} S_{\vec{k}x} \cdot S_{\vec{k}x}^{*} + F_{2\vec{k}} S_{\vec{k}x}' \cdot S_{\vec{k}x}^{*}' + f_{1\vec{k}} S_{\vec{k}y} \cdot S_{\vec{k}y}^{*} + f_{2\vec{k}} S_{\vec{k}y}' \cdot S_{\vec{k}y}^{*}', \tag{13'}$$

$$F_{1\vec{k}} = D + A\gamma_{1\vec{k}} + B\gamma_{2\vec{k}} + C\gamma_{3\vec{k}}, \quad F_{2\vec{k}} = D - A\gamma_{1\vec{k}} + B\gamma_{2\vec{k}} + C\gamma_{3\vec{k}}, \tag{13''}$$

$$f_{1\vec{k}} = D + a\gamma_{1\vec{k}} + b\gamma_{2\vec{k}} + c\gamma_{3\vec{k}}, \quad f_{2\vec{k}} = D - a\gamma_{1\vec{k}} + b\gamma_{2\vec{k}} + c\gamma_{3\vec{k}}, \tag{13'''}$$

where the commutation relation, $[S_{ix}, S_{iy}] \sim iS$, was used. Equation (13') is easily diagonalized and expressed as

42

$$N(E_{PF}^{(2)}-SD)=\sum_{\vec{k}}\{2S\sqrt{F_{1\vec{k}}f_{1\vec{k}}}(n_{\vec{k}}+\tfrac{1}{2})-SD+2S\sqrt{F_{2\vec{k}}f_{2\vec{k}}}(n'_{\vec{k}}+\tfrac{1}{2})-SD\}. \quad (14)$$

Equation (14) is just the Hamiltonian of PF, where $n_{\vec{k}}$ and $n'_{\vec{k}}$ are the occupation number of two spin wave modes, whose wave number is $\vec{k}$. The total number of $\vec{k}$ ammounts to $N/2$, and $\vec{k}$ are located in the first Brillouin zone of the simple cubic lattice.

When the three exchange parameters are in the domain, in which uudd is stable in low external applied field, at $h=h_{c2}$, $F_{2\vec{k}}$ becomes zero at some $\vec{k}$ which is parallel to one of the (111) axis. While $f_{2\vec{k}}$ is positive. This means that the present spin configuration (Eqs.(5) and (5')) is incomplete, for the purpose to obtain the phase diagram in finite magnetic field. Because the magnetic field is applied along the x-axis, and $S_{\vec{k}x}$ which is a linear combination of $S_{ix'}$, represents the deviation of the spins in the plane composed by $\vec{h}$ and $\vec{S}^0$.

Next we calculate the stability of uudd in finite magnetic field. The procedure is the same as that for PF. We assume that in the case of absence of the external fields, the spins of ^{3}He are aligned along the x-axis in uudd sequence, and are parallel to the z-axis. In this case Eq.(10) is used for up spin sites and Eq.(10') for down spin sites, but θ is different from that for PF, and will be shown later in Eq.(16). The energy of uudd in magnetic field is expressed as follows,

$$E_{uudd}=E_{uudd}^{M}(h)+\frac{1}{S}\left(\frac{\partial E_{uudd}^{M}(h)}{\partial\theta}\right)\frac{1}{N}\sum_{i}S_{ix}+E_{uudd}^{(2)}-SD_u \quad (15)$$

$$E_{uudd}^{M}(h)=2S^2[-J_2+2J_3+3K_P+K_F-2(2J_1+J_2+4J_3+3K_P)t^2-4K_Ft^4-\frac{h}{2S}t]$$

$$NE_{uudd}^{(2)}=D_u\sum_{i}(S_{ix}^2+S_{iy}^2)+\frac{1}{2}A^{+}\sum_{(nn)}^{\uparrow\uparrow}S_{ix}S_{jx}+\frac{1}{2}A^{-}\sum_{(nn)}^{\uparrow\downarrow}S_{ix}S_{jx}$$

$$+\frac{1}{2}a^{+}\sum_{(nn)}^{\uparrow\uparrow}S_{iy}S_{jy}+\frac{1}{2}a^{-}\sum_{(nn)}^{\uparrow\downarrow}S_{iy}S_{jy}+\frac{1}{2}B^{+}\sum_{(nnn)}^{\uparrow\uparrow}S_{ix}S_{kx}$$

$$+B^{-}\sum_{(nnn)}^{\uparrow\downarrow}S_{ix}S_{kx}+\frac{1}{2}b^{+}\sum_{(nnn)}^{\uparrow\uparrow}S_{iy}S_{ky}+b^{-}\sum_{(nnn)}^{\uparrow\downarrow}S_{iy}S_{ky}$$

$$+\frac{1}{2}C^{+}\sum_{(3\text{-rd})}^{\uparrow\downarrow}S_{ix}S_{\ell x}+\frac{1}{4}C^{-}\sum_{(3\text{-rd})}^{\uparrow\downarrow}S_{ix}S_{\ell x}+\frac{1}{2}c^{+}\sum_{(3\text{-rd})}^{\uparrow\uparrow}S_{ix}S_{\ell x}$$

$$+\frac{1}{4}C^-\sum_{(3-rd)}^{\prime\prime} S_{ix}S_{\ell x},$$

$$A^+=4[-J_1-(3K_P+K_F)\cos(2\theta)-2K_Ft^2],\quad a^+=A^+,$$

$$A^-=4\{(-J_1-3K_P-K_F)\cos(2\theta)\qquad\qquad a^-=4(J_1+3K_P+K_F+2K_Ft^2),$$

$$-K_F[\cos(2\theta)-\cos(4\theta)]\},$$

$$B^+=4[-J_2-(K_P+2K_Ft^2)\cos(2\theta)],\qquad b^+=4[-J_2-K_P\cos(2\theta)+2K_Ft^2],$$

$$B^-=2(-J_2+K_P+2K_F)\cos(2\theta),\qquad b^-=2(J_2-K_P-2K_F),$$

$$C^+=4[-J_3-\tfrac{1}{2}K_P\cos(2\theta)],\qquad\qquad c^+=C^+,$$

$$C^-=8(-J_3+\tfrac{1}{2}K_P)\cos(2\theta),\qquad\qquad c^-=8(J_3-\tfrac{1}{2}K_P),$$

$$D_u=-(a^++a^-+b^++b^-+c^++c^-),$$

where we put $\sin\theta=t$. $\sum_{(nn)}^{\prime\prime(N)}$ means the summation over the nearest neighbors parallel (antiparallel) spin pairs.

From the equation $\dfrac{\partial E^M_{uudd}(h)}{\partial\theta}=0$, we obtain the following equation:

$$4(2J_1+J_2+4J_3+3K_P)t+16K_Ft^3+\frac{h}{2S}=0. \qquad (16)$$

By making use of the following orthogonal transformation $E^{(2)}_{uudd}$ is easily simplified:

$$\vec{S}_{1\vec{k}}=\sqrt{\frac{2}{N}}[\sum_i^{corner}\cos(\vec{k}\cdot\vec{R}_i+\pi/4)\vec{S}_i+\sum_i^{b-c}\cos(\vec{k}\cdot\vec{R}_j+\pi/4)\vec{S}_j],$$

$$\vec{S}_{2\vec{k}}=\sqrt{\frac{2}{N}}[\sum_i^{corner}\cos(\vec{k}\cdot\vec{R}_i-\pi/4)\vec{S}_i-\sum_j^{b-c}\cos(\vec{k}\cdot\vec{R}_j-\pi/4)\vec{S}_j]. \qquad (17)$$

Or inversely,

$$\vec{S}_i=\sqrt{\frac{2}{N}}\sum_{\vec{k}}[\cos(\vec{k}\cdot\vec{R}_i+\pi/4)\vec{S}_{1\vec{k}}+\cos(\vec{k}\cdot\vec{R}_i-\pi/4)\vec{S}_{2\vec{k}}],$$

for corner site,

$$\vec{S}_j=\sqrt{\frac{2}{N}}\sum_{\vec{k}}[\cos(\vec{k}\cdot\vec{R}_j+\pi/4)\vec{S}_{1\vec{k}}-\cos(\vec{k}\cdot\vec{R}_j-\pi/4)\vec{S}_{2\vec{k}}], \qquad (18)$$

for body-center site,

where $\vec{S}_i$ and $\vec{S}_{n\vec{k}}$ are two-dimensional real vectors (S_{ix}, S_{iy}) and ($S^x_{n\vec{k}}$, $S^y_{n\vec{k}}$), and $\vec{k}$ is a vector in the first Brillouin zone of the simple cubic lattice. $E^{(2)}_{uudd}$ is then expressed as follows:

$$NE^{(2)}_{uudd} = \sum_{\vec{k}} [F^x_{\vec{k}}(S^x_{1\vec{k}},S^x_{2\vec{k}}) + F^y_{\vec{k}}(S^y_{1\vec{k}},S^y_{2\vec{k}})],$$

$$F^x_{\vec{k}}(x,y) = [D_x + (A^+ + A^-)\gamma_{1\vec{k}}]x^2 + 2(A^+ - A^-)\gamma'_{1\vec{k}}xy + [D_x - (A^+ + A^-)\gamma_{1\vec{k}}]y^2,$$

$$F^y_{\vec{k}}(x,y) = [D_y + (a^+ + a^-)\gamma_{1\vec{k}}]x^2 + 2(a^+ - a^-)\gamma'_{1\vec{k}}xy + [D_y - (a^+ + a^-)\gamma_{1\vec{k}}]y^2,$$

$$D_x = D + B^+\gamma^+_{2\vec{k}} + C^+\gamma^+_{3\vec{k}} + (B^-\gamma^-_{2\vec{k}} + C^-\gamma^-_{3\vec{k}}),$$

$$D_y = D + b^+\gamma^+_{2\vec{k}} + c^+\gamma^+_{3\vec{k}} + (b^-\gamma^-_{2\vec{k}} + c^-\gamma^-_{3\vec{k}}),$$

$$\gamma'_{1\vec{k}} = \sin(\tfrac{1}{2}k_x a) \cdot \cos(\tfrac{1}{2}k_y a) \cdot \cos(\tfrac{1}{2}k_z a),$$

$$\gamma^+_{2\vec{k}} = [\cos(k_y a) + \cos(k_z a)]/2, \qquad \gamma^-_{2\vec{k}} = \cos(k_x a),$$

$$\gamma^+_{3\vec{k}} = \cos(k_y a) \cdot \cos(k_z a), \qquad \gamma^-_{3\vec{k}} = \gamma^+_{2\vec{k}} \cdot \gamma^-_{2\vec{k}}. \tag{19}$$

Now "uudd" being stable" means that the two quadratic polynominal $F_{\vec{k}}x(x,y)$ and $F_{\vec{k}}y(x,y)$ are positive definite for any $\vec{k}$ in the first Brillouin zone of the bcc lattice. The conditions "uudd being stable" are given as follows:

$$D_x - \sqrt{(A^+ + A^-)^2\gamma_{1\vec{k}}^2 + (A^+ - A^-)^2\gamma'^2_{1\vec{k}}} \; > 0. \tag{20}$$

$$D_y - \sqrt{(a^+ + a^-)^2\gamma_{1\vec{k}}^2 + (a^+ - a^-)^2\gamma'^2_{1\vec{k}}} \; > 0. \tag{20'}$$

We define h_{c1} for a given set of exchange parameters, as the largest magnetic field strength for which the above conditions are satisfied. When the exchange parameters are in the domain in which uudd is stable at zero external field, as shown in Fig. 1, at $h = h_{c1}$, the condition (20') is broken for small $\vec{k}$ parallel to the x-axis. The condition (20') has the form $G_1(k_y^2 + k_z^2) + G_2 k_x^2 \geq 0$, for small $\vec{k}$, and $G_2 \geq 0$ is expressed as

$$(1 - t^2)(3K_P + K_F)[J_2 + 4J_3 - K_F - t^2(3K_P + K_F)] - [J_1 + 3(K_P + K_F)t^2]^2 \geq 0. \tag{21}$$

The condition (21), when t=0, coincides with that for spin wave of uudd being real, mentioned previously.
If the first order transition to PF is forbidden, uudd makes a second order phase transition to DS(100) at the applied field $h=h_{c1}$ perpendicular to spin plane.
The spin wave frequency of uudd in finite magnetic field is easily obtained from Eq. (19), using commutation relation $[S_{n\vec{k}}, S_{n\vec{k}}]{\sim}iS$. If $\vec{k}$ is reduced within the first Brillouin zone of uudd structure with the same method by which Eq. (13) is converted to Eq. (13'), we get four spin wave branches, as shown in reference 5.

In Figs. 3, 4 and 5, we show the domain for three cases, $\theta=-2.5$, -3.0 and -3.5 mK, in which both uudd and PF are stable or metastable at $h=0.4$ T. In these figures, as references, we show the lines $e_2=7$ mK^2, $e_3'=3.3$ mK^3, $T_N^M=1.0$ mK, and $h_c^M=0.4$ T where e_3' which is defined with the energy of the order state at $h=0$ as $E=E_0+(\pi^2/30)\times(k_BT)^4/e_3'$, is calculated by usual spin wave theory,[5] T_N^M is the transition temperature to uudd at $h=0$ calculated by the MFA, and h_c^M is the critical field where

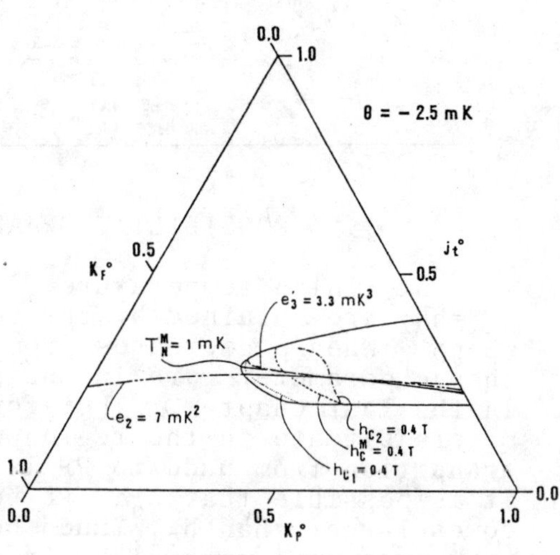

Fig.3 Domain for $\Theta=-2.5$ mK.

uudd makes a transition to PF, calculated by the MFA $[E_{uudd}{}^M(h_c{}^M) = E_{PF}{}^M(h_c{}^M)]$.

From these figures we can conclude that K_F must be nearly equal to K_P to explain all experimental results.

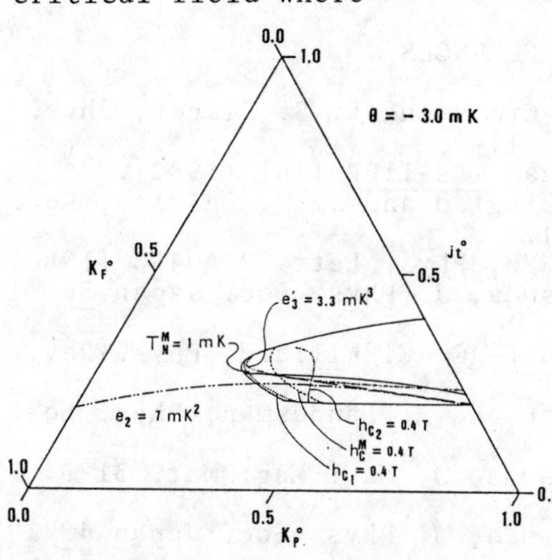

Fig.4 Domain for $\Theta=-3.0$ mK.

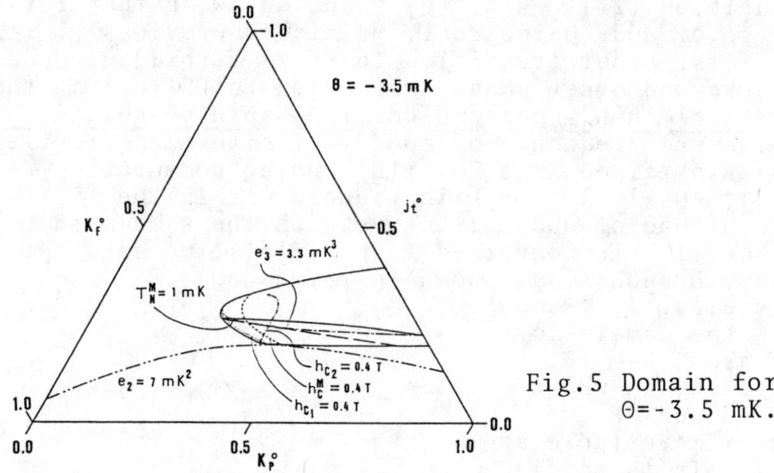

Fig.5 Domain for
$\Theta = -3.5$ mK.

POSSIBILITY OF APPEARANCE OF DS

At finite temperature, if the MFA is allowable, then $h_{c1}=h_{c2}$ are obtained by the replacement in Eq. (9), $|\vec{S}_i^0|$ $=S \rightarrow pS$. When p decreases from 1, h_{c2} increases if the exchange parameters are in the appropriate region as shown in the last chapter. Therefore even if the exchange parameters situate in the region where the first-order phase transition from uudd to PF is possible at low temperature, it is possible that h_{c2} (PF becomes stable at h_{c2}) becomes larger than h_{c1} (uudd becomes unstable at h_{c1}) when p becomes small near T_N. In such case, we shall observe a continuous magnetization-field curve, like that observed by Prewitt and Goodkind[10] near T_N.

REFERENCES

1 D. D. Osheroff, M. C. Cross and D. S. Fisher, Phys. Rev. Lett. 44 792 (1980).
2 D. D. Osheroff, Physica, 109-110B 1461 (1982).
3 M. Roger, J. H. Hetherington and J. M. Delrieu, Rev. Mod. Phys. 55 1 (1983).
4 D. D. Osheroff and C. Yu, Phys. Lett. 77A 458 (1980).
5 K. Iwahashi and Y. Masuda, J. Phys. Soc. Japan 50 2508 (1981).
6 J. H. Hetherington and F. O. C. Willard, Phys. Rev. Lett. 35 1442 (1975).
7 M. Roger, J. M. Delrieu and A. Landesman, Phys. Lett. 62 449 (1977).
8 K. Iwahashi and Y. Masuda, J. Mag. Mag. Mat. 31-34 733 (1983).
9 K. Iwahashi and Y. Masuda, J. Phys. Soc. Japan 49 1974 (1980).
10 T. C. Prewitt and J. M. Goodkind, Phys. Rev. Lett. 44 1699 (1980).

SCALING OF MAGNETIZATION IN bcc ^{3}He AT VARIOUS MOLAR VOLUMES

T. Shigi, T. Hata, S. Yamasaki and T. Kodama
Faculty of Science, Osaka City University, Osaka 558, Japan

ABSTRACT

Static magnetization in solid ^{3}He has been measured from 10 mK down to 0.3 mK for various molar volumes V (from 24.14 to 19.26 cm^3/mole). The results for bcc ^{3}He show that both the ordering temperature T_N and inverse of the maximum magnetization at T_N vary in proportion to $V^{16.5\pm1}$, that the magnetization in the ordered state is almost constant, and that the magnetization reduced by its maximum value is represented by a universal function of the reduced temperature T/T_N. The magnetization in hcp ^{3}He shows that its Weiss constant is very small, possibly less than 30 μK.

INTRODUCTION

Solid ^{3}He is a typical example of the so called "quantum solid" with a large zero point energy. The exchange interactions play an important role in this crystal and cause an nuclear ordering at about 1 mK under the melting pressure which is 3 orders of magnitude higher than expected from the nuclear dipole interaction. In 1974, Halperin et al[1] first found out the nuclear ordering in bcc ^{3}He by entropy measurements in a Pomeranchuk cell. Their results showed a sharp drop in entropy suggesting a first order transition. Since then, Kummer et al[2] measured the entropy in magnetic fields up to 12 kOe and discovered a different ordered state in bcc ^{3}He at melting pressure above 4 kOe. Prewitt and Goodkind[3] performed the static magnetization measurements below the ordering temperature T_N and found a sharp drop of the magnetization at T_N. Their results gave a clear evidence for the anti-ferromagnetic transition of first order. In 1980, Osheroff et al[4] succeeded in observing the anti-ferromagnetic resonance in the low field ordered phase of a bcc single crystal and proposed a new spin structure, up-up-down-down in the planes normal to the [100] direction.

In order to explain the above experimental results, low temperature properties of solid ^{3}He have to be described in terms of two, three and four atom exchange interactions[5]. The earlier pressure measurements at higher temperatures[6] showed that the exchange interactions themselves depend strongly on the molar volume. Therefore if these interactions should have different volume dependences, one could hope to make their mechanism clear separately by the experiments on molar volume dependence of thermodynamic quantities. So, we have measured the static magnetization of bcc ^{3}He through the ordering temperature for the molar volume from 24.14 cm^3/mole to 22.46 cm^3/mole. We have also extended this measurements to the hcp phase to obtain informations on its magnetic properties. In Fig.1, the region of the present experiments is indicated by the hatched area.

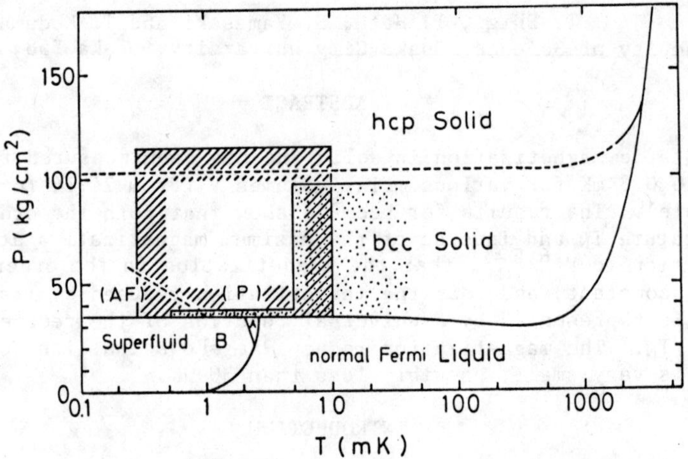

Fig.1. P-T phase diagram of ^{3}He. The hatched area:
this work, dotted area: experiments so far reported.

EXPERIMENTAL

The experimental arrangement was almost the same as that of
Prewitt et al[3] [7] (Fig.2). A sample of solid ^{3}He was formed in a
sintered silver powder. The size of this powder was found to be
2000 Å after sintering using a electron microscope, although its
nominal value was 700 Å. Cooling was achieved by nuclear demagnetization of copper (22 moles), which could maintain the cell temperature below 1 mK for more than two weeks[8]. The magnetization of
solid ^{3}He was measured statically with a SQUID magnetometer in the
trapped field of 26 mT. The cell temperature was determined by
means of the pulsed NMR of Pt wires at a resonant frequency of 250
kHz. This thermometer was calibrated against the superfluid ^{3}He-B
transition (Helsinki scale). The sample cell and the Pt wires were
mounted on a thick copper plate thermally connected to the nuclear
stage.

A sample of solid ^{3}He (^{4}He content less than 3 ppm) was made by
a blocked capillary method at about 2 K and annealed for one day.
The molar volume of the sample was determined from the pressure in
the cell using the P-V-T relation of solid ^{3}He on the melting
curve[9]. For the cell pressure measurements, we made use of the
system of ^{3}He magnetization measurements. The magnetic flux running
through the pick up coil changes linearly with pressure in the cell
because of the elastic expansion of the cell wall (thickness;1.5mm).
No hysterisis was observed in the relation between the SQUID output
and the pressure up to 150 bar. The molar volume was determined
within an accuracy of ±0.02 cm^3/mole. Although the SQUID system
picks up the change of pressure as well as the magnetization of

solid ^{3}He, these two signals become dominant at different temperature regions. The signal coming from the magnetization of solid ^{3}He is negligible around 2 K and the effect of pressure change is very small below 10 mK. In this temperature region, the total pressure increase due to the thermal expansion and the nuclear ordering of solid ^{3}He gives an error only less than 1 % for the magnetization measurements[10].

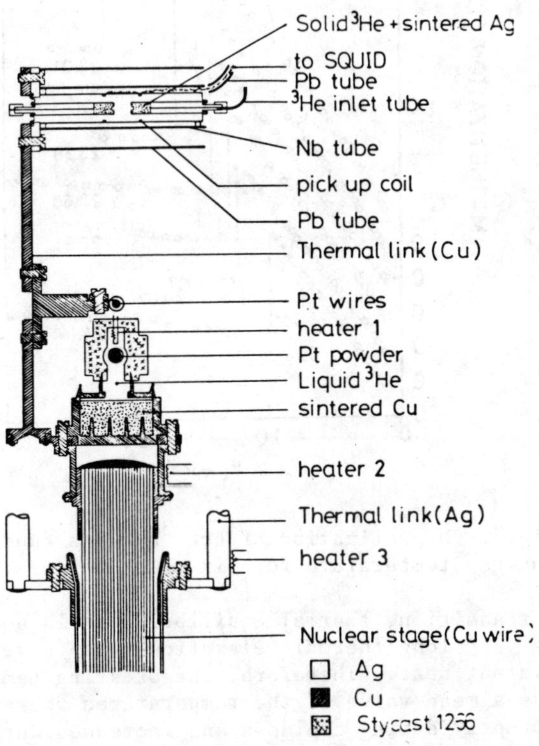

Fig.2. Main part of the experimental setup.

In the course of cooldown or warmup, the cell temperature was held constant untill thermal equilibrium was established between the ^{3}He sample and the Pt thermometer, and then both the temperature and the magnetization of solid ^{3}He were recorded. The typical thermal relaxation time was 15 min. in the ordered state and about 60 min. in the paramagnetic state at about 1 mK.

RESULTS AND DISCUSSIONS

Figure 3 shows the magnetization of bcc ^{3}He as a function of inverse temperature for six molar volumes. It was obtained by subtracting the magnetization of the cell body from a raw data so that the magnetization of bcc ^{3}He obeys the Curie-Weiss law above 5 mK[11].

50

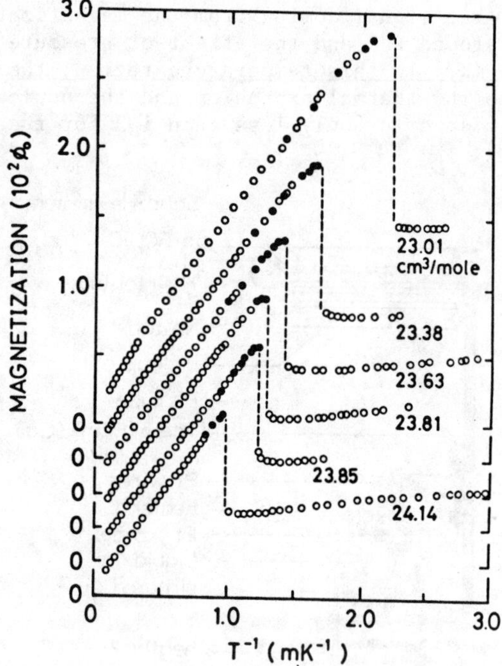

Fig.3. Magnetization of bcc ³He as a function of
inverse temperature for six molar volumes.

At the transition, thermal equilibrium could not be obtained
because of a very long thermal relaxation time (several days) related to a latent heat. Therefore, the ordering temperature was
determined as a mean value of the temperatures where magnetization
decreases abruptly during cooldown and increases during warmup. The
transition width was about 20 μK. Figure 4 shows the ordering temperature as a function of molar volume in logarithmic scale. The
ordering temperature varies in proportion to $V^{16.5\pm1}$, which is in
reasonable agreement with the previous measurements at high temperatures[6]. The ordering temperature on the melting curve was estimated by extraporation as 1.04 mK, which agrees well with that obtained by Osheroff et al[12].

The present experimental data of solid ³He inevitably involve
the magnetization of the ³He at the surface as well as that of bulk
solid ³He. As the sample is confined in very small pores of the

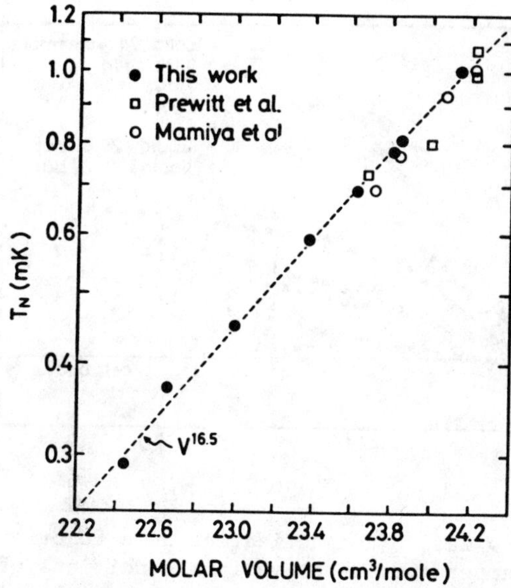

Fig.4. Molar volume dependence of the ordering
temperature of bcc ^{3}He plotted in logarithmic scale.
The broken line shows $V^{16.5}$. Solid circles: this work,
Open squares: Prewitt et al[3) 7)], Open circles: Mamiya
et al[10)].

sintered silver sponge, the amount of ^{3}He at the surface can not be
neglected. To examine the surface effect of the sample, magnetization measurements have been made under various conditions of ^{3}He.
Figure 5 shows the results. At first we consider the temperature
region below 2 mK, where the magnetization of the cell body is
constant. The Curie-Weiss behavior of the low and high pressure
liquid is probably related to the so called boundary magnetism[13)].
We would like to notice that the magnitude of this excess magnetization has considerable pressure dependence. Four samples were
measured for the mixture of liquid and solid ^{3}He. In this temperature region, the ratio of liquid and solid in the cell stays almost
constant when the temperature is changed along the melting curve.
The anti-ferromagnetic transition appears for all mixtures as well
as 100 % solid. The magnetization curve for three mixtures (30 %,
50 % and 75 % molar fraction of solid) are parallel for a broad temperature interval below T_N, and their temperature dependences are
almost the same as that of the high pressure liquid. As the difference of solid fraction results in the magnetization difference independent of temperature below T_N, we concluded that the magnetization
of bulk solid ^{3}He is almost constant below T_N and the apparent temperature dependence is due to the magnetization of the liquid at the
surface.

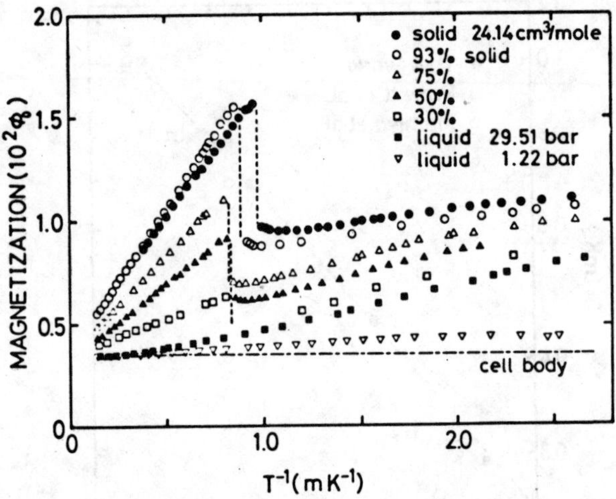

Fig.5. Raw data of magnetization as a function of inverse temperature under various conditions of the sample ^{3}He. The magnetization curves of the mixtures of liquid and solid ^{3}He are shown for the molar fraction of solid in the cell from 30 % to 93 %. Broken lines and a dotted broken line indicate the ordering in solid ^{3}He and the magnetization of the cell body respectively.

When the molar fraction of solid is increased to 93 %, the temperature dependence below T_N changes from that of lower solid fractions, and coincides with the behavior of 100 % solid near the melting curve. We see here that the behavior of the surface magnetization changes when the liquid portion in the cell decreases below a certain value. Furthermore, we would like to point out that the ordering temperature for three lower solid fractions is about 15 % higher than that of 100 % solid on the melting curve (Fig.6). For 93 % solid, T_N falls on a value in between. The 15 % higher ordering temperature corresponds to the melting curve of ^{3}He depressed by about 1 bar.

The surface magnetism above T_N is not clear, but this correction will tend to a small value at high temperatures. So, we estimated the surface magnetization as follows. It begins to appear at the temperature where the magnetization of liquid begins to have temperature dependence, and it increases with the same slope as that of solid ^{3}He below T_N. Assuming that this surface magnetization does not change for different molar volumes, we subtract this background from the raw magnetization data for all molar volumes.

For each sample, the magnetization starts to deviate from the high temperature Curie-Weiss behavior at $T \approx 5\,T_N$ and rises above the curve. At T_N, the magnetization drops by 61 % of its maximum value

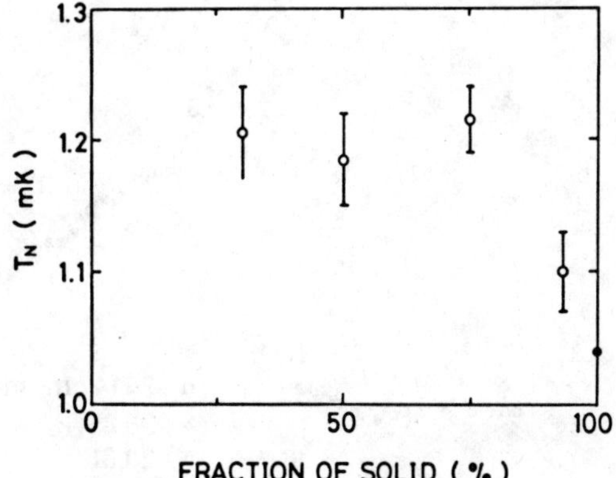

Fig.6. Ordering temperature as a function of solid fraction for the mixture of liquid and solid ^{3}He. Solid circle indicates the 100 % solid on the melting curve.

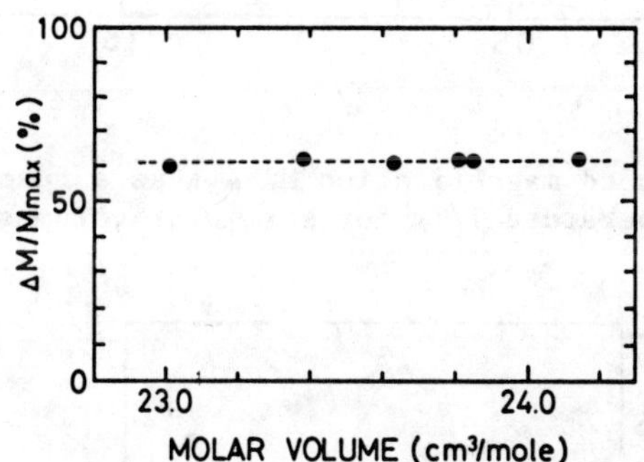

Fig./. Ratio of the magnetization drop ΔM to the maximum magnetization M_{max} at T_N as a function of molar volume.

for all molar volumes as shown in Fig.7. The maximum magnetization M_{max} was obtained by correcting the rounding near T_N. Based on the above facts, we analyzed all data to express the reduced magnetization M/M_{max} as a function of reduced temperature T/T_N. The results shown in Fig.8 indicate that M/M_{max} is represented by a universal function of T/T_N for all molar volumes. It is also important that $1/M_{max}$ is proportional to $V^{16.5\pm1}$ as shown in Fig.9, which is the same relation as T_N. Therefore, the magnetization of bcc ^{3}He is represented by a function of $V^{16.5}$ below $2(T/T_N)$. This leads to similar molar volume dependence of all the exchange interactions, which agrees with the recent calculations of exchange integrals by Avilov and Iordansky[14] and also by Roger[15].

Concerning the hcp ^{3}He, the magnetization has been measured for

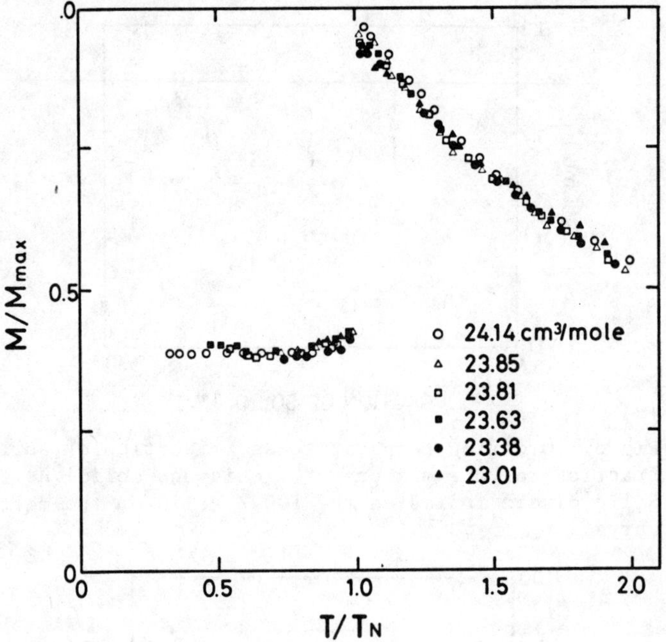

Fig.8. Reduced magnetization M/M_{max} as a function of reduced temperature T/T_N for six molar volumes.

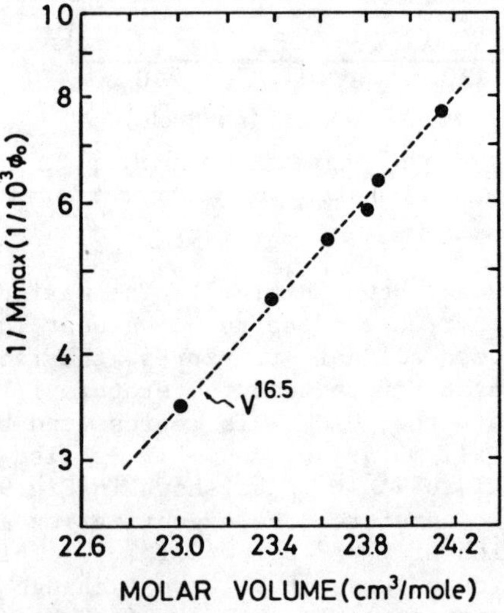

Fig.9. Inverse of the maximum magnetization in bcc ^{3}He as a function of molar volume plotted in logarithmic scale. The broken line shows $V^{16.5}$.

three different molar volumes. As shown in Fig.10, the results obey
the Curie-Weiss law, but the Weiss constants are so small that it is
impossible to determine even their sign. Their absolute values seem
to be less than 30 μK.

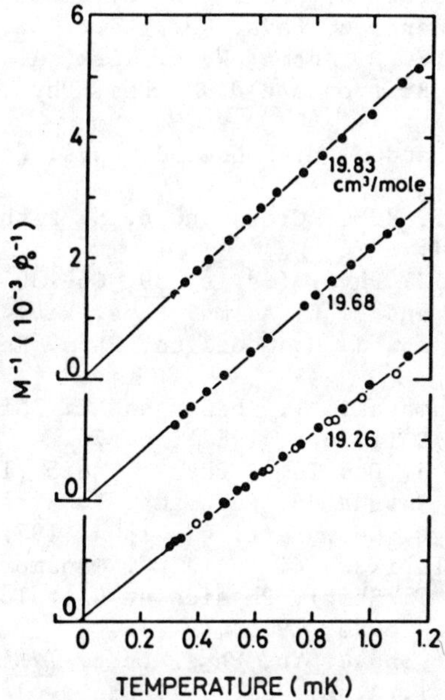

Fig.10. Low temperature part of inverse magnetization
of hcp ^{3}He as a function of temperature for three molar
volumes.

CONCLUSION

We have first studied systematically molar volume dependence of
the magnetization in bcc ^{3}He through the ordering temperature for a
wide region of molar volume, and found the relations

$$T_N \propto V^{16.5} \quad \text{and} \quad M_{max}^{-1} \propto V^{16.5}$$

We have also found that the magnetization in the ordered state is
almost constant. Based on the fact that the magnetization drops by
61 % of its maximum value at T_N, we have shown that the reduced magnetization M/M_{max} can be represented by a universal function of T/T_N
for all molar volumes. Data of the magnetization in hcp ^{3}He show
that its Weiss constant is very small, possibly less than 30 μK.

ACKNOWLEDGEMENTS

The authors wish to thank Dr. T. Mizusaki and Prof. A. Hirai
for the purity measurement of the sample ^{3}He. This work was partly

supported by Itoh Science Grant and also Grant-in-Aid for Scientific Research from the Ministry of Education, Science and Culture, Japan.

REFERENCES

1. W. P. Halperin, C. N. Archie, F. B. Rasmussen, R. A. Buhrman and R. C. Richardson, Phys. Rev. Lett. 32, 927 (1974).

2. R. B. Kummer, E. D. Adams, W. P. Kirk, A. S. Greenberg, R. M. Muller, C. V. Britton and D. M. Lee, Phys. Rev. Lett. 34, 517 (1975).

3. T. C. Prewitt and J. M. Goodkind, Phys. Rev. Lett. 39, 1283 (1977).

4. D. D. Osheroff, M. C. Cross and D. S. Fisher, Phys. Rev. Lett, 44, 792 (1980).

5. A. Landesman, J. Phys. (Paris) 39, C6-1305 (1978).

6. M. F. Panczyk and E. D. Adams, Phys. Rev. 187, 321 (1969).

7. T. C. Prewitt and J. M. Goodkind, Phys. Rev. Lett. 44, 1699 (1980).

8. T. Hata, S. Yamasaki, T. Kodama and T. Shigi, Proc. of ICEC 9 (Butterworth, Guildford, 1982), p.576.

9. E. R. Grilly, J. Low Temp. Phys. 4, 615 (1971).

10. T. Mamiya, A. Sawada, H. Fukuyama, Y. Hirao, K. Iwahashi and Y. Masuda, Phys. Rev. Lett. 47, 1304 (1981).

11. Y. Morii, K. Ichikawa, T. Hata, C. Kanamori, H. Okamoto, T. Kodama and T. Shigi, Physics at Ultralow Temperatures (Phys. Soc. of Japan, Tokyo, 1978), P.196.

12. D. D. Osheroff and C. Yu, Phys. Lett. 77A, 458 (1980).

13. A. I. Ahonen, T. A. Alvesalo, T. Haavasoja and M. C. Veuro, Phys. Rev. Lett. 41, 498 (1978).

14. V. V. Avilov and S. V. Iordansky, J. Low Temp. Phys. 48, 241 (1982).

15. M. Roger, Private Communication.

NEUTRON EXPERIMENTS ON SOLID ^{3}He

A. Benoit[*], J. Bossy, J. Flouquet[*], D. Rufin[*o] and J. Schweizer[+]

* CRTBT-CNRS, B.P. 166 X, 38042 Grenoble Cédex, FRANCE
o Air-Liquide, 38360 Sassenage, France
+ DN/RFG-CENG, B.P. 85 X, 38041 Grenoble Cédex, France

ABSTRACT

Neutron experiments on solid ^{3}He are described : the search for the (110) Bragg reflection in bulk and restricted geometries – the thermalization of the sample in presence of a neutron beam.

INTRODUCTION

Neutron diffraction is the only unambiguous way to determine a magnetic structure. In recent years, in the very low temperature physics the interest of, for example, nuclear magnetism[1] and electronuclear magnetism[2,3,4], competition between magnetism, superconductivity and non magnetic ground states[5,6] in many systems has motivated the experimentalists to performe neutron experiments.

For ^{3}He, important experiments have been performed in the liquid phase[7,8] in order to observe the excitation spectrum. The recent interest in the magnetically ordered phases of solid ^{3}He[9] strongly suggests that an attempt be made to observe the magnetic structure directly. The large zero field resonance frequency observed by Osheroff[10] proves that its symmetry is not cubic. The shape of the resonance spectrum indicates a planar anisotropy with spins aligned in (100) ferromagnetic planes. The uudd structure proposed may correspond to a sequence of (100) paramagnetic planes with spins pointing in the up up and down down directions. It will be the aim of neutron experiments to verify this proposal, since more complicated structures could fit the data as well. Furthermore, the new magnetic phase observed above $H_c \sim 4$ kOe[11] is still not identified.

It must be pointed out that the uudd structure proposed for solid ^{3}He has been previously observed in cerium intermetallic compounds such as CeSb[12] and CeBi[13]. Furthermore, as for solid ^{3}He, the magnetic Grüneisen coefficient

$$\Omega_m = - \frac{\partial \, \text{Log} \, T_N}{\partial \, \log V}$$

can be large : $\Omega_m \sim 40$ in CeSb[14]. Common points between these materials are the complexity of the exchange couplings. In CeBi and CeSb, the strong planar coupling may be associated with the strong mixing between 4f states and 5p holes of the valence band[14].

For many years, the scheme was that solid ^{3}He may be an example of an ordinary double exchange Heisenberg coupling. It is the failure of such an description which asks basic question. Recently a multiple exchange model has been proposed for the understanding of the experimental results[9].

58

After a short review of neutrons and their interaction with ^{3}He,
we describe the research for the nuclear Bragg reflections of the
bcc ^{3}He lattice. Measurements of the transmission of polarized neutrons through a weakly polarized solid ^{3}He target are described.

REVIEW

2.1. The high neutron-^{3}He absorption cross-section

Disadvantages : In ordinary neutron experiments performed at
very low temperature (T < 50 mK), the major problem is the thermalization of the sample. Liquid ^{3}He, which has an excellent thermal
conductivity, cannot be used since its neutron absorption cross
section σ_{abs} is enormeous :

$$\sigma_{abs}/\lambda = 2,962 \text{ barn/Å} \quad {}^{15}$$

By comparison those of aluminium, copper and silver are respectively
0.13, 2.2 and 36 barn/Å[16]. For a wavelength $\lambda = 1$ Å and a ^{3}He molar
volume of 24 cm^3, $\sigma_{abs}(^3$He) corresponds to a typical length of
0.1 mm. This requires the use of a thin sample. Furthermore, in
contrast to ordinary cases where σ_{abs} is low, the heating Q_o produced by the nuclear reaction

$$^1_o n + {}^3_2 He \rightarrow {}^1_1 H + {}^3_1 H + 1.2 \times 10^{-13} \text{ Joule}$$

is a major problem for the thermal contact of ^{3}He with the cooling
agent (a copper demagnetization stage) at very low temperature.
Assuming 60 % absorption of a number N_o of neutrons per second,
the time constant Δt below which the solid will stay in a magnetic
ordered state is governed by the drop ΔS of the magnetic entropy
at the first order transition[17], which occurs at $T_N \sim 1$ mK :

$$Q_o N_o \Delta t = T_N \Delta S$$

Using the experimental value of $\Delta S = 40$ % of the full magnetic
entropy (k Log 2), and $N_o = 10^4$ n/sec, one is left with
$\Delta t \sim 700$ sec.

2.2. Use of the high neutron ^{3}He – cross section : density measurement

However, the large value of σ_{abs} can be used for measuring
accurately the ^{3}He density. At high temperatures or for unpolarized
neutrons, the transmission N of N_o neutrons is equal to

$$N = N_o \exp - \alpha \qquad \text{with } \alpha = \frac{A\sigma_{abs}x}{V}$$

A, V and x are respectively Avogadro's number, the molar volume and
the thickness of the target. This relation can be used for the
molar volume determination using for the calibration, the pressure
molar volume relation of liquid ^{3}He phase, where the pressure can

be measured outside at room temperature.

2.3. Absorption of polarized neutrons in a polarized ^{3}He target : direct temperature measurement

For ^{3}He, the neutron capture occurs only for the singlet state
built by the nuclear I = 1/2 of ^{3}He and s = 1/2 of the neutron[18].
When the ^{3}He nuclei are polarized by an applied magnetic field H,
the transmitted neutrons N depend on the polarization f_n of the
^{3}He nuclei :

$$N = N_o \exp -\alpha(1 - f_n)$$

Far above T_N, for solid ^{3}He, f_n may be described by the well-known
Brillouin function :

$$f_n = B_I \left(\frac{\mu_n H}{kT}\right)$$

where μ_n is the ^{3}He magnetic moment. In the vicinity of T_N, a strong
departure from Brillouin behavior occurs. However, since magnetization experiments have been performed in solid ^{3}He[19,20], the study
of the transmission can give a direct measurement of the temperature
of the nuclei in the presence of the neutron irradiation. The measured flipping ratio (R_T) is the ratio of the transmission of
incoming neutrons polarized parallel (P_n = 1) to antiparallel with
respect to the nuclear polarization[21] :

$$R_T = \exp 2 \alpha f_n$$

2.4. Diffraction with polarized neutron : temperature of the container

The scattering cross section σ of the nuclei and the neutrons
depends of their respective polarizations (f_n, P_n) through the relation[22] :

$$\sigma = \bar{b}^2 + 2P_n \bar{b} IB f_n + (IBf_n)^2$$

$\bar{b}$, B are respectively the coherent and incoherent scattering lenghts.
In a diffraction experiment, by measuring the flipping ratio

$$R_D = \frac{\sigma^+}{\sigma^-} = \left(\frac{\bar{b} + IBf_n}{\bar{b} - IBf_n}\right)^2$$

another measurement of f_n can be obtained. This possibility has been
used to test that the cooling agent reaches a temperature lower
than T_N. A single crystal of Cu was chosen since f_n obeys a pure
Brillouin function in the mK range. The flipping ratio of the (111)
Bragg reflection of a single crystal of Cu (squeezed by a brass
screw onto the copper container of ^{3}He) was observed. Since
$IBf_n < \bar{b}$, the parameter R_D-1 gives T directly through the relation :

$$\frac{R}{D} - 1 = \frac{4IBf_n}{\bar{b}} \sim \beta\,\frac{H}{T}$$

with $\beta = 1.37 \times 10^{-8}$ K/Oe for copper[23].

2.5. Debye Waller attenuation - evaluation of a magnetic signal

The large zero point motion of the quantum crystals ^{3}He introduces another source of attenuation of the diffraction signal which is proportionnal to the Debye Waller factor e^{-2W}. In a Debye approximation, the Debye Waller factor is linked to the quadratic displacement of the atoms ΔX^2 through the relation[24]

$$e^{-2W} = \exp\left[-\frac{4\pi^2}{3}\frac{\Delta X^2}{d_{hk\ell}^2}\right]$$

where $d_{hk\ell}$ is the distance characteristic of $hk\ell$ planes. For $\lambda = 1$ Å, the first nuclear line $[110]$ will be attenuated by the factor 0.14, whereas the $[1/2\ 0\ 0]$ reflection, corresponding to the uudd structure is attenuated by 0.79. Taking into account the domain populations and the values of the coherent $(\bar{b} = 0.58 \times 10^{-12}$ cm) and incoherent $(B = 0.65 \times 10^{-12}$ cm) scattering amplitudes[25], the $[1/2\ 0\ 0]$ intensity of a uudd ground state may be an order of magnitude lower than the $(1\ 1\ 0)$ line.

Regarding the presumed weakness of the magnetic reflecions, the first attempts will be to observe the first $(1\ 1\ 0)$ nuclear reflection and to prove the possibility of cooling ^{3}He below T_N in the presence of the neutron beam.

HIGH TEMPERATURE MEASUREMENTS (K range) : RESEARCH OF THE (110) NUCLEAR BRAGG REFLECTION

These measurements were performed in order to detect the (110) Bragg reflections of bcc solid ^{3}He which have not been observed previously by neutron experiments.

Experiments have been performed either with a dilution refrigerator and a 1 K cryostat on different goniometers D1B and D2 of the high flux reactor of the Institut Laue Langevin (ILL) or goniometers of Siloe or Melusine reactors of the Nuclear Center of Grenoble.

The first idea was to apply to ^{3}He a technique which has been used successfully to solve the magnetic structure of $CeAl_2$[26] and $PrCu_5$[2]. The high sensitivity of the multidetector of the D1B spectrometer at ILL reactor allows one to make accurate differences between patterns measured on each side of T_N and thus to obtain directly the complete magnetic pattern.

In order to minimize the heating produced by the recoiling T and p particles in solid ^{3}He and to prevent ^{3}He to grow as a single crystal, ^{3}He was immersed in a sintered metallic powder (Cu,Ag). From atomic stopping cross section tables[27], the heating in ^{3}He is expected to decrease by 80 %.

The target was a boat-like copper container filled with copper or silver sintered powder. Their expective mean grain sizes are 1 μ and 700 Å. The thickness of the plate made by the powder is 0.6 mm, the effective ^{3}He thickness between 0.2 and 0.3 mm. A thin Cu foil and the filling capillaries were soldered with tin on the top of the container. Mechanical support was realized by two Al plates (figure 1). In this "restricted geometry", no powder pattern was observed. In a similar configuration, experiments performed on a solid ^{4}He target of 1 cm thickness failed to observe a powder pattern.

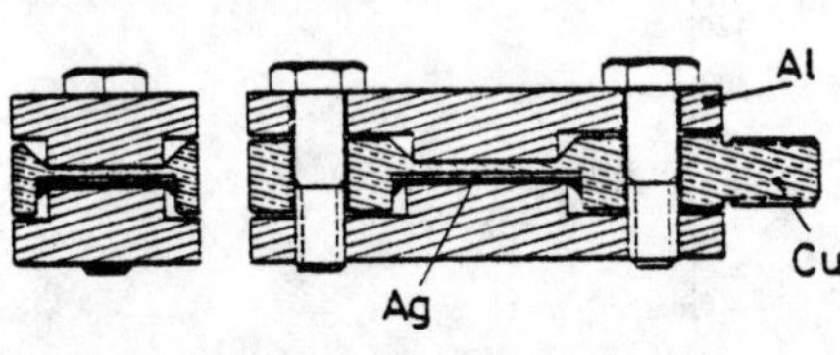

Fig. 1. ^{3}He target.

Experiments performed on a bulk ^{3}He sample of mean size 0.3 x 10 x 10 mm showed the (110) Bragg reflection of bcc ^{3}He. The surprising result was that the (110) single crystal reflection was also observed in restricted geometries with a magnitude comparable to that observed in a bulk geometry[28]. There seems to be no major difference between Cu and Ag. Systematic studies of the Bragg intensity were actually performed as a function of the packing fraction of the powder.

Figure 2a represents the rocking curve of solid ^{3}He observed in a bulk geometry (T = 1.22 K and p = 46.5 atm) and figure 2b, the signal after the crystal has melted. Figure 3 describes rocking curves obtained in a Cu restricted geometry. For each point, the counting time was 20 seconds.

EXPERIMENTS AT LOW TEMPERATURE (mK range)

4.1. Experimental set up

On the polarized neutron diffractometer DN2 at the Melusine reactor[21], a low temperature experiment was set up to check the thermalization of ^{3}He nuclei in a 700 Å powdered silver target (figures 4 and 5). A restricted geometry was chosen to optimize the cooling process despite the fact that, in a perfect single crystal of ^{3}He, the thermal conductivity may be high due to the fast spin diffusion given by the exchange coupling between the nuclear spins.

A polarizing field of H = 800 Oe was produced by SmCo$_5$ permanent magnets. Very low temperature were obtained by a dilution refrigerator in conjunction with a copper nuclear demagnetization stage, made of 20 moles of commercial copper wires. The initial and final demagnetization conditions lead to an estimated lowest temperature T_{Cu} = 0.3 mK on the copper nuclei. To check the feasibility of cooling down the ^{3}He target below its magnetic ordering temperature, two main temperature measurements had to be performed. The first one was designed to check whether the sample container (C) reaches the temperature of the copper nuclear demagnetization stage (DS), and the second served to verify that ^{3}He is below its Néel point. In an

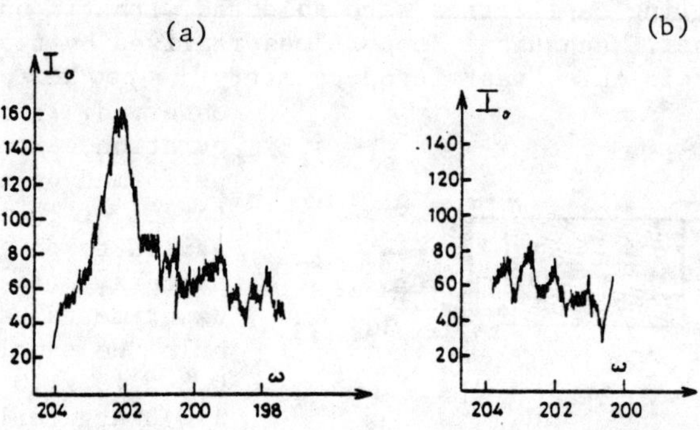

Fig. 2. Bulk geometry rocking curves of the solid phase (a) and of the liquid phase (b).

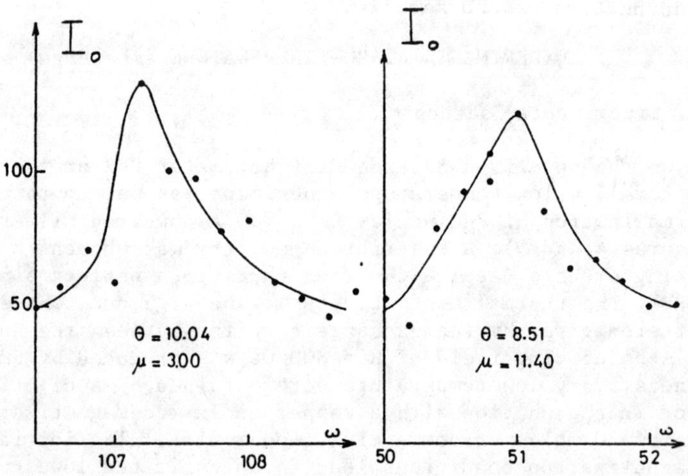

Fig. 3. Restricted geometry : rocking curve of the solid phase.

initial measurement, a copper single crystal was mounted on the sample container. After the end of the nuclear demagnetization process, the neutron beam was opened and reflection (111) of the monocrystal measured at regular time intervals. Temperatures down to 0.7 mK were obtained. For the experimental conditions, a counting time of one hour ($N_O = 10^4$ n/sec) was necessary to reach an experimental accuracy of 10 % on T. The radioactive heating due to the nuclear reaction was estimated to be $8 \cdot 10^{-10}$ W.

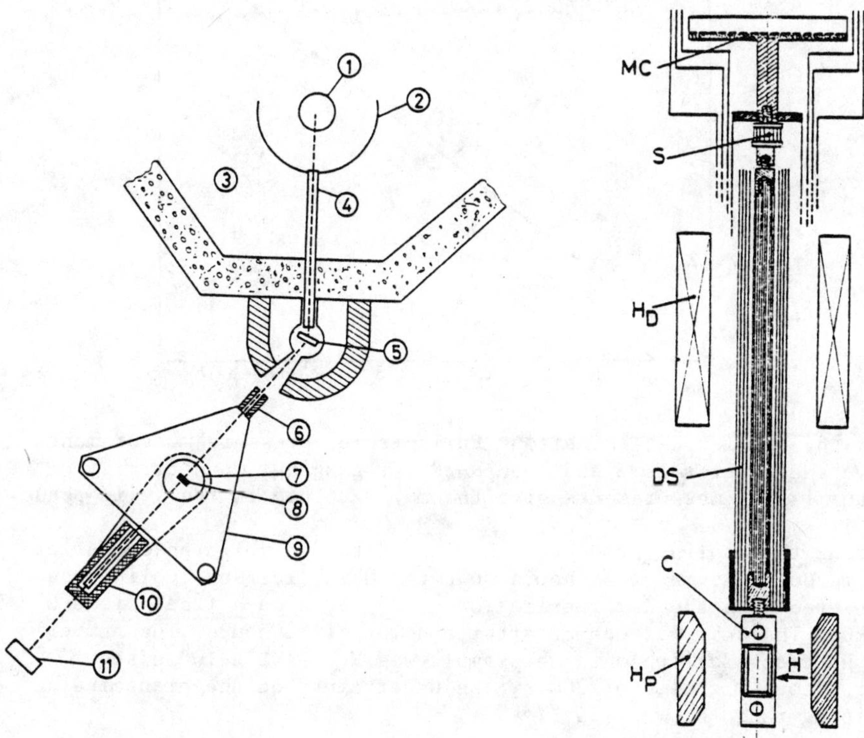

Fig. 4. Experimental set up
1) reactor's core 2) reactor 3) open pool 4) beam tube 5) monochromator 6) collimator 7) dilution refrigerator 8) ^{3}He cell 9) concrete block 10) detector and 11) stop beam.

Fig. 5. The experimental set up : MC, S, DS, H$_D$, C and H$_P$ are respectively the mixing chamber, the superconducting switch, the demagnetization stage, the demagnetization field, the container and the polarizing field.

The curve of the gauging of the molar volume is shown in figure 6. Using these data, the pressure in the target was increased up to the chosen density. The volume was continously measured

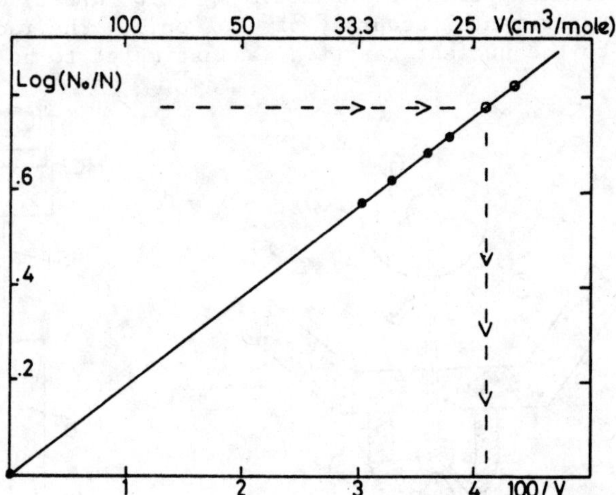

Fig. 6. Density determination. Full circle, pressure measurements give the molar volume and then N_O/N (V) gauging. Open circle, transmission measurements give the molar volume in the solid phase.

during the cooling procedure which lead to the solid phase. After a precooling time of 48 hours down to 10 mK, further cooling was achieved with the demagnetization of copper down a final field of 2 kOe. The beam was opened after a delay of 12 hours. The number of neutrons falling onto the sample was $N_O = 1.2 \times 10^4$ n/sec. For a counting time of 300 s, the uncertainty on the transmission fliping ratio R_T was 1.5 °/∘∘.

4.2. Results

According to the strength of the magnetic Grüneisen coefficient,[19] $\Omega_m \sim -17$. The Curie-Weiss temperature and the transition temperature decrease with the molar volume. Figure 7 reports, in arbitrary units, the susceptibility data corresponding to the two molar volume measured of 23.1 cm^3 and 24.2 cm^3 [19].

Figure 8 shows the corresponding flipping ratio as a function of the time of the neutron irradiation.

In the experiment at the molar volume of 23.1 cm^3, the maximum flipping ratio was observed immediately after the neutron beam was opened. Thus, this sample did not pass through the magnetic phase transition at 0.45 mK. On the other hand, for the sample at 24.2cm^3, where the phase transition occurs at 1 mK, the initial increase of the flipping ratio and thus of the susceptibility shows clearly

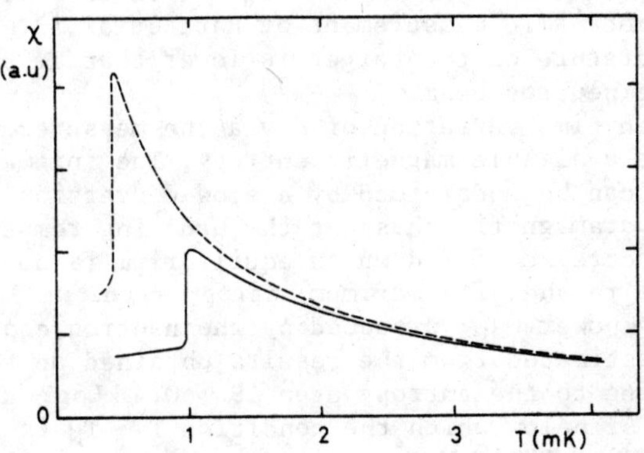

Fig. 7. Temperature variation of the susceptibility χ. In full and dotted lines respectively for V = 24.2 cm^3 and 23.1 cm^3 (see reference 19-20).

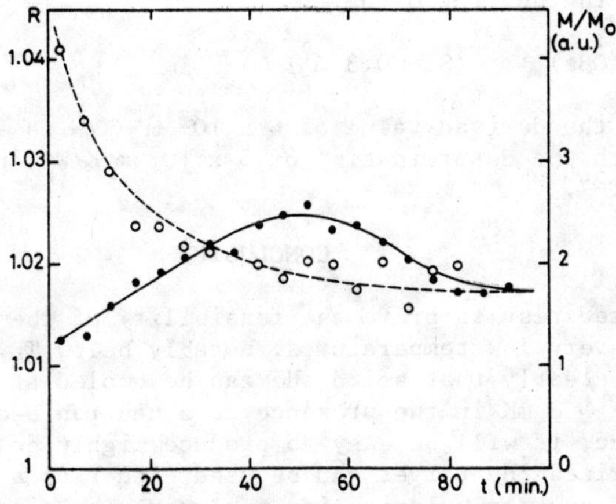

Fig. 8. Time dependence of the flipping ratio measured for V = 24.2 cm^3 (full circle) and for V = 23.1 cm^3 (open circle). The time origin corresponds to the opening of the neutron beam. M/M$_O$ is the corresponding variation of the ^{3}He magnetization.

66

that experiments below T_N can be maintained during a time of Δt = 50 min. after the opening of the neutron beam. The factor of 2 observed between the susceptibility measured at t = 0 (T $<$ T_N) and the susceptibility maximum (T = T_N) is in excellent agreement with the thermodynamic measurement of Hata et al.[19]. This proves that the temperature of the target is lower than T_N before the opening of the neutron beam.

The observed time variation of M via the measurement of R is related to the available magnetic entropy. The initial linear increase of R can be understood by a slow conversion of the ordered phase to the paramagnetic phase at the ordering temperature. After T_N, the slow decrease of R down to equilibrium is due to the fact that solid ^{3}He reaches its maximum entropy k Ln2 well above T_N[18].

The heating power $\dot{Q}_{3He}$ produced by the neutron capture on ^{3}He atoms can be estimated from the results obtained on the melting curve. According to the entropy drop $\Delta S \sim 0.4$ kLog2 at T_N and the measured time Δt below which the condition $T \leqslant T_N$ is realized, a value $\dot{Q}(^3He)$ = 8 x 10^{-10} W is derived through the relation :

$$\dot{Q}(^3He) \;=\; T\,\frac{\Delta S}{\Delta t}$$

The energy loss $\dot{Q}$ of the charged particles (proton and triton) produced by the nuclear reaction is $\dot{Q} \sim 1.3$ x 10^{-9} W. The comparable values of $\dot{Q}(^3He)$ and $\dot{Q}$ may show either parasitic heating coming with the beam (γ-rays, fast neutrons) or an overestimate of the role of the silver powder.

Extrapolating to infinite time, an equilibrium value of $T \sim 2$ mK seems to be reached. Thus, the Kapitza resistance R_K between ^{3}He and the silver powder can be estimated from the temperature difference δT and the heating $\dot{Q}(^3He)$:

$$\delta T \;=\; \frac{R_K}{S}\,\dot{Q}(^3He) \qquad (S = 0.3\ m^2)$$

At $T \sim 2$ mK, the derived value of 6 x 10^5 m^2 K/W is in excellent agreement with the determination of 5 x 10^5 m^2K/W reported by Mamiya et al.[29].

CONCLUSION

The reported results prove the feasibility of the study on solid ^{3}He at very low temperatures, notably below T_N. As this experiment shows clearly that solid ^{3}He can be cooled at an equilibrium temperature $T \sim 2$ mK in the presence of a neutron beam of N_0 = 10^4 n/sec, it will be easy to produce highly polarized ^{3}He. Thus, a polarized ^{3}He target can be used either as a neutron polarizer or as a neutron polarization analyser. The search for the magnetic lines will be the next attempt.

REFERENCES

1. Y. Roinel, V. Bouffard, G.L. Bachella, M. Pinot, P. Meriel,
   R. Roubeau, O. Avenel, M. Goldman and A. Abragam,Phys. Rev.
   Lett. 41, 1572 (1978).
2. A. Benoit, J. Flouquet, J.L. Génicon and J. Palleau, Physica
   108 B, 1103 (1981).
3. H. Bjerrum-Møller, J.Z. Jensen, M. Wulff, A.R. Mackintosh,
   O.D. Mac Masters and K.A. Jr Gschneidner, Phys. Rev. Lett. 49.
   482 (1982).
4. A. Benoit, J. Flouquet, K.A. Mc Ewen and W.G. Stirling, to be
   published.
5. S.K. Sinha, G.W. Crabtree, D.G. Hinks, H.A. Mook and O.A.
   Pringle, J. Magn. Magn. Mat. 31-34, 489 (1983).
6. A. Benoit, J.X. Boucherle, J. Flouquet, F. Holtzberg,
   J. Schweizer and C. Vettier, in "Valence Fluctuations in
   Solids", eds. L.M. Falicov, W. Hanke and M.B. Maple (North-
   Holalnd, Amsterdam, 197 (1981).
7. K. Sköld, C.A. Pelizzari, R. Kleb and G.W. Ostrowski, Phys.
   Rev. Lett. 37, 842 (1976).
8. See W.E. Stirling, J. de Phys. C6, 39, 1334 (1978).
9. M. Roger, J. Magn. Magn. Mat. 31-34, 727 (1983) and M. Roger,
   J.H. Hetherington and J.M. Delrieu, Rev. Mod. Phys. 55, 1
   (1983).
10. D.D. Osheroff, M.C. Cross and D.S. Fisher, Phys. Rev. Lett. 44,
    792 (1980).
11. See F.D. Adams, J.M. Delrieu and A. Landesman, J. de Phys.
    Lett. 39, L190 (1978).
12. J. Rossat Mignod, P. Burlet, J. Villain, H. Bartholin, T.S.
    Wang, D. Florence and O. Vogt, Phys. Rev. B16, 440 (1977).
13. H. Bartholin, P. Burlet, S. Quezel, J. Rossat Mignod and
    O. Vogt, J. Phys. C5-40, 130 (1979).
14. See J. Rossat Mignod, P. Burlet, S. Quezel, J.M. Effantin,
    D. Delacote, H. Bartholin, O. Vogt and D. Ravot, J. Magn. Magn.
    Mat. 31-34, 398 (1983).
15. J.Als. Nielsen and O. Dietrich, Phys. Rev. B133, 925 (1964).
16. G.E. Bacon, Neutron Diffraction, Ed. Clarendon Press, Oxford
    (1975).
17. W.P. Halperin, F.B. Rasmussen, C.N. Archie and R.C. Richardson,
    J. Low Temp. Phys. 31, 617 (1978).
18. L. Passel and R.I. Schermer, Phys. Rev. 150, 149 (1966).
19. T. Hata, S. Yamasaki, Y. Tanaka, T. Kodama and T. Shigi,
    Physica 107B, 201 (1981).
20. T. Hata, S. Yamasaki, M. Taneda, T. Kodama and T. Shigi, J.
    Magn. Magn. Mat. 737 (1983).
21. A. Benoit, J. Flouquet, D. Rufin and J. Schweizer, J. de Phys.
    Lett. 43, L431 (1982).
22. Y. Ito, C.G.Shull , Phys. Rev. 185, 961 (1969).
23. A. Benoit, J. Flouquet, D. Rufin and J. Schweizer, J. de Phys.
    C7, 43, 311 (1982).

24. See A. Guinier in "Théorie et technique de la radio-cristallographie", Ed. Dunod, Paris, 511 (1964).
25. Alfimenkov, "Report Joint Institute of Nuclear Research", Dubna, URSS (1980).
26. B. Barbara, J.X. Boucherle, J.L. Buevoz, M.F. Rossignol and J. Schweizer, Solid State Commun. $\underline{24}$, 481 (1977).
27. W. Whaling, Handbuch der Physik (Springer Verlag), Ed. by S. Flugge XXXIV, 193 (1958).
28. D. Rufin, Thesis Grenoble, 1982.
29. T. Mamiya, A. Sawada, H. Fukuyama, Y. Hirao and Y. Masuda, Physica 108 B, 847 (1981).

High Frequency NMR in bcc Solid ^{3}He:
Influence of the Defects

M.E.R. Bernier and G. Guerrier
DPhG/PSRM Cea Orme des Merisiers
91191 Gif/Yvette Cedex France

ABSTRACT

The spin lattice relaxation properties of solid ^{3}He have been extensively studied in the bcc phase at low magnetic fields. Recent high frequency measurements, reported in this paper, analysed in connection with former results provide a comprehensive study of the NMR spin lattice relaxation for temperatures ranging from the melting of the solid down to 15mK.

INTRODUCTION

We report NMR spin lattice relaxation measurements performed at a Larmor frequency of $\omega/2\pi$ = 199.84 MHz for temperatures ranging from the melting of the solid under study to 15 mK. The results are analysed in connection with the low frequency properties known from former studies to give a coherent picture of the spin lattice relaxation in the bcc phase of solid ^{3}He.

We shall discuss the various relaxation processes as they appear when lowering the temperature, putting some emphasis on (i) the correlation function describing the modulation of the dipolar interaction by the exchange of ^{3}He atoms, (ii) the low temperature relaxation properties as they appear to be dominated by the existence of defects in the solid, most likely dislocations.

RESULTS AND DISCUSSION

The spin lattice relaxation time T_1 has been measured in the bcc phase of solid ^{3}He for 24.94 cm^3 $\geqslant V_m \geqslant$ 22.67 cm^3. Before performing the measurements, the samples were annealed at temperatures close to the melting curve, a procedure which gives reproducible results.

The solid ^{3}He can be described by a sum of hamiltonians corresponding to the various systems of excitations, as appears in Fig. 1: Zeeman, exchange, lattice, vacancies, mass fluctuation waves or ^{4}He impurities, defects, dipolar interaction. During spin lattice relaxation experiments, the energy is transferred from the

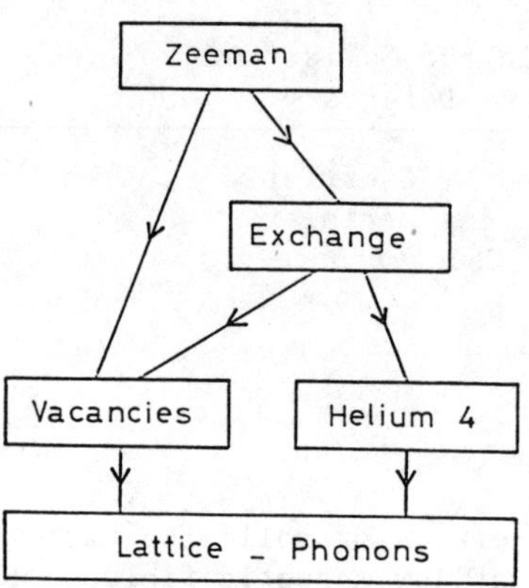

Fig. 1. Schematic diagram of the various systems of excitation ^{3}He. The lines show the couplings responsible for the spin lattice relaxation.

Zeeman to the lattice via the various systems of excitations as indicated by the links in Fig. 1. T_1 is a measure of the weaker of these links which forms a bottleneck for the relaxation. The relative strength of these links changes with temperature and a typical T_1 measurement is sketched as a function of T in Fig. 2. Our results at $\omega/2\pi$ = 199.84 MHz appear in Fig. 3. They are deduced from the recovery of the CW signal following a pulse of 90°. The temperature range can be divided into several regimes (Fig. 2) which we now consider in more detail, starting on the high in solid temperature side.

A. Region I:

T_1 is due to the modulation of the dipolar interaction by the vacancies which are strongly coupled to the lattice. T_1 exhibits a minimum as a function of the temperature when the average atomic jump frequency is of the order of the Larmor frequency. This minimum cannot be observed in our experiments as the solid does not exist in the corresponding temperature range. In a simple model, the relaxation time is given by:[2-3]

$$T_1 = T_{zv}^{-1} = \frac{2M_2 \tau_v}{3} \left(\frac{1}{1+\omega^2 \tau_v^2} + \frac{4}{1+4\omega^2 \tau_v^2} \right) \tag{1}$$

using a correlation function proportional to $\exp(-t/\tau_v)$ to describe the exchange of ^{3}He atoms with neighbouring vacancies. M_2 is the Van Vleck rigid lattice second-moment and τ_v a characteristic time for atomic jumps related to the concentration x_v and the atomic jump frequency ω_v of the vacancies: $\tau_v^{-1} \alpha x_v \omega_v$. All the existing measurements are consistent with a tunneling motion of vacancies whose concentration is thermally activated with a formation energy Φ: $\tau_v \alpha \exp(\Phi/T)$. The values of Φ deduced from the analysis for the two smaller molar volumes for which this regime is observed in a sufficiently large temperature range are in good agreement with former measurements:[4]

$V(cm^3)$	22.67	23.05
$\Phi(K)$	7.4 $\pm$ 0.2	7 $\pm$ 0.2

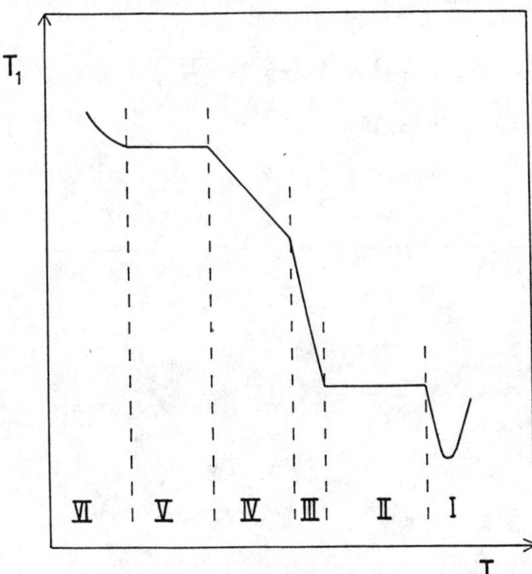

Fig. 2. Schematic variation of the spin lattice relaxation time T_1 as a function of the temperature. The temperature scale is divided into six regions corresponding to different mechanisms for the relaxation.

B. Region II:

When lowering the temperature, the Zeeman vacancy relaxation rate decreases exponentially with 1/T and the Zeeman energy is transferred to the exchange system by the modulation of the dipolar interaction by the quantum mechanical tunneling of the ^{3}He atoms, a process which is temperature independent. T_1 is thus independent of T as long as the exchange is strongly coupled to the lattice, which is true for T > 250 mK. Two or more atoms can be involved in the above mentioned exchange process. For a nearest neighbour Heisenberg hamiltonian

$$H_{ex} = -2 \, \Sigma_{i<j} \, J_{ij} \, \vec{I}_i \cdot \vec{I}_j$$

the Zeeman exchange relaxation time can be written:[2]

$$T_{ze}^{-1} = J_1(\omega) + 4 \, J_2(2\omega)$$

where the spectral densities J_n are the Fourier transforms of the correlation functions describing the modulation of the dipolar interaction by the exchange of ^{3}He atoms. Using a powder assumption the spectral densities are independent of n and, within the assumption of a Lorentzian correlation function,[5] can be written:

$$J(\omega/\omega_e) \, \alpha \, \exp \, (-\omega/\omega_e)$$

72

where $\omega_e^2 = M_4/(2M_2)$, M_2 and M_4 are the second and fourth Van Vleck moments of the rigid lattice. T_{ze} is given by:

$$T_{ze}^{-1} = \frac{\pi M_2}{3\omega_e} \left[\exp\left(-\frac{\omega}{\omega_e}\right) + 4 \exp\left(-\frac{2\omega}{\omega_e}\right)\right] \qquad (2)$$

$$\text{where } \omega_e = 3.36 \, J \qquad (3)$$

for the bcc lattice.

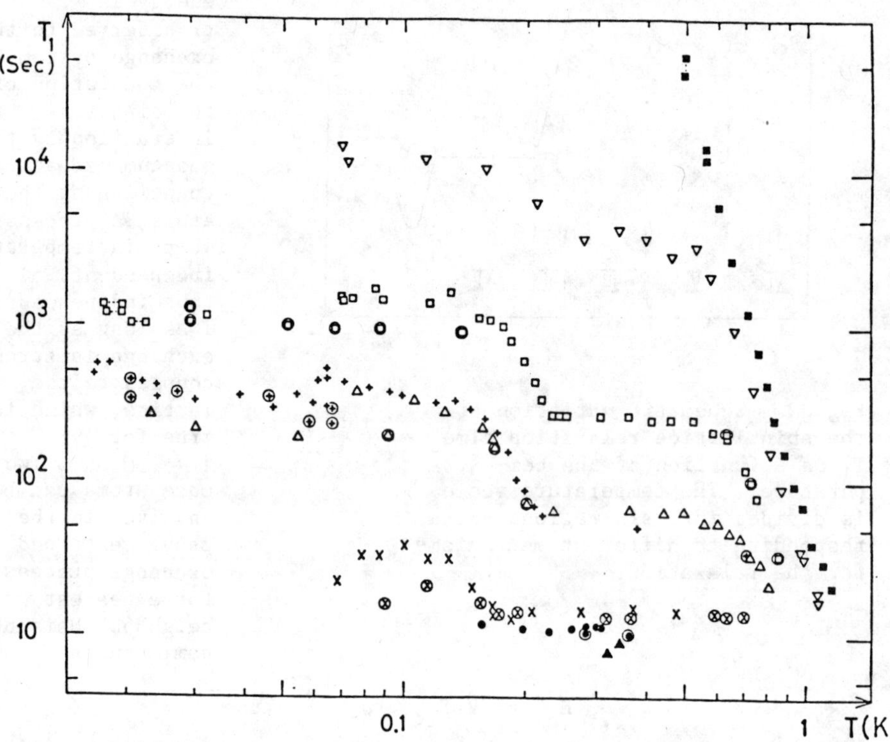

Fig. 3. Spin lattice relaxation time T_1 as a function of the temperature for several molar volumes. The samples are labeled I to VIII. Circled symbols correspond to measurements taken during warming, others taken during cooling: (I) ▲ $V = 24.94$ cm^3; (II) ● $V = 24.83$ cm^3; (III) x $V = 24.59$ cm^3; (IV) Δ $V = 23.91$ cm^3; (V) + $V = 23.91$ cm^3; (VI) $V = 23.52$ cm^3' (VII) ∇ $V = 23.05$ cm^3; (VIII) ■ $V = 22.67$ cm^3. At low temperature, whenever the relaxation was analysed as a sum of two exponentials we plotted only the shortest characteristic time which was also responsible for most of the relaxation.

A comparison of the experimental results with equation (2) is made in Fig. 4. $T_1 = T_1 (21/V)^2$ is used to remove the molar volume dependence of M_2 and reduced units are used to plot on the same universal curve both our results and former low-frequency measurements. The J values used for the analysis are taken from reference.[2]

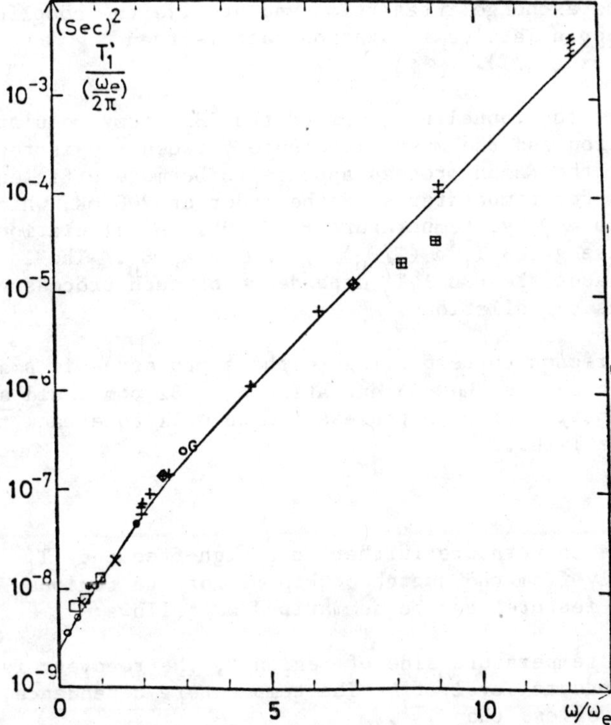

Fig. 4. Experimental values of $T_1'/(\omega_e/2\pi)$ against (ω/ω_e): symbols + refer to the present work and the full curve represents equation (2). Values of $T_1'/(\omega/\omega_e/2\pi)$ measured at different Larmor frequencies were extracted from the following references: ● Ref. 17; x Ref. 18; □ Ref. 19; O Ref. 20; ⊕ Ref. 21; G Ref. 22; ⊕ Ref. 23; the points ⊞ are estimates of relaxation times deduced from thermodynamic measurements in a field of 7T, Ref. 24.

The Lorentzian function describing the modulation of the dipolar interaction by the motion of ^{3}He atoms appears to remain a good approximation up to frequencies of 200 MHz, a fact which was questioned for some time.

It is worth pointing out that this analysis performed with a two particles exchange remains valid if we include three or four particles exchanges provided the molar volume dependence of the various exchange parameters is the same; ω_e will then appear as a functional form of the parameters instead of being given by equation (3). The preliminary measurements of the molar volume dependence of the parameters J_t, K_P, K_F corresponding to the various exchanges for three and four particles do satisfy this criterion.[6]

C. Regions III and IV:

Whereas the Zeeman-exchange coupling is temperature independent, the coupling of the exchange to the lattice is a strong function of T. A decrease in temperature reduces this coupling, making the exchange–lattice the bottleneck for the energy transfer to the lattice: T_1 is then a fast varying function of T. Two mechanisms compete for this relaxation:

--For small ^{4}He impurities concentrations x_4 the thermal vacancies modulate the exchange interaction and provide the coupling to the lattice. The spin lattice relaxation rate is then proportional to $x_v \alpha \exp(-\phi/T)$.

--If x_4 is large, the tunneling jumps of the ^{4}He atoms modulate the exchange interaction and the mass difference between the isotopes scatters the phonons; the Raman process appears to be more efficient than a direct process for temperatures of the order of 200 mK, which is to be compared with a Debye temperature $\theta_D \sim 20$K. A calculation of T_1^{-1} in the bcc phase gives $T_1^{-1} \alpha (T/\theta_D)^9 \Omega_D$, $\hbar\Omega_D = k_B \theta_D$. The T^9 dependence which replaces the usual T^7 dependence of such processes is due to lattice sums cancellations.[7]

The temperature region corresponding to these processes is small for our samples for which the ^{4}He concentration x_4 = 82 ppm would a priori allow a contribution of both processes. We will come back to this temperature range later.

D. Regions V and VI:

When lowering the temperature further, our high-frequency T_1 results differ slightly from the sketch of Fig. 2 for the regions V and VI. The observed features can be summarized as follows:

(i) On the high temperature side of region V, the recovery is exponential and T_1 tends to level off. The temperature dependence, if any, is weak, always less than T^{-1}.

(ii) When decreasing the temperature below T $\sim$ 100 mK the magnetization recovery is sometimes nonexponential. However, the initial part of the recovery cannot be properly described by a $t^{1/2}$ law as observed in the same temperature region at low frequency. Within the scattering of the data, the recovery with time can be analysed as the sum of two exponentials but with relative weights which can vary for successive measurements. However the two characteristic times never differ by more than a factor 2 and keep always the same order of magnitude. This behavior is to be contrasted with the one observed at higher temperatures where the recovery is always exponential and reproducible.

(iii) When decreasing the temperature in the neighbourhood of $T_{PS} \sim$ 60 mK we observe an abrupt drop of the recovery times which then increase slowly upon further cooling.

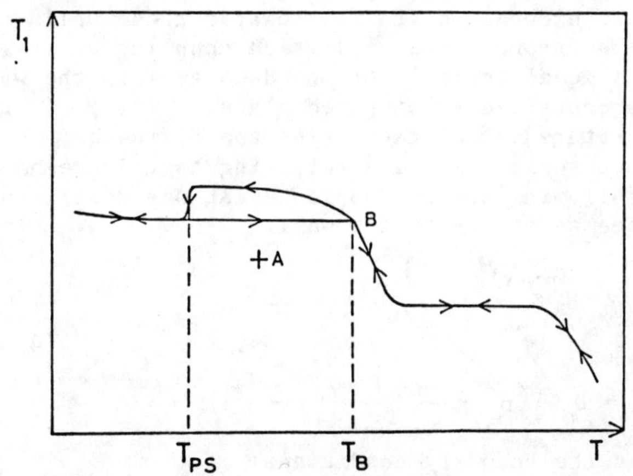

Fig. 5. Sketch of the general behaviour of T_1 as a function of T. At low temperatures, T < T_{PS}, whenever two characteristic times "T_1" can be measured, the shorter one, which accounts for most of the magnetization, is used. The point labeled A corresponds to a typical measurement made without annealing the sample.

(iiii) If the sample which has been cooled below T_{PS} is warmed up, the relaxation time T_1 exhibits a hysteric behavior schematically shown in Fig. 5. T_1 remains constant upon warming up to T_B. The hysteresis cycle has different amplitudes, depending on the rate of cooling through T_{PS} but keeps always the same general features.

Two remarks are in order:

--The measurement technique does not allow us to observe the behavior of the recovery for times shorter than a few seconds.

--The temperature labeled T_{PS} agrees with the phase separation temperature in solid ^{3}He-^{4}He mixtures for $x_4 = 82$ ppm, a value estimated after references 8-9.

Qualitative model for the relaxation on the low temperature "plateau":

The mechanism put forward to explain this relaxation assumes the existence of defects in the solid. Such defects are well known and have been studied in solid ^{4}He[10,11] and in a lower extent in solid ^{3}He.[12] Dislocations are known to exist in these solids. Their vibration frequencies vary from a few kHz to more than 100 MHz in solid ^{4}He, a typical vibration frequency for a dislocation of length $1 = 5\mu$ with a Burgers vector b ~ 3.5Å in a crystal of θ_D ~ 20K would be ν ~ $k_B\theta_D/\hbar$ b/1 ~ 28 Mhz. These frequencies overlap with the exchange frequencies. The dislocations are also strongly coupled to the lattice. In classical crystals they are known to build Frank networks[13] which form fairly stable structures which tend to pin the dislocations put in motion when annealing the crystals, thus forming walls of dislocations.

If we assume that the magnetic energy put into the solid is transferred by Zeeman and exchange diffusion towards these walls of dislocation lines where a strong exchange lattice coupling exists, we can directly transpose a model formerly proposed to explain the wall relaxation of solid ^{3}He contained in sintered glass.[14] If β, C, D stand for inverse temperatures, heat capacities and diffusion coefficients, with the subscripts z and e referring to the Zeeman or exchange systems, the following set of coupled equations describes the relaxation of the Zeeman energy to the walls:

$$\begin{cases} \dfrac{\partial}{\partial t}\,\beta_z = D_z\,\nabla^2\beta_z + \dfrac{1}{T_{ze}}\,(\beta_e - \beta_z) \\[2em] \dfrac{\partial}{\partial t}\,\beta_e = D_e\,\nabla^2\beta_e + \dfrac{C_z}{C_e}\,\dfrac{1}{T_{ze}}\,(\beta_z - \beta_e) \end{cases} \qquad (4)$$

when taking into account the boundary conditions:

(i) $t = 0$ $\beta_z = \beta_e = 0$

(ii) $t > 0$ $\beta_e = \beta_{lattice}$ on the walls

(iii) $\partial\beta_z/\partial x = \partial\beta_e/\partial x = 0$ in the center

where x is a position coordinate. Defining $D_t = C_z D_z + C_e D_e/C_z + C_e$ and using $D_e = 2D_z$, the results of a computer calculation can be summarized as follows:

a) If $C_z/C_e(T_{ze}) \ll \tau_D$ where τ_D is a diffusion characteristic time $\tau_D = \pi/4(V_o/A_o)^2 1/D_t$, A_o being the area of relaxing surface for a volume V_o, the recovery of the magnetization $M(t)$ for short times is well described by $[M(t)/M(\infty)]^2 = t/\tau_D$. This type of recovery corresponds to observations made in low fields experiments.[14-16]

b) If $C_z/C_e(T_{ze}) > \tau_D$ the situation is more complicated; $\partial/\partial t[M(t)/M(\infty)]^2$ increases with time towards τ_D^{-1} but may not in fact attain it; the maximum of the slope $\partial/\partial t[M(t)/M(\infty)]^2$ is denoted by $1/\tau$.
    $1/\tau$ shows a systematic dependence on $\left(C_z/C_e\,T_{ze}/\tau_D\right)$ and can be empirically fitted by:

$$\frac{\tau}{\tau_D} = 0.75 + 1.53\,\left(\frac{C_z}{C_e}\,\frac{T_{ze}}{\tau_D}\right)^{1/2} + 0.25\,\exp\left[-\,6.2\left(\frac{C_z}{C_e}\,\frac{T_{ze}}{\tau_D}\right)^{1/2}\right] \quad (5)$$

The computer calculation was made both in linear and spherical geometries with similar results.[14] In high field experiments $C_z/C_e\,T_{ze}/\tau_D \gg 1$ and using equation (2) for T_{ze}, equation (5) can be written as

$$\mathrm{Ln}\left[\tau\left(\frac{J}{2\pi}\right)\right] = \frac{\omega}{2\omega_e} + \mathrm{Const} \qquad (6)$$

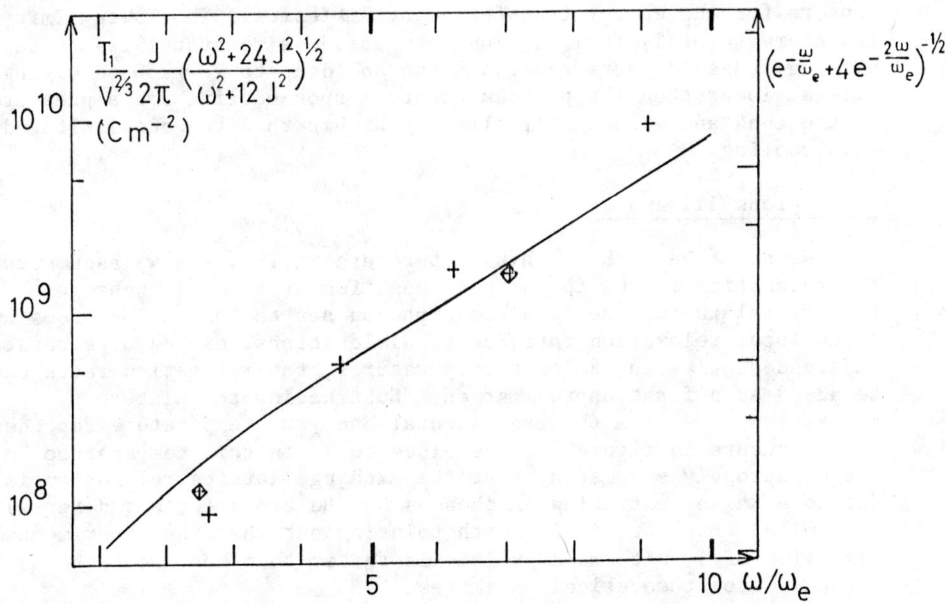

Fig. 6. Experimental value of
$Ln\{T_1/V^{2/3}(J/2\pi)[(\omega^2 + 24\ J^2)/(\omega^2 + 12\ J^2)]^{1/2}\}$ versus (ω/ω_e) compared to the expected value in a model of relaxation on walls of dislocation: in the solid line is $Ln[\exp(-\omega/\omega_e) + 4\ \exp(-2\omega/\omega_e)]^{-1/2}$. The symbols + correspond to this work and ⊕ to Ref. 23.

A comparison of the experimental results with equation (5) is given in figure 6 where the solid line represents equation (5). The constant term of equation (6) gives $V_o/A_o \sim 2 \times 10^{-3}$ cm $\sim 20\mu$, which is a typical length for dislocations in our sample. This value agrees with other estimates in solid ^{4}He and with the estimation of the distance between relaxation surfaces necessary to explain the low temperature relaxation in low field experiments.[15]

The small decrease in T_1 observed when lowering the temperature below T_{PS} is attributed to the creation of defects in the solid due to the migration of ^{4}He atoms. These atoms probably migrate through the dislocation lines which are known to be channels of easy diffusion in classical crystals. To anneal these defects it is necessary to warm the crystal up to T_B of figure 5, a temperature domain where the ^{4}He atoms seem to be able to migrate more rapidly through the crystal. Such hysteresis cycle has also been observed in sound attenuation measurements at 10 MHz.[12]

A decrease of T_1 is also observed at large molar volumes when the solid melts while lowering the temperature as shown in figure 3, where the hysteresis cycle of T_1 upon warming appears clearly. Upon melting, many defects appear in the crystal, they act as relaxation centers for the energy transferred by diffusion. The lower limit for the decrease of T_1 is T_{ze}. When the solid is recrystallized upon warming, these defects remain in the solid up to $T \gtrsim 150$ mK giving a plateau lower than the plateau observed upon cooling and a pressure in the constant volume cell slightly higher than the one monitored upon cooling.

E. Regions III and IV

We now come back to this temperature regime. If we assume that the relaxation is due to the superposition of a local exchange lattice relaxation due to a Raman phonon scattering on ^{4}He atoms and a non local relaxation rate due to dislocations, giving a temperature independent plateau at lower temperatures, the relaxation rates can be added as a first approximation. Subtracting the plateau relaxation rate from the experimental one, the temperature dependence of T_1 appears in figure 7 to be close to T^9 in this temperature range, strongly suggesting that the exchange lattice relaxation is due to a Raman scattering of phonons on ^{4}He atoms (or any defects) in the solid. However, it is worth pointing out that the experimental variation of T_1 with molar volume is faster than what would be expected from theoretical estimates.[1-7]

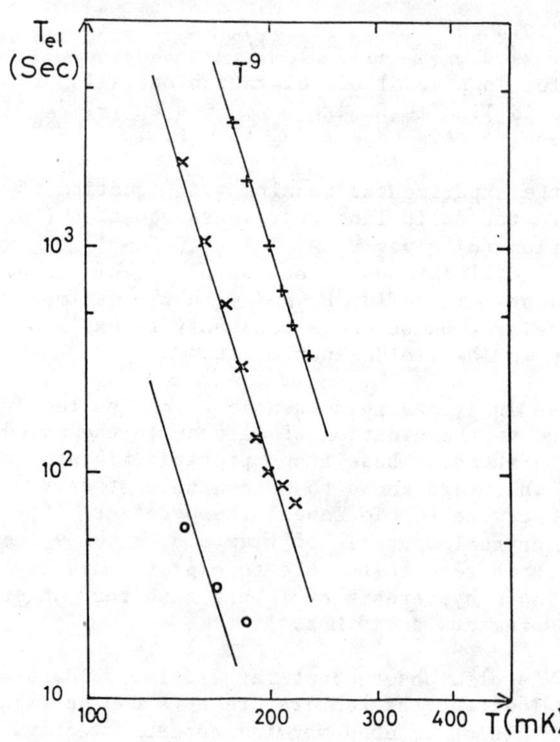

Fig. 7. Temperature dependence of the exchange lattice relaxation time T_{el} in the temperature region located between the Zeeman exchange plateau and the low temperature plateau. At a given temperature, the relaxation rate T_{el}^{-1} is taken as the difference between the experimental rate T_1^{-1} and the relaxation rate on the low temperature plateau $(T_1)_{Pl}^{-1}$. The symbol + correpsonds to $V = 23.52$ cm^3, $(T_1)_{Pl} = 1400$ sec; x corresponds to $V = 23.91$ cm^3, $(T_1)_{Pl} = 335$ sec; 0 corresponds to $V = 24.59$ cm^3, $(T_1)_{Pl} = 32$ sec.

F. Other features:

Erratic pressure jumps of maximum amplitude ⩰ 0.7 mb are observed for the whole temperature range (15 mk ⩽ T ⩽ 400 mK). They happen at random, towards higher and lower pressure without changing the average pressure at a given temperature. They do not interfere with the measured relaxation properties of the crystal and do not produce measurable heating. This phenomenon, observed on a macroscopic scale, is attributed to the presence of twins in the solid and to the reversible motion of these defects.[13]

CONCLUSION

The spin lattice relaxation mechanisms of bcc solid ^{3}He appear to be understood for temperatures ranging from the melting of the solid down to 15 mK and Larmor frequencies $\omega/2\pi$ ⩽ 200 Mhz.

The qualitative difference appearing in the correlation functions describing the modulation of the dipolar interaction by exchange in the bcc (Lorentzian function) and hcp (Gaussian function) phases tends to support the idea of different topological exchanges in these two phases but a theoretical calculation seems presently too difficult to succeed.

REFERENCES

1.   M.E.R. Bernier and G. Guerrier, Physica B + C, June 1983.

2.   A. Landesman, Ann. Phys. 9 (1975) 69
      A. Landesman, Ann. Phys. 8 (1973) 53
      R.A. Guyer, R.C. Richardson, L.I. Zane, Rev. Mod. Phys. 43 (1971) 532

3.   N. Bloembergen, E.M. Purcell and R.V. Pound, Phys. Rev. 73 (1948) 679
      H.C. Torrey, Phys. Rev. 92 (1953) 962
      H.C. Torrey, Phys. Rev. 96 (1954) 690
      H.A. Resing and H.C. Torrey, Phys. Rev. 131 (1963) 1102
      M. Eisenstadt and A.G. Redfield, Phys. Rev. 132 (1963) 635

4.   N. Sullivan, G. Deville and A. Landesman, Phys. Rev. 11 (1975) B 1858

5.   W.C. Thomlinson, J.F. Kelly, R.C. Richardson, Phys. Letters. 38A, (1972) 531

6.   T. Mamiya, A. Sawada, H. Fukuyama, Y. Hivao, K. Iwashashi and Y. Masuda, Phys. Rev. Lett. 47, 1304 (1981)

7.   M. Bernier and A. Landesman, Journal de Physique C5a (1971) 213

8.   D.O. Edwards, A.S. McWilliams and J.G. Daunt, Phys. Rev.
     Letters 9 (1962) 195

9.   M.F. Panczyk, R.A. Scribner, J.R. Gonano and E.D. Adams, Phys.
     Rev. Letters 21 (1968) 594

10. M.A. Paalanen, D.J. Bishop, H.W. Dail, Phys. Rev. Letters 46,
     664 (1981)

11. D.J. Bishop, M.A. Paalanen and J.D. Reppy, Phys. Rev. B 24
     (1981) 2844

12. I. Iwasa, N. Saito, H. Suzuki, Journal de Phys. (Paris) C5-37
     (1981)

13. J. Friedel, Dislocations, Pergamon Press -- Oxford (1964).
     Les Dislocations -- Monographies de Chimie Physique -- Gauthier
     -- Villars -- Paris (1956)

14. R.P. Giffard, J.P. Stagg, W.S. Truscott and J. Hatton, J. Low
     Temp. Phys. 31 (1978) 817

15. R.P. Giffard, D. Phil. Thesis Oxford (1968)

16. E.R. Hunt, R.C. Richardson, J.R. Thomson, R.A. Guyer, and H.
     Meyer, Phys. Rev. 163 (1967) 181

17. M. Bernier, Thesis, Univ. of Paris (1971)

18. M. Bernier, J. Low Temp. Phys. 3 (1970) 29

19. M. Bernier, G. Deville, J. Low Temp. Phys. 16 (1974) 349

20. R.C. Richardson, E. Hunt, H. Meyer, Phys. Rev. 138 (1965) A
     1326

21. M. Bassou Thesis, Univ. of Paris (1981)

22. D. Thoulouze, G. Bonfait, Y. Chabre, Journal de Physique C7, 41
     (1980) 111

23. M. Bernier, A.S. Greenberg, G. Guerrier, unpublished

24. W.P. Kirk, E.D. Adams, Proc. 13th Int. Conf. Low Temp. Phys. 2
     (1972) 149

PRE-TRANSITIONAL EFFECTS FOR VACANCY FORMATION AND MOBILITY IN BCC ^{3}HE

M. Chapellier, M. Bassou, M. Devoret,
J.M. Delrieu and N.S. Sullivan*

Université d'Orsay, 91405 Orsay, France, and
Centre d'Etudes Nucléaires, 91191 Saclay, France

While the properties of vacanices (formation enthalpy H, mobility μ, bandwidth ....) have been measured systematically for relatively low molar volumes ($V_m < 23$ cm^3/mole) using both x-ray[1] and NMR techniques[2] only a few studies[1] have been carried out for high molar volumes close to the minimum of the melting curve. Since the vacancy formation energy $\Phi(V_m)$ determined at high densities decreases rapidly for increasing V_m and extrapolates to values close to the estimated bandwidths[3] for V_m 24.9 cm^3/mole, reliable determinations of $\Phi(V_m)$ and the mobility are needed as a function of density close to V_m^{max} where one anticipates that a reformulation of the description of the ground state may be necessary when the vacancy tunnelling matrix elements become comparable to the formation energy. Recent measurements of the nuclear spin lattice relaxation times have enabled us to determine these properties, and assuming that the entropy of formation is always zero, we find that both $\Phi(V_m)$ and log μ vary as $\left(V_m - V_m^{max}\right)^{1/2}$ for $23.0 < \overline{V_m} < 24.9$ cm^3/mole. This dependence can be understood within the context of a mean field treatment of the delocalized vacancies in terms of density fluctuations. In this approach the vacancies form an energy band with gap Δ $\left(V_m - V_m^{max}\right)^{1/2}$ and coherence length $\xi \sim 1/\Delta$.

The mobility $\mu \propto \exp(a/\xi)$. This conclusion would be different if one allows for an entropy of formation $S(V_m)$ for the vacancies.[4] NMR measures $\mu \exp(H(V_m)/k_BT)$ where the enthalpy $H(V_m)$ = $\Phi(V_m) - TS(V_m)$ and the T-independent factor is $\mu \exp S(V_m)$ and not simply μ. The striking V-dependence observed for this factor (4 decades for $23.0 < V_m < 24.9$ cm^3/mole) could also be understood in terms of an increase in the entropy of formation on approaching V_m^{max}.

1. S.M. Heald, Ph.D. Thesis, University of Illinois at Urbana-Champaign, 1976. (unpublished)
2. N.S. Sullivan, G. Deville and A. Landesman, Phys. Rev. B11, 1858 (1975).
3. J.H. Hetherington, Phys. Rev. 176, 231 (1968).
4. This has also been pointed out by J.H. Hetherington and P. Sokol (private communications).

*Present address: Dept. of Physics, University of Florida, Gainesville, FL 32611

SPIN STRUCTURE OF SOLID ^{3}He BELOW 1 mK NEUTRON DIFFRACTION*

P.R. Roach, S.K. Sinha, B.K. Sarma, M. Vrtis,
Y. Takano,[a] M. Misawa,[b] K. Sköld,[c] R. Kleb and Z. Sungaila

Materials Science and Technology Division
Argonne National Laboratory, Argonne, Illinois 60439

and

J.B. Ketterson and W.P. Halperin

Physics Department, Northwestern University,
Evanston, Illinois 60201

An ultralow temperature neutron diffraction facility has been constructed at the Intense Pulsed Neutron Source (IPNS-1), with the purpose of resolving the magnetic spin structure of ordered solid ^{3}He ($T_s \sim 1.1$ mK). We have successfully grown large single crystals of ^{3}He and (110) Bragg reflection peaks from solid ^{3}He ($T \sim 20$ mK) have been observed with a signal to noise ratio of 10. In the absence of neutron flux, the cryostat has cooled below 0.5 mK.

*Work supported by the U.S. Department of Energy.
[a]Present address: University of Tokyo, Japan.
[b]Present address: Tohoku University, Japan.
[c]Permanent address: Studsvik, Sweden.

MOLAR VOLUME DEPENDENCE OF THE PRESSURE OF SOLID ^{3}He AT VERY LOW TEMPERATURE

T. Mamiya, A. Sawada, H. Fukuyama, K. Iwahashi, and Y. Masuda
Nagoya University, Nagoya, Japan

The pressure of solid ^{3}He has been measured as a function of temperature T between 0.3 and 50 mk at molar volumes between 24.19 and 23.31 cm^3. The entropy discontinuity obtained from the pressure jump at the ordering transition turned out to be almost independent of molar volumes, being about .40Rℓn2 in the studied range of molar volumes. Spin wave velocity decreases from 8.5 to 5.0 cm/s as molar volume decreases from 24.19 to 23.51 cm^3. Pressure measurements above the ordering temperature reveal that besides the known $1/T$ term there are significant terms proportional to $-1/T^2$ and $+1/T^3$. The reduced pressure $(P-P_o)$ V/T_N versus T_N/T at various molar volumes does not follow a universal function with a maximum deviation of 13%, where $P-P_o$ is the pressure associated with nuclear magnetism, V is molar volume, and T_N is the ordering temperature. The Kapitza resistance between solid ^{3}He and sintered silver in the pressure cell was also measured. The Kapitza resistance for all molar volumes investigated varies as $2 \times 10^6/T$ (mk) m^{2}K/W and it is almost the same magnitude as the Kapitza resistance between liquid ^{3}He and silver powder.

ORIENTATIONAL ORDERING IN SOLID H_2 AND D_2 AT INTERMEDIATE J=1 CONCENTRATIONS

James R. Gaines and Paul E. Sokol*

Department of Physics
Ohio State University
Columbus, Ohio 43210

ABSTRACT

At temperatures below 4K for zero pressure the previously free rotation of the J=1 excitations in solid H_2 or D_2 becomes hindered. The hindered rotor is then orientationally ordered. This orientational ordering is discussed for concentrations of J=1 molecules $x < 0.55$.

The largest number of experiments that have probed the orientational ordering have used NMR techniques. These experiments as well as approaches using Monte Carlo simulations, x-ray diffraction, and specific heat measurements will be discussed. The bulk of experimental evidence suggests that the freezing of the orientational degrees of freedom proceeds continuously with no definite transition temperature but the evidence for a transition of a cooperative nature with a well defined transition temperature will also be presented.

INTRODUCTION

Orientational ordering of J=1 molecules, at concentrations x, below the critical concentration (X_c = 0.55) for the hcp to fcc phase transition (originally called the "lamba transition in hydrogen") has been studied by means of nuclear magnetic resonance (NMR) since the first experiments of Sullivan and Pound[1] in 1972. In 1976, Sullivan[2] ruled out the possibility that the ordering for intermediate concentrations (x < .55) was produced by an axial crystalline field and speculated that this orientational ordering might be analogous to the ordering of spins in a spin-glass. In 1978, Sullivan et al.[3] interpreted their latest NMR data on H_2 alloys with x < 0.55 in terms of a <u>quadrupole glass</u> transition with a well defined transition temperature. A key concept, frustration, was introduced in this paper. The energetically favoured "Tee" configuration of adjacent quadrupoles cannot be realized on an hcp lattice or any 3D lattice. At sufficiently high concentrations, the orientational fluctuations drive a phase transition to an fcc lattice on which the molecular orientations are ordered in a long range periodic Pa_3 configuration. At reduced concentration (x < 0.55),

*Present Address: Department of Physics, University of Illinois, Urbana, Illinois 61801

this frustration is no longer able to drive a lattice structure change but it is thought to be a major contributing factor to a quadrupole glass transition.

No one disputes that orientational ordering of the quadrupole moments occurs as the temperature is reduced. There is mounting evidence and consensus from the NMR experiments for the viewpoint that the ordering occurs continuously with no well defined transition temperature.[4,5,6] I endorse that point of view for the range of concentration below x = 0.29, but data taken at O.S.U.[7] suggest that cooperative behavior does exist in the ordering for concentration between 0.29 < x < 0.55.

CURRENT STATUS OF THE NMR EXPERIMENTS

In the most recent experiments at Harvard[6], it is found that: (1) For H_2, no unambiguous order parameter can be found in the temperature dependent NMR lineshape changes to assign a definite transition temperature to the ordered state. (2) For D_2, relaxation studies (T_1) for concentrations below x = 0.3 show no indication of a transition to a quadrupole glass state.

The most recent Duke experiments[5] covering the H_2 concentration range x < 0.55 can be summed up by their remark: "Over the investigated temperature and concentration region, there is no hint of a well-defined orientational ordering phase transition in the observed NMR spectrum... Furthermore, the absence of hysteresis or other non-equilibrium phenomena indicates that there is no transition region characteristic of the glass transition in vitreous silica or other glasses."

The most recent experiments of Sullivan and coworkers, employing c.w. lineshape studies, T_1 measurements, and stimulated echoes do not indicate the well-defined critical behavior expected for a phase transition. Instead a smooth cooperative freezing of the orientational degrees of freedom is indicated.

INFORMATION FROM X-RAY DIFFRACTION STUDIES, SPECIFIC HEAT
MEASUREMENTS, AND MONTE CARLO SIMULATIONS

Vital information about the nature of the orientational ordering at low temperatures is available from sources other than the NMR experiments. These sources are: specific heat measurements[8,9]; x-ray diffraction studies[10]; and Monte Carlo simulations.[11,12,13]

The x-ray results[10] demonstrated that the lattice structure in H_2, in the zero pressure region of the T-x phase diagram conjectured to be the quadrupolar glass phase, was hcp and not fcc as commonly thought. Furthermore, it didn't matter whether the samples were aged into this region from the long range ordered fcc phase or brought into the glass region from the hcp disordered phase--the lattice

86

structure was still hcp. The inconsistencies of the various NMR
determinations of the transition temperature were pointed out and the
most likely explanation of these inconsistencies ruled out.

The recent measurements of the specific heat by Haase[9] were
initiated in the hope of finding a first or second order phase
transition in this concentration region. The measurements do not
show any cusp-like behavior. The specific heat, C, decreases roughly
as $T^{1.5}$ and the ratio C/T shows a rounded peak in the vicinity of
0.2K to 0.4K that decreases in magnitude with decreasing concentration. Searches for long thermal relaxation times have not detected
any and the only remanence effect observed was associated with the
isothermal conversion of a sample from the fcc phase to the hcp
phase.

In the "computer experiments" of Klenin and coworkers, there is
strong indication that the thermal history (e.g. the cooling rate
from the disordered state or whether the sample is prepared by isothermal conversion) is important. Geometric effects arising from the
frustration of the quadrupoles appear to be more important than
dilution effects. For instance, the sharp transition to an ordered
state for x=1 is broadened as x is reduced but there is no drastic
change in the behavior of the order parameters until the two dimensional percolation limit of about .55 is reached.

The main thrust of the work by Devoret and Esteve is the investigation of whether the freezing of the motion of the quadrupoles
is a smooth approach toward a unique ground state or a strongly
cooperative phenomena. Their Monte-Carlo simulations support the
idea that the freezing is a cooperative phenomenon.

THE ARGUMENT FOR A PHASE TRANSITION

The argument for a transition to an orientationally ordered
phase at intermediate concentrations rests on the existence of dips
or cusps in the spin-rotation relaxation time (T_1) and the time constant that characterizes the spin echo dephasing (T_E) for $0.29 < x <$
0.52. These dips or cusps are not seen for concentrations below
x = .33 so the limit 0.29 is obtained by an extrapolation. The
increase in relaxation rate responsible for a cusp or dip in T_1 is
thought to come from critical fluctuations of the order parameters.

Experiments at O.S.U. were performed on D_2 alloys with J=1 concentrations ranging from 0.52 to 0.02. D_2 was chosen over H_2 because
of its much lower J=1 to J=0 conversion rate. The NMR probe for
these alloys was about 1% oH_2 impurity that was added to the
samples. The quadrupole moment of the oH_2 molecule responds to the
fields created by the quadrupole moments of the J=1 pD_2 molecules so
the relaxation time, T_1, of these impurity oH_2 molecules depends on
the amplitude of the spectral density function of the pD_2 rotational
correlations evaluated at the Larmor frequency (30 MHz) of the

impurity NMR probe. The H_2 NMR is then monitored to extract information about the pD_2 rotational correlations. When oH_2 was added to solid D_2 with $x = 0.83$, the relaxation rate peaked at the order-disorder transition temperature T_{do} (slightly below T_{od} as determined by specific heat measurements[14]) and the lineshape change followed that of the D_2 NMR.

In our initial experiments, we concentrated on the range of alloys that can be made from gas mixtures of nD_2 ($x = .33$) and D_2 with $x = 0.022$ (the 20K equilibrium value). Of particular concern was the low concentration experiment as it could provide an estimate for a lattice caused by inclusion of mass 2 defects in the mass 4 lattice. No such effects were observed on the H_2 NMR. Spin echo measurements using a 90°-90° sequence were used to obtain the lineshape while a saturating 90° pulse followed by a 90°-90° echo sequence was used to measure T_1. Where it was possible to use a 180°-90° sequence to measure T_1, the values obtained agreed well with those obtained with the three pulse sequence.

In addition to the proton resonance in these samples from the added oH_2 impurity, there was an unwanted signal from HD impurities present in the nD_2 gas. The HD resonance lineshape was a gaussian that was temperature independent. The second moment measured for this gaussian agreed well with that calculated using standard formuli for broadening due to like and un-like spins. The broadening of oH_2 spins due to <u>intermolecular interactions</u> is characterized by the same second moment as that of the proton spin on the HD molecule. If the composite echo signal due to HD and oH_2 is multiplied by $\exp(+M_2t^2/2)$ where the HD free induction decay signal is proportional to $\exp(-M_2t^2/2)$, one obtains a constant baseline (the HD signal) and a time dependent signal due to oH_2 molecules—that is now broadened just by intramolecular interactions. This procedure is illustrated in Fig. 1 where the top trace shows an echo obtained in a sample with $x = .33$ at .035K. The gaussian intermolecular broadening correction has been made on the lower trace and a constant representing the HD signal subtracted. This correction amounts to the removal of about 1 KHz linewidth from the data. From the remaining echo signal, the fourier transform gives the absorption lineshape from which moments or the derivative can be computed for comparison with cw measurements.

Our sample with $x = .33$ did show a dip in T_1 and a large increase in linewidth or splitting of the derivative peaks as the temperature was lowered below 1K. For the other samples studied with $x < .33$, no dip in T_1 was observed, and although the NMR line did broaden with reduced temperature, no sudden increase in splitting was observed that could indicate a phase transition had occurred.

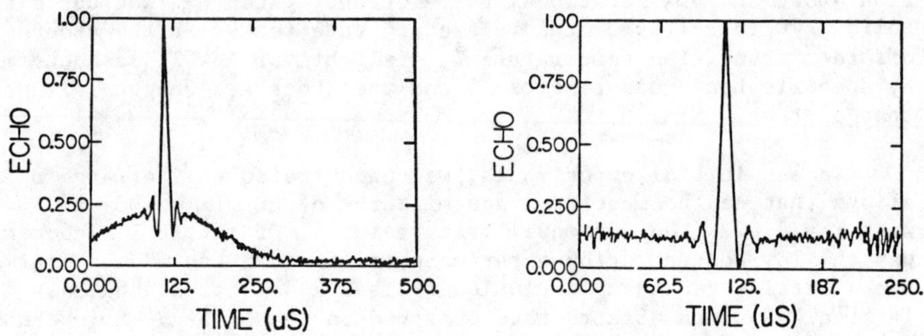

Fig. 1. Top Trace: Echo for x=0.33 at T+0.035K. Bottom Trace: HD
signal removed and intermolecular broadening removed as described in
the text.

The situation is quite different for samples we tested with
concentrations from .42 to .52. These samples had been prepared some
fifteen years ago by David White for the specific heat measurements
reported by Grenier and White[14]. Since the samples had been enriched
in their J=1 concentration by selective absorption, there was no HD
impurity detectable.

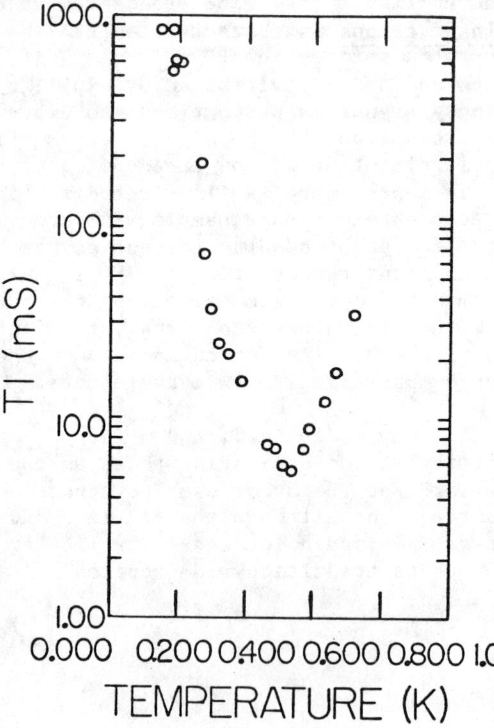

Fig. 2. Measured spin
lattice relaxation time vs.
temperature for x=0.52.

At 30 MHz we measured
the spin-lattice relaxation
time, recorded the echo, and
measured the echo decay
time. T_1 was observed to
drop from about 230ms at 4K
to 2ms at .56K for the
sample with x = .52. This
is shown in Fig. 2. At
fixed pulse separation, the
echo height was observed to
drop and then increase again
as the temperature was
raised from below 0.1K to
1K. The echo decay constant
passes through a minimum
near 0.3K before recovering
to its high T value. The
scaled relaxation rates are
shown in Fig. 3, where the
dip obtained for x = .33 is

given as well. To accomodate this rate on the same graph with the data for x = .52, it must be scaled by a factor 8. A scale factor of 4 is used for the x = .42 data. A high temperature "background" rate of 4 s^{-1} has been subtracted from the data on the high temperature side of the cusp in order to make the data reflect the temperature dependent fluctuations.

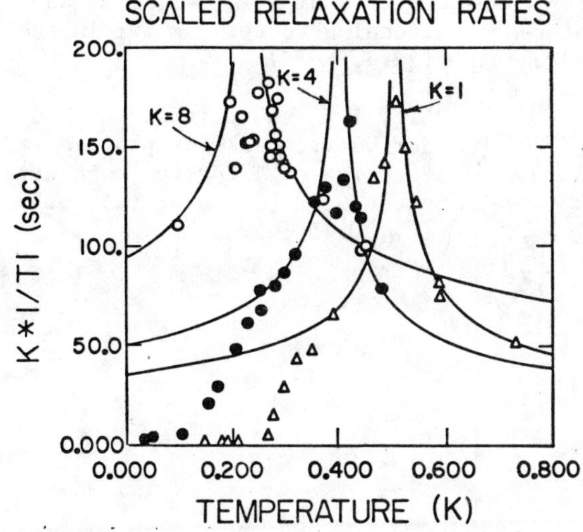

Fig. 3. Scaled Relaxation Rates, KT^{-1} for x=0.52 (Δ), x=0.42 (o), x = .33 (o). Measured rates after background subtraction are multiplied by K=4 for x=0.42 and K=8 for x=0.33.

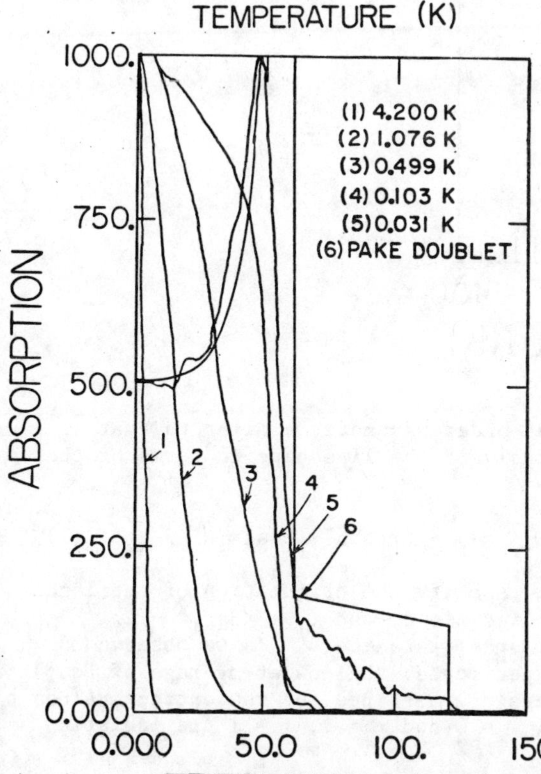

Fig. 4. Lineshapes for the sample with x=0.52 at the temperatures, 4.2K (1), 1.076K (2), 0.499K (3), 0.013K (4), 0.031K (5).

Lineshapes for the x = .52 samples, obtained by fourier transforming the echo response, are shown in Fig. 4. A very dramatic increase in width and a marked lineshape change occur between the extreme tempera-tures. The cusp in T_1 occurs at about 0.55K for this concentra-tion. While such change are visually interesting, it has re-mained a difficult problem to quantify these changes and decide from them, if a transition has taken place,

90

and if so, at what temperature.

In the CW lineshape studies of many groups, the splitting ($\Delta\nu$) in frequency units between the maximum and minimum of the derivative of the lineshape is plotted as a function of temperature. We have computed the derivative from our lineshape and show the splitting as a function of temperature for concentrations of .52, .42, .33, and .25 in Fig. 5. The temperature corresponding to the maximum in the relaxation rate is indicated on the figure.

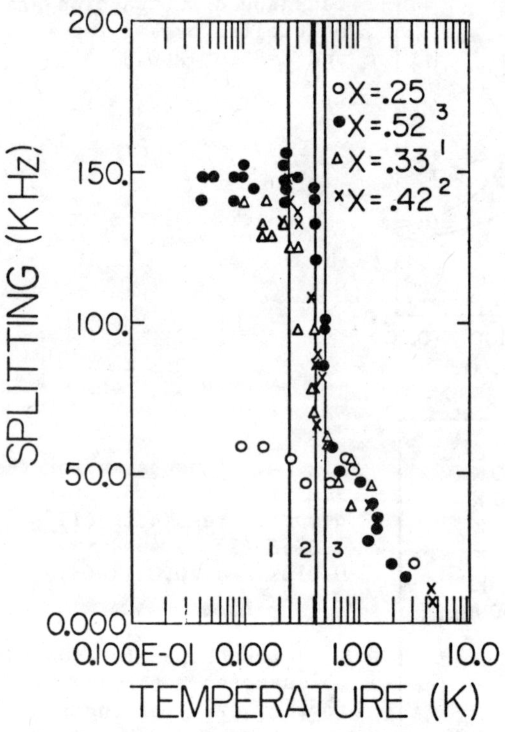

Fig. 5. The splitting of the derivative peaks is plotted as a function of T for the concentrations 0.52 (o), 0.42 (x), 0.33 (δ), 0.25 (o).

In an attempt to find an order parameter similar to that calculated by Klenin[15], we analysed the lineshape in terms of the expression for the frequency[3]

$$\Delta\omega_i = \frac{3d}{4}\left[\sigma_i(3\cos^2\theta_i - 1) + 3\eta_i\sin^2\theta_i\cos2\phi_i\right] \tag{1}$$

where d = 57.7 kHz, θ_i and ϕ_i specify the orientation of the local axis with respect to the applied field, and $\sigma_i = \langle 3j_{zi}^2 - 2\rangle$, $\eta_i = \langle J_{xi}^2 - J_{yi}^2\rangle$ are local order parameters. As we observe no angular dependence in our experiments, the powder average of Eq. 1 is appropriate. The powder average lineshapes for the extreme values of the order parameters $\sigma = -2$, $\eta = 0$ and $\sigma = 1$, $\eta = 1$ are indistin-

guishable so we treated Eq. 1 as though $\eta = 0$. With this approximation, the lineshape for a perfectly ordered powder specimen ($\sigma_i = -2$)
is a Pake doublet with maximum frequency of 3d. Designating the
center frequency of the lineshape function as $\omega = 0$, we can define.

$$Q = \frac{\int_0^\infty \omega I(\omega)d\omega}{\int_0^\infty \omega I_{PD}(\omega)d\omega} \qquad (2)$$

with $I(\omega)$ the observed lineshape, $I_{PD}(\omega)$ the Pake Doublet lineshape
with $\sigma = -2$. Q is thus representative of the average local quadrupole moment. Values of Q calculated from the above expression are
given for $x = .52$, $x = .42$, $x = .33$, and $x = .25$ as shown in Fig. 6.

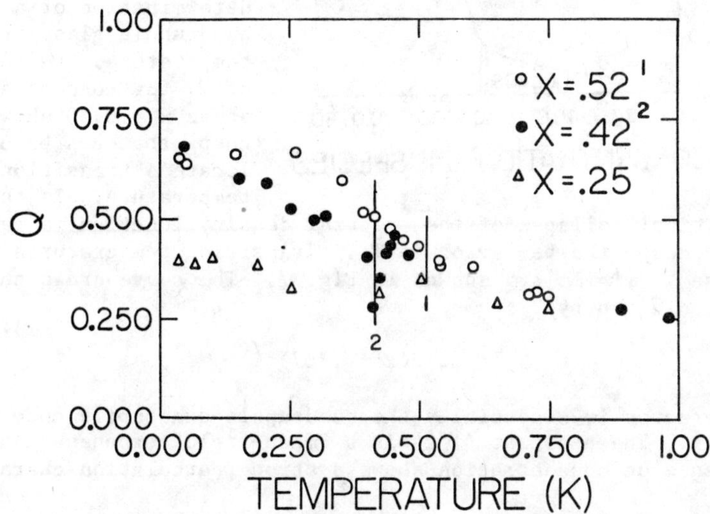

Fig. 6. The quantity Q defined in Eq. 2 is plotted as a function of
T for the concentrations 0.52 (o), 0.42 (o), 0.25 (△).

A global comment on the NMR results is that there is no obvious
way to extract the static order parameter from the lineshape
results. Whether some parameter such as Q, the second moment M_2, or
the splitting of the derivative peaks is used, the problem in
determining a transition temperature arises because there are both
static and dynamic contributions to the lineshape. Above T_c, the
linewidth contains a fluctuation component that increases as T_c is
approached. Below T_c, the fluctuation component diminishes (possibly
symmetric in its behavior about T_c) and the static order parameter

increases. Without a detailed theoretical picture, it is not possible to untangle these components. For example, in the disordered phase at 4.2K, some static ordering exists as we are able to produce solid echoes at this temperature.

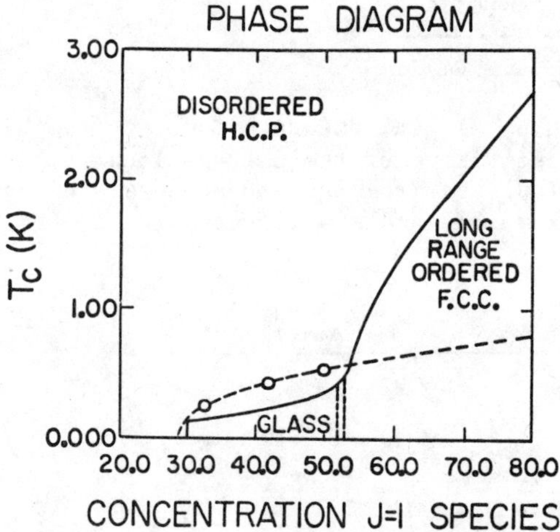

Fig. 7. Transition temperatures determined from the minimum in T_1 are plotted as a function of concentration. The curve through the points is given in Eq. 3.

In conclusion, NMR lineshape changes do not yield an unambiguous determination of a quadrupole glass transition temperature. Measurements of T_1 for concentrations above x = 0.30 show dips or cusps that can be used to locate a transition temperature. In this case, the critical collapse of the spectral density function is responsible for the cusp-like be-havior in T_1. Transition temperatures obtained from the T_1 minima are showns in Fig. 7. The curve drawn through the points is given by:

$$T_c = 1.12(x - .287)^{1/2} \tag{3}$$

The prefactor is essentially the quadrupole-quadrupole coupling constant. The exponent (1/2) is a "mean-field" exponent, and the dependence on concentration shows a strong percolation character.

THE CASE AGAINST A PHASE TRANSITION

Minima in T_1 can occur when the correlation frequency characteristic of the fluctuating local magnetic field becomes comparable to the Larmor frequency. Such minima can occur in a system without a phase transition occurring. If an exponential correlation function[16] is used, the predicted minimum value of T_1 for 30 MHz is 0.2ms and it occurs for ω_o/ω_c = 0.8. For lower frequencies, the minimum value of T_1 is proportional to the frequency. For a gaussian correlation function[17] (which is more realistic at these concentrations) the minimum in T_1 occurs when ω_o/ω_c = 0.67 giving a value of T_1 at the minimum of 0.17ms for 30 MHz. For lower frequencies, the minimum value of T_1 is also proportional to the frequency.

If the minimum in T_1 is of this ordinary variety, then combined with the smooth behavior observed through the lineshapes and linewidths, there is no way to assign a definite transition temperature to the orientational ordering observed by NMR techniques.

Preliminary results on the frequency dependence of the relaxation time have indicated that the temperature at which the minimum occurs is not changed for frequencies down to 8 MHz nor does the minimum values of T_1 scale with the frequency. In our experiment, the minimum occurs for $\omega_o/\omega_c = 1/25$ with a magnitude of ten times that calculated for an ordinary minimum.

If, despite these difficulties in identifying the minimum in T_1 with an ordinary minimum (e.g. the temperature dependence, frequency dependence, and magnitude are not consistent with conventional models), we can use the temperature at which the minimum occurs to obtain the value of $\omega_c = \omega_o$. Furthermore, at a lower temperature, the echo decay time has a minimum that corresponds to $\omega_c \simeq d$. The motional narrowing model can then be used to extract the correlation frequency as a function of temperature. This approach parallels that of Sullivan and Esteve.[18] Independent of model, the measured echo decay time, $\ln(T_{echo})$, can be plotted as a function of T^{-1} in order to display thermally activated behavior. Results of $x = 0.52$ are shown in Fig. 8. The slopes of the lines shown are 4.1K and 0.15K. The positive slope indicates that the system has made a transition from the regime $\omega_c > d$ to $\omega_c < d$. The NMR measurements of T_1 and the spin echo response, interpreted in this manner, thus probe the time scale of the fluctuations.[18] While the correlation frequency decreases rapidly with decreasing temperature, freezing out certain specific motions at definite temperatures, there is no single temperature that characterizes the ordering. This situation is comparable to that encountered in real glasses where only a transition region can be defined.

SUMMARY

Dramatic changes in the NMR lineshape and increases in the linewidth are observed with decreases in temperature for H_2 and D_2 alloys with $x < 0.55$. Lack of detailed theory prohibits the determination of a static order parameter in the presence of a large fluctuation contribution to the linewidth.

Dips in the relaxation time T_1 or cusps in the relaxation rate are observed for $x > 0.30$ but not for concentrations below this-- indicative of percolation behavior. If the dips are interpreted as ordinary T_1 minima, then the NMR transient responses can be used to obtain the time scale of the rotational fluctuations. Such an interpretation of the measurements indicates a drastic slowing down of the rotational fluctuations analogous to the behavior observed in real glasses. If the dips are not interpreted as conventional T_1 minima, then they can be used to assign a unique transition temperature to the orientationally ordered phase for $0.30 < x < 0.55$.

94

The spin echo response indicates that the correlation frequency becomes comparable to the dipolar frequency at T = 0.028K for x = 0.52. Well below this temperature (e.g. at T = 0.035K), the lineshape is similar in appearance to that of a Pake doublet with $|\sigma|$ 1.5 whereas a perfectly ordered powdered samples would yield $|\sigma| = 2$. The linewidth contributions can be considered to be "static" at this temperature. At 0.1K, the line is bell shaped (as observed by others). Above 0.28K, fluctuations make significant contributions to the lineshape and linewidth making the determination of a state order parameter difficult.

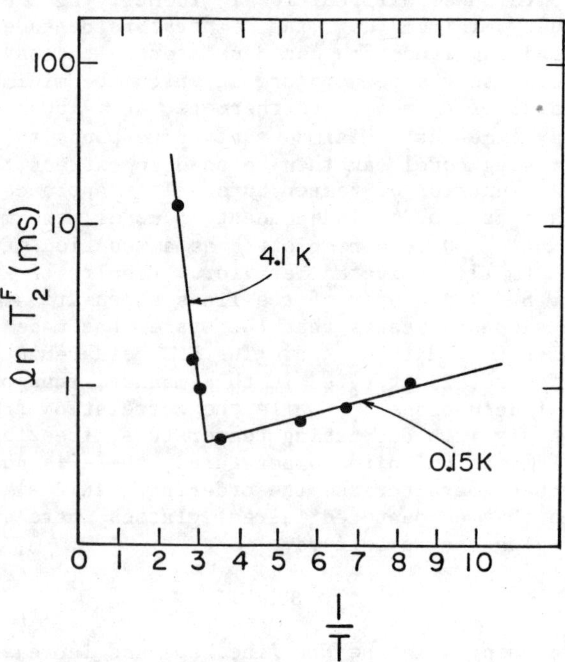

Fig. 8. The echo decay time T_F is plotted as a function of T^{-1}. The high temperature slope yields an activation energy of 4.1K, while the low temperature slope yields an activation energy of 0.15K.

ACKNOWLEDGEMENT. The support of the National Science Foundation (Grant Number DMR 8106166) is gratefully acknowledged.

REFERENCES

1. N.S. Sullivan and R.V. Pound, Physics Letters 39A, 23 (1972).

2. N.S. Sullivan, Journal de Physique 37, 209 (1976).

3. N.S. Sullivan, M. Devoret, B.P. Cowan and C. Urbina, Phys. Rev. B17, 5016 (1978).

4. S. Washburn, R. Schwiezer and H. Meyer, Solid State Commun. 35, 623 (1980).

5. S. Washburn, M. Calkins, H. Meyer and A.B. Harris, J. of Low Temp. Phys. 49, 101 (1982).

6. D. Candela, S. Buchman, W.T. Vetterling and R.V. Pound, Phys. Rev. B27, 3084 (1983).

7. W.T. Cochran, J.R. Gaines, R.P. McCall, P.E. Sokol and B. Patton, Phys. Rev. Lett. 45, 1576 (1980).

8. O.D. Gonzalez, D. White and H.L. Johnston, J. Phys. Chem. 61, 773 (1957).

9. D.G. Haase, Solid State Commun. 44, 469 (1982).

10. J.V. Gates, P.R. Granfors, B.A. Fraass and R.O. Simmons, Phys. Rev. B19, 3667 (1979).

11. M.A. Klenin and S.F. Pate, Physica 197D, 185 (1981).

12. M.A. Klenin, Phys. Rev. B26, 3969 (1982).

13. M. Devoret and D. Esteve, preprint.

14. G. Grenier and D. White, J. Chem. Phys. 40, 3015 (1964).

15. M.A. Klenin, Phys. Rev. Lett. 42, 1549 (1979).

16. J.R. Gaines, Y.C. Shi and J.H. Constable, Phys. Rev. B17, 1028 (1978).

17. A.B. Harris and E. Hunt, Phys. Rev. Lett. 19, 845 (1966).

18. N.S. Sullivan and D. Esteve, Physica 107B, 189 (1981).

SPIN-LATTICE RELAXATION IN DEUTERIUM AT REDUCED PARA CONCENTRATION

R. V. Pound, D. Candela, S. Buchman, and W. T. Vetterling

Lyman Laboratory of Physics, Harvard University, Cambridge Ma. 02138

ABSTRACT

We have measured NMR spin-lattice relaxation times in solid D_2 for para-D_2 concentrations $0.24<X<0.31$ and temperatures $40mK<T<700mK$. No evidence is seen for the sharp features in t_1 vs T that were reported by others. A model for cross-relaxation has been developed which allows our data for solid D_2 to be compared directly with earlier data for solid H_2.

Electric quadrupole-quadrupole (EQQ) interactions between the J=1 molecules in solid hydrogen and solid deuterium lead to a number of interesting phenomena in which the molecular axes become preferentially aligned along certain directions.[1] The J=0 molecules in these solids do not participate in this quadrupolar ordering, due to their spherically symmetric rotational wavefunctions. The fraction X of J=1 molecules may be varied by preparation or aging.

For $X>0.55$ there is a transition at a few kelvins into the Pa_3 phase, which has long range orientational order. While hydrogen and deuterium freeze into hcp crystal structure, the phase transition upon cooling is accompanied by a crystallographic transformation to fcc structure. This is apparently driven by the low orientational energy of the Pa_3 phase.

If the J=1 fraction X is sufficiently low (below <u>approximately</u> 0.5), the crystal structure is believed to remain hcp at all temperatures.[2] In this case, NMR experiments provide evidence of orientational ordering at temperatures near 300mK. It has been hypothesized[3] that this ordering is analogous to that which occurs in a spin-glass, since the EQQ Hamiltonian combines the features of frustration and randomness (due to the random occupation of lattice sites with J=1 molecules). However, definitive experimental evidence of spin-glass behavior in the solid hydrogens has been lacking.

We have recently completed measurements of the nuclear spin-lattice relaxation time t_{1N} in solid deuterium for $0.24<X<0.31$ and $40mK<T<700mk$, at a Larmor frequency of 46MHz. (The subsipt N is used to denote the narrow component of the NMR line due to J=0 molecules.) The only measurements reported earlier of t_{1N} at temperatures definitely low enough to enter the ordered region,[4] had shown a distinctive feature at X=0.33, T=170mK, which is near the hypothesized "quadrupolar glass" transition. In particular, it was found that the relaxation showed evidence of two distinct time constants, both of which had discontinuities in their derivatives with respect to temperature at about 170mK. These two time constants are shown as curves (d) and (e) in Fig. 1. The rapid increase in t_{1N} below this temperature was attributed to spin-glass-like orientational ordering.

Our measurements of the relaxation show no evidence of two distinct time constants, and the single time constant we measure shows a broad, shallow minimum as a function of temperature. These measurements are shown as open circles (X=0.30) and open triangles (X=0.24) on Fig. 1. As can be seen on this figure, there is no indication of a well-defined transition into the ordered state in the neighborhood of 170mK.

Earlier measurements of t_{1N} in solid deuterium at X=0.29 for T>150mk[5] are shown as curve (a) on Fig. 1. Although these measurements did not extend to sufficiently low temperatures to exclude a transition, it should be noted that the mimimum in t_{1N} reported in Ref. 5 is deeper and at a lower temperature than the minimum we measured. This is consistent with the assumption of an orientational correlation time which gradually increases with decreasing temperature, because the measurements of Ref. 5 were taken at a Larmor frequency of 3MHz and thus were sensitive to orientational fluctuations at a lower frequency than in our measurements. The measurements of Ref. 4 do not fit into this pattern, as is apparent from Fig. 1.

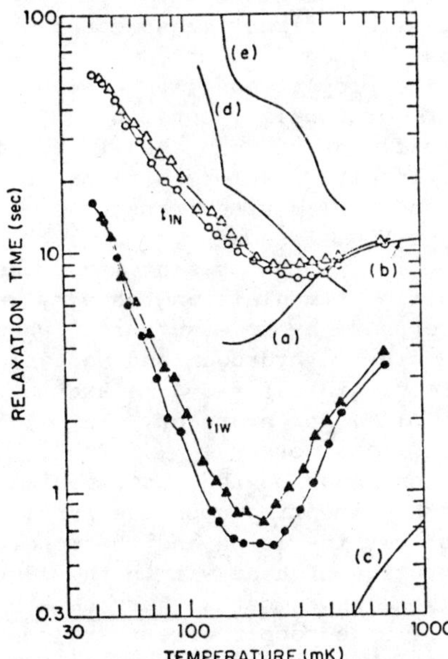

Fig. 1. Measured nuclear spin-lattice relaxation times t_{1N} for the J=0 line (open circles and triangles) and inferred relaxation times t_{1W} for the J=1 line (filled circles and triangles). Solid lines (a)-(e) show previously reported measurements. See text for explanation of symbols.

All of the deuterium relaxation measurements we have discussed so far, have been performed upon the narrow component of the NMR line which is due to molecules with J=0. The J=1 molecules contribute a wide component to the NMR line, which is difficult to observe in CW experiments (the width is due to intramolecular quadrupole interactions with the deuteron, which vanish for J=0). However, the J=0 molecules we observe relax primarily via cross-relaxation with the J=1 molecules,[6] and thus can serve as a useful probe of orientational fluctuations in the J=1 system. We have developed a simple theory for cross-relaxation similar to that developed in Ref. 6, which allows us to infer the relaxation time t_{1W} of the wide J=1 line from the relaxation time t_{1N} we measure for the narrow J=0 line. The result of our calculation is:

$$t_{1W} = 16X^2 g_o{}^3 t_{1N}{}^2 / [5(1-X)]^2 t_2{}^4,$$

where g_o is the density of the J=1 line at its center (in units of inverse frequency) and t_2 is the transverse relaxation time computed from the intermolecular dipole-dipole interactions which are responsible for spin diffusion within the J=1 line. The solid circles and triangles in Fig. 1 show the t_{1W} we compute from our measured t_{1N}. These may be compared with a direct measurement of t_{1W} for X=0.33 using pulse techniques[6], which is plotted as curve (c) in Fig. 1. It can be seen that the proportionality between t_{1W} and t_{1N}^2 is verified, but that the proportionality constant differs from the predicted value by a factor of more than five. This is not surprising, since this proportionality factor is based on a very rough theory of spin diffusion and varies with transverse relaxation time as t_2^{-4}. (Curve (b) on Fig. 1 shows t_{1N} as reported in Ref. 6; it agrees very well with the present measurements and those of Ref. 5.) The J=1 relaxation times also agree very well with hydrogen J=1 relaxation times reported by some authors,[7] when properly scaled by the relative strengths in hydrogen and deuterium of the intramolecular interactions responsible for J=1 relaxation. Others have reported <u>sharp</u> minima in $t_1(T)$ for hydrogen,[8] in apparent disagreement with Ref. 7 and this work.

In view of the general lack of consensus on the experimental aspects of solid hydrogen, it is worth commenting upon the particular difficulties this substance poses for refrigeration and thermometry. These result from the continuous evolution of heat within the sample as J=1 molecules convert into J=0 molecules. Most workers have used sample chambers filled with brushes of fine copper wire to reduce the thermal gradient between sample and refrigerator, which is mainly due to the Kapitza boundary resistance at the sample-wire interface. For example, the deuterium measurements we report were taken using a sample chamber containing 1,500 thermally anchored copper wires, while the chamber we use for hydrogen measurements contains 12,000 wires. Despite these precautions, sample to chamber temperature differences can still be large.

The measurements shown in Fig. 2 illustrate this problem for hydrogen. In this figure, curve (a) is a temperature calibration of the carbon resistance thermometer affixed to the mixing chamber of our refrigerator. The data points show the sample temperature determined by NMR thermometry as a function of the resistance of the same carbon thermometer, for three different values of the ortho-concentration X. The solid curves (b)-(d) show a single-parameter fit to the functional form expected if the Kaptiza resistance dominates the thermal impedance between sample and refrigerator.

In light of such measurements we feel that NMR thermometry on the sample itself is the only reliable means of determining the temperature of solid hydrogen or deuterium at dilution refrigerator temperatures. Figure 2 suggests that it is not possible to cool a hydrogen sample of intermediate ortho-concentration much below 100mK. Since the heat liberated by a deuterium sample is only about 1/30 that liberated by a hydrogen sample of the same size (at X=0.5), thermal contact problems are not quite so severe in solid deuterium. The minimum achievable temperature is expected to scale as the fourth root of the heat load. The lowest temperature to which we were able

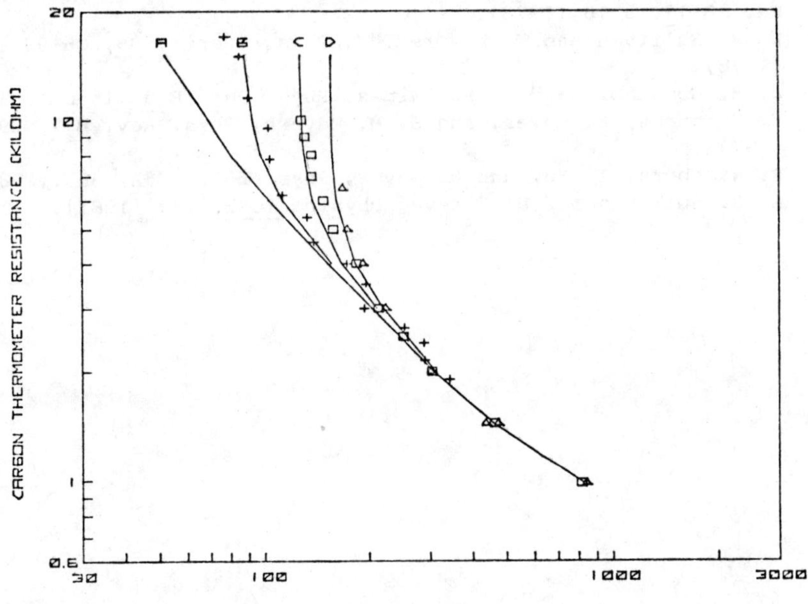

Fig. 2. Temperature calibration for resistance thermometer attached to the mixing chamber of our apparatus (A), and expected carbon thermometer resistance as a function of hydrogen sample temperature for three different ortho-concentrations (B, X=0.20.   C, X=0.44. D, X=0.67). The data points show sample temperatures as measured by NMR thermometry for the corresponding ortho-concentrations.

to cool deuterium samples with X=0.3 was 40mK, which is roughly consistent with this law. Since the temperature of the hypothesized quadrupolar glass transition was more than four times this value, thermal gradients within our sample may be excluded as a possible explanation for our failure to observe a sharp transition.

In conclusion, our relaxation measurements in solid deuterium failed to confirm the observation of a well defined transition into an ordered state, as had been reported in Ref. 4. Our methods for thermal contact to the sample and for thermometry ensured that the broad minimum we observed in $t_1(T)$ was truly characteristic of the sample, despite the substantial problems posed by spin-conversion heating.

This work was supported by NSF grant DMR 79-07023.

REFERENCES

1. I. F. Silvera, Rev. Mod. Phys. 52, 393 (1980).
2. J. V. Gates, P. R. Granfors, B. A. Fraass, and R. O. Simmons, Phys. Rev. B 19, 3667 (1979).

3. N. S. Sullivan, M. Devoret, B. P. Cowan, and C. Urbina, Phys. Rev. B 17, 5016 (1978).
4. N. S. Sullivan and M. Devoret, J. Phys. (Paris) 39, C6-92 (1978).
5. J. H. Constable and J. R. Gaines, Phys. Rev. B 3, 1556 (1971).
6. F. Weinhaus, H. Meyer, and S. M. Meyers, Phys. Rev. B 7, 2948 (1973).
7. S. Washburn, I. Yu, and H. Meyer, Phys. Lett. 85A, 365 (1981).
8. N. S. Sullivan and D. Esteve, Physica 107B, 189 (1981).

THERMODYNAMIC PROPERTIES OF SOLID H2 AT
INTERMEDIATE ORTHOHYDROGEN CONCENTRATION

D. G. Haase and L. R. Perrell
Physics Department, North Carolina State University
Raleigh, North Carolina 27650

ABSTRACT

We have measured the specific heat of solid hydrogen samples
grown under vapor pressure for orthohydrogen molar concentrations
0.65 > X > 0.2 and between 1 and 0.15 Kelvin. A thermal relaxation
technique was used for the measurements to negate the heating effect
of orthohydrogen conversion and to allow examination of long time
heat release. The specific heat results showed no transitions or
remanence which might be associated with a glass phase, with one exception. It was found that all relaxations were exponential with a
single time constant regardless of sample thermal history. Measurements were made of samples isothermally converted at T = 0.3K from
the ordered fcc phase to the hcp phase, in which case a gradual disorientation of the rotational moments is seen despite the relatively
sharp fcc to hcp structural transition.

INTRODUCTION

By far the majority of work performed to date on the so-called
glass phase of solid hydrogen has been in the form of NMR experiments.[1] One major attraction of the solid H2 system is its possible
usefulness as a model for other, less conceptually simple, types of
glasses. Therefore, we have measured thermodynamic properties of
solid H2 at low temperatures and reduced orthohydrogen concentrations in an explicit attempt to seek correspondences to the
properties of other systems such as spin glasses or structural
glasses. While the NMR results at other laboratories have provided
the impetus for further interest in solid H2, the thermodynamic
results provide a complementary view of the system.

We have measured the specific heat of solid H2 samples grown at
low pressures and having orthohydrogen molar concentrations 0.65 > X
> 0.20. Our data, in the range 1.0 K > T > 0.15 K, includes the
so-called quadrupolar glass phase as well as the molecularly ordered
fcc (Pa3) phase.[2] The slow conversion of the quadrupolar orthohydrogen molecules to the spherical parahydrogen state allows the opportunity to measure the properties of the solid as the quadrupolar
concentration is continuously and isothermally decreased. An unwanted side effect is the significant heating caused by this species
conversion. We have, therefore, used a thermal relaxation technique
to determine this specific heat of our samples.[3] A schematic of the
sample cell is given in Figure 1. The cell was filled with a sintered copper sponge of nominal 100 micron pore size to reduce surface resistance and thermal gradients within the sample. The cell
was connected by short copper wires to the dilution refrigerator

mixing chamber. Thus the sample could be heated to 10 to 30 mK above the equilibrium temperature, the heat turned off and the time relaxation of the cell temperature observed and recorded. All of the heater control, temperature recording, and data analysis were handled by a small computer. The orthohydrogen concentrations of the samples were determined by the application of the usual conversion rate law and thermal analysis of the sample gas after the experiment. Because of the sample internal heating it was found difficult to cool and measure samples grown at pressures high enough to maintain constant volume at 1 K, thus filling the sample cell. Thus all of the data shown here involves samples grown at the triple point and which partially filled the sample cell. Two samples were grown at 100 Bars, however, and their Cv results above 0.5 K agree qualitatively with the vapor pressure results.

One advantage of the relaxation technique is that we can seek long time relaxations of glassy defects in a manner similar to that described by Zimmerman, et al,[4] in structural glasses. However, in our measurements on solid H2 and D2 (33% para-D2)[3] we have found no non-exponential thermal relaxations or relaxation having a second long time exponential time constant indicative of defect relaxation. The experimental relaxations were determined by the size of the copper thermal link from the cell to the mixing chamber, and were in the reange of 5 to 120 seconds. In a search for thermal history effects the specific heat was measured after the sample was quickly cooled to particular temperature and when it was "annealed" at that same temperature for a period of hours. Within the resolution of

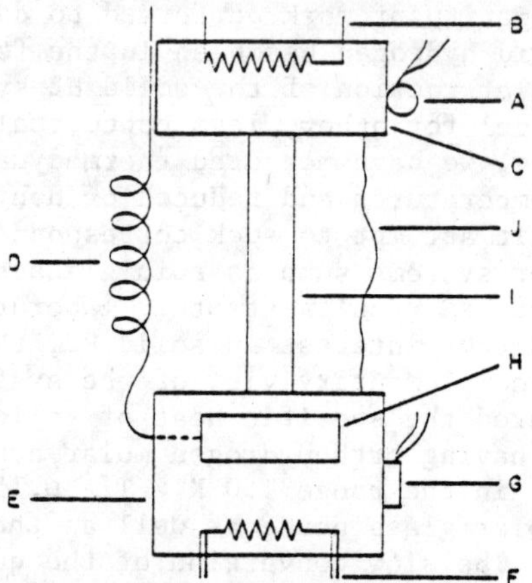

Fig. 1. Schematic of sample cell used for specific heat measurements on solid H2. Cell volume = 0.162 cm^3, surface area > 50 cm^2. A - Mixing chamber thermometer. B - Automatically controlled heater. C - Mixing chamber. D - Copper-nickel fill capillary, 0.010" I.D. E - Sample cell, OFHC copper. F - Sample cell heater, 4 lead. G - Thermometer, Speer 220 ohm slab, 0.5 mm thick. H - Sample volume filled with 100 micron sintered copper powder. I - Cell support, thin walled stainless steel tube. J - Thermal link, 0.03 cm. diameter copper wires.

our measurements no time or preparation dependent effects were
noted, with one exception discussed below.

Cv VERSUS T AT CONSTANT ORTHOHYDROGEN CONCENTRATION

The results of our Cv measurements for several orthohydrogen
concentrations are shown in Figure 2. None of these samples were
cooled into the rotationally ordered fcc phase and the data were
reproducible upon warming and cooling. The onset of glassy
behavior is normally considered to occur at about 200 - 300 mk for
these concentrations.[5] Consistent with previous $\partial P/\partial T|v$ data from
this laboratory, the specific heat results show no evidence of a
first or second order phase transition in the hcp solid. There is,
however, a rounded maximum in Cv at about 0.7 to 0.8 K in the
hisher orthohydrogen concentration samples. Plots of Cv/T showed
peaks at about 0.3 to 0.4 K for all of the concentrations with
that peak broadening and moving to lower temperatures as X decreased.

Comparison of the Cv data with our previous $\partial P/\partial T|v$ data

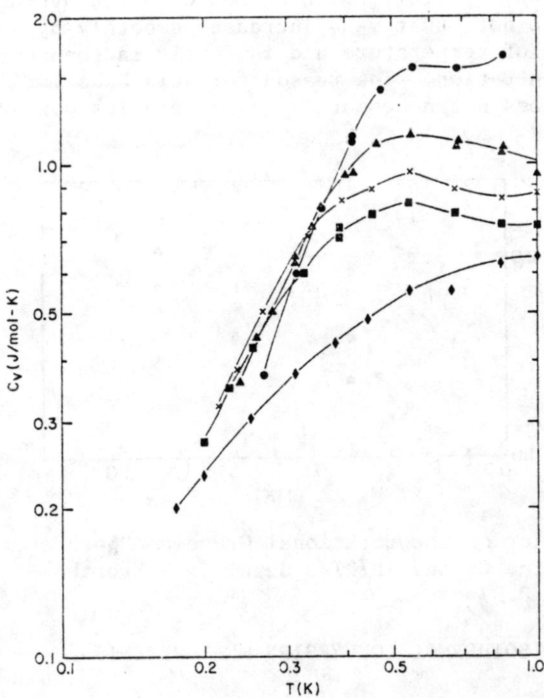

Fig. 2. Cv versus T for solid H2 samples grown under vapor pressure.
Points displayed are averages of several consecutive measurements.
None of these samples were cooled into the fcc phase. ● - X(ortho)
= 0.58. ▲ - X = 0.45 ✗ - X = 0.38 ■ - X = 0.32 ◆ - X = 0.20

104

allows a test of the Gruneisen relation:

$$V \; \partial P/\partial T)_v = \gamma_L \; C_{v,L} + \gamma_{ROT} \; C_{v,ROT}$$

where γ is the Gruneisen parameter and the L and ROT subscripts
indicate the contributions due to the lattice vibrations and
molecular rotations, respectively. The Gruneisen relation is a
result of equilibrium thermodynamics, specifically dependent upon
the interchangeability of volume and temperature derivatives of the
free energy. In a non-equilibrium solid it is possible that these
conditions would not hold and there has been a small controversy of
an analogous violation of the Maxwell relations in spin-glasses.[6]
In solid H2 below 1 K the lattice specific heat is negligible[7]
compared to the rotational contribution thus simplifying our
analysis. Although all the results shown here are strictly Cp,
since the samples are at vapor pressure, we have noted no significant difference from Cv data of a sample grown at a pressure of
100 Bars.

A plot of γ_{ROT} as calculated from our Cv and $\partial P/\partial T|v$ [8]
results, using V = 22.9 cm^3/mol is shown in Fig. 3. The theoretical value of γ_{ROT} is 1.677 which agrees with the average of these
results. We do note that γ_{ROT} increases smoothly by about 40%
over the range of temperature and is fairly independent of orthohydrogen concentration. The reason for this behavior is not understood but it does not necessarily indicate a loss of ergodicity in
the solid.

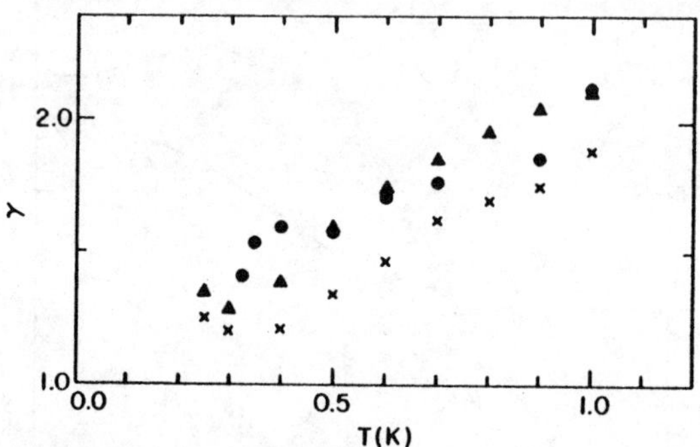

Fig. 3. Behavior of the rotational Gruneisen parameter obtained by
comparison of the Cv and $\partial P/\partial T|v$ data. $\bullet$ - X(ortho) = 0.55 $\blacktriangle$ - X
= 0.45 $\times$ - X = 0.37

ISOTHERMAL CONVERSION MEASUREMENTS

One phenomenon of interest in the field of spin-glasses is the
field cooling experiment in which a sample cooled in an externally
applied field shows remanence effects after the field is removed.[9]
In solid H2 a conceptually similar experiment can be performed by
cooling the sample into the rotationally ordered fcc phase and then

isothermally converting the sample into the glass phase. The fcc structure acts as the analogue of the external field, allowing the formation of a rotationally ordered structure, which due to frustration effects is not available in the hcp phase.

We first made an isothermal measurement of Cv as a function of time, i.e., orthohydrogen concentration, after the sample was initially cooled into the ordered phase. The results of this experiment are shown in Fig. 4. At T = 0.3 K the primary contribution to the specific heat is the orthohydrogen molecular rotation. The Cv data shows a gradual change in the rotational specific heat with decreasing orthohydrogen concentration. This may be interpreted as a gradual disordering of the molecular moments from the initial ordered configuration in the fcc phase. This contrasts with the sharp transition seen in the P(X) data of Haase, et al,[10] but is not contradictory. Ramm, et al,[10] have shown (and we have corroborated that) for X > 0.6 the pressure change on cooling from the hcp to the fcc phase is positive. However, the Cv anomaly seen at higher X is very small or non-existent at X > 0.6. In agreement with Harris, et al,[12] we take this to mean that the P(X) data indicates the change in crystalline phase but is not so

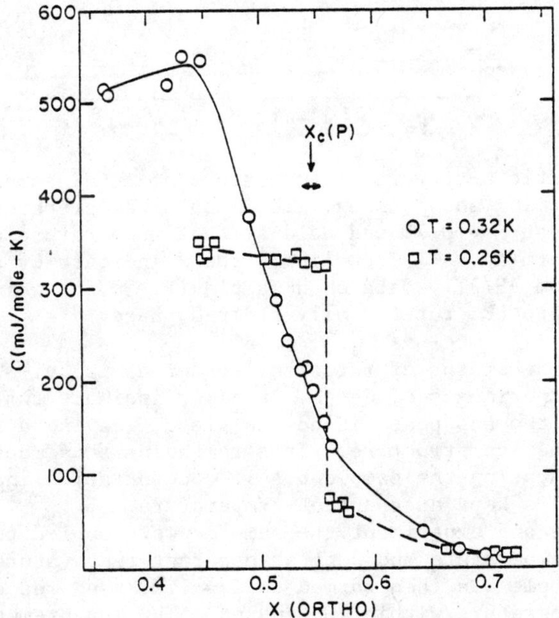

Fig. 4. Specific heat versus orthohydrogen molar concentration X for two samples. The 0.32 Kelvin sample was held at that temperature during the entire run. The 0.26 Kelvin sample was heated to 1 K and quickly recooled twice: at X - 0.557 and X - 0.446. Xc(P) denotes the location and width of the pressure transition observed in Ref. 10.

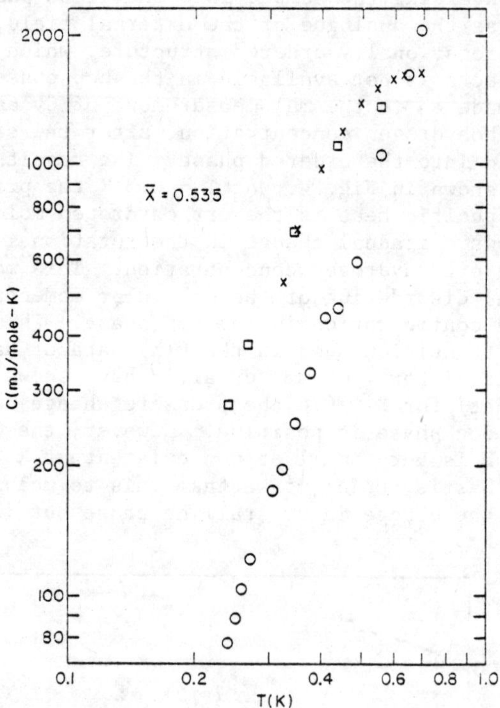

Fig. 5. Specific heat versus temperature of an H2 sample of X =
0.535 upon warming and cooling. The sample was prepared by cooling
to T = 0.24 K at X = 0.70 and held at that temperature until X =
0.55. O - warming. □ - cooling. The **x** indicate Cv points
calculated from $\partial P/\partial T|v$ data on a sample (X - 0.55) which had not
been cooled into the rotationally ordered phase.

sensitive to the status of rotational order as is Cv. The x-ray
diffraction experiments of Gates, et al,[13] indicate that the
conversion to the hcp phase is not complete, leaving a small
fraction of the fcc structure. This remaining fcc fraction is
recovered by heating the sample but is concentration independent
if the sample is kept at constant temperature.

In the second experiment the samples were cooled to the
measurement temperature and kept at constant temperature until X =
0.55. The sample was then warmed to T = 1.0 K and recooled to the
measuring temperature within 20 minutes. The measurement of Cv(X)
proceeded as shown in Fig. 4. The effect of the warming and
recooling of the sample was to increase Cv by a factor of roughly
3 at X = 0.53. The amount of disorder shows a behavior similar to
that of the previous sample for X > 0.49.

We interpret these data to mean that the orthohydrogen rotational moments retain the long range order over some period of time
even after conversion into the hcp crystal phase, provided that the

sample is kept at low temperature. Heating the sample induces thermal disorder and the loss of the memory of the rotational order. Removal of the fcc structure, like removal of the ordering magnetic field on a spinglass, allows the moments to slowly reorient to a disordered configuration. Since the fcc to hcp phase transition involves a sliding of crystal planes perpendicular to the c axis, it is reasonable to conjecture that in-plane rotational order is maintained in the crystalline transition.

A further illustration of this behavior is shown in Fig. 5. A sample of X = 0.55, which had been prepared in the ordered fcc phase, was heated to 0.7 K and then recooled over a period of four hours, while Cv was measured as a function of temperature. The Cv data does not indicate a sharp phase transition upon heating, yet there is a large hysteresis upon recooling. The cooling curve is in reasonable agreement with Cv calculated from the $\partial P/\partial T|v$ data[8] for a sample which was never cooled into the ordered phase.

CONCLUSION

We have made careful and extensive measurements of the specific heat, and in previous work $\partial P/\partial T|v$, in solid H2 at low temperatures and reduced orthohydrogen concentrations. We have not found any evidence of long relaxation or equilibrium times or of thermal history effects in the solid for characteristic measurement times on the order of seconds. This result differs from some of the NMR results but we note that NMR is a probe of local order while we have examined global properties. There is no evidence in our data of a distinct phase transition at low temperatures, or of a violation of the Gruneisen relation.

On the other hand, in samples isothermally converted at T = 0.3 K from the rotationally ordered fcc phase into the hcp phase there is a lons time disorientation of the molecular rotational moments occurring in contrast to a relatively sharp crystal structure transition. Although it is possible that fcc residue in the hcp structure may affect this remanent rotational order, there is theoretical evidence[14] that indicates geometric frustration in the hcp structure can account for such long time relaxation.

ACKNOWLEDGEMENTS

We acknowledge many helpful conversations with Dr. M. A. Klenin and Dr. H. Meyer. This research was supported by NSF grant DMR 79-23202 and by a contract with the Office of Naval Research.

REFERENCES

1. J. R. Gaines, this conference.
2. I. F. Silvera, Rev. Mod. Phys. 52, 393 (1980).
3. D. G. Haase, Solid State Commun. 44, 469 (1982).
4. J. Zimmerman and G. Wever, Phys. Rev. Lett. 46, 661 (1981).
5. N. S. Sullivan and D. Esteve, Physics 107B, 189 (1981).

W. T. Cochran, J. R. Gaines, R. P. McCall, P. E. Sokol, and
B. R. Patton, Phys. Rev. Lett. <u>45</u>, 1576 (1980).

6. Various authors, Phys. Rev. Lett. <u>49</u>, 238-242 (1982) and
    references therein.

7. J. F. Jarvis, H. Meyer, and D. Ramm, Phys. Rev. <u>178</u>, 1461
    (1969).

8. D. G. Haase and A. M. Saleh, Physica <u>107B</u>, 191 (1981).

9. J. A. Mydosh and G. J. Niewenhuys in <u>Ferromagnetic Materials</u>,
    vol. I, edited by E. P. Wohlfarth (North Holland, Amsterdam,
    1980), p. 71.

10. D. G. Haase, J. O. Sears, and R. A. Orban, Solid State Commun.
    <u>35</u>, 891 (1980).

11. D. Ramm, H. Meyer, and R. L. Mills, Phys. Rev. <u>B1</u>, 2763 (1970).

12. A. B. Harris, S. Washburn, and H. Meyer, J. Low Temp. Phys. <u>50</u>,
    151 (1983).

13. J. V. Gates, P. R. Granfors, B. A. Fraas, and R. O. Simmons,
    Phys. Rev. <u>B19</u>, 3667 (1979).

14. M. A. Klenin, this conference and to be published.

ON THE ORIENTATIONAL ORDERING IN THE QUADRUPOLAR GLASS PHASE OF SOLID ORTHO-PARA HYDROGEN MIXTURES*

N.S. Sullivan
Department of Physics, University of Florida
Gainesville, Florida 32611

ABSTRACT

Several research groups have in recent years studied the
orientational ordering of orthohydrogen for intermediate concentrations $0.20 < x < 0.55$ for which the orientational degrees of
freedom of the ortho molecules become "frozen" in an apparently
glass-like phase on cooling from high temperatures. This behavior is
related to fundamental questions concerning the underlying physics of
glass formation in general. The collective freezing of a dilute
random array of molecular rotators represents a particularly
elementary example of the combined effects of frustration and disorder which are thought to be common to many glass formers. A clear
definition of what is meant by the term "freezing" and "glass" can be
given in terms of autocorrelation functions for the variables
describing the molecular orientations. These functions can be related to recent NMR experiments which explore the molecular dynamics
as "freezing" occurs.

INTRODUCTION

One of the most fascinating and challenging problems of
contemporary solid state physics has been that of achieving an
understanding of the spin glass phases. These have been observed in
dilute magnetic alloys when the interactions between the spins are in
competition with each other due to the nature of the interactions
between the spins and because of the disorder. No long range
periodic order (FM or AFM) exists but these phases do exhibit special
properties which distinguish them from the completely disordered
(paramagnetic) phases. New concepts such as frustration have had to
be developed to understand the origin of spin glasses and to treat
the problem of competing interactions in random dilute systems in
general. This has led to the realization that one is dealing with a
new physical problem that is related to a wide class of systems –
dilute magnetic alloys, anti-ferroelectrics, molecular crystals,
electric dipolar glasses and others.

The quadrupolar glass phase observed in solid hydrogen falls
into this category and provides a particularly clear example which
sheds light on the underlying physics in these systems. The
orthohydrogen molecules (orbital angular momentum J=1) are (unlike

* This work has been carried out in collaboration with M. Devoret
and D. Esteve at the Centre D'Etudes Nucleaires de Saclay, France.

110

para-molecules with J=0) non-spherical and they tend to orient to
minimise their interactions with neighbouring ortho molecules. The
dominant anisotropic interaction is the electric quadrupole-
quadrupole interaction and the problem is that of ordering an array
of electric quadrupoles located randomly at the sites of an hcp
lattice. This leads to a fascinating enigma common to many glass
formers. While two isolated quadrupoles are in their lowest energy
state when mutually perpendicular, it is impossible to arrange all
molecules mutually perpendicular on any 3D lattice. This topological
incompatibility between the lowest energy configurations and the
lattice symmetry is called frustration. In order to realise a
periodic orientational structure a compromise must be sought for –
the lattice undergoes a structural change to an fcc lattice on which
the molecules orient according to a 4-sublattice Pa_3 structure[1] with
the molecules aligned parallel to one of the body diagonals in each
sublattice. The bond energies for this structure are -2Γ as opposed
to -4Γ for the isolated tee configuration. ($\Gamma = 6Q^2/25\, a_o$ where Q is
the electric quadrupole moment and a_o the molecular separation.) The
price paid to achieve a periodic structure is high, although for the
cubic structure the bond energies are all equivalent.

The frustration is even more severe for quadrupoles on an hcp
lattice. The calculated ground state is a PC_{21} configuration[2] in
which some molecules are closely parallel to their neighbours (i.e.
highly frustrated) while others are almost perpendicular (i.e. non-
frustrated). Compared to the cubic structure the frustration is no
longer uniformly distributed. This inhomogeneity sets up additional
constraints in the system, renders nucleation of the long range order
especially difficult and the distribution of quadrupoles on the hcp
lattice can be considered to be superfrustrated. Any periodic
orientational ordering, if it can be nucleated, will be very unstable
with respect to site dilution (substitution of a quadrupole bearing
ortho molecule by a spherical para molecule). This is illustrated in
figure 1 which shows the invariance of the orientations of the
quadrupoles with respect to site dilution for a non-frustrated
lattice (figure 1a) as opposed to the reorientations that occur for
the frustrated triangular lattice (figure 1b).

Fig. 1. Schematic representation of the effect of substitutional
disorder on quadrupolar arrays.

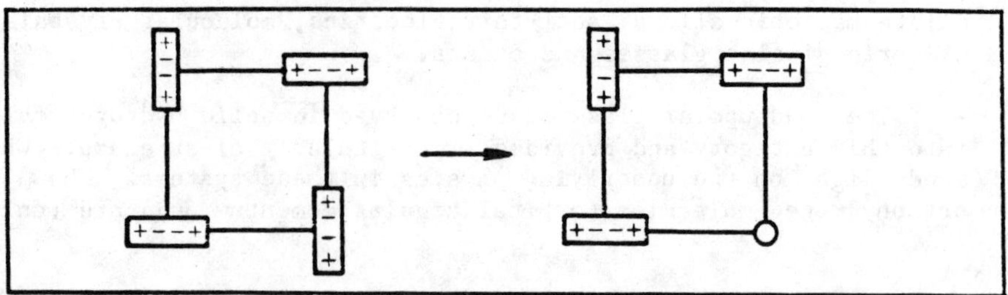

(a)   Non-frustrated (square lattice)

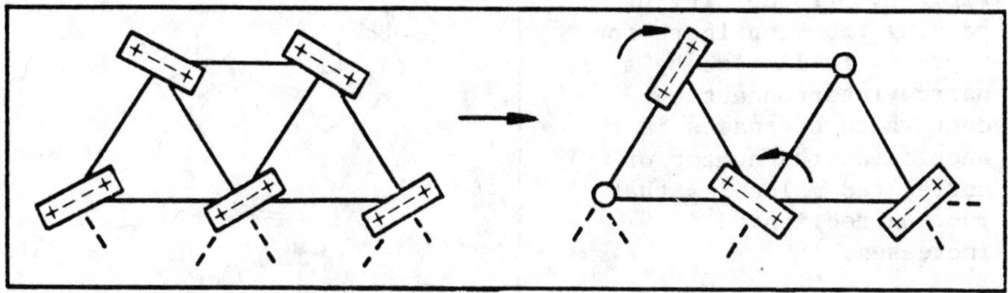

(b)   Frustrated (triangular lattice)

What is even more striking is that for the randomly diluted
frustrated lattice, the lowest energy configurations are not
unique.   There are several closely lying energy configurations, but
in order to pass from one configuration to another, one needs to
reorient a large number of molecules.   This is a direct consequence
of the superfrustration for quadrupoles on an hcp lattice and can
lead to a blocking effect for the molecular orientations as one cools
from the high temperature phase.   In order to pass from one global
configuration such as C_1 in figure 2a to a lower energy state C_2 we
need to find paths in configuration space that have low potential
barriers V_{12} separating the two states.   Since this involves the
simultaneous reorientation of large correlated clusters, the
"distance" ξ in configuration space becomes very long.   This is shown
schematically in figure 2b where the probability of passing from C_1
to C_2 is represented by a narrow duct (with cross-section
$\sim 1/N^{\alpha}$, $\alpha \approx 1$).   The problem of reaching the lowest energy
configurations depends on this connectivity in configuration space.
For highly frustrated systems with a large number of low lying energy
configurations, the system can become trapped in configuration space
when the probability of exploring all configurations becomes so small
that one cannot reach the lowest energy states on any reasonable time
scale.   The system becomes non-ergodic at a temperature which depends
on the experimental conditions - e.g. cooling rate, thermal history,
etc.   Disordered frustrated systems are inherently non-ergodic and
the qualitative description given here describes the behaviour of
many glass formers.

Fig. 2a. Low potential
barriers V_{12} between
configurations can be
found by reorienting
simultaneously large
clusters of molecules
corresponding to large
separations in
configuration space.

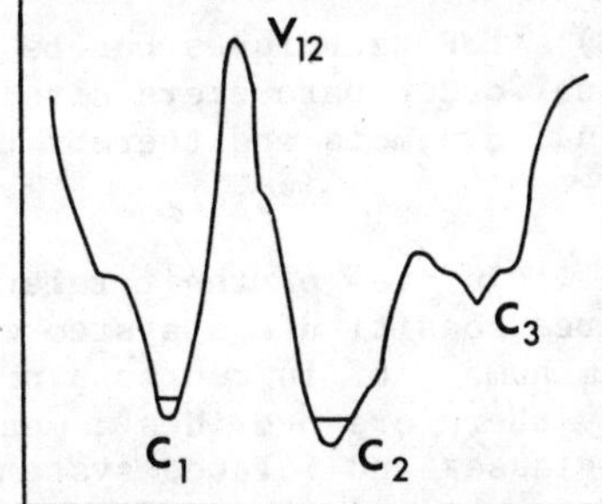

112

Fig. 2b. Probability of passing from configuration C_1 to C_2 is limited by a narrow interconnecting duct which decreases in section as the number of correlated molecules that must be reoriented increases.

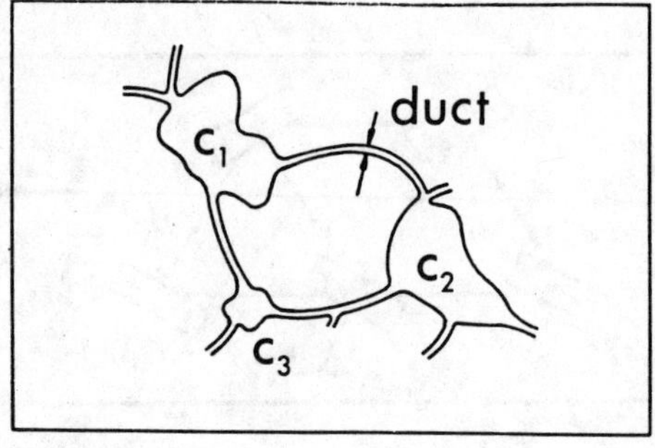

The canonical spin glasses (e.g. CuMn, AuFe ...) with long range interactions oscillatory functions of spatial separation) do exhibit subtle phase transitions (with critical exponents for the non-linear susceptibility) and one may ask if they belong to the general class of glass formers. This has not been answered, but it should be kept in mind that the mathematical description[4] of the ∞-range spin glass requires an infinite number of order parameters which can be parameterised in terms of a variable related to the time scale t being considered.[5] The thermodynamic limit is valid for long t but linear response is lost, while one does observe linear response for short t. The detailed dynamical behaviour therefore plays a major role for the long range spin glasses but a clear physical interpretation of the behaviour of these systems has not been given.

The interest in solid hydrogen ortho-para mixtures as an orientational glass former is at least four-fold:

(i)   The fundamental interactions are short ranged and well known. First principles statistical mechanics methods can be used where applicable. (These methods have been used with remarkable success for high ortho concentrations[6] but conventional methods fail for frustrated systems in the presence of disorder.)

(ii)   The qualitative features of the combined effects of frustration and disorder are clear. For ortho concentrations $0.20 < x < 0.55$, long periodic configurations are not observed and one finds instead a glassy phase in which the molecular orientations are frozen randomly from one site to another.

(iii)   NMR techniques can be used to determine the local orientational order parameters directly, to follow the molecular orientational dynamcis and thereby test various models for the glass formation.

(iv)   In view of their relative simplicity, solid H_2 ortho-para mixtures consititute a system which has what many believe to be the minimum number of ingredients necessary for glass formation and their study therefore provides a means of testing current physical models of glasses and related systems that are conceptually less simple. We now need to pass from this qualitative discussion to a

more quantitative description of the orientational ordering in order
to interpret the results of recent experiments.

ORDER PARAMETERS AND NMR EXPERIMENTS

In order to be able to give precise definitions of what we mean
by the terms "freezing" and "glass" we need to define the orientational order parameters. These parameters specify the degrees of
freedom of a spin-1 object and go beyond the specification of the
polarization $\langle J_z \rangle$ of a vector (which would sufficient for spin $\frac{1}{2}$).
In our case, in the absence of any interactions which break time
reversal symmetry, the vectorial components $\langle J_x \rangle$, $\langle J_y \rangle$, $\langle J_z \rangle$ vanish
and we must consider higher order components – the quadrupole moments
representing the components of a second rank tensor $Q_{\alpha\beta}$ which are
the expectation values of the operators

$$Q_{\alpha\beta} = \frac{1}{2}\left(J_\alpha J_\beta + J_\beta J_\alpha\right) - \frac{1}{3} J^2 \tag{1}$$

α, β refer to an arbitrary set of Cartesian axes (x,y,z). Of the 9
components $Q_{\alpha\beta}$ only 5 are independent. If we choose 3 local axes
x_i, y_i, z_i for the ith molecule which coincide with the principle
axes of the quadrupole tensor (analogous to the principal axes of an
ellipsoid) only two intrinsic quadrupolar parameters remain. These
are

$$\sigma_i = \frac{3}{2}\left\langle Q_{zz}(i) \right\rangle \quad \text{and} \quad \eta_i = \left\langle Q_{xx}(i) - Q_{yy}(i) \right\rangle. \tag{2}$$

The orientation of a molecule is described by a probability
distribution in space which may look like a cigar (figure 3a) if the
molecule is aligned preferentially along an axis, or a disc (figure
3b) if it prefers to lie in a plane. In general the probability
distribution will be intermediate between these two extremes. The
local axes (x_i, y_i, z_i) are the principal axes for the
distribution, σ_i measures the mean <u>alignment</u> along z_i and η_i the
<u>eccentricity</u> of the distribution (or departure from axial
symmetry). (Following convention the axes are chosen so that
$|\sigma| \geq \sqrt{3}\ \eta \geq 0$.)

Fig. 3 Representations of the
probability distributions for
the molecular orientations for
preferential alignment (a)
along an axis or (b) in a
plane.

(a)   (b)

114

In the long range periodically ordered configurations the
molecules are aligned along well defined axes with $|\sigma| = 2$
and $\eta = 0$; it is only the local axes (x,y,z) that vary from one
sublattice to another. In the glass phase at low temperatures, not
only do the local triads (x,y,z) point in different directions at
different sites, but the intrinsic quadrupolar parameters also vary
from site to site (see figure 4). This is known from studies of the

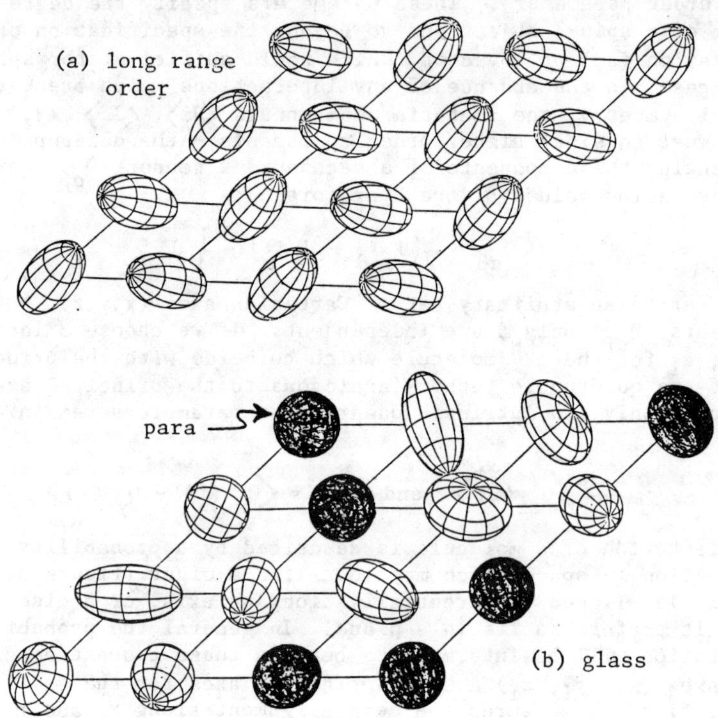

Fig. 4. Comparison of the orientational ordering for (a) a plane
section of the Pa_3 structure and (b) the quadrupolar glass phase.

CW NMR specta whose lineshapes are determined by the intramolecular
dipole-dipole interaction which depends directly on the values of the
local order parameters. If the molecular motion is slow compared to
D = 57 kHz (the strength of the intramolecular dipolar interaction)
each molecule contributes a doublet with frequency separation[7]

$$\Delta \nu_i = \frac{3}{2} D \left[\sigma_i \, P_2(\cos\theta_i) + \frac{3}{2} \eta_i \, \sin^2 \theta_i \, \cos 2\phi_i \right] \qquad (2)$$

(θ_i, ϕ_i) are the polar angles specifying the orientation of the
applied magnetic field with respect to the molecular reference frame
(x_i, y_i, z_i). The NMR spectra have been analysed assuming axial
symmetry ($\eta = 0$ at each site) and one then finds for the glass phase a
broad distribution $P(\sigma)$ of order parameters with $P(\sigma) \propto |\sigma|$ to a good

approximation. This analysis breaks down if the molecular motion is fast or comparable to 100 kHz for which the NMR lineshape becomes motionally narrowed. There has been some inconsistency in the literature with analyses of spectra in terms of a distribution of quasi-static order parameters[8] and interpretations of the results of pulsed NMR experiments in terms of molecular reorientations occurring at rates faster than 10^5 Hz.[9]

If one studies the evolution of the system as a function of temperature, no distinct abrupt change in $P(\sigma)$ inferred from the NMR spectra (assuming quasi-static order parameters) as a means for detecting possible transitions to the glass phase. This is due to the fact that the local symmetry is broken even in the completely disordered high temperature paraorientational phase as soon as a neighbouring quadrupole (ortho molecule) is replaced by a spherical (para) molecule. This effect does not exist for spin glasses (in the absence of an external field). The order parameters resulting from this dilution effect alone which breaks the cubic symmetry of the environment of a given ortho molecule can be calculated in the high temperature limit and is given by

$$\sigma = 3x(1-x)(5\Gamma/3T)^4 \tag{3}$$

for an ortho fracton X.

For this reason we have studied the nuclear spin-lattice relaxation times which measure the spectral densities of the molecular fluctuations at the chosen Larmor frequency ω_o (typically $10^7 - 10^8$ Hz). The modulation of the intramolecular dipolar interactions (strength D) and the spin-rotation coupling $C\underline{I}\cdot\underline{J}$ leads to a relaxation rate

$$T_1^{-1} = C^2 \sum_m J_{1m}(\omega_o) + 9D^2/5 \sum_m m^2 J_{2m}(m\omega_o) \tag{4}$$

where $J_{LM}(\omega_o)$ are the spectral densities of the autocorrelation functions of the irreducible operator equivalents $\Theta_{LM}(t)$ of the spherical harmonics $Y_{LM}(t)$ describing the molecular orientations. (The Θ_{LM} are simply related to the Cartesian components given above:

$$\Theta_{20} = 1 - \frac{3}{2} J_z^2, \quad \Theta_{2\pm2} = \frac{\sqrt{3}}{8}(J_\pm)^2 \text{ and} \tag{5}$$

$$\Theta_{2\pm1} = \mp \sqrt{\frac{3}{8}} (J_\pm J_z - J_z J_\pm).$$

If we assume that the dominant mechanism is due to fluctuations of σ with respect to a mean value $\bar{\sigma}$ one can show that the autocorrelation functions are given by

$$G_{2m}^{(i)}(t) = D_{mo}^2(\chi_i) D_{mo}^{2*}(\chi_i) \langle \Theta_{20}(t) \Theta_{20}(0) \rangle_i$$

$$= (2 - \bar{\sigma}_i - \bar{\sigma}_i^2) F_{2m}(\chi_i) \mathscr{G}(t) \tag{6}$$

116

where the form $\mathcal{G}(t)$ needs to be determined from some microscopic
model. The $D_{mo}(\chi_i)$ are rotation matrix elements calculated for each
site for a transformation from the laboratory reference frame
(determined by the applied magnetic field) to the local molecular
reference frame (x_i, y_i, z_i). χ_i denote the Eulerian angles for the
transformation. $F(\chi_i)$ is a purely geometrical factor determined by
these angular factors. The above expression shows the strong
dependence of T_1 of a given isochromat in the spectrum on the value
of the local order parameter $\bar{\sigma}_i$. This explains the observed
dependence of T_1 on the component of the spectrum studied in the
glass phase.[10] This is seen when the spectral diffusion from one
isochromat ν^i in the spectrum to another ν^j is blocked. This occurs
because the intermolecular dipolar interactions $(d^{inter} \approx 1 \text{ kHz})$ is
too weak to assure energy conservation – flip-flops mediated by
$d^{inter} I_i^+ I_j^-$ cannot conserve energy for two isochromats $(\nu_i^i - \nu_i^j)$ whose
order parameters differ by more than 1% since $D^{intra} \approx 100 \, d^{inter}$.
Since the diffusion rates fall off as $1/R_{ij}^6$ with spatial separation
the spectral diffusion should be very effectively quenched in the
glass phase since the probability of finding equivalently oriented
molecules is only non-neglible for well separated pairs. This is to
be contrasted with the case of the Pa_3 configuration for which
molecules on the same sub-lattice are parallel and all molecules have
the same order parameter to a close approximation.

A more important feature is the observation[11] of a fairly sharp
minimum in T_1 at a temperature close to those for which earlier
studies reported rapid changes in the wings of the CW lineshapes.[7]
Fig. 5. Temperature
dependence of relaxation
time in quadrupolar glass
phase.

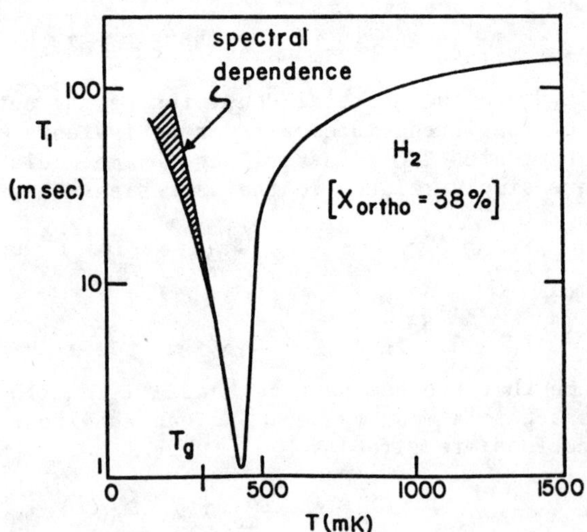

The value of T_1 at the minimum is not inconsistent with that calculated using the above expression and a simple Gaussian or even exponential form for $\mathcal{G}(t)$ provided one allows for the dependence on the order parameters σ. Such an analysis leads to a strong temperature dependence for $\mathcal{G}(t)$ with the characteristic time τ_c varying by several orders of magnitude over a small temperature interval (~ 30 mK) at approximately 400 mK for x=38%. (See figure 5)

This temperature dependence has also been studied by the group at Ohio[12] who adopted the ingenious technique of using orth-hydrogen impurities to probe the orientational ordering of ortho-para D_2 mixtures in the quadrupolar glass regime. The temperature dependence that they observed was so strong that they described their observations in terms of a cusp-like behaviour for the relaxation rate T_1^{-1} at a well defined temperature $T_g(x)$ marking a transition to the glass phase. More systematic measurements are needed to distinguish between a real transition (i.e. some discontinuity in the chosen autocorrelation functions or their derivatives as a function of temperature) and a smooth dynamical freezing of the molecular motion over a narrow temperature interval. Very different conclusions have been drawn from recent studies at Duke[10] (and also at Harvard[13]) for large samples of H_2 (and D_2), respectively. In these experiments only a very smooth evaluation of T_1 as a function of temperature was observed and these groups find no evidence for a transition (smooth or abrupt) to a glass phase. The difference between their experiments and those at Saclay[11] and Ohio[12] is that the latter groups grew their samples on extremely fine copper wire (~ 100 microns) with the view of minimising the thermal gradients that can occur in the samples due to the heat released by the ortho-para conversion. Bulk samples without any Cu wire were used at Duke while very large samples (~ 1 cm^3) grown on relatively large wires were studied at Harvard. It is difficult to attribute the differences to thermal gradients for the D_2 studies[13] but there is another possibility. It has been shown that the cubic phase can be stabilized by epitaxial growth on copper and gold surfaces[14] and the striking differences between the bulk samples and those on grown in confined geometrics on fine wires may be due to difference in lattice constraints imposed by the two different experimental technqiues. Further experiments that address this question (e.g. neutron diffraction studies) are needed to resolve this issue.

In order to test the hypothesis that the transition to the glass phase is a smooth but rapid freezing of the molecular dynamics, we have developed an NMR technique based on the observation of stimulated echoes to detect slow molecular reorientations.[15] The idea was to choose the experimental method so that the stimulated echo amplitudes would be damped by the molecular motion when this is reduced to the range $10^4 < \omega_m < 10^5$ Hz and see if this occurs at a temperature slightly lower than that of the minimum observed for T_1 (which occurs for $\omega_m \sim 10^7$–10^8 Hz). The results shown in figure 6 do indeed indicate a reduction in the echo amplitude for $T < T\left(T_1^{min}\right)$.

118

Fig. 6. Motional damping
of stimulated echoes in the
glass phase.

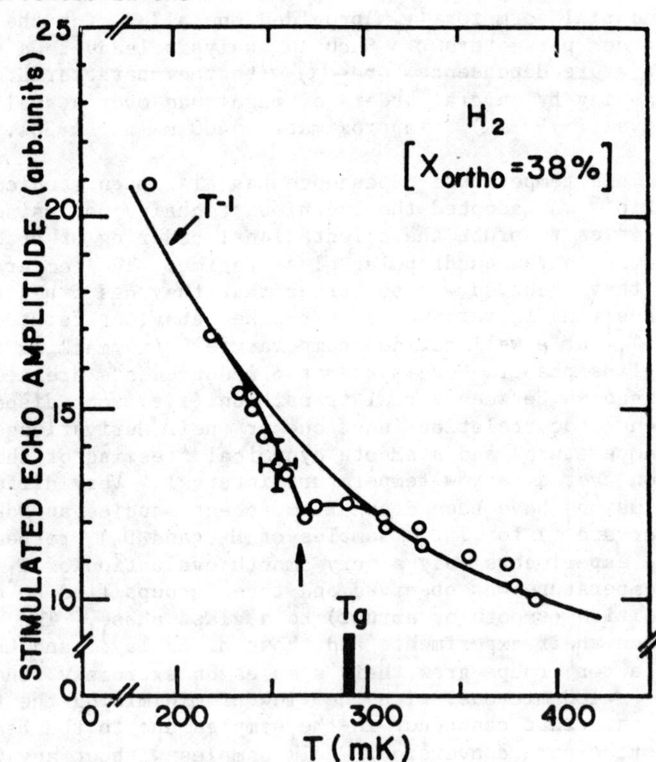

The reduction is small ($\sim$ 30%) and is attributed to the intrinsic
spatial inhomogeneity of the molecular dynamics in the glass phase.
(The dip in the echo amplitude is regarded as the superposition of a
large number of sharp minima occurring at slightly different temperatures.)

OVERALL PICTURE

Using the above results we can now give a precise definition of
the terms "freezing" and "glass" used in connection with quadrupolar
glass phase of solid hydrogen. We consider the single particle autocorrelation function introduced above for the irreducible tensorial
operators Θ_{2m}:

$$G_i(t) = \frac{1}{2} \left(\sum_m [\Theta_{2m_i}(t) \Theta_{2m_i}^+(0) + \Theta_{2m_i}(0) \Theta_{2m_i}^+(t)] \right)_T \qquad (7)$$

This is a scalar and it is independent of the choice of reference
frame $G(0) = 5/2$ while

$$G_i(\infty) = \sum_m |\langle \Theta_{2m_i} \rangle|^2 = \sigma_i^2 + \eta_i^2; \text{ i.e. } G_i(\infty) \text{ measures the magnitude of}$$

the _total quadrupolarization_ $q_i = \left(\sigma_i^2 + \eta_i^2\right)^{1/2}$ of molecule i.
(σ and η are evaluated in the local principal reference frame.)
Experimentally, we observe the spatial average $\overline{G}(t) = \frac{1}{N} \sum_i G_i(t)$ and
the general behaviour is sketched below. T_1 studies probe changes in
$G(t)$ for short times 10^{-6}–10^{-8} s, CW studies for 10^{-4}–10^{-5} s
(determined by D^{intra}) and stimulated echo methods have been useful
for the interval 10^{-2}–10^{-3} s. It is only legitimate to interpret the
CW spectra in terms of a sum of Pake doublets[1] if $G(t)$ is flat for
$0.1\ D^{-1} < t < 10\ D^{-1}$.

When we claim that the molecular orientational motion is frozen
for a time scale t_{sc} we mean that

    (i) $G(t)$ is flat over the time scale t_{sc}, and

    (ii) $G(t_{sc}) > G(\infty) = \overline{q_{eqm}^2}$

where $\overline{q_{eqm}^2}$ is the statisical thermal equilibrum value of the mean
square quadrupolarization taking into account _all_ configurations. In
the glass model in which the system is frustrated from exploring all
the lowest energy configurations on any experimental time scale we
have $G(t_{expt}) > \overline{q_{eqm}^2}$. Both properties (i) and (ii) need to be
established to confirm the glass model. NMR studies alone cannot
prove (ii) directly. There are however indirect indications that
(ii) may be valid. These are provided by observations of hysteresis
following long thermal cycles. The configurations realized can vary
from one cycle to another and small irreversible effects are observed
on warming following an initial cooldown. These are seen in both T_1
measurements[15] and line shape studies. In the 300–450 mK temperature interval we found that if one resets the temperature (following
an initial measurement at a known temperature), one must wait a long
time (30–45 min) before steady consistent values of T_1 ($\sim$ msec)
are observed despite the fact that the magnitude of the NMR signal
indicates that the temperature was completely stabilized in less than
1 min. Furthermore, if the lineshape is carefully recorded in this
temperature region and the sample is then cooled, held below 100 mK
for approximately 1 hour and warmed to precisely the initial
temperature, we find that the functional form of the lineshape is _not_
the same. One needs to warm an additional 60–80 mK to observe the
same form (after normalising total areas to account for the trivial
temperature dependence T^{-1}). These effects are subtle, but they do
provide the first evidence of the remanence effects that would be
expected for a glass-like freezing of the orientational degrees of
freedom of the ortho molecules.

REFERENCES

1. For a general review of the orientational ordering ortho-para H_2
   mixtures see J.R. Gaines in these proceedings (and reference
   therein).

2. H.M. James, Phys. Rev. $\underline{167}$, 862 (1968).

3. Site dilution occurs naturally via ortho → para conversion which allows one to study the behaviour for several concentrations by simply aging a given sample.

4. G. Parisi and G. Toulouse, J. de Phys. (Paris) Lett. $\underline{41}$, 361 (1980); H. Sompolinsky, Phys. Rev. Lett, $\underline{47}$, 935 (1981).

5. N. Bontemps, J. Rajchenbach and R. Orbach, J. de Phys. (Paris) Lett. $\underline{44}$, L47 (1983).

6. N.S. Sullivan, J. de Phys. (Paris) $\underline{37}$, 981 (1976).

7. N.S. Sullivan, M. Devoret, B.P. Cowan and C. Urbina, Phys. Rev. $\underline{B17}$, 5016 (1978).

8. S. Washburn, M. Calkins, H. Meyer and A.B. Harris, J. Low Temp. Phys. $\underline{49}$, 101 (1982).

9. I. Yu, S. Washburn, M. Calkins and H. Meyer, to be published.

10. S. Washburn, I. Yu and H. Meyer, Phys. Lett. $\underline{85A}$, 365 (1981).

11. N.S. Sullivan and D. Esteve, Physica $\underline{107B}$, 189 (1981).

12. W.T. Cochran, J.R. Gaines, R.P. McCall, P.E. Sokol and B.R. Patton, Phys. Rev. Lett. $\underline{45}$, 1576 (1980).

13. R.V. Pound in these proceedings and references therein.

14. A.E. Curzon and A.I. Mascal, Brit. J. Appl. Phys. $\underline{16}$, 130 (1965).

15. N.S. Sullivan, D. Esteve and M. Devoret, J. Phys. C. (Solid State) $\underline{15}$, 4895 (1982).

COMPUTER STUDIES OF ORIENTATIONAL ORDERING
IN THE HCP PHASE OF SOLID HYDROGEN

M. A. Klenin
Physics Department, North Carolina State University
Raleigh, North Carolina 27650

ABSTRACT

Computer experiments have been carried out on a simple model
for solid hydrogen in the "glass phase." It is found that although
a transition may occur to a three-dimentional orientationally
ordered phase, this requires very long Monte Carlo run times. The
data suggest that the ordering which occurs on real experimental
time scales corresponds to local symmetry breaking. It is driven
by the multiple order parameter interactions intrinsic to the hcp
lattice, even in the absence of dilution.

INTRODUCTION

There has been a great deal of interest recently in the
so-called "orientational glasses." These include molecular solids
in which either quadrupolar or dipolar intermolecular interactions
may dominate. They are characterized experimentally by a quenching
of the orientational degrees of freedom which occurs without an
observable onset of long range order.[1] The discussion here is
applicable to the rather diverse quadrupolar systems (ortho-
parahydrogen, ortho-paradeuterium, as well as N_2-Ar alloys) and
has been developed to distinguish between two different aspects
of the problem -- namely the respective roles of dilution-induced
local symmetry breaking and intrinsic frustration effects associated with the lattice structure of the experimental systems in
their "glassy" regimes. It should be noted that most recent work
on the subject suggested an analogy with the spinglass problem,
emphasizing both frustration and disorder induced through dilution
of the quadrupolar species, but without examining the role of these
various ingredients separately. In real materials, the quadrupolar
glass regime appears to coincide in all cases with a cubic-
hexagonal structural phase transition, rather than with a well
defined quadrupole concentration. This fact suggests that the
frustration effects created by the hcp lattice are important in
their own right, with or without the additional complications
created by dilution. In addition, experiments which study concentration dependence of the "glass" transition indicate a lower
critical concentration of approximately 28 percent quadrupole
concentration. This lies well above the percolation threshold for
the lattice, and suggests that the ordering, whatever its form, is
not describable in terms of dilution alone.

The model used here for computation is a simple one. All data
are obtained for an EQQ (electric quadrupole-quadrupole)[2] interaction between classical rotators on a rigid lattice. The standard

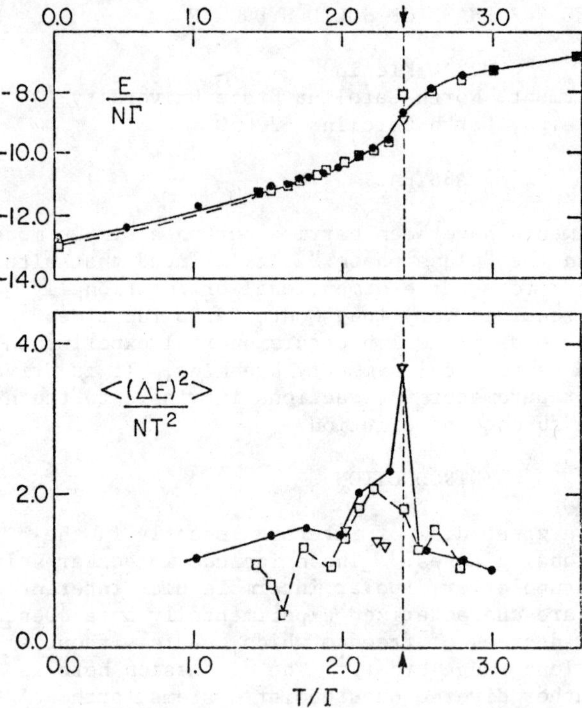

Fig. 2. Internal energy per quadrupole vs. temperature (upper curve) and specific heats. Circles represent the warming curve, squares, an intermediate timescale cooling curve. Triangles represent a cooling curve annealed at $T = 2.4\Gamma$. The arrows indicate the transition temperature, where annealing was carried out.

called "short" or "intermediate", it is comparable to run times necessary for attainment of equilibrium in some of the well studied nonpathological systems.) To study the time-scale on which equilibrium is achieved, the same initial configuration was "annealed" at $T/\Gamma = 2.4$ for 2×10^4 MCS/S. The primary effect of annealing on the bulk properties is restoration of the second-order nature of the transition (see Fig. 4a) and in reintroducing a sharp specific specific heat peak. Annealing experiments below about $T = 2.0\Gamma$ appear to give results identical to those of the warming runs. Furthermore, below this temperature, times greater than about 3000-4000 MCS/S appear to be adequate to stimulate cooling from the annealed array. If annealed in the vicinity of $T = 2.4\Gamma$ the system undergoes an apparently ordinary phase transition over a narrow temperature range.

The energy curves obtained under the various annealing schedules are virtually indistinguishable. More detailed analysis of the system configuration requires the computation of order parameters. Data from the "equilibrium" runs of Fig. 2 -- i.e. from both the warming run and annealed cooling runs -- are displayed in Fig. 3. The large positive values of ψ_1, and $\bar{\psi}_1$, are consistent with a well-defined transition into the LRO ground state. The apparent transition temperature is consistent with

Because the glassy properties of the system are of interest
and because earlier computer experiments indicate characteristics
of slow relaxation, particular attention was paid to size effects
(which are directly related to defect formation) and to run time
effects. In addition, a number of initial configurations were
used at various temperatures, and comparisons were made for slow
warming, slow cooling, and quenching runs. These comparisons
included calculations both of various order parameters and their
susceptibilities and of internal energies and specific heat.
Sample size varied between 128 and 2520 lattice sites; beyond
about 500 sites final configurations appeared to be independent
of sample size. Because of practical limitations most of the
data shown are taken for 576 or 720 sites. Non-periodic
boundary conditions were used to avoid prejudicing the sublattice
structure, and all the bulk parameters were computed using only
sites interior to the sample.

The accessibility of the long-range order configurations was
investigated by means of a "ground state" search from equilibrium
states at temperatures in the transition region. This "quenching"
process invariably produced a configuration different from that
of the LRO ground state. The ground state search was conducted
in a manner analogous to the local field technique used for
classical Heisenberg spins.[7] In the present case one may compute
a quadrupolar field tensor

$$h_{\alpha\gamma}(\vec{r}_i) = \frac{1}{2}\,\Gamma \sum_{\substack{i \\ \alpha'\gamma'}} K_{ij}^{\alpha\gamma\alpha'\gamma'}\, Q_j^{\alpha'\gamma'} \tag{6}$$

RESULTS AND DISCUSSION

find the principle axes, and rotate the quadrupole moment at site i
so that it is aligned along the direction for largest negative
eigenvalue of $h_{\alpha\gamma}(\vec{r}_i)$, thus minimizing the energy. It should be
noted that the present model yields a non-degenerate local field
tensor.

The calculated values of energy and specific heat are
summarized in Figure 2. Data are displayed for three sorts of
runs. The warming curve (closed circles) was prepared at T = 0
in the lowest-lying LRO configuration. Each run consisted of
between 4000 and 5000 MCS/S, and the final configuration was used
as the initial configuration at the next higher temperature. The
temperature intervals varied over the transition region as shown,
and increase in run times produced no discernible effect on the
data. The open squares represent several cooling runs beginning
from a random initial array at $T/\Gamma = 5.0$, and reducing the
temperature in steps of 0.5Γ to the transition region. The energy
curves are virtually identical except very near $T/\Gamma = 2.4$. At
these relatively short run times, the transition appears to be
first order, and the specific heat peak, clearly observable
in the warming runs, is truncated. Below this temperature the
statistical noise in the specific heat persists to the lowest
temperatures measured. (Although the MC run time is here

124

with the Q_{im} the (five) independent linear combinations of the
$Q_i^{\alpha\gamma}$. Normalization is chosen such that $\frac{1}{N} \Sigma_{jm}\ Q_{im}^2 = 1$, and thus
the value $\psi = 1$ indicates complete ordering into the given refer-
ence state.

For a number of reasons it is convenient to use reference
states which differ slightly from the configurations shown. The
discussion below is based on order parameters ψ_1, $\bar{\psi}_1$, ψ_0. The
first two are closely related to the herringbone structures; ψ_0
was the "P" structure as its reference states. One may identify
the configurations of Fig. 1 as follows:

(1) $\psi_o \neq 0$; $\psi_s = \bar{\psi}_s = 0$ for all s>0: the system is in the P
     configuration

(2) $\psi_s > 0$, $\bar{\psi}_s > 0$ or $\psi_s < 0$, $\psi_s < 0$; and $\psi_o = 0$:
     H1 structure obtains with axis $\vec{k}_s$

(3) $\psi_s > 0$, $\psi_s < 0$ or $\psi_s < 0$, $\psi_s > 0$; and $\psi_o = 0$:
     H2 structure obtains with axis $\vec{k}_s$

(4) $|\psi_s| > 0$, $\psi_s = 0$ or $|\bar{\psi}_s| > 0$, $\psi_s = 0$; and $\psi_o = 0$:

     Long range herringbone order occurs in each plane with
     axes of successive planes aligned. The stacking rule
     determining H1 or H2 structures is broken, and stacking
     occurs randomly.

Other sets of order parameter values correspond to coexistence
of the various structures or to some unidentified long range
ordered structure.

The associated susceptibilities may be obtained from order
parameter fluctuations in the usual way:

$$\chi_\alpha \equiv \frac{1}{NT} \left\langle \left[\Delta\ (N\psi_\alpha) \right]^2 \right\rangle \tag{3}$$

where $\left\langle (\Delta x)^2 \right\rangle = \left\langle x^2 \right\rangle - \left\langle x^2 \right\rangle$; angular brackets indicate
averages over MC samples, and T is the temperature.

Internal energy per quadrupolar "spin" and specific heat are
given as

$$E = \frac{\Gamma}{2N} \sum_{\substack{\alpha\gamma \\ \alpha'\gamma' \\ ij}} \left\langle Q_i^{\alpha\gamma} Q_i^{\alpha'\gamma'} \right\rangle K_{ij}^{\alpha\gamma\alpha'\gamma'} \tag{4}$$

and the specific heat:

$$C = \frac{1}{NT^2} \left\langle \left[\Delta(NE) \right]^2 \right\rangle \tag{5}$$

Thus one has direct access to these quantities; they do not depend
on the choice of order parameter definition.

provide a function of the $Q_i^{\alpha\gamma}$ which takes into account any sublattice decomposition which occurs. The known low-energy states are shown in Fig. 1. The two "herringbone" structures, H1 and H2, are four-sublattice structures, and are distinguished by the relative orientation of molecules in the inequivalent lattice planes. The P (pinwheel) structure lies at slightly higher energy, but is associated with a higher entropy, since it is invariant under rotations about the crystal c-axis. These structures form the basis for the order parameter definitions used here.

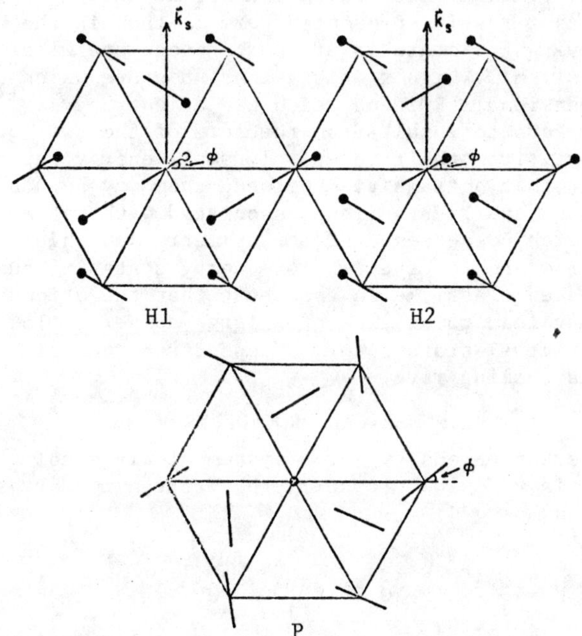

Fig. 1. Possible four- and eight-sublattice arrangements for classical quadrupoles on the hcp lattice (after Ref. 3). H1 and H2 are nearly degenerate and are associated with a lower internal energy than the P structure. The vector $\vec{k}_s$ indicates the direction of doubling of the unit cell. Solid circles represent the ends of the molecules pointing upward from the paper.

The explicit form for the order parameters are similar to those used in spinglass simulations. Each involves an "overlap" between the state observed at a particular stage of the MC run and a particular "reference" state, such as those shown above -- i.e. the order parameter $\psi\alpha$ is

$$\psi\alpha = \frac{1}{N} \sum_{jm} Q_{jm}^{(\alpha)} \, Q_{jm} \tag{2}$$

126

MC rejection procedure is used. The only non-standard aspect of
the computation is development of order parameter definitions in
such a way as to permit discussion of local and global symmetry
breaking and to distinguish among states which are very nearly
degenerate. The numerical results suggest a resolution of the
apparent discrepancies reported between experiments using purely
local order parameter probes (such as NMR) and those using bulk
sample measurements.

There is one major difference between the present system and
systems exhibiting spinglass behavior. In both of the pure rigid
crystalline forms the quadrupolar glasses possess a long-range
ordered (LRO) configuration which lies at an energy minimum.[3-6]
One of the major results presented here is that in the hcp phase,
the lowest lying LRO state is at least degenerate with (and
possibly at slightly higher energy than) a state which displays
only two-dimensional LRO, and which has an entropy $\propto N^{1/3}$.
This fact appears to result from the form of the two lowest lying
configurations, from their relatively small energy separation and
from the fact that a transition between them may be achieved
locally through the fluctuations associated with yet another order
parameter, which possesses a higher symmetry and which is easily
accessible from each of the two low-energy states. Thus the
system is indeed glasslike in the sense that the orientational
transition may lead to any one of a large number of low tempera
ture and the actual state observed depends on such experimental
parameters as cooling rate.

NUMERICAL PROCEDURES

The interaction energy for a system of classical quadrupoles
interacting via a Coulomb force may be written in Cartesian
coordinates as

$$H = \frac{1}{2} \Gamma \sum_{\substack{\alpha\gamma \\ \alpha'\gamma' \\ ij}} Q_i^{\alpha\gamma} K_{ij}^{\alpha\gamma\alpha'\gamma'} Q_j^{\alpha'\gamma'} \tag{1}$$

when the $Q_i^{\alpha\gamma}$ are the second rank tensors associated with the
molecular orientation at site i; α,γ label Cartesian components.
The fourth rank coupling tensor is given by

$$K_{ij}^{\alpha\gamma\alpha'\gamma'} = \frac{\partial^4}{\partial r_{i\alpha} \partial r_{i\alpha'} \partial r_{i\gamma} \partial r_{i\gamma'}} \left(\frac{1}{|\vec{r}_i - \vec{r}_j|} \right)$$

This form for the Hamiltonian is particularly convenient for
Monte Carlo work. It should be borne in mind however that there
exist only five independent components for expectation values of
the $Q_i^{\alpha\gamma}$ at each site, and thus five independent components for
the local order parameter. Order parameter definition must

the specific heat peak and annealing point indicated in the
previous figure. There is no evidence for the multiple LRO
transitions predicted by mean field theory, but in the transition
region a significant magnitude is observed for the order parameter ψ_o which describes the P-structure, and which has no
overlap with either ψ_1, or $\bar{\psi}_1$. As one would expect, given the
form of the LRO ground state, ψ_o vanishes as T goes to zero.
The values plotted are averages over several thousand MCS/S.
Thus although one should describe the behavior ψ_o as dominated
by fluctuation effects in this regime, the fluctuations occur
over a relatively long time scale. The temperature range of the
effect is of the order of the transition temperature itself.

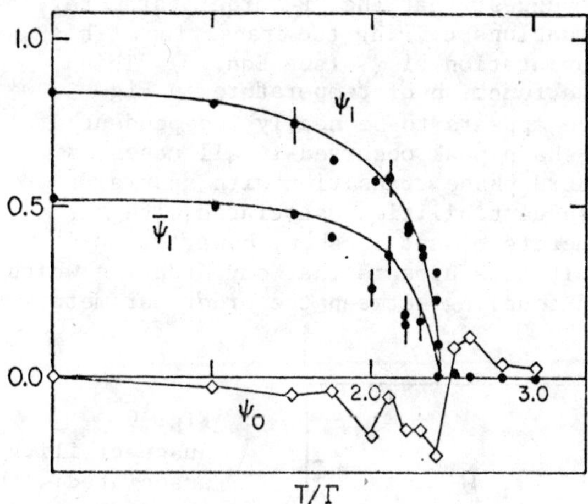

Fig. 3. Order
parameters for runs
at "equilibrium" in
Fig. 2. ψ_1 and $\bar{\psi}_1$
indicate formation
of a "herringbone"
pattern. ψ_0 indi-
cates a tendency to
order in the "P"
structure.

Fig. 4. Order para-
meters for the
"quenched" runs. The
dotted lines indicate
the results of a
ground state search.

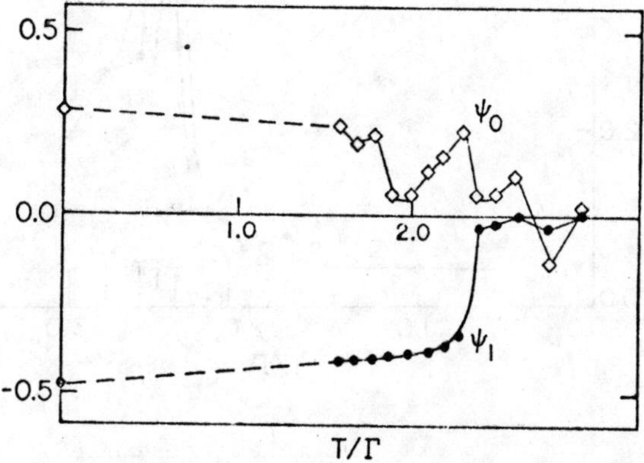

128

The behavior of the non-equilibrium arrays may be contrasted
with the above. Under intermediate timescale cooling conditions,
the system finally orders in planar arrays ($\psi_1 \neq 0$; the absolute
sign is unimportant, since only relative signs of ψ_1 and $\bar{\psi}_1$ enter
into a description of the symmetry), and the transition occurs at
the same temperature as the LRO transition. However, there are
two major differences in the order parameter descriptions of the
two cases. Here $\psi_0 \neq 0$ persists to T = 0. The dotted lines on
the figure indicate the result of a ground state search from the
point shown, and the data points indicate the results of several
runs. Clearly the spatial form of the configurations differs
from the LRO results, despite virtually indistinguishable energy
values. Figures 4 and 5 suggest that the "P" order parameter is
associated with the fluctuations driving the transition. This
is corroborated by the computation of χ_0 (see Eqn. 3) This
quantity is displayed as a function of temperature in Fig. 5.
The temperature dependence appears to be nearly independent of
annealing schedule. The sharp peak observed in all cases is
characteristic of a standard phase transition with spatial
symmetry breaking. The susceptibilities associated with the
"correct" LRO order parameters ψ_1 and $\bar{\psi}_1$ show, however, no
evidence of an anomaly. It thus appears that any ordering which
results occurs because of coupling between the order parameters
of the system.

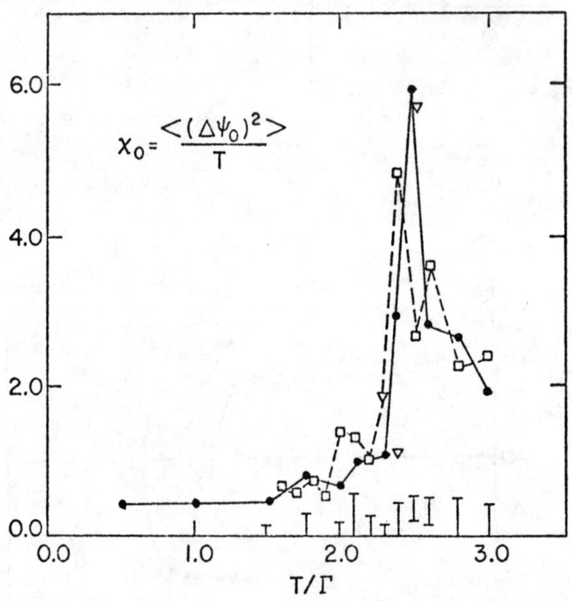

Fig. 5.
Susceptibility
associated with
the order para-
meter ψ_0 versus
temperature.
Symbols are as
in Fig. 1. The
bars lying below
$\chi = 0.5$ indicate
the range of
magnitude of the
susceptibilities
associated with
the "correct"
symmetry of the
system.

Further corroboration of this is obtained from visual inspection of the arrays themselves. For $T \simeq 2.0\Gamma$ both regions of P ordering and H1 / H2 ordering are found. The P structure can apparently be formed easily from either herringbone configuration. Tentatively, one may conclude that this represents a purely local fluctuation in the system, but long-range interactions between fluctuations is being investigated. It is found at $T = 0$ that the "quenched" runs differ from the "annealed" runs in that the former contain orientational stacking faults -- i.e. each triangular plane appears to order completely in a herringbone, and all planes have a common axis ($\vec{k}_s$ in Fig. 1). Successive planes may however represent H1 or H2 structures with nearly equal probability. Thus it seems that a number of order parameters may appear, each of which involves a local symmetry breaking, but all of which taken together destroy this symmetry breaking on a global scale. In all of the above data this is attributable to the hcp structure alone. This distinction between local and global properties is consistent with observed experimental behavior.

Some preliminary order parameter data have also been obtained to study the role of dilution by non-interacting sites. In Fig. 6 data are displayed for a simulation of $T = 0$ downconversion. The non-zero order parameters are plotted for both quenched and LRO conditions at full concentration χ. Run #1 was started from the LRO array (the plot indicates ψ_1 and $\bar{\psi}_1$ as a function of χ), while Run #2 was started from the "quenched" array of Fig. 4 (ψ_1 and ψ_0 are plotted).

There are several noteworthy features of these curves: (1) either of the structures obtained is at least metastable at at $T = 0$, even on down conversion; (2) the high concentration data in all cases follows the standard percolation probability curve for site dilution; (3) complete order parameter suppression appears to occur for $\chi \simeq 0.3$, rather than at the percolation threshold,[8] and this feature appears to correspond to experiment. More detailed analysis is being carried out to determine the effects of run times, sample size, and annealing schedule.

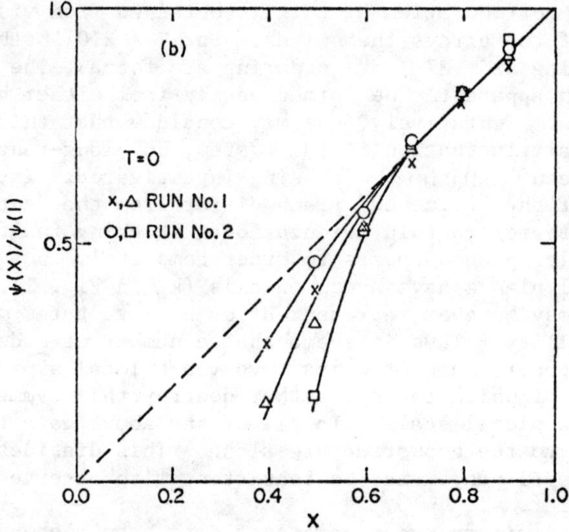

Fig. 6. Order parameters as a function of concentration. X and O
represent the extent of two-dimensional ordering (ψ_1 of Figs. 3
and 4) △ and □ are necessary to complete the description of
the X = 1.0 configurations and are related to interplanar
correlations. In each case, the concentration-dependent order
parameter has been normalized to its value in the initial array
of the down conversion run.

CONCLUSION

All of the above may be summarized as follows:

(1) The structure of the hcp lattice itself leads to a
     complicated interaction among order parameters of the
     system. This results in a number of very nearly
     degenerate states whose ordering is two-dimensional,
     rather than three-dimensional in nature.
(2) Despite the absence of long-range order under quenched
     conditions, the symmetry-specifying order parameters
     of classical Landau theory provide a useful framework
     for discussion of local ordering and fluctuation
     effects.
(3) Dilution effects do not appear to be purely percolative
     in nature; one may conjecture that fluctuation
     effects lead to suppression of order well above the
     percolation threshold.
(4) Experimental results seem to be consistent with the
     qualitative picture drawn here. Real systems appear
     to correspond to the non-adiabatic simulations, and
     thus one may expect rather different behavior, depend-
     ing on whether microscopic or macroscopic experimental
     probes are used.

ACKNOWLEDGEMENTS

I am grateful to D. G. Haase, H. Meyer and R. G. Palmer for comments and criticism. This research was supported by NSF grant DMR 79-23202.

REFERENCES

1. For a review and further references see J. R. Gaines, this conference.
2. Orientational interactions for H_2 and D_2 are discussed by I. F. Silvera, Rev. Mod. Phys. $\underline{52}$, 393 (1980); those for N_2, by T. A. Scott, Phys. Rep. $\underline{2}$, 1 (1976).
3. H. M. James, Phys. Rev. $\underline{167}$, 862 (1968).
4. A. I. Kitaigorskii and K. V. Mirskaya, Soviet Phys. Crist $\underline{10}$, 121 (1965).
5. H. Miyagi and T. Nakamura, Progr. Theort. Phys. (Kyoto) $\underline{37}$, 641 (1967).
6. A. B. Harris, Phys. Rev. $\underline{B2}$, 3495 (1970).
7. L. R. Walker and R. E. Walstedt, Phys. Rev. $\underline{B22}$, 3816 (1980).
8. V. Shante and S. Kirkpatrick, Adv. Phys. $\underline{20}$, 325 (1971).

On the Quasiparticle Interaction in liquid ^{3}He

E. Krotscheck

Institute for Theoretical Physics
University of California
Santa Barbara, CA 93106

ABSTRACT

Microscopic models to understand the quasiparticle interaction and the spectrum of elementary excitations in liquid ^{3}He are reviewed. It is argued that progress in this field is likely to come from a union of the major techniques of many-body theory; perturbation theory; variational methods and Monte-Carlo methods.

A critical analysis of ground-state theories and especially the approximations underlying the variational method is offered with particular concern for their consequences for the excitation spectrum and the quasiparticle interaction. Infinite orders of perturbation theory are necessary to obtain a satisfactory description. The Babu-Brown model for the quasiparticle interaction, which is shown to incorporate a topological subset of the hypernetted chain diagrams, provides the necessary guidance. However, variational wave functions alone do not adequately describe the energy-dependence of the general vertex function. An improvement, which permits the microscopic calculation of the "direct" interaction, is offered.

The simplest numerical applications of the theory for the effective mass and the spin-susceptibility of liquid ^{3}He provide substantial improvements upon results given by earlier, purely variational, approaches.

1. INTRODUCTION

Our understanding of the ground-state of liquid ^{3}He at zero temperature has improved greatly in recent years. It is now feasible to solve the many body Schrödinger equation by direct Monte-Carlo integration.[1] A substantial amount of raw material is already available in these calculations for studies of excited states. To begin with, one thinks of the response of a quantum liquid to adiabatic external perturbations, most efficiently expressed through Landau's quasiparticle interaction. Other interesting subjects are the response to time-dependent external fields, collective excitations and their damping, temperature-dependent effects, and, probably the most difficult problem, the possibility and the structure of a superfluid phase.

Attempts to understand these phenomena from the bare two-body interaction are not new. Neither are claims that the physically interesting quantities can or cannot be derived "from first principles". Without attempting to analyze the historical developments, it seems clear that none of the supposedly microscopic attempts in the literature can withstand a critical analysis.

It should be said at the outset that this paper will not present a fully microscopic, quantitative theory for the quasiparticle interaction either. Rather, we analyze the problem in the light of present ground-state technology, and investigate suitability of the available theoretical approaches. The basic assumption behind the present study is that a direct numerical treatment, in the sense of the Monte-Carlo ground-state calculations, will not be available in the foreseeable future. Therefore we have to resort to the available perturbative and variational theories in their widest sense, analyze their strengths and weaknesses, and try to reach our goal with the help of "the best of both worlds". To some extent, we must review the development of ground-state formalisms to see where they work, why they work, and where they fail.

The study of the development of ground-state theories is *per se* historically interesting. Serious ground-state calculations with strong interactions started with the development of Brueckner's summation of ladder diagrams[2] and its refinements through the Coupled-Cluster Theory[3]. For some time it was believed that Brueckner theory and its descendants were adequate to explain the ground-state of dense quantum liquids. However, numerical application[4,5] to liquid ^{3}He give typical "ground-state" energies of 0.15 K at a density of 0.006 A^{-3}, as opposed to the experimental value of -2.52 K at a density of 0.0166 A^{-3}. As Kümmel *et al*[5] point out, it is easy to obtain agreement with experiments by introducing terms simulating three body effects, but fourth-order Bethe-Faddeev theory is needed to show convergence. The formidable effort necessary for nuclear matter calculations[6] underscores the difficulty of this path vividly. Turning to our more immediate concern, namely the quasiparticle interaction, Babu and Brown[7] showed that the *minimum* addition to Brueckner-Hartree-Fock type theories needed to obtain a satisfactory description is a "self-consistent" summation of RPA diagrams.

The next generation of ground-state theories, more successful than perturbation-theory methods, starts from a variational *ansatz* for the wave function constructed to incorporate, in an average sense, the geometrical correlations between two particles. The starting point of a variational theory is the choice of a model wave function $|\Psi_0\rangle$ generated from the wave function $|\Phi_0\rangle$ of the corresponding noninteracting system through

$$|\Psi_0\rangle = F|\Phi_0\rangle / \langle\Phi_0|F^\dagger F|\Phi_0\rangle^{\frac{1}{2}}, \tag{1.1}$$

where F is a *correlation operator*, usually of the Feenberg form

$$F = \exp\left\{\sum_{i<j} u(r_{ij}) + \sum_{i<j<k} u(\vec{r}_i, \vec{r}_j, \vec{r}_k) + \ldots\right\}. \tag{1.2}$$

As far as numerical applications are concerned, I will restrict myself here to the two-body part of the Feenberg function (1.2). The presence of three-body correlation factors[8] may lead to some quantitative changes. The main conclusions, however, which are concerned with effects *not* included in the local correlation operator (1.2), will remain unchanged.

For the explicit ansatz (1.2) for the wave function, one may derive cluster expansion and summation techniques, whose diagrammatic richness goes far beyond the Bethe-Goldstone and RPA theory, which permit an approximate, but self-consistent, summation of ring, ladder, and self-energy diagrams. These techniques are generally referred to as Hypernetted-Chain (HNC) theories. The application of the Hypernetted Chain method[9] for Fermi systems (FHNC) leads

134

to the correct prediction of both short- and long ranged interparticle correlations in a unified picture. In the Hypernetted-Chain *approximation* (i. e.
neglecting certain sets of "elementary" diagrams), the theory predicts[10,11] a
minimum of the energy around -1.1 K at a density of 0.012 A^{-3}. Comparison with
the exact Monte-Carlo evaluation of the energy expectation value[1] with the same
two-body correlation factor reveals that the Hypernetted-Chain *approxmiation*
misses roughly .5 K, i. e. 10% of the total binding energy from local two-body
correlations. The remaining 1 K binding energy can be attributed about equally
to static three body correlations and momentum dependence.

The crucial breakthrough in ground state theories came, however, neither
from perturbative nor from variational calculations. The Gordian knot was
finally untied by the direct greens function Monte Carlo integration of the many
body Schrödinger equation[1].

Compared with BBG theory approaches, variational methods have the
advantage that the origin of the "missing" energy is easily understood. I will
therefore base the further discussion on the variational method, but must also
appeal to perturbation theory to keep track of those effects that are only poorly
described by a variational wave function.

Owing to the relative success of variational calculations it is occasionally
believed that one can obtain a complete and accurate description of a quantum
liquid by going beyond the Feenberg form (1.2) and inventing a correlation
operator F "good enough" to describe the desired effect. Such a statement, while
in its generality trivial, requires specification. The success of the variational
method rests on the fact that it allows a self consistent summation of ring,
ladder, and self energy diagrams and provides therefore simultaneously a
correct treatment of short and long ranged correlations. Apart from recent progress in Boson systems[12], the same diagrammatic richness has not yet been
reached in perturbative treatments of the many body problem. Of course, if the
"sufficiently realistic" correlation operator F is so complicated that such self
consistent summations are no longer feasible, the variational approach looses
its justification. To take advantage of both the wide arrays of diagrams summ-
able with simple, local correlation functions and the accurate description of
state-dependence in perturbation theory, it is useful to have a theory which
allows conveniently to step back and forth between the different pictures.

The theory that most successfully combines the variational approach and
perturbative methods in the many body problem is known as the correlated
basis functions (CBF) theory[13] CBF theory seeks the improvement of the ground
state wave function and, at the same time, a description of excited states in a
non-orthogonal basis $\{|\Psi_m>\}$ of wave functions generated by the correlation-
operator F acting on a corresponding basis of $\{|\Phi_m>\}$ of independent particle
wave functions:

$$|\Psi_m> = F|\Phi_m>/<\Phi_m|F^\dagger F|\Phi_m>^{\frac{1}{2}}. \tag{1.3}$$

Superficially, one may think of CBF theory as perturbation theory in the non-
orthogonal basis (1.3), and regard the algorithms to generate the required
matrix elements within the basis of correlated wave functions[14] as a "black box".
A deeper analysis of this perturbation theory with non-orthogonal wave functions
reveals CBF theory as a comprehensive and flexible theory of effective interactions useful in precisely those place where conventional perturbative treatments
reach their practical limitations. Thus, I will base the further discussion on the
Correlated Basis Functions Theory, which contains both, "one-shot" variational
treatments and purely perturbative approaches, as special cases.

2. CORRELATED WAVE FUNCTIONS AND CBF THEORY

The theory of variational wave functions and the connected CBF perturbation theory has acquired a substantial amount of technology. It is not the aim of this paper to review these developments. Nonetheless, the techniques of the FHNC/CBF theories are not commonplace, and a jump into the middle of the action would leave this paper inaccessible to many. We therefore offer an outline of the essentials which emphasizes plausibility arguments and a discussion of the physical relevance of each a step at the expense of formal derivations.

2.1 Expectation values

The variational description of a quantum liquid usually starts with a diagrammatical analysis of the energy expectation value

$$H_{00} = \langle \Psi_0 | H | \Psi_0 \rangle. \tag{2.1}$$

The purpose of this analysis is twofold: First, it allows the summation of wide classes diagrams generated by a set of so-called "elementary" diagrams via the integral equations of (F)HNC theory. Second, it provides exact relations between the partial distribution functions, (which are accessible without further approximation by Monte-Carlo techniques) and the basic building blocks necessary to construct the effective interactions of CBF theory[14].

2.2 Optimization

For the calculation of the energy expectation value, it is necessary to specify the pair correlation function, the three body factor, etc.. Due to the variational principle, the energy expectation value is an insensitive quantity. A rough guess determined, for example, by a healing distance constraint (low order constraint variation, LOCV[15]) is usually adequate if one is interested in the ground state energy alone. More advanced methods determine the pair correlation function by minimization of the ground state energy[10].

$$\Omega(r) = \frac{\delta H_{00}}{\delta u(r)} = 0. \tag{2.2}$$

The optimization of the pair correlation function is necessary for the correct description of the phonon contributions to the static structure function. It is also required to make the connection between variational and perturbative theories and to accelerate the convergence of the subsequent CBF- perturbation series. Numerical instabilities of the variational problem (2.2) reflect physical instabilities of the system: An instability against long-wavelength density fluctuations, for example, at low densities indicates droplet formation[16]. At high densities one finds an instability against fluctuations of finite wavelength indicating an impending instability against the formation of a quantum crystal[17].

2.3 Off-diagonal quantities

The energy expectation value H_{00} is just one matrix element to be calculated between variational wave functions. To derive a systematic perturbation theory we must extend the theory to the calculation of general matrix elements of the Hamiltonian and the unit operator. Let us start with the basic definitions and notations.

(Off)-diagonal matrix elements of the unit operator and the Hamiltonian occur in the combinations

$$N_{mn} = \langle \Psi_m | \Psi_n \rangle \tag{2.3}$$

and

136

$$H'_{mn} = <\Psi_m | H - H_{00} | \Psi_n>, \tag{2.4}$$

where $|\Psi_m>$ and $|\Psi_n>$ are correlated states identified by the sets of occupation numbers $\{m\}$ and $\{n\}$. An especially important object is the combination[14]

$$W_{mn} = H'_{mn} + \tfrac{1}{2}(H_{mm} + H_{nn})N_{mn} \tag{2.5}$$

that can be identified with the full d-body vertex function in an energy-averaged representation, where d is the number of orbitals in which m and n differ. We note also that the diagonal matrix elements H'_{mm} for a one particle, one hole state $|\Psi_m>$ are the single particle energies of CBF theory. Their analytic structure is reminiscent of a Hartree-Fock spectrum[14] but their physical interpretation is different.

Additional features of the single particle spectrum and the off diagonal matrix elements will be discussed below as needed.

2.4 CBF Perturbation Theory

The generation of a perturbation series in this non orthogonal, correlated basis of wave functions is a moderately straightforward task[18] and leads to a perturbation series of a structure similar to Rayleigh Schrödinger perturbation theory with multi-particle interactions. The only awkward feature is the existence of unlinked diagrams canceling in different orders of the perturbation series. The demonstration[19,20] that these "catastrophic" diagrams cancel order by order was one of the most important steps in the early developments of CBF theory. More advanced theories[21,22] avoid this technical problem by construction and derive integral equations independent of the correlation operator which generate classes of diagrams similar to, but more elaborate than, the integral equation techniques of conventional perturbation theory. As special cases we mention the CBF generalization of Bethe Goldstone or Galitzki ladders or the RPA ring diagrams within the general framework of the Correlated Coupled Cluster theory[21] Study of these infinite summations, when carried out together with the analysis of variational wave functions, provides important insight into the approximations underlying the variational ansatz (1.1). The formal manipulations are complicated[22] since cluster expansions and CBF-diagrams must be considered at the same time. I wish to restrict myself therefore to the simplest case, where the non local parts of the CBF effective interactions are approximated by a local hole state average. This simplification also permits the most transparent identification of the correlations introduced by the CBF perturbation series. Owing to the immediate relevance of ring diagrams for the theory of the quasiparticle interaction we concentrate on this subset of the CBF perturbation expansion.

The sum of all CBF ring diagrams may be cast in the form

$$E_{RING} = E_{opt} + E_{RPA} - E_{RPA}^{coll}. \tag{2.6}$$

The first term, E_{opt} indicates the fact that the CBF perturbation series takes care of the optimization of the *local* two-body correlations. It has the form

$$E_{opt} = -\tfrac{1}{2}\int \frac{d^3q}{(2\pi)^3} \frac{\hbar^2 q^2}{4mS(q)} \left\{ \sqrt{1 + \frac{4m\tilde{\Omega}(q)}{\hbar^2 q^2}} - 1 \right\} \tag{2.7}$$

Here, $S(q)$ is the static structure factor. Note that $\tilde{\Omega}(q)$ is the Fourier transform of the functional variation of H_{00} with respect to the pair correlation function , (c.f. eq. (2.3)). The term ΔE_{opt} is zero if optimal two-body correlations are used to generate the CBF series.

The second term is the well-known[23] sum of the RPA ring-diagrams

$$E_{RPA} = -\frac{i}{2}\int\frac{d^3q\,d\omega}{(2\pi)^4}\left\{\log[1-U(q)\Pi_0(q,\omega)] + U(q)\Pi_0(q,\omega)\right\}, \qquad (2.8)$$

in which $\Pi_0(q,\omega)$ is the bare particle-hole propagator, i. e. the Lindhard function. The third term, E_{RPA}^{coll}, is identical with the expression (2.8), but with the full Lindhard function replaced by its "collective" approximation

$$\Pi_0^{coll}(q,\omega) = \frac{\hbar^2q^2/m}{\omega^2-(\hbar^2q^2/4mS_F(q))^2}. \qquad (2.9)$$

Here, $S_F(q)$ is the static structure factor of a non interacting system, and the particle hole interaction $U(q)$ is of the form

$$U(q)=\tilde{\Omega}(q)S^{-2}(q)+\frac{\hbar^2q^2}{4m}(S^{-2}(q)-S_F^{-2}(q)). \qquad (2.10)$$

In the case $\tilde{\Omega}(q) = 0$, this $U(q)$ is precisely the particle hole interaction needed to reproduce S(q) via the fluctuation dissipation theorem from the density response function if the particle hole propagator is replaced by its collective approximation (2.9). The equations are readily extended to the spin channel. The particle-hole interactions for both, normal and spin-polarized ^{3}He are shown at experimental equilibrium density in fig. 1.

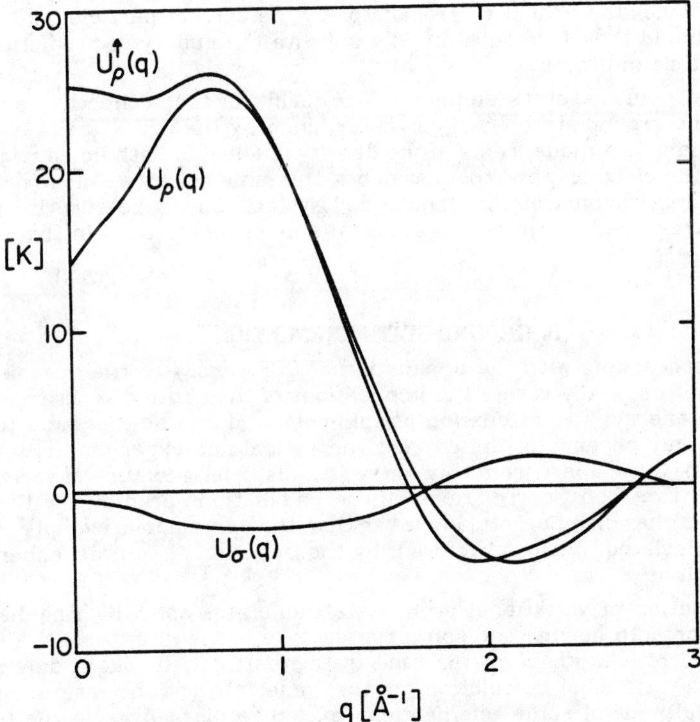

Fig. 1: The local particle hole interactions are shown for normal ^{3}He in the density channel, $U_\rho(q)$, and in the spin channel, $U_\sigma(q)$, at experimental equilibrium density $\rho = 0.0166A^{-3}$. Also shown is the density channel potential for the spin polarized liquid at the same density, $U^\uparrow(q)$.

138

These potentials correspond to the pseudopotentials introduced by Aldrich and Pines[24].

We see that the summation of CBF ring-diagrams has two effects:

(i)   It provides the first step in the optimization of the two-body correlation factor by the inclusion of virtual phonons[25] and

(ii) It replaces the "collective" particle-hole propagator by the Lindhard function. If the two-body correlations have been optimized already, only the propagator corrections survive, and the sum of all CBF ring-diagrams vanishes when the particle-hole propagator is replaced by its collective approximation.

This clarification of the correlations introduced by the variational wave functions and by perturbation is one of the keys to a satisfactory microscopic theory of the quasiparticle interaction. The simple variation of the energy expectation value H_{00} with respect to the quasiparticle occupation number must clearly lead to wrong results, especially since it is not clear how to "undo" the frequency sums. An "energy - averaged" particle-hole propagator can be acceptable for ground-state calculations. But it is wrong if one looks at a specific $\omega \to 0$ limit. This is the basic explanation why purely variational calculations of Fermi-Liquid parameters are usually poor and experience substantial corrections in CBF theory[26,27]. It offers also the cure: The inclusion of infinite order CBF ring diagrams replaces the collective propagator by the correct particle hole propagator, and should therefore substantially improve the quality of predictions for the quasiparticle interaction.

A second point deserves emphasis: The quality of the "collective" approximation for the ground-state energy is determined by the existence of a strong collective mode. This mode exists in the density channel in both liquid ^{3}He (zero sound) and the electron gas (the plasmon). The same collective mode does not exist in the spin-channel of the unpolarized system, and is no surprise that a "collective" treatment of these correlations is much worse than in the density channel.

3. GROUND-STATE ENERGETICS

Before proceeding with the application of CBF theory to the quasiparticle interaction, let us briefly review the implications of the above discussion for the ground-state energy. The discussion of spin-polarized ^{3}He at the same time is relevant not only because of the current theoretical and experimental interest in this system. Quite apart from this, there is a disturbing feature of variational theories with two- body correlations: These calculations predict[28,10] that the spin polarized phase is energetically lower than the spin unpolarized phase. The problem is alleviated, but not removed, by the inclusion of spin dependent two body correlations[29].

A compilation of variational ground state energies and CBF ring-diagram corrections for both normal and spin polarized ^{3}He is shown in table 1. The two body potential of Aziz *et. al.*[30] was used in this calculation. Slight differences with a recent variational calculation for the same interaction[11] are due to an improved treatment of some "elementary" exchange diagrams (in the framework of the above discussion, the first-order insertions in the particle-hole propagator). As usual, the purely variational results of the normal phase lie above the ones for the polarized system. The comparison with variational Monte-Carlo calculations[28,31] show that the neglect of elementary "dd - diagrams" is somewhat more severe for the polarized system, and leads to an underestimate of the energy difference between the two phases of the about 0.3 K.

Ground-state energy of normal and polarized ^{3}He							
ρ [A^{-3}]	normal phase				polarized phase		
	E_{var}	E_{pp}	E_{sp}	E_{tot}	E_{var}	E_{pp}	E_{tot}
0.0100	-1.01	-0.15	-0.25	-1.41	-1.08	-0.06	-1.14
0.0112	-1.11	-0.19	-0.28	-1.58	-1.20	-0.07	-1.27
0.0130	-1.15	-0.25	-0.32	-1.62	-1.24	-0.11	-1.35
0.0142	-1.09	-0.30	-0.34	-1.73	-1.18	-0.14	-1.32
0.0148	-0.98	-0.32	-0.36	-1.66	-1.11	-0.16	-1.27
0.0166	-0.55	-0.41	-0.39	-1.35	-0.76	-0.21	-0.97
0.0180	-0.00	-0.48	-0.43	-0.91	-0.31	-0.26	-0.58
0.0200	1.02	-0.59	-0.48	-0.06	0.65	-0.34	0.31

Table 1: Variational results for the ground-state energy (E_{var}), propagator corrections in the density channel (E_{pp}) and spin- correlations (E_{sp}) for normal (Cols 2-5) and spin-polarized liquid ^{3}He. E_{tot} is the sum of the distinct contributions in each case.

Propagator corrections for normal ^{3}He are somewhat larger that those for the spin-polarized phase, which suggests that the spin-polarized system should be more dominated by a zero-sound mode. They are, for the normal phase of the same order of magnitude as the effect that is occasionally[32] interpreted as "backflow". In a more conventional language, one would identify backflow with the current- current coupling term in the quasiparticle interaction. This is *not* included in the local approximations for the particle-hole interactions shown in fig. 1. The results of table 1 in the density channel are the minimum *additional* binding energy to be expected from momentum dependent correlation due to the inclusion of the correct particle-hole propagator in the ring diagrams.

More interesting than the propagator corrections in the density channel are the spin correlations. These are not present in the spin-polarized system. For the unpolarized system, since no optimized spin correlations were used, the ring diagram corrections include also the "single operator chain" contributions from collective[29] (or state-averaged) spin correlations. It is illuminating to note that the use of the "collective" approximation for the particle-hole propagator in the spin-channel underestimates the energy gained from spin- correlations by about 30%. The inclusion of the correct particle-hole propagator in the spin channel stabilizes the normal phase, even if we allow for the fact that the FHNC "elementary" diagrams in the spin-polarized phase are somewhat bigger.[31] One would also expect that three body correlations are weaker in the polarized system. Once a main source of the spurious instability of the correlated state against spin fluctuations has been identified, it should be no surprise that the quasiparticle interaction calculated in the variational/CBF framework is substantially improved over purely variational estimates.

4. SINGLE-PARTICLE SPECTRUM

It is a long-standing experience that purely variational calculations[13,27] predict an effective mass of a ^{3}He atom in liquid ^{3}He of about 0.7 in sharp contrast with the experimental evidence[33,34,35,36] from specific heat measurements that m^*/m should be between 2 and 3 at zero vapor pressure. It is also known that low-order CBF corrections to m^*/m are able to resolve most of the discrepancy[26,11]. The physical origin of the large effective mass of a ^{3}He atom near the Fermi surface has been most clearly identified by Brown $et.$ $al.$[37] as due to the coupling of single particle excitations to the low lying mode of spin fluctuations. Low order CBF calculations of the kind mentioned above[26,11] include most of this effect, but, in a system as dense as liquid ^3He, it is hard to estimate the convergence of the CBF expansion. Therefore, it is useful to analyze the single particle spectrum from a different point of view[38] which illuminates the underlying problems more clearly and may also be extended further away from the Fermi surface.

In a study similar to the analysis of ring diagrams[22] one can show[39] that there are exact cancellations between the CBF corrections to the hole-line insertions and contributions to the CBF single-particle spectrum,

$$u_{CBF}(k) = \frac{\delta H_{00}}{\delta n(k)}. \tag{4.1}$$

We avoid the details of the largely formal manipulations leading to this result in favor of local model of the particle-hole interaction (2.10) introduced above. In this approximation, the self-energy is of the familiar form

$$\Sigma(k,E) = U_0 + u_F(k) + i \int \frac{d^3q\,d\omega}{(2\pi)^4} G^0(\vec{k}-\vec{q},E-\omega)\,U^2(q)\Pi(q,\omega) \tag{4.2}$$

where $\Pi(q,\omega)$ is the (spin) density response function. Here

$$\Pi(q,\omega) = \frac{\Pi_0(q,\omega)}{1-U(q)\Pi_0(q,\omega)}. \tag{4.3}$$

and $u_F(k)$ is the Fock term of the particle hole interaction (2.10). U_0 is a constant related to the chemical potential. $u_F(k)$ should be identified with the contribution to the self energy originating from the G-matrix with the additional provision that multiple phonon exchange effects are included. The momentum-dependence of this term is weak and similar to the momentum dependence of the CBF spectrum.

The second term in eq. (4.2) is, of course, the most important one. It is the ordinary RPA expression for the self-energy[23] in terms of the particle hole interaction introduced above. It is the only term that has a non trivial energy dependence and off shell extension at this stage of the development.

Results for the momentum dependence of the on-shell effective mass

$$\frac{m^*}{m} = \left[1 + \frac{m}{\hbar^2 k}\frac{d\Sigma(k,e(k))}{dk}\right]^{-1} \tag{4.4}$$

at different densities are shown in fig. 2.

The results of the present calculation may be summarized as follows:

(i)   Fig. 2 shows at all densities the expected sharp enhancement peak of the effective mass around the Fermi momentum, arising from the attractive interaction in the spin channel. The width of the enhancement peak $\delta k \approx 0.2 k_F$ is largely independent on density and the strength of the interaction.

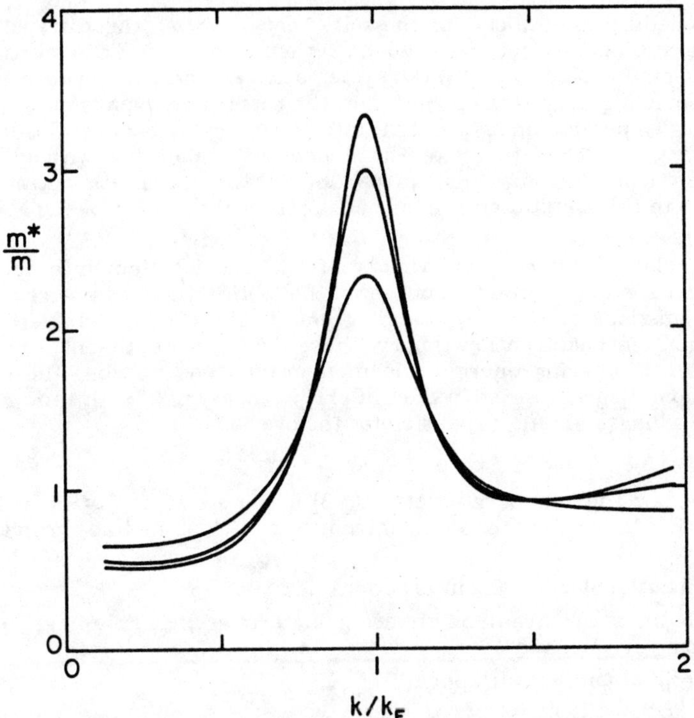

Fig. 2: The effective mass in normal ^{3}He is shown as a function of momentum for three densities $\rho = 0.0148\ A^{-3}$, $\rho = 0.0166\ A^{-3}$, and $\rho = 0.0180\ A^{-3}$. The maximum of the effective mass increases with density.

(ii) The actual size of the enhancement peak depends very sensitively on the strength of the spin-interaction. An increase in this strength by a few percent drives the system to a Landau-instability, a decrease by a few percent lets the effective mass drop well below 2. It is hard to see why the local approximation for the particle-hole interaction and the simple particle-hole propagator should be so accurate. With slight adjustments of parameters, or a judicious choice of diagrams, any sensible empirical result can easily be reproduced. The enhancement of the effective mass can be understood qualitatively, but the description lacks a mechanism to ensure stability of the system.

For completeness I should mention also that there is a second enhancement peak[38] of the effective mass at or above $2\,k_F$. This secondary enhancement is due to the coupling of single-particle excitations to the zero sound mode. It is necessary to understand the sharp drop of m^* above k_F and the existence of a stable zero-sound mode at large momentum transfers[40] in a unified picture. The actual size and location of the second enhancement peak depends sensitively on zero sound dispersion relation. The probably best estimate of the effect is the one by Friman *et al.*[38] using the Aldrich-Pines pseudopotential. The effective masses shown in fig. 2 should be regarded with some caution above about $1.5\,k_F$.

142

5. QUASIPARTICLE INTERACTION

The procedure used above for the self-energy involved three steps, namely (1) the generation of a variational model, (2) the summation of CBF perturbation diagrams to infinite order, and (3) the systematic cancellation between topologically corresponding diagrams appearing in the respective expansions. The topological analogies provide enough insight into the interplay between FHNC theory and perturbation theory to make the connections plausible without further lengthy analytical manipulations. In particular, they facilitate a transparent presentation to the uninitiated reader of the physical meaning of the accessible technology.

To formulate a microscopic theory of the quasiparticle interaction one must consider the diagrammatic content of the (F)HNC theory in more detail. To this end, let us look at some typical diagrams in the two-particle vertex function, see fig. 3. In perturbation theory, these diagrams represent functions of four 4-momenta, obeying energy- and momentum conservation. The topologically equivalent diagrams arise in the CBF/FHNC theory of effective interactions. To generate a diagrammatic expansion for the overlap

$$N_{ph,p'h'} = <ph' \,|\, N(1,2) \,|\, hp'>_a,$$

one must expand the Feenberg function (1.2) in powers of $h(r) = e^{u(r)}-1$. Using this correlation factor $h(r)$, one may introduce a "Goldstone-like" graphology in which

(i) $h(r)$ takes the place of the interaction,

(ii)   "particle lines" are drawn as upward going arrows and "hole-lines" as downward going arrows, and

(iii)   *no energy denominators* appear.

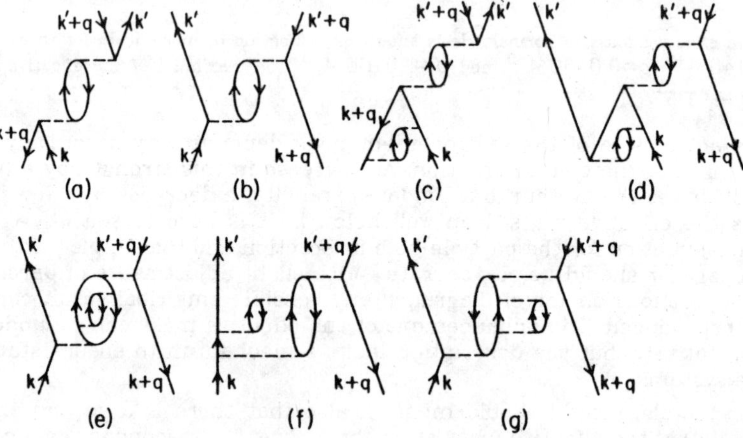

Fig. 3: Some typical diagrams contributing to the general two particle vertex. Note that diagrams (a) and (c) vanish in the Landau limit.

Typical diagrams contributing to the overlap (5.1) assume topologically the same form as shown in fig. 3. However, all summations over internal particle and hole lines become trivial and lead simply to the Slater function

$$l(rk_F) = A^{-1} \sum_{k<k_F} e^{i\vec{k}\cdot\vec{r}}. \tag{5.2}$$

Qualitatively, one can therefore interpret the correlation factor $h(r)$ as a screened interaction divided by an averaged energy denominator. The details of energy-averaging can be studied for special cases. The rule for the ring-diagrams is the collective approximation for the propagator given above, where also the corresponding result for the factorizable insertions is mentioned.

It is precisely these simplifications of the propagator that makes possible the vast summation of diagrams within the FHNC theory. As an example, consider the "chain-diagram" shown in fig. 3a. The exchange term to this diagram is shown in fig. 3b. These exchange diagrams are, within the FHNC summations, linked to other chain diagrams in various ways (Figs. 3c and 3d) to form structures driving other chain diagrams (Fig. 3e). In addition to these chain-summations, "parallel" connections (Fig. 3f) and "factorizable self-energy insertions" are included.

The diagrams shown in Figs. 3a-3d are topologically identical to those considered by Babu and Brown[7] in their analysis of the quasiparticle interaction. Diagrams of the type shown in fig. 3e are lumped in the "direct" interaction, while the hole line insertions (Fig. 3e) still await consideration.

From a topological point of view we can therefore identify the Babu-Brown model with a special subset of the FHNC or parquet-diagram[12] theories. Yet there are two important differences:

(i) The Babu-Brown model treats the energy-dependence of the propagator in the intermediate states exactly, whereas FHNC uses a collective approximation. This point is crucial for the quasiparticle interaction, and is the source of the incorrect prediction of most Landau parameters by a purely variational theory.

(ii) The Babu-Brown model requires the "direct" interaction as an external input, to be provided, for example, by a G-matrix calculation[41] In liquid helium, the latter procedure does not work well[7] since the multiple-phonon exchanges dominate the "direct interaction". More recent implementations of the theory[42] resort therefore to parametrized approximations for this quantity.

An alternate and completely microscopic derivation of the direct interaction is offered by the FHNC/CBF theory. This quantity may most simply be derived by topological analogy as outlined above, while mathematically rigorous derivation may be achieved by the generalization of the recent analysis of ring diagrams[22] which formed the basis of the considerations of the preceding section.

At this point of the discussion it should be plausible that one can work "backward" from the full FHNC theory of the quasiparticle interaction, identify what should be regarded as the "direct" interaction, and then do the summations suggested by Babu and Brown with the *correct* propagator. For those with a background in the FHNC/CBF theory, I give the local approximation for the "direct" interaction, which is

$$d(q) = X_{dd}(q) - \frac{\hbar^2 q^2}{4m} X_{dd}(q). \tag{5.3}$$

As a start towards the numerical implementation of the general procedure outlined above, the RPA-diagrams in the cross-channel (Fig. 3b) were calculated

144

for $l = 0$. The results for F_0^a are shown in table 2. Note that *no* effective mass correction was applied, and no z-factors due to the renormalization of the quasiparticle pole strength are included. For comparison the results in purely variational approximation (i.e. without the corrected propagator) are also shown.

Variational and propagator-corrected Fermi liquid parameter F_0^a							
$10^3 \rho$ [A^{-3}]	10.0	11.2	13.0	14.2	14.8	16.6	18.0	20.0
present work	-.43	-.44	-.45	-.46	-.47	-.48	-.49	-.49
variational	-.73	-.81	-.85	-.85	-.86	-.91	-.95	-.98

Table 2: Predictions for F_0^a from the present calculation (row 2) and the purely variational calculation of ref. 27 (row 3) as a function of density.

The results shown in Table 2 are certainly not the final answer for microscopic calculations of F_0^a. At least three more corrections are to be applied. These are

(i) Effective mass corrections

(ii) Renormalization of the quasiparticle pole strength by z-factors, and

(iii) Propagator Corrections in the particle-particle and the hole-hole channel.

The corrections (i) and (ii) suffer presently still from substantial numerical uncertainties, they will both tend to make F_0^a more negative. Note that the correction (ii) is essentially responsible for the instability of the normal Jastrow state against spin-density fluctuations. The propagator corrections (iii) are positive[27] and also non-negligible.

6. SUMMARY

Let me summarize the extent to which our understanding of the quasiparticle interaction can presently be based on a purely microscopic theory. Generally, summaries attempt to impress the reader with excellent agreement with experimental results. Many of these impressive demonstrations have a short lifetime. In view of this it seems better to offer an honest survey of w at has been accomplished, and which problems are still ahead of us.

Let me outline first what has been accomplished: First of all, the ground state wave function is by now very much pinned down. In the theory of the quasiparticle interaction in ^{3}He, two major problems have been resolved which precluded hitherto even a *qualitatively* correct prediction of the effective mass and the magnetic susceptibility. The resolution of both problems lies ultimately in the inclusion of the correct particle-hole propagator.

Two other major problems remain. One of them is the role of nonlocalities. Without addressing this problem, it will not be possible to understand the difference between F_1^s and F_1^a. The second problem is the question of what stabilizes the single-particle spectrum. Changing the local interaction in the spin-channel by 10% can lead to a drop of the effective mass ratio below 2, or lead to a Landau-instability, i.e. F_1^s <-3.

What one ideally would like to have is a machine capable of generating the quasiparticle interaction from a given bare two-body force, and no other ingredients. Such a machine exists today for the ground-state energy in the form of the Greens-functions Monte Carlo technique. It works, although one might wish it to work more efficiently. The analogous machine is not yet available for the excited states of a quantum liquid. To generate a quantitative solution of the problem will require the concerted effort of different schools of many-body theory.

(i) The self-consistent summation of RPA and ladder diagrams which is, for example, performed in the FHNC theory and may also be possible by the extension of parquet-diagram techniques[12] is a necessity: Ladder diagrams are needed to eliminate the core-repulsion. Generalized RPA diagrams are required to satisfy Pauli-principle sum rules and to include long- wavelength excitations. The iteration of ring- diagrams into the ladders is needed to include multiple-phonon exchange processes, which are essential to generate a qualitatively correct "direct" interaction in the Babu-Brown model, they are needed, as well, to guarantee the correct stability behavior of the variational problem (2.2)[16]

(ii) Integral equation techniques do have the disadvantage that they rely ultimately on approximations for "elementary" diagrams, whereas such diagrams are not negligible in the helium liquids, and the series of elementary diagrams shows poor convergence[43] The advantage of the FHNC analysis of variational wave functions is that it allows us in principle to work "backwards" from Monte-Carlo computations of (partial) distribution functions to generate the diagrammatic quantities needed for the construction of the effective interactions. The local average of the particle-hole interaction (eq. (2.10)) in the density and in the spin channel and, to some extent the "direct" interaction (5.3), can be constructed from present Monte Carlo data. The extension to partial distribution functions would give access to nonlocalities through the structural relations[14] for the off-diagonal quantities (2.4). The definite advantage of FHNC techniques over diagrammatic perturbation theory is the relation between the driving terms of the FHNC integral equations and quantities accessible via Monte- Carlo integration.

(iii) Variationally based theories produce the wrong energy dependence of all off-shell quantities and must be improved in that respect. The most natural way to do this is to go back to perturbative many-body theory, study the theorems proven there, find if they are satisfied within the variational theory and correct, if necessary, possible deficiencies within the CBF expansion. For the problem of the quasiparticle interactions, the Babu-Brown theory is the obvious place to start. I have presented here what may be regarded as the first step in that direction. While the results are encouraging, this first step will certainly not be the last word. I should also point out that the sets of diagrams suggested by the Babu-Brown theory are only the minimum diagrammatic structures necessary for a qualitatively satisfactory model of the quasiparticle interaction. Experience with "elementary" diagrams in variational theories suggests that these sets of diagrams may very well be not sufficient.

This work was supported, in part, by the Deutsche Forschungsgemeinschaft through a Heisenberg Fellowship, and by the National Science Foundation under Grant No. PHY77-27084. Valuable comments and suggestions by J. W. Clark and A. D. Jackson are gratefully acknowledged.

References

1. M. A. Lee, K. E. Schmidt, M. H. Kalos and G. V. Chester, *Phys. Rev. Lett.* **46** p. 728 (1981).

2. K. A. Brueckner, *Lecture notes of the Les Houches Summer School*, Dunod, (1959).

3. F. Coester and H. Kümmel, *Nucl. Phys.* **17** p. 477 (1960).

4. E. Ostgaard, *Phys. Rev.* **180** p. 263 (1969).

5. H. Kümmel, K. H. Lührmann and J. G. Zabolitzky, *Phys. Rep.* **36** p. 1 (1978).

6. B. D. Day, *Phys. Rev C* **24** p. 1203 (1981).

7. S. Babu and G. E. Brown, *Ann. Phys.* **78** p. 1 (1973).

8. C. C. Chang and C. E. Campbell, *Phys. Rev.* **13** p. 3779 (1976).

9. E. Krotscheck and M. L. Ristig, *Nucl. Phys. A* **242** p. 389 (1975).

10. E. Krotscheck, R. A. Smith, J. W. Clark and R. M. Panoff, *Phys. Rev. B* **B24** p. 6383 (1981).

11. E. Krotscheck and R. A. Smith, *Phys. Rev. B*, (1983). (in press)

12. A. D. Jackson, A. Lande and R. A. Smith, *Phys. Rep.* **86** p. 55 (1982).

13. E. Feenberg, *Theory of Quantum Fluids*, Academic, New York (1969).

14. E. Krotscheck and J. W. Clark, *Nucl. Phys.* **A328** p. 73 (1979).

15. J. S. Bell and E. J. Squires, *Adv. in Phys.* **10** p. 211 (1961).

16. E. Krotscheck, *Phys. Rev. A* **15** p. 397 (1977).

17. L. Castillejo, A. D. Jackson, B. K. Jennings and R. A. Smith, *Phys. Rev. B* **20** p. 3631 (1979).

18. J. W. Clark, "Variational Theory of Nuclear Matter," pp. 89 in *Progress in Nuclear and Particle Physics*, ed. D. H. Wilkinson, Pergamon, Oxford (1979). Vol. 2

19. C.-W. Woo, Ph. D. Thesis, Washington University 1966.

20. C.-W. Woo, *Phys. Rev.* **151** p. 138 (1966).

21. E. Krotscheck, H. Kümmel and J. G. Zabolitzky, *Phys. Rev. A* **22** p. 1243 (1980).

22. E. Krotscheck, *Phys. Rev. A* **26** p. 3536 (1982).

23. A. L. Fetter and J. D. Walecka, *Quantum Theory of Many - Particle Systems*, Mc Graw Hill, New York (1971).

24. C. H. Aldrich III and D. Pines, *J. Low Temp. Phys.* **32** p. 689 (1978).

25. C. C. Chang and C. E. Campbell, *Phys. Rev. B* **15** p. 4238 (1977).

26. H.-T. Tan and E. Feenberg, *Phys. Rev.* **176** p. 360 (1968).

27. E. Krotscheck, R. A. Smith and J. W. Clark, *Lecture Notes in Physics* **142** p. 270 (1981).

28. C. Lhuiller and D. Levesque, *Phys. Rev. B* **23** p. 2203 (1981).

29. J. C. Owen and G. Ripka, *Phys. Rev. B* **25** p. 4914 (1982).

30. R. A. Aziz, V. P. S. Nain, J. C. Carly, W. J. Taylor, and G. T. McConville, *J. Chem. Phys.* **70** p. 4330 (1979).

31. R. M. Panoff, private communication

32. K. E. Schmidt and V. R. Pandharipande, *Phys. Rev. B* **19** p. 2504 (1979).

33. J. Wilks, *The Properties of Liquid and Solid Helium*, Claredon, Oxford (1967).

34. J. C. Wheatley, *Rev. Mod. Phys.* **47** p. 415 (1975).

35. D. S. Greywall and P. A. Busch, *Phys. Rev. Lett.* **49** p. 146 (1982).

36. D. S. Greywall, *Phys. Rev. B* **26** p. 2747 (1982).

37. G. E. Brown, C. J. Pethick and A. Zaringhalam, *J. Low Temp. Phys.* **48** p. 349 (1982).

38. B. L. Friman and E. Krotscheck, *Phys. Rev. Lett.* **49** p. 3536 (1982).

39. E. Krotscheck, J. W. Clark and A. D. Jackson, to be published.

40. K. Skjold, C. A. Pelizzari, R. Kleb, and C. E. Ostrowski, *Phys. Rev. Lett.* **37** p. 842 (1976).

41. O. Sjöberg, *Ann. Phys.* **78** p. 39 (1973).

42. T. L. Ainsworth, K. S. Bedell, G. E. Brown and K. F. Quader, *J. Low Temp. Phys.* **50** p. 317 (1983).

43. R. A. Smith, *Phys. Lett. B* **63** p. 369 (1976).

QUASIPARTICLE SCATTERING AMPLITUDE FOR NORMAL LIQUID ^{3}He

M. Pfitzner and P. Wölfle*
Physik-Department der Technischen Universität München,
Garching, Federal Republic of Germany

ABSTRACT

A new parametrization of the quasiparticle scattering amplitude
(QSA) in terms of eigenfunctions of the exchange operator is introduced. The first six eigenfunctions provide an excellent fit of the
transport coefficients and the known Landau parameters. In the second part we calculate the QSA from a Bethe-Salpeter equation with an
irreducible kernel derived from the density polarization potential
and the f-sum rule as well as the single particle spectrum. Transport properties and Landau parameters evaluated with this QSA are
compared with experiment.

INTRODUCTION

It is well known that the macroscopic properties of a strongly
interacting normal Fermi liquid are to a very good approximation
those of a gas of elementary excitations, the Landau quasiparticles[1].
The interaction of these quasiparticles is described in terms of the
scattering amplitude (QSA) t for binary collisions. The QSA contains
the full information on the Landau parameters, and hence the thermodynamic properties, and it determines the transport properties in the
normal phase[1]. Even for the pair-correlated state the transport
properties[2], the transition temperature T_c and the strong coupling
corrections[3] may be expressed in terms of the normal state QSA.
Following the pioneering work of Abrikosov and Khalatnikov[4] and
the later work by Dy and Pethick[5] and others, recently there have
been several attempts to derive a more accurate QSA. Assuming that
the QSA may be calculated from an effective potential in Born approximation, Sauls and Serene[6] determine the first coefficients of a
Legendre expansion of the effective potential by fitting to normal
state experimental data, while Levin and Valls[7] calculate the effective potential from spin fluctuation theory. While it was in general
possible to find good fits to the transport parameters by
the Legendre expansion[6] at $L \leq 2$, accurate description of the Landau
parameters required $L \leq 3$, or 7 adjustable parameters[6].
Within a second, more microscopic approach, Bedell and Pines[8]
attempt to calculate the QSA from a Bethe-Salpeter equation, using
the information on polarization potentials obtained from the dynamical theory of quantum liquids by Aldrich and Pines[8]. There is a general difficulty with this procedure in that the solution of the Bethe-
Salpeter equation for a given quasiparticle interaction does not
automatically satisfy exchange symmetry. This defect is corrected in
ref. 8 by an arbitrary symmetrization procedure. Furthermore, the

---

*Also at Max-Planck-Institut für Physik und Astrophysik, Munich,
Federal Republic of Germany.

assumption of identifying the current density polarization potential
of ref. 9 with the p-wave component of the quasiparticle interaction
in ref. 8 is not justified, in particular if the effective mass is a
strongly varying function of momentum. Recent results of microscopic
theories[10,11,12] show a rapidly decreasing single particle effective
mass.

In section 2 we present a new parametrization of the QSA in
terms of eigenfunctions of the exchange operator, discuss its potential merits as compared to the effective potential approximation and
show explicit results. In section 3 the method of constructing exchange symmetric solutions of the Bethe-Salpeter equation developed
in our earlier work[13] is briefly reviewed. Section 4 is devoted to a
derivation of the quasiparticle interaction function from the quantum kinetic equation, the polarization potentials of Aldrich and Pines,
sum rule arguments and the single particle spectrum of refs. 10-12.
The results of a preliminary evaluation of the QSA are used in section 5 to calculate the QSA, higher Landau parameters, the transport
properties and the pair interaction coefficients.

PARAMETRIZATION OF SCATTERING AMPLITUDE

At low temperatures, the QSA of a Fermi liquid is a function of two
variables. These have been traditionally chosen[4] to be the angles
(θ,ϕ), where $\cos\theta = \hat{p}_1 \cdot \hat{p}_2$ and $\cos\phi = [(\hat{p}_1 \cdot \hat{p}_1{}') - (\hat{p}_1 \cdot \hat{p}_2{}')]/[1 - (\hat{p}_1 \cdot \hat{p}_2)]$.
Here $\vec{p}_1, \vec{p}_2 (\vec{p}_1{}', \vec{p}_2{}')$ are the momenta of the incoming (outgoing) quasipart-
icles. We find it more convenient to use instead the variables
(μ, Q), where Q is the normalized momentum transfer q, $Q = q/p_F$, and
$\mu = \cos\theta_L = \cos(\hat{p} \cdot \hat{p}')$ (see Fig. 1). Considering only spinconserving

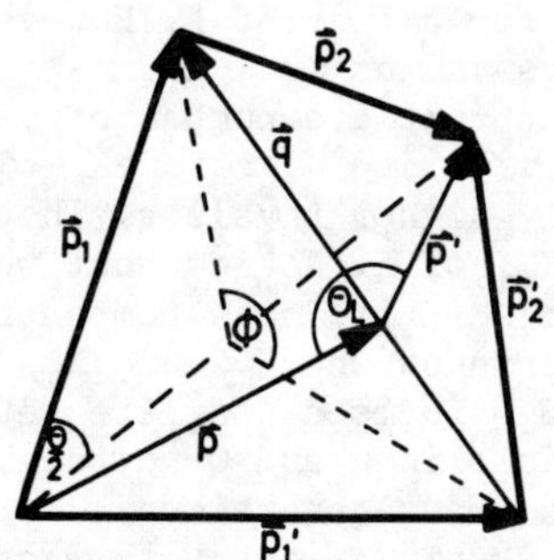

Fig. 1. Representation of
the variables (θ,ϕ) and (q,θ_L)
of the scattering amplitude.

interactions, there are two independent spincomponents of the QSA,
which may be chosen as the spin symmetric (antisymmetric) combinations $A^s(A^a)$ or as the singlet (triplet) components $A^0(A^1)$. Here the
A's are dimensionless QSA's obtained by multiplying the dimensional
QSA by a factor density of states $N(\epsilon_F) = (m^* p_F/\pi^2)$.

Since the QSA is essentially a two particle wavefunction, the Pauli principle requires the following exchange symmetry

$$A^\circ(\mu,Q) = A^s - 3A^a = A^\circ(\bar{\mu},\bar{Q})$$

$$A^1(\mu,Q) = A^s + A^a = -A^1(\bar{\mu},\bar{Q}), \tag{1}$$

where $(\bar{\mu},\bar{Q})$ are obtained from (μ,Q) by interchanging momenta $\vec{p}_1 \leftrightarrow \vec{p}_2$ (see eq.(17b) of ref. 13). A general parametrization of the QSA is thus obtained by expanding in a complete set of eigenfunctions of the exchange operator, with eigenvalues ± 1.

The eigenfunctions of the exchange operator, defined by

$$\hat{X}\, X_{m\pm}(\mu,Q) = X_{m\pm}(\bar{\mu},\bar{Q}) = \pm\, X_{m\pm}(\mu,Q) \tag{2}$$

have not yet been determined in general form, as far as we know. However, we have solved[13] the simpler problem of finding the eigenfunction $X_k^{(L)}$ within a subspace of total function space defined by

$$\hat{X}^{(L)}\, X_k^{(L)}(\mu,Q) = \sigma_k X_k^{(L)}(\mu,Q),$$

where

$$X_k^{(L)}(\mu,Q) = \sum_{\ell=0}^{L} x_{k,\ell}^{(L)}(Q) P_\ell(\mu) = \vec{x}_k^{(L)}(Q)\cdot\vec{P}(\mu), \tag{3}$$

$P_\ell(\mu)$ are the Legendre polynomials and $\hat{X}^{(L)}$ is the projection of the exchange operator onto the subspace. This requires solving a set of $L+1$ coupled integral equations for the $x_{k,\ell}^{(L)}(Q)$, which can be done analytically[13]. The $X_k^{(L)}(\mu,Q)$ are bipolynomials of degree L in μ and of arbitrary degree in Q^2. For L=0 and L=1 the $x^{(L)}$ have been explicitly given in ref. 13. In general the eigenvalues (and eigenfunctions) are different for different L. However, eigenfunctions $x^{(L)}$ belonging to the eigenvalues $\sigma_o = 1$ and $\sigma_1 = -1$ are reproduced for $L'>L$ by adding the appropriate number of vector components $x_{k,\ell}^{(L)} = 0$, $\ell>L$) and these may therefore be determined by considering successively higher L values. They may be identified with the eigenfunctions $X_{m\pm}$ of $\hat{X}$. Notice that the action of the full exchange operator $\hat{X}$ on $\vec{x}_k^{(L)}$ produces components $\ell>L$, for $\sigma_k \neq \pm 1$.

The set of eigenfunctions X_{m+} and X_{m-} containing powers not higher than μ^L and Q^{2L} forms a complete set of exchange (anti-) symmetric bipolynomials in μ and Q^2 of degree L and may be used to expand the QSA in the most general way. For example, for L=2 there are altogether six eigenfunctions with eigenvalue $\sigma_o = 1$ and $\sigma_1 = -1$, respectively, given explicitly by

$$\vec{x}_{o+} = \begin{pmatrix} 1 \\ 0 \\ 0 \end{pmatrix}; \quad \vec{x}_{1+} = 2^{-1/2}\begin{pmatrix} y_1 \\ -3z \\ 0 \end{pmatrix}; \quad \vec{x}_{2+} = (9/7)^{1/2}\begin{pmatrix} y_2 \\ 2zy_3 \\ 0 \end{pmatrix}; \tag{4}$$

$$\vec{x}_{3+} = (5/7)^{1/2}\begin{pmatrix} y_2 \\ -\frac{3}{2}zy_3 \\ \frac{7}{2}z^2 \end{pmatrix}; \quad \vec{x}_{o-} = (3/2)^{1/2}\begin{pmatrix} y_1 \\ z \\ 0 \end{pmatrix}; \quad \vec{x}_{1-} = \begin{pmatrix} y_2 \\ -\frac{3}{2}zy_3 \\ -\frac{5}{2}z^2 \end{pmatrix};$$

where
$$z = 1 - Q^2/4 \; ; \quad y_1 = P_1^{(1,0)}(Q^2/2-1) = \tfrac{3}{4}Q^2-1 \; ;$$

$$y_2 = P_2^{(1,0)}(Q^2/2-1) = \tfrac{5}{8}Q^4-2Q^2+1 \; ; \quad y_3 = P_1^{(3,0)}(Q^2/2-1) = \tfrac{5}{4}Q^2-1 \; ;$$

Here only the first three components have been written (all other components are zero). The functions $P_n^{(\alpha,\beta)}(x)$ are the Jacobi polynomials. The functions $X_k^{(L)}(\mu,Q)$ satisfy the orthonormality relations

$$\int_0^2 dQ \int_{-1}^1 d\mu \, \frac{Q}{2}\left(1-\frac{Q^2}{2}\right) X_k^{(L)}(\mu,Q) X_{k'}^{(L)}(\mu,Q) = \delta_{kk'} \tag{5}$$

Notice that the factors $(1-Q^2/4)^\ell$ appearing in the ℓ-th component of $X_{m\pm}$ ensure that $X_{m\pm}(\mu,Q)$ is independent of μ for $Q = 2$.

Expansion of the QSA in terms of the 6 eigenfunctions given in (4) requires as input as many experimental quantities. The result of such a fit is nevertheless interesting for comparison with other theories and is holding a certain potential for prediction of further experimental quantities. We have determined the QSA by requiring that the Landau parameters $F_0^{s,a}$, F_1^{s}, F_2^{s} are reproduced exactly and by performing a least squares fit to the transport quantities shear viscosity η, thermal conductivity κ, and spin diffusion coefficient D. For the sets of data, based on Wheatley's[14] m^*/m values and on Greywall's[15] m^*/m values, respectively, the fit was accurate to better than one percent.

The resulting QSA components $A_\ell^{s,a}(Q)$ for $\ell=0,1$ (based on Wheatley's data) are shown in fig. 2. Also shown is the result of fitting the effective potential form

$$A^{0,1}(\vec{p}_1,\vec{p}_2;\vec{p}_1'\vec{p}_2') = V^{0,1}(\vec{p}_1-\vec{p}_1') \pm V^{0,1}(\vec{p}_1-\vec{p}_2') =$$
$$= \sum_{\ell=0}^{L} V_\ell^{0,1}\left[P_\ell(\hat{p}_1\cdot\hat{p}_1') \pm P_\ell(\hat{p}_1\cdot\hat{p}_2')\right], \tag{6}$$

with L = 3, given in ref. 6.

For general L there are $(2L+1)$ eigenfunctions in the effective potential approximation, as compared to $(1/2)(L+1)(L+2)$ for the complete expansion. As seen in fig. 2 the overall behavior in the two approaches is similar for $A_1^{s}(q)$, but there are significant differences in $A_0^{a,s}(q)$ and $A_1^{a}(q)$. One may express the effective potential form of the QSA in terms of eigenfunctions $X_{m\pm}$. In the case L = 2 there is one additional eigenfunction, proportional to $(0,-y_3z,z^2)$, outside the subspace spanned by the five effective potential functions. In the above fits we found the coefficient of this function to be comparable to the other coefficients, especially for Greywall's m^*/m data.

In the (θ,ϕ) representation, the eigenfunctions of the exchange operator take the simple form $Y_{\ell\ell'}(\theta,\phi) = P_\ell(\cos\theta)P_{\ell'}(\cos\phi)$, with ℓ' even (odd) for exchange symmetric (antisymmetric) functions. The

152

subspace of bipolynomials of degree L in $\cos\theta$ and $\cos\phi$ is spanned
by the $(L+1)^2$ orthogonal functions $Y_{\ell\ell'}$ with $\ell,\ell'\leq L$. However, these
functions are less convenient for expanding the Bethe-Salpeter equa-
tion (see below), because they do not have the property, neccessary
for the QSA, to become independent of ϕ for $\theta\to0$.

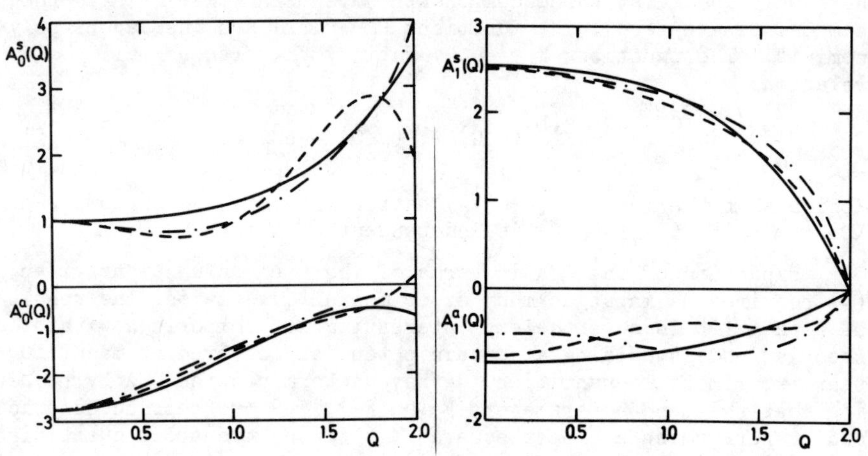

Fig. 2. Comparison of s- and p-wave components of the scattering
amplitude from ref. 6 (dashed lines), ref. 8 (dashed-
dotted lines) and this work (solid lines).

INTEGRAL EQUATION FOR SCATTERING AMPLITUDE

From the Bethe-Salpeter equation for the complete vertex part
the following integral equation connecting the particle-hole irreducible vertex part, or quasiparticle effective interaction,
$F^{s,a}(\mu,Q)$, with the scattering amplitude $A^{s,a}(\mu,Q)$, may be derived:

$$A^{s,a}(\hat{p}\cdot\hat{p}';Q) = F^{s,a}(\hat{p}\cdot\hat{p}';Q) - \int\frac{d\Omega_{p''}}{4\pi} F^{s,a}(\hat{p}\cdot\hat{p}'';Q) \tag{7}$$

$$\chi(\hat{p}''\cdot\hat{q};Q)\ A^{s,a}(\hat{p}''\cdot\hat{p}';Q)$$

Here it has been assumed that $A^{s,a}_{pp}(q)$ and $F^{s,a}_{pp}(q)$ depend only weakly
on $|\vec{p}|$ and $|\vec{p}'|$ and this dependence has been neglected. The integral
over p'' in the sum on intermediate states may then be done, yielding
the particle-hole propagator

$$\chi(\hat{p}'' \cdot \hat{q}; Q) = - \alpha^{s,a}(Q) \frac{1}{m^* p_F} \int_0^\infty dp'' p''^2 \; \frac{f^0_{\vec{p}''+\vec{q}/2} - f^0_{\vec{p}''-\vec{q}/2}}{\varepsilon_{\vec{p}''+\vec{q}/2} - \varepsilon_{\vec{p}''-\vec{q}/2}} \tag{8}$$

The quasiparticle weight factors $\alpha^{s,a}(Q)$ are introduced to allow for a q-dependent distribution of spectral weight among quasiparticles and high energy multi-particle-hole excitations. $\alpha^{s,a}(Q)$ is expected to decrease from 1 at Q=0 to a value $\alpha \sim 0,5 - 1$ at $Q = 2$.

Given the effective interaction
$$F^{s,a}(\mu,Q) = \sum_{\ell=0}^{L} F_\ell^{s,a}(Q) P_\ell(\mu),$$

the integral equation (7) may be solved in closed form[13]. For L = 1, in particular, one finds
$$A_o^{s,a}(Q) = \frac{F_o^{s,a}(Q)}{1 + \chi_o(Q) F_o^{s,a}(Q)} \tag{9}$$

$$A_1^{s,a}(Q) = \frac{F_1^{s,a}(Q)}{1 + \frac{1}{2} \chi_1(Q) F_1^{s,a}(Q)} \; ,$$

where
$$\chi_n(Q) = \int_0^1 d\mu \; \chi(\mu;Q)(1-\mu^2)^n \; .$$

For L>1, the different even and odd (in ℓ) components are mixed separately. However, eqs. (9) remain to be a useful first approximation due to the fact that the functions $F^{s,a}$ for $\ell \geq 2$ are decreasing rapidly.

As already indicated, the QSA calculated from the integral equation (7) will possess exchange symmetry only for rather special choices of the effective interaction $F^{s,a}(\mu,Q)$. In our earlier work[13] we have proposed to restore exchange symmetry for a given effective interaction having only s- and p-wave components, for example, by adding higher ℓ components in a systematic way. This may be done by expanding the input function $A_{in}^{0,1}(\mu,Q)$ in the set of eigenfunctions $X_k^{(1)}(\mu,Q)$ of the exchange operator $\hat{X}^{(1)}$, as

$$A_{in}^{0,1}(\mu,Q) = \sum_k a_k^{0,1} X_k^{(1)}(\mu,Q). \tag{10}$$

An exchange symmetric QSA with the same ℓ = 0,1 components as $A_{in}^{0,1}$ may be constructed by adding a function $B^{0,1}(\mu,Q) = \sum_k b_k^{0,1} X_k^{(1)}(\mu,Q)$, symmetrizing as

$$A^{0,1}(\mu,Q) = \sum_k \frac{1}{2}(a_k^{0,1} + b_k^{0,1}) \left[X_k^{(1)}(\mu,Q) \pm X_k^{(1)}(\mu,Q) \right] \; , \tag{11}$$

and requiring that $b_k^{0,1}$ be chosen such that the s- and p-wave components of (10) and (11) be identical. Using the eigenfunction property of $X_k^{(1)}$ this is equivalent to putting

154

$$\frac{1}{2}(a_k^{0,1} + b_k^{0,1}) = a_k^{0,1}/(1 \pm \sigma_k) \tag{12}$$

Equation (12) determines the symmetrized scattering amplitude (11) completely.

It is seen that the solution exists only if the expansion (10) does not contain a term with $\sigma_k = -1$ or $\sigma_k = 1$. In other words, the input QSA must be orthogonal to all exchange antisymmetric (singlet case) and exchange symmetric (triplet case) eigenfunctions, respectively, implying the integrability conditions

$$\int_o^2 dQ \int_{-1}^1 d\mu Q(1-Q^2/4) \; A_{in}^{0,1}(\mu,Q)X_{m\pm}(\mu,Q) = 0. \tag{13}$$

It remains to specify the eigenfunctions $X_k^{(1)}(\mu,Q)$ and eigenvalues σ_k. Quite generally, the components of the eigenvectors $X_k^{(L)}(Q)$ (see (3)) are given by

$$x_{k,\ell}^{(L)}(Q) = \alpha_{k,\ell}(1-Q^2/4)^\ell P_n^{(2\ell+1,0)}(Q^2/2-1), \quad \ell = 0,1,..,L \\ k = 0,1,....$$

where $n = [(k+1)/(L+1)] - \ell$ ($[x]$ is the largest integer contained in x) and the $P_n^{(\alpha,\beta)}(x)$ are Jacobi polynomials of degree n (note that $P_n^{(\alpha,\beta)}(x) \equiv 0$ for n<0). The coefficients $\alpha_{k,\ell}$ are found by solving a system of (L+1) linear homogeneous equations. The eigenvalues are determined by the zeros of the determinant. For L=1 one finds

$$\sigma_k = 2 \{(-1)^K(K+3)(K-1)+(-1)^k[(K+3)^2+3(K+2)K]^{1/2}\}/[K(K+1)(K+2)] \tag{15}$$

where K is the largest integer contained in (k+1)/2.

An exchange symmetric QSA may now be calculated by the following procedure. For given effective interaction $F^{s,a}(\mu,Q)$, containing L+1 angular momentum components, eq. (7) is solved for $A^{s,a}(\mu,Q)$, using eqs. (9) and their generalizations. The resulting QSA will not be exchange symmetric in general. Expanding $A^{0,1}(\mu,Q)$ in terms of $X_k^{(L)}(\mu,Q)$ one finds the coefficients $a_{k,\ell}^{0,1}$. Ensuring that the integrability conditions (13) are satisfied (otherwise the given effective potential $F^{s,a}(\mu,Q)$ is not compatible with exchange symmetry) the symmetrized QSA is obtained from eqs. (11) and (12). If the added higher ℓ components are sizeable, the effective potential calculated from the symmetrized QSA by solving (7) in the reverse direction may deviate from the input values. In this case the procedure should be reiterated including the higher ℓ components until convergence is achieved.

EFFECTIVE INTERACTION

In this section we discuss the relation of the effective interaction appearing in eq. (7) with the single particle spectrum, in particular in regions further away from the Fermi surface, and with the density response function. Experimental information on the single particle spectrum can be inferred from the temperature dependence of the specific heat and other thermodynamic quantities. The density response function is measured directly in neutron scattering experiments. As shown in ref. 1, the integral equation (7) for the QSA follows from the quantum kinetic equation

$$\delta f_{\vec{p}}(q,\omega) = \frac{f^0(\varepsilon_{\vec{p}+\vec{q}/2})-f^0(\varepsilon_{\vec{p}-\vec{q}/2})}{\omega-\varepsilon_{\vec{p}+\vec{q}/2}+\varepsilon_{\vec{p}-\vec{q}/2}} \left[\delta\varepsilon_{\vec{p}}(q,\omega)+U^{ext}_{\vec{p}}(q,\omega)\right], \qquad (16)$$

where the change in quasiparticle energy is given by

$$\delta\varepsilon_{\vec{p}}(q,\omega) = \sum_{\vec{p}'} f_{\vec{p}\,\vec{p}'}(q)\ \delta f_{\vec{p}'}(q,\omega).$$

Here the quasiparticle interaction function $f_{\vec{p}\vec{p}'}(q)$ is assumed to be independent of ω. In fact the scattering matrix is given by the energy change of the quasiparticle $\vec{p}$ induced by adding a quasiparticle in state $\vec{p}'$, for $\omega \to 0$, or $t_{\vec{p}\vec{p}'}(q) = \delta\varepsilon_{\vec{p}}(q,0)/\delta f_{\vec{p}'}(q,0)^1$.

Keeping only density and current density molecular fields, one may define interaction functions $f^S_0(q)$ and $f^S_1(q)$ by

$$\delta\varepsilon_{\vec{p}}(q,\omega) = f^S_0(q)\delta n(q,\omega) + f^S_1(q)\vec{p}\cdot\vec{J}(q,\omega), \qquad (17)$$

where $\delta n(q,\omega) = \sum_{\vec{p},\sigma}\delta f_{\vec{p}\sigma}(q,\omega)$ and $\vec{J}(q,\omega) = \sum_{\vec{p}\sigma}\vec{p}\delta f_{\vec{p}\sigma}(q,\omega)$ are the changes in particle density and momentum density, respectively. Analogous relations may be introduced for the spinantisymmetrie quantities.

From eqs. (16) and (17) the density response function $\chi(q,\omega) = \delta n(q,\omega)/U^{ext}(q,\omega)$ may be calculated as

$$\chi(q,\omega) = \frac{\chi_0}{1-f_0\chi_0-f_1\dfrac{\chi_1^2/\chi_0}{1+f_1(\chi_1^2/\chi_0-\chi_2)}}, \qquad (18)$$

where

$$\chi_n(q,\omega) = \sum_{\vec{p}}\frac{f^0(\varepsilon_{\vec{p}+})-f^0(\varepsilon_{\vec{p}-})}{\omega-\varepsilon_{\vec{p}+}+\varepsilon_{\vec{p}-}}(\vec{p}\cdot\hat{q})^n, \qquad (19)$$

A result similar to eq. (18) has been given in ref. 16. In the high frequency limit, $\omega/qv_F \gg 1$, eq. (18) takes the form used by Aldrich and Pines[9],

$$\chi(q,\omega) = \frac{\chi_{sc}(q,\omega)}{1-\left[f^S_q + (\omega^2/q^2)f^V_q\right]\chi_{sc}(q,\omega)}, \qquad (20)$$

156

if one identifies χ_{sc} with χ_0, f_q^S with $f_0^S(q)$ and

$$f_q^V = \frac{m^2\gamma_0^2(q)f_1^S(q)}{1+ mn\gamma_0(q)f_1^S(q)} \quad , \tag{21}$$

where

$$\gamma_0^{-1}(q) = \frac{m}{n}\,\frac{1}{q^2}\,2\sum_p\, f^O(\varepsilon_p)[\varepsilon_{\vec{p}+\vec{q}}-2\varepsilon_{\vec{p}}+\varepsilon_{\vec{p}-\vec{q}}] \quad . \tag{22}$$

For quadratic spectrum, $\varepsilon_p = p^2/2m^*$, γ_0 is a constant, $\gamma_0 = m^*/m$. The polarizationpotential f_q^V is seen to be related to the screened current density potential, rather than the unscreened potential, as assumed by Bedell and Pines[8].

Imposing the f-sum rule on the density response function (18), equivalent to $\lim_{\omega\to\infty} \chi(q,\omega) = (n/m)(q^2/\omega^2)$, we find the following relation between the effective interaction function $f_1^S(q)$ and the single particle spectrum

$$f_1^S(q) = \frac{1}{nm}\left[1 - \gamma_0^{-1}(q)\right] \quad , \tag{23}$$

which reduces to the usual effective mass relation in the limit $q \to 0$.

For the strong variation of the single particle effective mass m_p^* with p found in recent theories[10-12], implying a fall off of m^* to the bare mass m at $p \cong$ $_F$, the function $\gamma_0^{-1}(q)$ is found to increase drastically from $\frac{m}{m_0^*}$ at q=0 to a value ~ 1 at $q \cong 2p_F$

This implies via eq. (23) a rapid decrease of $f_1^S(q)$ to values much smaller than $f_1^S(0)$ at $q = 2p_F$. The corresponding decrease in f_q^V is not quite as rapid due to the approximate cancellation of part of the q-dependence by the screening denominator as long as $[mn\gamma_0(q)f_1^S(q)] > 1$. Also the factor $\chi_0(q,\omega)$ multiplying f_q^V in eq. (20) is increasing roughly proportional to $\gamma_0^{-1}(q)$.

We may now relate the effective interaction $F^{S,a}(\mu,Q)$ with the polarization potentials of Aldrich and Pines[9]. From eqs. (17) and (20) one has

$$F_0^{S,a}(\mu,Q) = N(\varepsilon_F)\, f_q^{S,a} \quad . \tag{24}$$

From eqs. (17), (20), and (23), as well as using the fact that $p = p' = p_F(1-Q^2/4)^{1/2}$ for quasiparticles on the Fermi surface, one may conclude

$$F_1^S(Q) = 3\frac{m^*}{m}(1-Q^2/4)(1-\gamma_0^{-1}(Q)) \tag{25}$$

There is no analogous relation for the spin antisymmetric function $F_1^a(Q)$. Eq. (25) expresses $F_1^s(Q)$ in terms of the single particle energies ε_p.

The analysis presented in this section is based on the assumption that the quasiparticle picture holds for energies and momenta far outside the Landau regime. This assumption appears to be justified to some extent by recent results[12]. In fact the imaginary part of the single particle selfenergy is found to increase with energy less rapidly than anticipated previously. Nevertheless, there will be a contribution from multiparticle-hole excitations in eq. (16), which we have neglected for the present. Such a contribution may be expected to smoothen the rapid q-dependence of F_1^s as given by (25). The qualitative effect of the single particle spectrum on the interaction function, however, should remain.

MODEL CALCULATION OF SCATTERING AMPLITUDE

We have carried out a preliminary calculation of the QSA along the lines described in the last two sections and in ref. 13. To be definite, we have used a model form for the single particle effective mass similar to that given in ref. 10,

$$\frac{m_p^*}{m} = 1 + \frac{1}{3}F_1^s \frac{1-(2p/q_o)^2}{\left[1+(2p/q_o)^2\right]^2} \quad , \tag{26}$$

with $q_o = 3.5\ p_F$. At $p = 2p_F$, the effective mass ratio is down to a value <1.

Instead of evaluating the functions $\gamma_o(q)$ in eq. (22) and $\chi(\mu,Q)$ in eq. (8) by performing the full p-integration, we have for the present approximated these integrals rather crudely by

$$\gamma_o(q) \cong m_q^*/m$$

and

$$\chi(\mu,Q) \cong \frac{m_{q/2}^*}{m_o^*} \ \alpha^{s,a}(Q) \ \left[1-(1-\mu^2)Q^2/4\right]^{1/2} \tag{27}$$

Eqs. (27) are equivalent to replacing the single particle spectrum by $p^2/(2m_q^*)$ and $p^2/(2m_{q/2}^*)$ in the case of $\gamma_o(q)$ and $\chi(\mu,Q)$, respectively. This is motivated by the observation that the average p-value in the two integrations is near $p\cong q$ and $p\cong q/2$, respectively. A more accurate evaluation will be presented in a forthcoming paper.

The spinsymmetric $\ell=0$ component of F, $F_0^s(Q)$, has been identified with the density polarization potential of ref. 9 as described in detail in our paper[13].

The spinantisymmetric functions $A_0^a(Q)$ and $A_1^a(Q)$ and hence, by eq. (9), $F_0^a(Q)$ and $F_1^a(Q)$ are determined by imposing the q = 0 value of $F^a(Q)$ and exploiting the four integrability conditions (13) for $A_0^a(Q)$ and $A_0^a(Q)$, taking into account that $A_{0,1}^s(Q)$ is already known via eq. (9). The five conditions suffice to determine the coefficients of the five lowest eigenfunctions $X_k^{(1)}(\mu,Q)$ in the expansion (10). For the spinsymmetric function $A^s(\mu,Q)$ we have determined the coefficients of the 20 lowest eigenfunctions. For this calculation we have put the quasiparticle weight factors $\alpha^{s,a}(Q) = 1$.

The resulting components $A_{0,1}^{s,a}(Q)$ of the symmetrized QSA resemble the general picture shown in fig. 2. The deviations are of the same kind as those of the result of ref. 13, shown in fig. 5 of ref. 13, and could be a consequence of the crude approximations of $\chi(\mu,Q)$ and $\gamma_0(q)$.

We have evaluated the transport parameters viscous relaxation time τ_η, spin diffusion time τ_D, thermal relaxation time τ_κ and quasiparticle lifetime τ_N^0 as defined, e.g., in ref. 13, as well as the Landau parameters F_1^a, $F_2^{s,a}$ and the pair interaction parameters G_ℓ (see ref. 13). The results for two sets of m*/m data are collected in the table.

TABLE IA

P[bar]	$\tau_\eta T^2$	$\tau_D T^2$	$\tau_\kappa T^2$	$\tau_N^0 T^2$	F_1^a	F_2^s	F_2^a	G_0	G_1	G_2
0 exp.	1.72	0.43	0.51			~0				
0 th.	1.22	0.53	0.52	0.59	-0.86	0.05	-0.05	0.86	-0.21	-0.05
21 exp.	0.88	0.19	0.29	0.30		~1				
21 th.	0.75	0.25	0.25	0.27	-1.01	0.82	0.51	0.72	-0.30	0.04
34 exp.	0.72	0.19	0.22	~0.25		~1				
34 th.	0.60	0.21	0.20	0.23	-0.89	0.20	1.45	0.35	-0.39	0.06

TABLE IB

P[bar]	$\tau_\eta T^2$	$\tau_D T^2$	$\tau_\kappa T^2$	$\tau_N^0 T^2$	F_1^a	F_2^s	F_2^a	G_0	G_1	G_2
0 exp.	1.58	0.40	0.47		-0.55	~-0.4				
0 th.	1.07	0.46	0.47	0.49	-1.09	2.74	-0.74	1.47	-0.10	0.06
21 exp.	0.80	0.18	0.26	~0.3	-0.99	~0.5				
21 th.	0.70	0.23	0.23	0.23	-1.26	5.61	-0.49	1.46	-0.17	0.08
34 exp.	0.66	0.18	0.22	~0.25	-0.99	~0.5				
34 th.	0.66	0.20	0.21	0.20	-1.18	3.65	0.10	1.09	-0.25	0.09

Relaxation times and interaction parameters from experiment and theory based on m*/m values of Ref. 14 (A) and Ref. 15 (B). The experimental values are taken from Ref. 17(τ_η), Ref. 18(τ_D), Ref. 14(τ_κ), Ref. 19(τ_N), Ref. 15,20 (F_1^a) and Ref. 21 (F_2^s).

The agreement with experiment is reasonable but not completely satisfactory. In particular the calculated τ_η at low pressure is too small and F_2^s is consistently large for Greywall's data. As has been discussed in ref. 13, large values of $F_2^{s,a}$ may result from the rather drastic truncation of the eigenfunction expansion of $A_{0,1}^a(Q)$ in the present model. In general, quantities calculated by averaging over μ and Q can be expected to be more accurately given than those involving averages along a line in the (μ,Q) plane, like F_ℓ and G_ℓ. There is clearly more to be done in terms of completing the input information and refining the calculation.

CONCLUSION

In this paper we have tried to determine the QSA for normal Fermi liquids in two ways. In the first part we have discussed the general and complete expansion of the QSA in terms of eigenfunctions of the exchange operator $X_{m\pm}(\mu,Q)$, which, in particular, have the correct behavior near the degenerate points Q=2 and θ=0. The result of fitting an eigenfunction expansion of the QSA with the six lowest eigenfunctions to experiment is shown in fig. 2. It should provide an accurate representation of the low ℓ components ($\ell\leq2$) of the QSA. With more accurate data available on the Landau parameters F_1^a, $F_2^{s,a}$ as well as further transport data in the superfluid phase, one might be able to determine also the four ℓ=3 eigenfunction coefficients.

In the second part of the paper we have addressed the question of calculating the QSA from a semimicroscopic theory. There are some indications[12] that the quasiparticle picture might be applicable well beyond the Landau regime. Under these premises one may take a straightforward extension of Landau's integral equation for the QSA to momentum transfers q up to $2p_F$ as a starting point. The input quantities of this equation are the irreducible kernel, or effective interaction $F(\mu,Q)$, and the particle-hole propagator. We discuss in section 4 how the lowest spinsymmetric angular momentum components of F may be related to the polarization potential[9] (ℓ=0) and the single particle spectrum (ℓ=1). There is no hope to get information on the higher ℓ components of the effective interaction for the foreseeable future, except possibly on the Landau parameters for ℓ=2. On the other hand, exchange symmetry of the QSA severely restricts the variation of $F(\mu,Q)$. We have attempted to reconstruct the μ-dependence of $F(\mu,Q)$ out of the Q-dependence for the lowest ℓ-components $F_{0,1}(Q)$, by imposing exact exchange symmetry on the QSA. This program has been partially successful, as may be seen from the table. However, as could have been expected, the reconstruction depends rather sensitively on the input functions, $F_{0,1}^s(Q)$, and, in particular, the particle-hole propagator. It is hoped that more accurate input information on $F_0^s(Q)$ and ε_p will eventually lead to good agreement with available data and will allow to predict quantities involving components up to ℓ=3.

REFERENCES

1. For a review see: G. Baym and C.J. Pethick, in "The Physics of Liquid and Solid Helium", Part II, K.H. Bennemann and J.B. Ketterson, eds. (Willy, New York, 1978).
2. D. Einzel and P. Wölfle, J.Low Temp.Phys. 32, 19 (1978); P. Wölfle and D. Einzel, J.Low Temp.Phys. 32, 39 (1978); D. Einzel, to be published.
3. D. Rainer and J.W. Serene, Phys.Rev. B13, 4745(1976); J.W. Serene and D. Rainer, Phys.Rev. B17; J.Low.Temp.Phys. 34, 589 (1979); Phys.Rep., to be published.
4. A.A. Abrikosov and I.M. Khalatnikov, Rep.Progr.Phys. 22, 329 (1959).

5. K. Dy and C.J. Pethick, Phys.Rev. 185, 373 (1969).

6. J. Sauls and J.W. Serene, Phys.Rev. B24, 183 (1981).

7. K. Levin and O. Valls, Phys.Rev. B20, 105, 120 (1979).

8. K. Bedell and D. Pines, Phys.Rev.Lett. 45, 39 (1980).

9. C.H. Aldrich III and D. Pines, J.Low Temp. Phys. 32, 689 (1978)

10. G.E. Brown, C.J. Pethick, and Ali Zaringhalam, J.Low Temp. Phys. 48, 349 (1982).

11. S. Fantoni, V.R. Pandharipande, and K.E. Schmidt, Phys.Rev. Lett. 48, 878 (1982).

12. B.L. Friman and E. Krotscheck, Phys.Rev.Lett. 49, 1705 (1982).

13. M. Pfitzner and P. Wölfle, J.Low Temp.Phys. 51, (1983), 535.

14. J.C. Wheatley, Rev. Mod. Phys. 47, 415 (1975).

15. D.S. Greywall, Phys.Rev. B 27, 2747 (1983).

16. S. Babu and G.E. Brown, Ann.Phys. (N.Y.) 78, 1 (1973).

17. J.M. Parpia, D.J. Sandiford, J.E. Berthold, and J.D. Reppy, Phys.Rev.Lett. 40, 565 (1979).

18. D.F. Brewer, D.S. Betts, A. Sachrajda, and W.S. Truscott, Physica 108B, 1059 (1981).

19. J.C. Wheatley, in "Progress in Low Temperature Physics", Vol. 7, D.F. Brewer, ed. (North-Holland, Amsterdam, 1978).

20. L.R. Corrucini, D.D. Osheroff, D.M. Lee, and R.C. Richardson, J.Low Temp.Phys. 8, 229 (1972).

21. K. Nara, I. Fujii, K. Kaneko, and A. Ikushima, Physica 108B, 1203 (1981).

SPECIFIC HEAT OF ^{3}He IN THE FERMI LIQUID REGION

M. C. Mayberry, William E. Fogle[*] and Norman E. Phillips
Department of Chemistry, and Materials and Molecular Research
Division, Lawrence Berkeley Laboratory, University of California,
Berkeley, California 94720

ABSTRACT

A CMN thermometer has been calibrated by nuclear orientation
thermometry at low temperatures and by He vapor pressure thermometry
at high temperatures. The calibration agrees well with the NBS
temperature scale between 100 and 200 mK. Specific heat data on ^{3}He
in the Fermi liquid region obtained with this thermometer are in good
agreement with recent measurements at Bell Laboratories. It is
argued that discrepancies with other data arise from differences in
the underlying temperature scales.

INTRODUCTION

The effective mass for ^{3}He that can be derived from the specific
heat in the normal or Fermi liquid region is of fundamental importance to an understanding of the superfluid states as well as the
normal state. Its value, however, is not well established because
the existing specific heat data vary by up to 40% in magnitude (see
Table I for examples) and also exhibit significantly different temperature dependences. The pioneering investigation of the Fermi
liquid properties by Wheatley and colleagues produced extensive
specific heat data[1] (hereinafter referred to as the La Jolla data
although the measurements were actually made at several different
laboratories) which were generally accepted as fairly close to correct. More recent measurements at Helsinki[2] gave substantially
lower values for the specific heat and a different temperature dependence. Both sets of data have been approximately verified by
independent measurements at other laboratories: in the case of the
La Jolla data by Hébral et al.[3] (Grenoble) in measurements near
the melting pressure; in the case of the Helsinki data by Zeise et
al.[4] (Cornell-I) in measurements at pressures throughout the liquid
range. Furthermore, Greywall[5] (Bell Laboratories) has reported intermediate values of the specific heat that agree with data obtained
by Halperin et al.[6] (Cornell-II) near the melting pressure.

The disparities in the specific heat data have led to speculat-
ion that surface contributions to the specific heat, which might be
different in the different experimental arrangements, are significant. On the other hand, the values for the ratio of the heat
capacity just below the superfluid transition to that just above the
transition and for the ratios of the heat capacities at different
pressures are much more consistent with one another (see Table II).
This suggests certain types of systematic errors, the most probable

[*]Present address: National Bureau of Standards,
Washington D.C. 20234

TABLE I. Heat capacity of liquid ^{3}He at 10 mK, 0 and 28 atm pressure, as reported from five laboratories.

Laboratory	Zero Pressure		28 atm	
	$\frac{C}{RT}$ (K^{-1})	Deviation from mean	$\frac{C}{RT}$ (K^{-1})	Deviation from mean
La Jolla[a]	2.94	+18%	4.41	+11%
Grenoble[b]	----	----	4.70	+18%
Helsinki[c]	2.11	−16%	3.33	−16%
Cornell-I[d]	2.20	−12%	3.30	−17%
Bell Laboratories[e]	2.74	+10%	4.11	+ 3%
Cornell-II[f]	----	----	4.00	+ 1%

[a]Reference 1 [d]Reference 4
[b]Reference 3 [e]Reference 5
[c]Reference 2 [f]Reference 6

TABLE II. Heat capacity ratios for liquid ^{3}He, as reported from four laboratories. C_A and C_N are the specific heats in the superfluid A and normal phases, respectively, and T_A is the equilibrium temperature.

Laboratory	$\left|\frac{C(28\ atm)}{C(0\ atm)}\right|$ 10 mk		$\left|\frac{C_A}{C_N}\right|_{T_A}$, 33 atm	
	Value of ratio	Deviation from mean	Value of ratio	Deviation from mean
La Jolla[a]	1.50	−1.7%	2.85	−4.0%
Cornell-II[b]	----	----	3.00	+1.0%
Helsinki[c]	1.58	+3.5%	3.06	+3.0%
Bell Laboratories[d]	1.50	−1.7%	----	----

[a]References 1 and 7 [c]Reference 2
[b]Reference 6 [d]Reference 5

of which is temperature scale error. Furthermore, while there is no obvious correlation between the reported specific heat and the surface area of CMN or sinter, there is a correlation between specific heat and temperature scale: the Grenoble and La Jolla measurements were based on a Curie-Weiss relation,

$$\chi = \frac{C}{T - \Delta} \, , \qquad\qquad (1)$$

for a CMN thermometer with $\Delta = 0$; the Cornell-I data were based on
a temperature scale derived from the Helsinki phase diagram; the
Bell Laboratories temperature scale is anchored to the same thermodynamic temperature measurements that are the basis of the Cornell
-II data. These considerations suggest the importance of additional
measurements on temperature scales that can be independently related
to the thermodynamic scale. New measurements of this type are reported here and, in addition, the earlier data are re-examined from
the point of view of the possibility that the discrepancies arise
primarily from temperature scale differences.

MEASUREMENTS

With one exception (Cornell-II) the temperature scales used in
the specific heat measurements described above depend on the use of
Eq. (1) for the susceptibility of a cylinder, with diameter equal to
height, of powdered CMN or LCMN (La-diluted CMN). These scales are
generally tied to the He vapor pressure scales at high temperatures,
but an independent low-temperature determination of T is required to
obtain the value of Δ, which is assumed to be a constant and is of
the order of 0 ± 1 mK. The scale used in our measurements was of
this type with CMN as the working substance and Δ determined by
$Co^{60}Co$ NO (nuclear orientation) thermometry. Independent NO temperatures were derived from both "on-axis" and "off-axis" γ-ray intensities. Fifteen comparisons at about 17 mK gave an average
difference of 0.11% with a standard deviation of 0.77%. Temperatures
obtained with a $Co^{60}Co$ NO thermometer have been tested, to comparable accuracy, by comparison with a noise thermometer at NBS.[8] The
agreement between the on-axis and off-axis temperatures indicates
the absence of certain possible experimental errors that would arise
from misalignment, closure domains etc. A total of five different
CMN thermometers using either grease or ^{3}He to provide thermal contact and either ac or dc measuring techniques were investigated. In
all but one of these the CMN was from a sample that had been used
for thermodynamic measurements that gave $\Delta = 0.75$ mK for a spherical
single crystal,[9] and the values of Δ for the powdered samples were
0.95 mK (to within a variation of about 0.1 mK apparently associated
with effects of different superconducting coils and flux-trapping or
shielding tubes). One of these thermometers was used in the specific
heat measurements. (For the one CMN thermometer that used CMN from
a different source, apparently with excess water of hydration,
$\Delta \approx 0$ mK). The temperature scale has been compared[10] with an NBS
SRM 768 fixed point device[11]. At the Ir, $AuAl_2$, and $AuIn_2$ points
($\sim$ 99-200 mK) discrepancies with the NBS scale are 0.35% or less,
and at the Be point ($\sim$ 23 mK) 1.3%, which is reasonable given the
1.7% transition width. At the W point ($\sim$ 16 mK) the scale is 1.1
mK higher than suggested by the original NBS calibration and 0.6 mK

higher than the revised value. However, our value of the W point (16.26 mK) is in better agreement with the Helsinki values[12] (15.86, 16.29 mK) determined by Pt NMR thermometry, and with the Bell Laboratories value[13] (15.97 mK) determined relative to the Cornell-II thermodynamic scale[6]. If, as now seems to be the case, the W points have the same value for all the fixed point devices, these comparisons are particularly relevant to the specific heat data.

The specific heat data taken to date were obtained in a cell that was designed primarily for thermometer intercomparisons and did not permit measurements at pressures above the vapor pressure. The working thermometer, the CMN thermometer already described, contained approximately 0.1 cm^3 of CMN that passed a 37μ sieve and was immersed in the ^3He sample. Thermal contact to the Cu cell body was made through sintered Cu powder. It was possible to fill the cell to various levels above the sinter (with 5 to 15 cm^3 of liquid ^3He) and thus to determine separately the heat capacity of the cell, with ^3He filling the sinter, and that of liquid above the sinter. As expected for such an arrangement, the thermal equilibrium times were long, but good thermal isolation made it possible to establish linear temperature drifts between successive heat capacity points. There is thus no reason to expect errors from this source greater than those indicated by the precision of the data. The data are also provisional in that they cover a relatively limited temperature interval, but they do permit a choice between the values reported earlier.

Our data extend from 13 to 170 mK and are shown in Figs. 1 and 2 as C/RT where C is the molar heat capacity. Below 30 mK, C/RT is constant, as expected for a simple Fermi liquid, and approximately equal to 2.68 K^{-1}. Figures 1 and 2 also include other experimental data, represented by solid curves, for comparison. Curve (b) in Fig. 1 represents the Bell Laboratories data as reported[5] on a modified temperature scale with Δ chosen to give a constant C/T in the low temperature limit. Above 30 mK the maximum deviation of our data from that curve is 0.5%. Below 30 mK the deviation increases to 2%, but our data are in essentially exact agreement with the Bell Laboratories data recalculated on the original temperature scale (i.e., on the scale on which the W point is 15.97 mK and Δ was chosen to correspond to the Cornell-II thermodynamic scale). Similarly, modifying our scale to give agreement with the NBS W point (rather than the NO thermometer) raises the lowest temperature points from 2% below curve (b) to 2% above. Thus, within limits imposed by uncertainties of a few percent in the temperature scales, our data agree with the Bell Laboratories data.

COMPARISON WITH OTHER DATA

Extrapolation of our data ($C/RT=2.68$ K^{-1}) to 10 mK permits a comparison with the zero-pressure data in Table I. Assuming that all of the measurements are of the same quantity, i.e., the bulk liquid specific heat, the relatively good agreement between our data and the Bell Laboratories data, which are on independently established temperature scales, suggests that these measurements are most likely to be correct. (As noted above, the Helsinki and Cornell-I data are

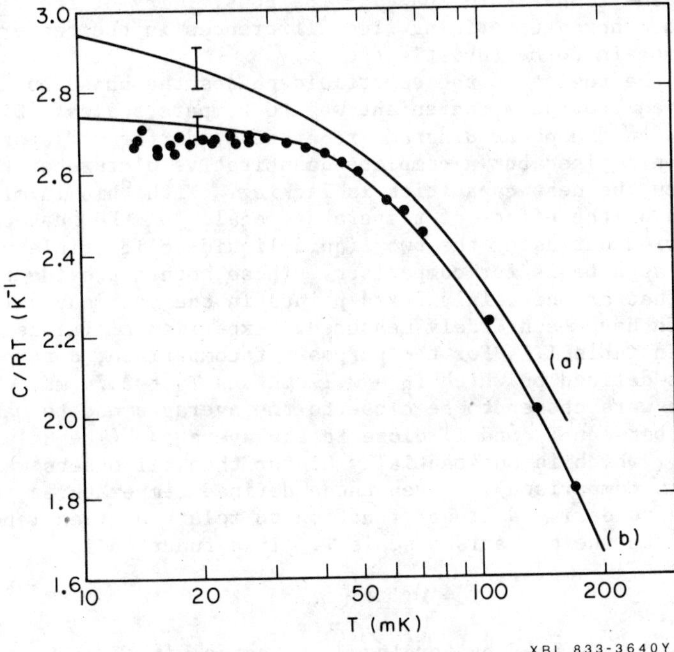

Fig. 1.   C/RT for liquid ^{3}He in the Fermi Liquid region at zero pressure. Solid curves represent reported data from La Jolla (a) and Bell Laboratories (b). Error bars indicate approximate scatter of reported data. Our data points are shown by circles.

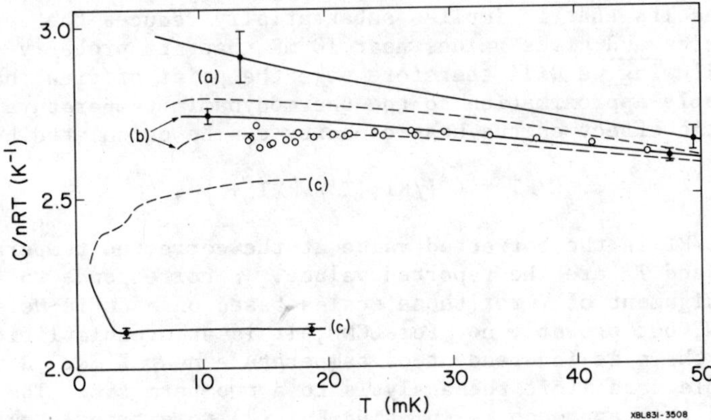

Fig. 2.   C/RT for liquid ^{3}He in the Fermi Liquid region at zero pressure. Solid curves represent reported data from La Jolla (a), Bell Laboratories (b), and Helsinki (c). Error bars indicate approximate scatter of reported data. Dashed curves represent corrected data from the corresponding laboratories. Our data points are shown by circles.

on the same temperature scale). The possibility of the discrepancies with other data arising from differences in the temperature scales remain to be investigated.

In the region of the superfluid phases the phase boundaries themselves provide a convenient way to compare scales. Discrepancies between the phase diagrams reported by different laboratories have been noticed but a complete quantitative picture of the possible effect on the heat capacities is lacking. With this in mind, an analysis of the effect of temperature scales on the heat capacities was carried out using the two liquid-liquid-solid triple points, T_A and T_B, as a basis for comparison. These points provide two data points that are certainly fixed points in the thermodynamic sense and which have been widely measured. Experimental values are collected in Table III. For the purpose of comparison, a temperature scale is defined on which T_B = 2.17 mK and T_A = 2.74 mK. These temperatures were chosen to be close to the averages and to make the difference between T_B and T_A close to the average. (The Helsinki value of T_A-T_B, which is substantially higher than all others was omitted from this comparison). Given these defined temperatures it is possible to derive a linear function to relate a given laboratory scale T' to the comparison scale T. This function is:

$$T = \lambda T' + \tau , \tag{2}$$

with λ and τ defined by forcing T' to agree with T at T_A and T_B. Its form is justified by the prominent role of Eq. (1) in all temperature scales under consideration except that used for the Cornell-II measurements. The comparison scale which is based on somewhat arbitrary choices for T_A and T_B was originally intended only as a basis for comparison between laboratories, but the adjustment of the specific heat results that it implies substantially reduces the reported discrepancies _and_ gives values near 10 mK that are probably correct. In the following we will therefore take the point of view that it is a reasonable approximation to the thermodynamic temperature. Given λ and τ the effect on the heat capacity can be calculated by:

$$C/RT = (C'/RT')T'/\lambda\left(\lambda T'+\tau\right) , \tag{3}$$

where C/RT is the corrected value at the corrected temperature T and C'/RT' and T' are the reported values. τ corresponds to a change in the assignment of Δ for those scales based on a Curie-Weiss law. For CMN, but probably not for LCMN, it is an oversimplification to assume the τ is independent of temperature near 2 mK. However, the available data limit the analysis to a two-term fit. The major weakness of this approach is in using Eq. (2) to extrapolate temperature scale corrections derived between 2 and 3 mK to substantially higher temperatures. In Figs. 2 and 3 the corrected specific heat data are shown by the dashed curves.

The Cornell-II scale is a thermodynamic scale based on measurements of the heat of melting of ^{3}He and gives temperatures relative to T_A which was determined by the criterion that the entropy of the liquid is Rln2 at the minimum of the melting pressure. Using the re-

TABLE III. Experimental values of ^{3}He triple point temperatures.

Laboratory	T_A (mK)	T_B (mK)	T_A-T_B (mK)	Type of scale
La Jolla[a]	2.725	2.155	0.570	LCMN, zero sound
Cornell-II[b]	2.75	2.18	0.57	Latent heat
Helsinki[c]	2.79	2.16	0.63	LCMN, Pt NMR
Orsay[d]	2.75	2.20	0.55	Pt NMR
Florida[d]	2.68	2.10	0.58	Latent heat
Average	2.739	2.159	0.580	

[a]Reference 14
[b]Reference 6
[c]Reference 2
[d]Reference 15

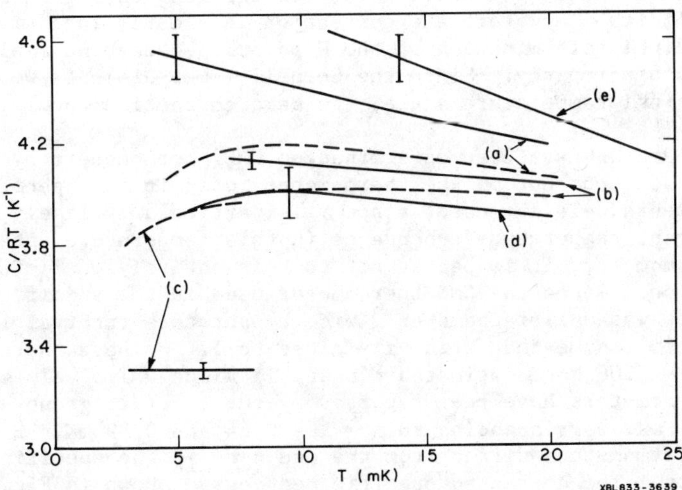

Fig. 3 C/RT for liquid ^{3}He in the Fermi Liquid region at 28 atm. Solid curves represent reported data from La Jolla (a), Bell Laboratories (b), Helsinki (c) Cornell-II (d), and Grenoble (e). The Cornell-I data would be within a few percent of the Helsinki data. Error bars indicate approximate scatter of reported data. Dashed curves represent corrected data from the corresponding laboratories, except that the correction for Cornell-II is too small to be shown.

ported values of T_A and T_B gives $\lambda = 1.00$, and $\tau = -0.01$ mK. The Cornell-II specific heat data were taken along the melting pressure but have been corrected by the authors to 28 atm to give a comparison with the La Jolla data. The pressure dependence of the heat capacity measured by the La Jolla group was assumed to be correct in making the adjustment. The use of the pressure dependence from another laboratory would change the Cornell-II values by less than 1% (see Table II). The correction to C' is less than 1% and is not shown in Fig. 3. This is the one set of data for which there is no justification for the form of Eq. (2), but the derived correction is negligible.

The Helsinki scale is based on a Pt NMR thermometer which was used to calibrate a more sensitive LCMN thermometer. From the data in Table III, λ = 0.905 and τ = +0.21 mK. Although the correction was determined by the reported values of T_A and T_B alone, it expresses correctly the relation of the entire Helsinki phase diagram[2] to that determined recently[14] at La Jolla. The correction to the specific heat, the largest of those considered, is primarily due to the large deviation of λ from one. In essence, while the measured T_A and T_B are within 1% of values determined in some other laboratories, T_A-T_B is 10% higher than the average value. The ripple in the corrected data is possibly due to an error in picking the smooth curve used here to represent the reported data. The correction to the Helsinki scale eliminates the difference between the Helsinki and La Jolla phase diagrams and brings the specific heat data into much more reasonable agreement with other data, strongly suggesting that it has some validity. However, the correction is clearly inconsistent with the Helsinki determination of the W point. We have no explanation for this discrepancy. Since the Cornell-I measurements were based on the Helsinki temperature scale, the same correction applies to that data.

T_A and T_B have been determined on CMN thermometers by the La Jolla group, but not on the thermometers used in the specific heat measurements. In any case, since Δ is particularly likely to have a significant temperature dependence in this temperature interval for a CMN thermometer, it is better not to rely entirely on T_A-T_B for the correction. Since the CMN thermometer used in the specific heat measurements was calibrated over a wide temperature interval it is reasonable to assume that λ is very close to 1. We have therefore assumed λ = 1.00, and estimated τ using T_A alone. Two values of T_A on CMN thermometers have been reported by the La Jolla group -- 2.10 mK and 2.35 mK, corresponding to τ = Δ = 0.64 and 0.39 mK. Both of these thermometers differ from the one used in the specific heat measurements. The corrected specific heat curve shown in Fig. 2 corresponds to Δ = 0.4 mK (taking Δ = 0.5 mK would lower the 10 mK value of C/RT by an additional 1% and give a constant C/RT at lower temperatures). The Grenoble specific heat data were obtained with a similar thermometer, and a similar correction seems appropriate. For this thermometer the value of T_A suggests τ = 0.6 mK.[17] The corrected data have not been shown in Fig. 3, but the correction would be about 50% greater than that shown for the La Jolla data.

The Bell Laboratories temperature scale was based on an LCMN thermometer (10 mole %) below 15 mK and a CMN thermometer above 15 mK. These thermometers were apparently assumed to have the same value of Δ and that value was originally chosen to make an extrapolation of measurements of the melting pressure (above 8 mK) fit the

Cornell-II data at T_A. The value of Δ was later reduced by 0.3 mK
to make the lowest-temperature specific heat data linear in temperature. Since the original scale was in good agreement with the Cornell
-II scale (as shown by comparison of the melting pressures) and
temperatures on the later scale were adjusted downwards by a constant
0.30 mK, comparison with λ and τ for the Cornell-II scale gives
$\lambda = 1.00$ and $\tau = 0.29$ mK.

On the corrected scale the overall spread in specific heat data
at 10 mK is reduced to 10% (from 33%) at zero pressure, and to 6%
(from 28%) at 28 atm. The remaining discrepancies in the specific
heats are of the order of those in the specific heat ratios (Table
II) and therefore reflect other systematic errors in the data. Furthermore, the widely divergent La Jolla and Helsinki data converge
on the Cornell-II and Bell Laboratories data, for which the adjustments were small, and on our data. The latter three scales are based
on both high- and low-temperature calibrations on the thermodynamic
scale while the first two are based on only one or the other. The
Helsinki scale is derived from Pt NMR thermometry between 2 and 7 mK.
The CMN scale used in the La Jolla measurements was based on high
temperature calibrations against the He vapor pressure scales. Our
scale is tied to the He vapor pressure scales at high temperatures
and to NO thermometry at low temperatures. The Cornell-II and Bell
Laboratories scales are complementary: The Cornell-II scale gives the
melting pressure-thermodynamic temperature relation from 1 to 23 mK
with the major uncertainty being a scaling factor which depends on
T_A. The Bell Laboratories CMN scale is based on the He vapor pressure scales at high temperature with Δ chosen to give agreement with
the Cornell-II scale at T_A. The excellent agreement between melting
pressures measured on the two scales provides confirmation of both.
Although these considerations do not prove that the specific heat
discrepancies are entirely due to temperature scale errors, they do
show that obvious discrepancies between the scales in the 2-3 mK
range can account for the specific heat differences when extrapolated
to higher temperatures in a plausible way.

This work was supported by the Director, Office of Basic Energy
Sciences, Material Sciences Division of the U.S. Department of Energy
under contract Number DE-AC03-76SF00098.

REFERENCES

1. W. R. Abel, A.C. Anderson, W. C. Black, and J. C. Wheatley,
    Phys. Rev. 147, 111 (1966); A. C. Mota, R. P. Platzeck, R. Rapp,
    and J. C. Wheatley, Phys. Rev. 177, 266 (1969).
2. T. A. Alvesalo, T. Haavasoja, and M. T. Manninen, J. Low Temp.
    Phys. 45, 373 (1981).
3. B. Hébral, G. Frossati, H. Godfrin, and D. Thoulouze, Phys.
    Lett. A 85, 290 (1981).
4. E. K. Zeise, J. Saunders, A. I. Ahonen, C. N. Archie, and R. C.
    Richardson, Physica 108B, 1213 (1981).
5. Dennis S. Greywall, Phys. Rev. B 27, 2747 (1983).
6. W. P. Halperin, F. B. Rasmussen, C. N. Archie, and R. C.
    Richardson, J. Low Temp. Phys. 31, 617 (1978).

7. R. A. Webb, R. J. Greytak, R. T. Johnson, and J. C. Wheatley, Phys. Rev. Lett. $\underline{30}$, 210 (1973).

8. R. J. Soulen, Jr. and H. Marshak, Cryogenics $\underline{20}$ 408 (1980).

9. R. A. Fisher, E. W. Hornung, G. E. Brodale, and W. F. Giauque, J. Chem. Phys. $\underline{58}$, 5584 (1973).

10. W. E. Fogle, E. W. Hornung, M. C. Mayberry, and Norman E. Phillips, Physica $\underline{109 \text{ \& } 110B}$, 2129 (1982).

11. R. J. Soulen, Jr., NBS Special Publication 260-62 (1979). (USGPO, Washington, D.C. Stock No. 002-002-02047-8).

12. E. Lhota, M. T. Manninen, J. P. Pekola, A. T. Soinne, and R. J. Soulen, Jr., Phys. Rev. Lett. $\underline{47}$, 590 (1981).

13. D. S. Greywall and P. A. Busch, J. Low Temp. Phys. $\underline{46}$, 451 (1982).

14. D. N. Paulson, M. Krusius, J. C. Wheatley, R. S. Safrata, M. Koláč, T. Těhal, K. Svec, and J. Matas, J. Low Temp. Phys. $\underline{34}$, 63 (1979).

15. Taken from a review by R. C. Richardson, Physica $\underline{90B}$, 47 (1977).

16. R. T. Johnson, D. N. Paulson, C. B. Pierce, and J. C. Wheatley, Phys. Rev. Lett. $\underline{30}$, 207 (1973).

17. D. Thoulouze, private communication.

A MICROSCOPIC THEORY OF FULLY SPIN-POLARIZED ^{3}He*

H. R. Glyde
Physics Department
University of Delaware, Newark, Delaware 19711

S. I. Hernadi
Physics Department
University of Ottawa, Ottawa, Ont., Canada K1N 6N5

ABSTRACT

The ground state energy (E), Landau parameters (F) and single particle energy spectrum ($\epsilon(k)$ and m^*) in fully spin polarized liquid ^{3}He (^{3}He$^\uparrow$) are calculated directly from the bare interatomic potential within the Galitskii-Feynmann T-matrix and Hartree-Fock (GFHF) approximations. The E agrees well with variational calculations, the F with model calculations and the $\epsilon(k)$ and m^* with results expected from nuclear matter. This suggests the effective interaction in ^{3}He$^\uparrow$ is dominated by hard core repulsion and Fermi statistics and that these components of the full interaction can be well described from first principles by a GF T-matrix.

INTRODUCTION

Sparked by the pioneering papers of Lhuillier and Laloë[1] and of Castaing and Nozières[2] there has been a renewed interest[3] in spin-aligned or spin-polarized Fermi systems. Examples of spin-polarized Fermi systems are spin-polarized ^{3}He[3,4−8] (^{3}He$^\uparrow$), deuterium[3,8−13] (D$^\uparrow$) and dilute solutions[3,14−16] of ^{3}He$^\uparrow$ immersed in liquid ^{4}He. In ^{3}He$^\uparrow$ the net nuclear spin of $\frac{1}{2}$ in each atom is polarized along a given direction (the 2 electrons on each atom having opposite spins). In D$^\uparrow$ the net electron spin of $\frac{1}{2}$ on each atom is polarized and the net nuclear spin of 1 may or may not be polarized. Spin-aligned Bose systems such as H$^\uparrow$ are also of great current interest (see Ref. 9 for recent references).

In this paper we wish to use fully spin-polarized ^{3}He (which we denote by ^{3}He$^\uparrow$) as a system in which to test a Brueckner-Hartree-Fock (BHF) theory[17−19] of dense Fermi systems. While the BHF theory arguably provides the best general description of nuclear matter[20] its success in more dense Fermi liquids such as ^{3}He has been limited.[21−23] For this test we use, in fact, the Galitskii-Feynman-Hartree-Fock (GFHF) theory[23−26] in which the Galitskii-Feynman (GF) T-matrix replaces the Brueckner T-matrix. To identify the factors needed in a microscopic description of the effective interaction in ^{3}He$^\uparrow$ we turn first to normal liquid ^{3}He.

In normal ^{3}He the effective interaction between atoms has three components: (1) a strong repulsion between atoms at close approach due to the hard, repulsive core of the bare He-He potential (2) statistical correlations (mainly repulsion) between atoms

---

*Supported by NSERC (Canada) and University of Delaware

having like spin due to the Pauli exclusion principle and (3) molecular field like interactions between pairs of atoms induced via the collective dynamical excitations in the liquid (density excitations (zero sound) and spin fluctuations).[8,27-30] Since none of these components is small a first principles description of the total effective interaction has proved very difficult. A GFHF theory takes account of hard core repulsion and statistical correlations only. Its limited success in normal ^{3}He may therefore be due to the neglect of induced interactions or, in addition, to a limited description of the hard core and statistical components.

Fully spin polarized ^{3}He$^\uparrow$ offers a system in which to test this question. Firstly, with all spins aligned, spin fluctuations are 'frozen out' and there can be no spin fluctuation induced interaction. Secondly, with all spins aligned, the Fermi statistical correlations operate between all atoms. This increases statistical correlations and may make the interaction induced via the density excitations (dynamical correlations) relatively less important. With induced interactions reduced, a GFHF theory may provide a reasonably successful first principles description of ^{3}He$^\uparrow$ – a question we now explore.

GFHF THEORY

The ground state energy of ^{3}He$^\uparrow$ in the GFHF approximation is[25]

$$E = \sum_1 \frac{P_1^2}{2m} n_1 + \frac{1}{\Omega} \sum_{1,2} [\Gamma(12,12) - \Gamma(21,12)] n_1 n_2 \tag{1}$$

The sums are over momentum states (spin $\uparrow$ only) from 0 to $P_F^\uparrow = (6\pi^2 N/\Omega)^{1/3}$, $n_1 = \theta(P_F^\uparrow - P_1)$ is the step function and

$$\Gamma(34;12) = U_0(34;12) + \frac{1}{\Omega} \sum_{5,6} U_0(34,56) G_2^{HF}(56, E_{12}) \Gamma(56,12) \tag{2}$$

is the GF T-matrix. Γ describes the interaction between a pair of particles scattering from initial momentum states 1 and 2 to final states 3 and 4 via intermediate states 5 and 6. Γ depends on energy via the Fourier transform of the two particle Green function

$$G_2^{HF}(56, E_{12}) = \left[\frac{(1-n_5)(1-n_6)}{E_{12} - \epsilon_5 - \epsilon_6 + i\eta} - \frac{n_5 n_6}{E_{12} - \epsilon_5 - \epsilon_6 - i\eta} \right] \tag{3}$$

and for the 'on energy shell' values of Γ needed in (1), $E_{12} = \epsilon_1 + \epsilon_2$.

We define $\epsilon(p)$ from the single particle Green function $G_1^{HF}(1,\omega)$ and invoke the 'on energy shell approximation' ($\omega = \epsilon_1$) so that

$$\epsilon(1,\epsilon_1) = \frac{P_1^2}{2m} + \Sigma(1,\epsilon_1) \tag{4}$$

where

$$\Sigma(1,\epsilon_1) = \frac{1}{\Omega} \sum_2 [\Gamma(12,12) - \Gamma(21,12)] n(2) \tag{5}$$

The $\epsilon(1,\epsilon_1) \equiv \epsilon_1$ then depends only on momentum p_1. This approximation is equivalent to Landau's definition,[27-29] $\epsilon_1 \equiv \delta E/\delta n_1$ (since on shell energies are used in E)

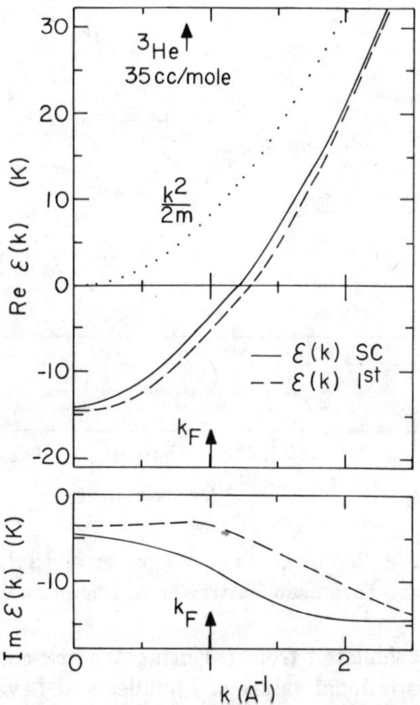

FIG. 1. The single particle energy spectrum $\epsilon(k)$ in ^{3}He↑ at $\Omega = 35$ cc/mole;
——— self consistent, — — — after one iteration
beginning with $\epsilon(k) = k^2/2M$ as input to the T-matrix.

but neglecting rearrangement terms. (In unpolarized ^{3}He we found[23] rearrangement contributed less than 5% to ϵ_1). We retain both real and imaginary parts of Σ so ϵ_1 is complex. Since all spins are up, only triplet interactions exist and only odd angular momentum (L) components of Γ enter,

$$[\Gamma(12,12) - \Gamma(21,12)] = 2a_o = 2\sum_{L\,odd}(2L+1)\Gamma_L.$$

In this scheme, $\epsilon(p)$ is a continuous function [19] of p having no gap at p_F. In previous BHF calculations [21,22] in ^{3}He a gap in $\epsilon(p)$ at p_F was introduced. Here Γ and ϵ were solved iteratively until consistent using the Beck[31] potential as input U_0.

GROUND STATE ENERGY

The self consistent SPE spectrum obtained by iterating Eqs. (2), (3) and (4) until consistent is shown in Fig. 1 for ^{3}He↑ fixed at $\Omega = 35$ cc/mole. Also shown in Fig. 1, as a dashed line, is the first interaction value of $\epsilon(k)$, obtained when simply free particle (kinetic) energies are used as input to Γ. The similarity of the first and final values of $\epsilon(k)$ shows that self consistency is achieved rapidly.

174

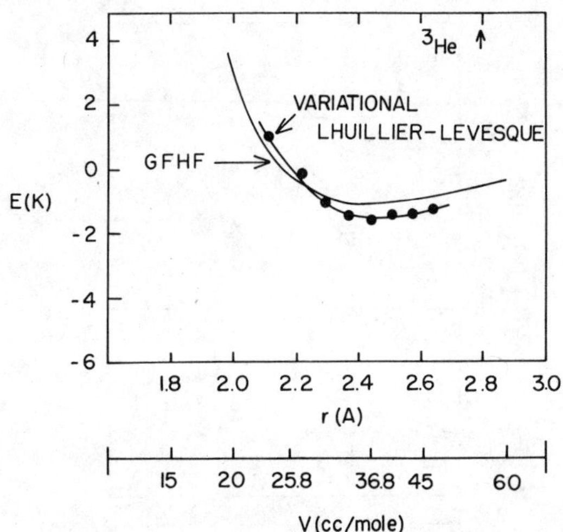

Fig. 2. The ground state energy of ^{3}He†;
GFHF = Galitskii-Feynmann-Hartree-Fock, Lhuillier-Levesque is Ref. 6.

The GSE of ^{3}He† calculated from (8) using the self-consistent $\Gamma^{\dagger\dagger}$ is shown in Fig. 2 along with the variational values of Lhuillier and Levesque[6] (LL). The GFHF minimum is $E_0 = -1.2$K at volume $\Omega_0 = 58.4$Å$^3 = 35.1$ cc/mole compared to the LL result $E_0 = -1.56$K at $\Omega_0 = 65.2$Å$^3 = 37.8$ cc/mole. The combined errors in both calculations are large enough that the saturation values of Ω_0 nearly overlap. Also the GFHF E_0 is quite sensitive to the specific definition of the SPF spectrum $\epsilon(k)$. For example, if we ignore the imaginary part of $\epsilon(k)$ entirely, which is generally the choice in nuclear matter[18, 19] we obtain $E_0 = -1.5$K. However, we believe retaining the Im $\epsilon(k)$ is more faithful to the present GFHF scheme.

The inverse compressibility is

$$\kappa^{-1} = \Omega\left(\frac{\partial^2 E}{\partial \Omega^2}\right) = 55 \quad \text{atm} \tag{6}$$

at saturation, or $(n\kappa)^{-1} = 24$K $(n = \Omega_0^{-1})$, and first sound velocity is $c_1 = (mn\kappa)^{-1/2} = 260$ m/sec. The compressibility κ varies greatly along the GSE curve and is therefore not very precise. The κ^{-1} increases rapidly at smaller values and decreases to effectively zero at larger volumes. The present κ^{-1} at saturation is $\approx 20\%$ greater than that of LL $((n\kappa)^{-1} = 17.4$K) and twice the observed κ^{-1} in normal ^{3}He $((n\kappa)^{-1} = 12.1$K).

Clearly the agreement between the variational and GFHF GSE in ^{3}He† is good. This allows us cautiously to infer that the effective interaction in ^{3}He† is dominated by pair hard core repulsion and Fermi statistic correlations. This is supported by LL, who showed that the addition of three-body correlations in ^{3}He† to the variational wave function did not substantially lower the GSE, and by CBF results[8].

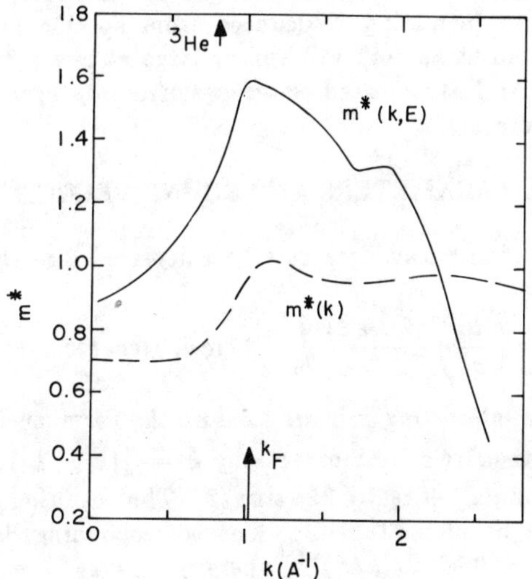

FIG. 3. Effective mass in ^{3}He↑ at saturation, $\Omega = 35$ cc/mole.

EFFECTIVE MASS

The effective mass is obtained from the SPE spectrum as

$$m^*(k,E) \equiv \left[\frac{M}{k}\left(\frac{\partial \epsilon}{\partial k}\right)\right]^{-1} = \frac{1 - \left(\frac{\partial \Sigma}{\partial E}\right)_k}{1 + \left(\frac{\partial \Sigma}{\partial T_k}\right)_E}$$

including both the energy (ω) and momentum (k) ($T_k = k^2/2M$) dependence of the self energy. This effective mass is often[19,20] split up into a product of a 'k mass'

$$m^*(k) = \left[1 + \left(\frac{\partial \Sigma}{\partial T_k}\right)_E\right]^{-1}$$

and an 'E mass'

$$m^*(E) = 1 - \left(\frac{\partial \Sigma}{\partial E}\right)_k$$

where clearly $m^*(k,E) = m^*(k)m^*(E)$. In Fig. 3 we show the 'k mass' and total effective mass at saturation for ^{3}He↑. The $m^*(k)$ is approximately independent of k showing a small increase at $k_F^\downarrow$ while the total $m^*(k,E)$ shows a marked enhancement at k_F. This enhancement is clearly contained in $m^*(E)$ and is due to the energy dependence of the self energy Σ. In normal ^{3}He we obtain a similar enhancement in $m^*(k,E)$ at k_F (and slightly above k_F) which is approximately 50% larger than that shown in Fig. 3.

An enhancement of $m^*(E)$ at k_F has been obtained within the BHF in nuclear matter calculations[19]. It has also been obtained within correlated Basis Function (CBF) theory[32] in normal liquid ^{3}He. An enhancement of m^* in ^{3}He has been proposed by

Brown et al[33] and Fantoni et al[34] as an explanation for the observed sharp decrease in the apparent m^* with increasing T deduced from specific heat measurements. Physically, if m^* is peaked at k_F, m^* will appear large at low T because states near k_F only will be excited. As T is increased states away from k_F become excited and the average apparent m^* decreases.

LANDAU PARAMETERS AND SOUND VELOCITIES

The spin triplet Landau parameters may be calculated directly from the GF T-matrix using

$$F_L^{\uparrow\uparrow} = \left(\frac{dn}{d\epsilon}\right)^{\uparrow} \frac{(2L+1)}{2} \int_0^{\pi} d\theta \, \sin\theta P_L(\cos\theta) 2a_o(\theta) \tag{7}$$

Here the momenta of the interacting pair are fixed on the Fermi surface, $|k_1| = |k_2| = k_F$, and θ is related to their relative momenta by $k = \frac{1}{2}(\vec{k}_1 - \vec{k}_2) = k_F \sin\theta/2$ with Center-of-Mass momentum $\vec{k}_1 + \vec{k}_2 = 2k_F \sin\theta/2$. The resulting $F_L^{\uparrow\uparrow}$ at saturation ($\Omega_0 = 35$ cc/mole) are listed in Table 1. The corresponding density of states is $(dn/d\epsilon)^{\uparrow} = 3/(4\Omega_0 \epsilon_F^{*\uparrow}) = 0.00145 \, (k\text{Å}^3)^{-1}$, where $\epsilon_F^{*\uparrow} = \epsilon_F^{*\uparrow}/m^{*\uparrow} = 8.89$K, which is a factor of 10 smaller than the value in unpolarized ^{3}He $((dn/d\epsilon) = 0.015 \, (k\text{Å}^3)^{-1})$. The $F_L^{\uparrow\uparrow}$ in Table 1 are small both because the even L components of Γ_L are missing and because the normalizing $(dn/d\epsilon)^{\uparrow}$ in (7) is small.

$F_o^{\uparrow\uparrow}$	$F_1^{\uparrow\uparrow}$	$F_2^{\uparrow\uparrow}$	$F_3^{\uparrow\uparrow}$	$F_4^{\uparrow\uparrow}$	$F_5^{\uparrow\uparrow}$	$m^{*\,\uparrow}$
-0.33 ± 0.2	-0.24 ± 0.05	-0.06	0.35	0.15	0.1	0.92
$3.0 \pm 0.5^*$						

TABLE 1.

Landau parameters and $m^* = (1 + F_1^{\uparrow\uparrow}/3)$ calculated directly from the GF T-matrix using Eq. (7). The starred value of $F_o^{\uparrow\uparrow}$ is obtained by fitting the Fermi liquid relation $(n\kappa)^{-1} = (2\epsilon_F^{*\uparrow}/3) \; (1 + F_o^{\uparrow\uparrow})$ to $(n\kappa)^{-1}$ obtained from the GSE.

We may also calculate $F_o^{\uparrow\uparrow}$ from the GSE by fitting the Fermi liquid relation $(n\kappa)^{-1} = (2\epsilon_F^{*\uparrow}/3) \; (1 + F_o^{\uparrow\uparrow})$ to the GS $(n\kappa)^{-1} = 24$K. This gives $F_o^{\uparrow\uparrow} \approx 3.0$ which includes effects of dynamic correlation (rearrangement contributions) to second order. The differences between $F_o^{\uparrow\uparrow}$ calculated from κ^{-1} and from (7) shows that dynamic correlations are still important in ^{3}He$^{\uparrow}$. The $F_o^{\uparrow\uparrow} \approx 3.0$ and $m^{*\uparrow} = 0.92$ agree well with the values $F_o^{\uparrow\uparrow} = 1.82$ and $m^{*\uparrow} = 0.82$ obtained by Bedell and Quadar[5]. This $m^{*\uparrow} = 0.92$ agrees well with the $m^*(k_F)$ in Fig. 3, but the $m^*(k_F, E)^{\uparrow} \approx 1.5$, which is a higher order value, is substantially larger. Unfortunately, all the Landau parameter values here depend on $m^{*\uparrow}$. We have used $m^{*\uparrow} = 0.92$ but it could be larger. (c_1 is independent of m^*).

Using the value of $m^{*\uparrow} = 0.92$ we find a Fermi velocity $v_F^{*\uparrow} = 230$ m/sec which is 4 times that in ^{3}He. With $F_o^{\uparrow\uparrow} \approx 3.0$, we find a ratio of zero sound to Fermi velocity of $s \approx 1.2$ giving $c_o^{\uparrow} \approx 280$ m/sec. The $c_o^{\uparrow} = v_F^{*\uparrow} s$ is dependent on $m^{*\uparrow}$, decreasing to $c_o^{\uparrow} = 240$ m/sec if we use $m^{*\uparrow} = 1.5$ ($v_F^{*\uparrow}$ decreases, s increases).

DISCUSSION

The present and previous[4-6,8] calculations suggest that both the kinetic energy and potential energy are greater in magnitude in ^{3}He$^\uparrow$ than in unpolarized ^{3}He. The increase in KE slightly outweighs the drop in PE to give higher total GSE in ^{3}He$^\uparrow$. The GFHF and variational predictions of the GSE agree well. ^{3}He$^\uparrow$ is predicted to be stiffer than ^{3}He, having both a larger first and zero sound velocity. On the other hand the Landau parameters (e.g. $F_o^{\uparrow\uparrow}$ and $m^{*\uparrow}$) are predicted to be smaller in ^{3}He$^\uparrow$ than in ^{3}He. These results suggest the effective interaction in ^{3}He$^\uparrow$ is dominated by short range (hard core) and Fermi statistics correlations. The molecular field-like induced interactions (via density correlations), which contribute strongly to the Landau parameters and are significant factors in the GSE[35] in normal ^{3}He, appear to be substantially less important in ^{3}He$^\uparrow$. Spin fluctuations are, of course, entirely frozen out in ^{3}He$^\uparrow$.

The reasonable predictions of the present GFHF theory for the GSE and Landau parameters in ^{3}He$^\uparrow$ suggests it can describe the short range and Fermi statistical components of the total effective interaction well. The GFHF theory might therefore form the basis of a successful microscopic theory of normal ^{3}He if combined[36] with a description of the induced components of the interaction.

In this regard, it is interesting to note that many calculations find the GSE of ^{3}He$^\uparrow$ and of normal ^{3}He are nearly equal, the calculated difference being very sensitive to details of the calculation. This suggests that normal ^{3}He is close to being unstable to ferromagnetic phase formation.[4]

---

[1] C. Lhuillier and F. Laloë, J. Phys. (Paris) **40**, 239 (1979).

[2] B. Castaing and P. Nozières, J. Phys. (Paris) **40**, 257 (1979).

[3] J. Phys. (Paris), Colloq. **41**, C7 (1980), *Proceedings of the Colloque International CNRS on Spin-Polarized Quantum Systems*, Aussois, France, 21-26 April 1980.

[4] R. A. Guyer and M. D. Miller, Phys. Rev. **B21**, 3917 (1980), **B18**, 3521 (1978).

[5] K. S. Bedell and K. F. Quader, Phys. Lett. to be published.

[6] C. Lhuillier and D. Levesque, Phys. Rev. **B23**, 2203 (1981).

[7] G. Schumacher, D. Thoulouze, B. Castaing, Y. Chabre, P. Segransan, and J. Joffrin, J. Phys. (Paris) **40**, L143 (1979); H. Godfrin, G. Frossati, A. S. Greenberg, B. Hebral and D. Thoulouze, Phys. Rev. Lett. **44**, 1695 (1980); H. Godfrin, G. Frossati, B. Hebral and D. Thoulouze, J. Phys. (Paris) **41**, c7-275 (1981).

[8] E. Krotscheck, R. A. Smith, J. W. Clark, and R. M. Panoff, Phys. Rev. B **24**, 6383 (1981).

[9] I. F. Silvera and J. T. M. Walraven, Phys. Rev. Lett. **45**, 1268 (1980); G. H. van Ypren, A. P. M. Matthey, J. T. M. Walraven, and I. F. Silvera, Phys. Rev. Lett. **47**, 800 (1981).

[10] M. D. Miller, L. H. Nosanow, and L. J. Parish, Phys. Rev. B **15**, 214 (1976); M. D. Miller and L. H. Nosanow, Phys. Rev. B **15**, 4376 (1977); C. E. Hecht, Phys. Rev. B **23**, 3547 (1981).

[11] R. D. Etters, J. V. Dugan, Jr., and R. W. Palmer, J. Chem. Phys. **62**, 313 (1975).

[12] T. K. Lim, Phys. Rev. B **25**, 2057 (1982).

[13] R. M. Panoff, J. W. Clark, M. A. Leo, K. E. Schmidt, M. H. Kalos and G. V. Chester, Phys. Rev. Lett. **40**, 1675 (1982).

[14] A. E. Meyerovich, J. Low Temp. Phys. **47**, 271 (1982).

[15] E. P. Bashkin and A. E. Meyerovich, Adv. Phys. **30**, 1 (1981).

[16] D. S. Greywall and M. A. Paalanen, Physica, **109**, 1575 (1982), Phys. Rev. Lett. **46**, 1292 (1981).

[17] K. A. Brueckner, C. A. Levenson and H. M. Mahmand, Phys. Rev. **95**, 217 (1954), K. A. Brueckner, Phys. Rev. **97**, 1853 (1955), **100**, 36 (1955), H. A. Bethe and J. Goldstone, Proc. Roy. Soc. Lond. **A238**, 551 (1957).

[18] B. D. Day, Rev. Mod. Phys. **50**, 495 (1978).

[19] J. P. Jeukenne, A. Lejeune and C. Mahaux, Phys. Reports **25C**, 83 (1976).

[20] C. Mahaux, in *The Many Body Problem, Jastrow Correlations Versus Brueckner Theory*, edited by R. Guardiola and J. Ros (Springer Lecture Notes in Physics, Berlin, 1980).

[21] K. A. Brueckner and J. L. Gammel, Phys. Rev. **109**, 1040 (1958).

[22] E. Østgaard, Phys. Reb. **187**, 391 (1969) and references cited there.

[23] H. R. Glyde and S. I. Hernadi, Phys. Rev. B (in press).

[24] V. M. Galitskii, Sov. Phys. JETP **7**, 104 (1958).

[25] A. L. Fetter and J. D. Walecka, *Quantum Theory of Many-Particle Systems*, (McGraw-Hill, New York, 1971).

[26] R. F. Bishop, H. B. Ghassib and M. R. Strayer, Phys. Rev. **A13**, 1570 (1976).

[27] P. Nozières, *Theory of Interacting Fermi Systems* (Benjamin, New York, 1964).

[28] L. D. Landau, Zh. Eksp. Teor. Fiz. **30**, 1058 (1956) (Sov. Phys. JETP **3**, 920 (1957), Z. Eksp. Teor. Fiz. **32**, 59 (1957), (Sov. Phys. JETP **5**, 101 (1957).

[29] G. Baym and C. Pethick, *The Physics of Liquid and Solid Helium*, Part II edited by K. H. Bennemann and and J. B. Ketterson (Wiley Interscience, New York, 1978).

[30] A. J. Leggett, Rev. Mod. Phys. **47**, 331 (1975).

[31] D. E. Beck, Mol. Phys. **14**, 311 (1968).

[32] E. Krotscheck and R. A. Smith, Phys. Rev. B (in press).

[33] G. E. Brown, C. Pethick and A. Zaringhalam, J. Low Temp. Phys. **48**, 349 (1982).

[34] S. Fantoni, V. R. Pandharipande and K. E. Schmidt, Phys. Rev. Lett. **48**, 878 (1982).

[35] K. E. Schmidt, M. A. Lee, M. H. Kalos and G. V. Chester, Phys. Rev. Lett. **47**, 807 (1981); **46**, 728 (1981); K. E. Schmidt and V. R. Pandharipande, Phys. Rev. **B19**, 2504 (1979).

[36] S. Babu and G. E. Brown, Ann. Phys. **78**, 1 (1973).

NUCLEAR POLARIZATION OF ^{3}HE GAS
AT LOW TEMPERATURES BY OPTICAL PUMPING

M. Leduc, P.J. Nacher, S.B. Crampton (*) and F. Laloë
Laboratoire de Spectroscopie Hertzienne de l'E.N.S. (**)
24, rue Lhomond, 75231 Paris Cedex 05

ABSTRACT

A colour centre laser system has been developed that can produce over 60% nuclear polarization of ^{3}He atoms in the gas phase at room temperature. We discuss in this article the efficiency of the optical pumping technique in ^{3}He as a function of the pumping intensity and other important parameters. At low temperatures, the low rate of the metastability exchange collisions limits the maximum nuclear polarization to relatively small values.

To obtain higher nuclear polarizations at low temperatures, we have developed a polarization transfer technique in which the optical pumping is done in ^{3}He gas at room temperature and the polarization is transported by spin diffusion to a second region of gas at liquid helium temperatures ($1.7 \leqslant T \leqslant 4.2$ K). Loss of polarization at low temperatures is strongly reduced by molecular hydrogen which freezes on the cold surfaces. The efficiency of the polarization transfer was measured and found nearly complete under good conditions. Nuclear polarizations up to 50% were obtained in a gas at T = 4.2 K, with a number density n of order 10^{18} cm^{-3}.

I. INTRODUCTION

The motivation for this work is to be able to study the quantum transport properties of spin polarized ^{3}He

----

(*) Permanent address : Williams College,
                        Williamstown, Mass. 01267
(**) Laboratoire associé au C.N.R.S., L.A. n°18

180

(^{3}He $\uparrow$) at low temperatures [1]. The antisymmetrization
principle when applied to binary collisions between ^{3}He
atoms predicts interesting effects in highly polarized
^{3}He at temperatures near 1 K, such as enhanced viscosity
and heat conduction, the appearance of spin waves, and the
coupling of heat conduction and spin diffusion [2]. In
addition, nuclear polarization of liquid ^{3}He [3] could
permit observation of changes of the ^{3}He phase diagram
with nuclear polarization [1].

Several alternative methods can be used to create
high nuclear polarization in ^{3}He at low temperatures. An
example is the method of fast melting of the solid suggested by Castaing and Nozières [4] and demonstrated by
several groups [1][5][6]. Another possible technique is
based on the coupling between nuclear spins of ^{3}He on the
surface of a solid and electronic spins of this solid
[7][8]. One can also use another method, optical pumping,
which was first applied to ^{3}He by   F.D.
Colegrove, L.D. Schearer and G.K. Walters [9]. The recent
development of colour centre lasers [10] providing the λ =
1.08 μ transition of He makes it possible to obtain large
nuclear polarizations in gaseous ^{3}He [11]. Optical pumping therefore seems an attractive possibility to study
some of the unusual properties in ^{3}He   at low temperatures.

There are in fact other potential applications of
high nuclear polarizations by optical pumping in gaseous
^{3}He :

* polarized targets for the study of nuclear reactions
  [12] [13].
* neutron polarizers [14].
* polarized beams of electrons [15][16][17], ions [18]
  or metastable helium atoms [19].
* neutral polarized beams for fusion reactors [20].
* Zeeman maser [21] or magnetometer [22].
* polarization of dilute solutions of ^{3}He in ^{4}He [23].

In the next sections, we shall present the predictions made for the limits of the optical pumping method
of polarizing ^{3}He nuclei [24]. Experimental results are
then reported, both at room temperature and at 4.2 K and
below.

II. LASER OPTICAL PUMPING OF ^{3}HE :
KINETICS AND EXPERIMENTS AT ROOM TEMPERATURE

Optical pumping is a method by which angular momentum is transferred from photons to matter through absorption of polarized photons by the atoms, followed by re-emission through spontaneous emission. In the case of the ^{3}He atom, whose ground state 1^1S_0 can carry only nuclear magnetism (the ^{3}He nuclear spin is 1/2), it is not possible to optically pump directly from the ground state. However, an indirect optical pumping method is possible [9], where the angular momentum is transferred through the 2^3S_1 metastable state of the helium atom. The idea is to populate a number of 2^3S_1 states by a discharge in ^{3}He gas and to optically pump these 2^3S_1 atoms by the $2^3S_1-2^3P$ line at $\lambda = 1.08\ \mu$. The orientation achieved in the 2^3S_1 state is transferred to the reservoir of 1^1S_0 ground states by $2^3S_1-1^1S_0$ metastability exchange collisions [25].

In these collisions, the nuclear spins of the atoms are unaffected, so that the nuclear orientations of the 2^3S_1 and 1^1S_0 states become strongly coupled. By using discharge lamps as light sources for the optical pumping, it is possible to obtain nuclear polarizations M of the order of 15 to 20% in a ^{3}He gas at pressures of the order of 1 Torr [26]. Higher values of M are predicted and actually observed with a laser.

II.a. **Kinetics of pumping**

Three atomic levels are involved in the pumping process, the ground state 1^1S_0, the metastable state 2^3S_1 (which has two hyperfine sublevels $F = 1/2$ and $F = 3/2$) and the 2^3P excited state (which has five sublevels split by fine and hyperfine interactions having the same order of magnitude). These levels are shown in figure 1. The 2^3S-2^3P line at $1.08\ \mu$ has 9 components, labelled C_1 to C_9, whose frequency difference spreads over 40 Ghz. The laser is supposed to be monochromatic and in coïncidence with one of the 9 above mentioned C components of the line. One can derive rate equations for the evolution of the 6 atomic populations of the magnetic sublevels of the 2^3S_1 state and of the 2 nuclear sublevels of the ground state. One takes into account the effect of different physical processes : the coupling of atoms with the infra-red photons,

182

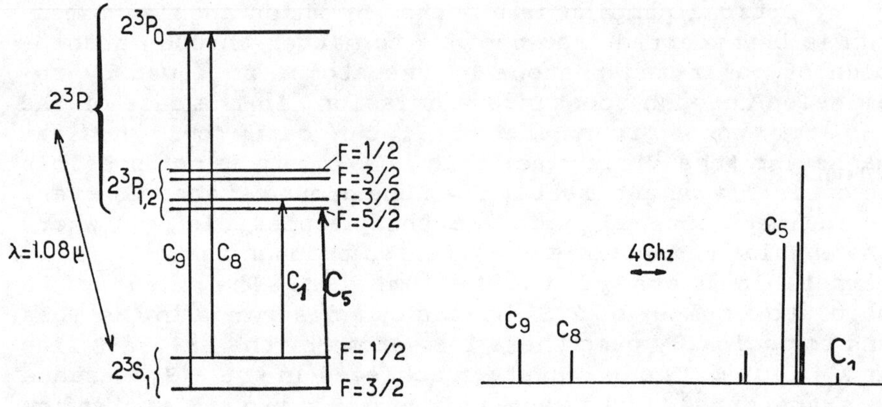

Figure 1 : Level scheme of the 2^3S and 2^3P states of the ^3He atom. Components of the 2^3S - 2^3P line at λ = 1.08 μ.

the metastability exchange collisions and all the other relaxation phenomena affecting the orientation of all three levels. A calculation of the absorption and emission of photons was undertaken in two extreme cases : no disorientation of the 2^3P state by collisions before reemission and total collisional disorientation. The first assumption is valid at very low gas pressure. The second one holds at high pressure and assumes a total thermalization of populations in the 2^3P before reemission by spontaneous emission; in this case, the pumping process is only due to depopulation [27][9]. The experimental conditions correspond to an intermediate situation which is difficult to study in detail. The calculation of the metastability exchange collisions between 2^3S_1 and 1^1S_0 atoms is based on two assumptions [25] : the nuclear spins of both atoms are unaffected and the electronic spin of the 2^3S_1 atom is transferred to the outgoing metastable atom without any change. Other relaxation phenomena are described phenomenologically. A relaxation probability $1/T_r$ is assigned to the ground state atoms, taking into account both the wall collisions and the various relaxing collisions in the plasma.

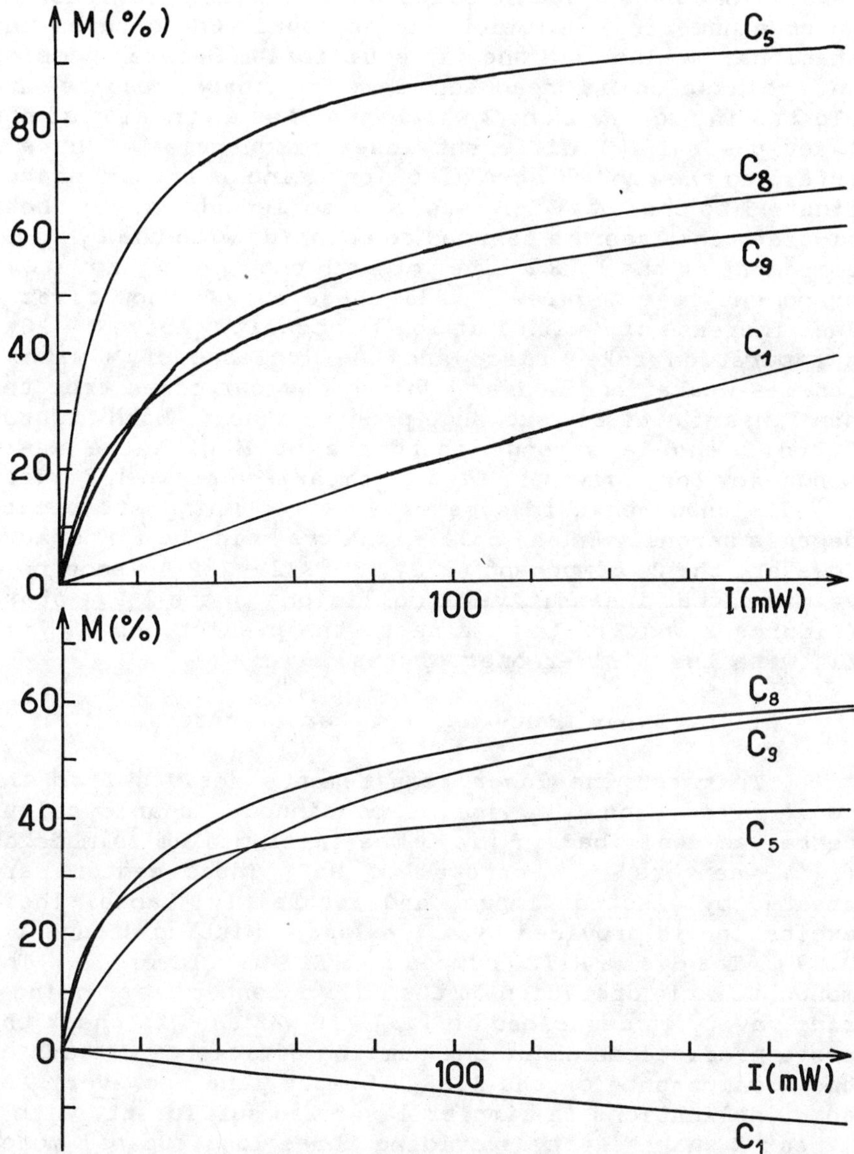

Figures 2 and 3 : Calculated ^{3}He polarization M as a function of the laser intensity I for the laser frequency incoïcidence with several C components of the 1.08 μ line (see fig. 1).
<u>Figure 2</u> : With zero disorientation of the 2^3P state.
<u>Figure 3</u> : With total disorientation of the 2^3P state.

184

When adding the contributions of all these processes, one gets a set of coupled non linear equations for which a numerical solution can be found. One obtains the stationary value of M when an equilibrium between pumping and relaxation is reached. Some of these results are plotted in figures 2 and 3 which give M as a function of the laser power I for different laser frequencies. Figure 2 refers to the case of zero disorientation of the 2^3P state, figure 3 to the total 2^3P state disorientation. In both figures, the laser is assumed to coïncide with the C_1 component of the 1.08 μ line, or with the C_5 or with the C_9 component (see figure 1). All these curves show first a fast increase of M with I at low I intensity. Above M = 20%, a saturation takes place and the increase of M with I becomes weaker and weaker. This behaviour comes from the non linearity of the exchange process itself. Both figures 2 and 3 show a strong dependence of M with the laser frequency for a given I. Also a comparison between figures 2 and 3 shows that in some cases the pumping efficiency depends strongly on the collisional rate in the 2^3P state. However, the C_9 component (2^3S_1, F = 3/2 - 2^3P_0) leads to M values rather insensitive to collisions in the 2^3P state (figures 2 and 3). In this case, the predictions of [24] fit with the earlier ones of [26].

II.b. **Experiments at room temperature**

To match the laser requirements derived from the above predictions, we used a continuous tunable colour center as described in [10]. The laser medium is made of $(F_2^+)^*$ centers in a crystal of NaF. These centers are created by electron impact and are fairly stable. Their excitation is provided by a dye laser emitting around λ = 0.89 μ. The dye is HITC, pumped by a Kr^+ laser. The monochromatic operation of the colour center laser using a ring cavity is described in [28]. It was used to check the above predictions about the pumping efficiency of each of the 9 components of the 1.08 μ helium line. However, for some applications, a simpler laser was sufficient, with a linear X shape cavity providing three longitudinal modes distant of roughly 150 Mhz. This laser beam, circularly polarized, is sent to a ^3He cell of approximatively 100 cc filled at a pressure ranging between 0.1 and 5 Torr. The magnetic field B_0 of about 4 Gauss is parallel to the laser beam. The detection of the ^3He nuclear polarization M is optical : one monitors the circular polarization rate of the λ = 668 nm line emitted by the helium discharge; these signals are calibrated, the accuracy being better than 10% [29].

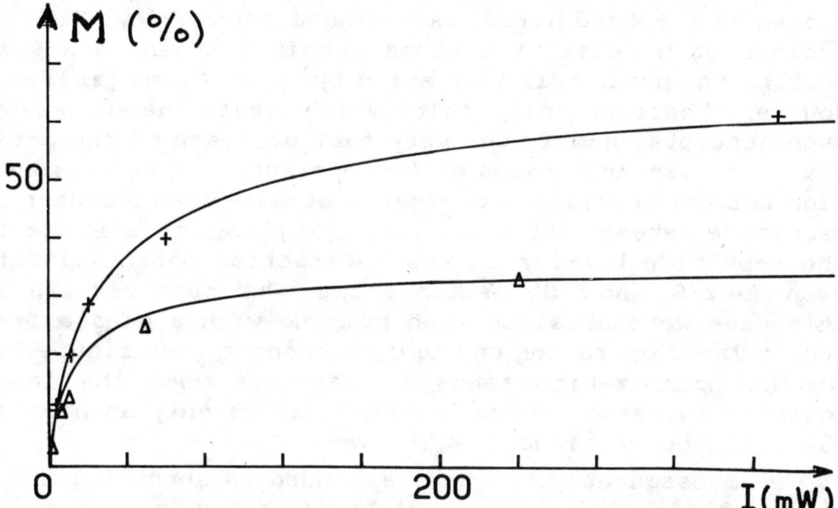

Figure 4 : ^{3}He polarization as a function of light intensity. Crosses are experimental results obtained with a multimode laser, triangles with a single-mode laser. The solid curves are the theoretical predictions in both cases.

Values of M were measured as a function of the laser power I. The highest value of M was 70% for I $\simeq$ 300 mW on the C_8 component (2^3S_1, F = 1/2 $\leftrightarrow$ 2^3P_0 and for a ^{3}He pressure of the order of 0.3 Torr. M slowly decreases when the pressure is raised above 0.3 Torr, as the density of metastable atoms cannot be increased in the same proportion without altering violently the discharge features. In figure 4 is plotted M as a function of I for pumping on the C_9 transition (2^3S_1, F = 3/2 $\rightarrow$ 2^3P_0).
Two sets of results are presented : some are measurements made with the single frequency laser, others with the 3 mode laser. The difference between the 2 cases comes from the number of metastable atoms directly excited by the laser within a Doppler atomic profile (it is larger with a 3 modes than a 1 mode structure). The fit between experimental points and results of the model is good, as seen in figure 4.

III. ^{3}HE ORIENTATION AT LOW TEMPERATURE

Similar experiments were repeated using ^{3}He gas cooled to 4.2 K and below, taking advantage of low ^{3}He nuclear spin relaxation rates obtained around 4.2 K by coating the pyrex cell with solid hydrogen films [30][31]. However, there is a difficulty which limits the success of such attempts, namely the very fast decrease of the metastability exchange rates with temperature. The cross section for this collision decreases by more than 2 orders of magnitude between 300 K and 4.2 K [32][33]; this is due to the repulsive barrier in the interaction potential between the 2^3S_1 and 1^1S_0 helium atoms. Our best results in this case were obtained when pumping with a single frequency laser operating on the C_5 component (see figure 1). The ^{3}He polarization tends to saturate when the laser power is increased above 20 mW. M values only as high as 25% could be obtained in this way.

Subsequently, we have extended to about 50% the M values obtainable in ^{3}He gas at liquid helium temperatures by an indirect technique involving laser optical pumping of room temperature gas and transport of the polarization to a low temperature gas by spin diffusion. The present device consists of two pyrex cells having different sizes and connected by a long thin pyrex tube (figure 5). The larger cell is at room temperature. Its shape, volume and gas pressure are chosen for optimum optical pumping of the ^{3}He gas. Its volume is approximatively 100 cc; a RF discharge populates the metastable states; it is illuminated by a vertical laser beam. The smaller cell is immersed in a liquid helium bath, whose temperature can presently be varied only between 4.2 and 1.7 K. Its volume ranges from 0.3 to 10 cc. The exchange time constants of atoms between top and bottom are important parameters of the double cell experiment. They determine the diameter of the connecting tube (see figure 5), given the 70 cm length. In one experiment, the tube had internal diameter 3 mm and the lower cell had a 10 cc volume : this provided a build-up time constant of $\sim$ 25 minutes for the polarization of the lower cell. For smaller bottom cells, smaller connecting tubes were used. The double cell was filled with ^{3}He at appropriate pressures so as to obtain approximatively 0.5 Torr of ^{3}He in the upper part when the bottom part is at low temperature. Hydrogen was added in order to coat the lower part of the cell interior. Tempe-

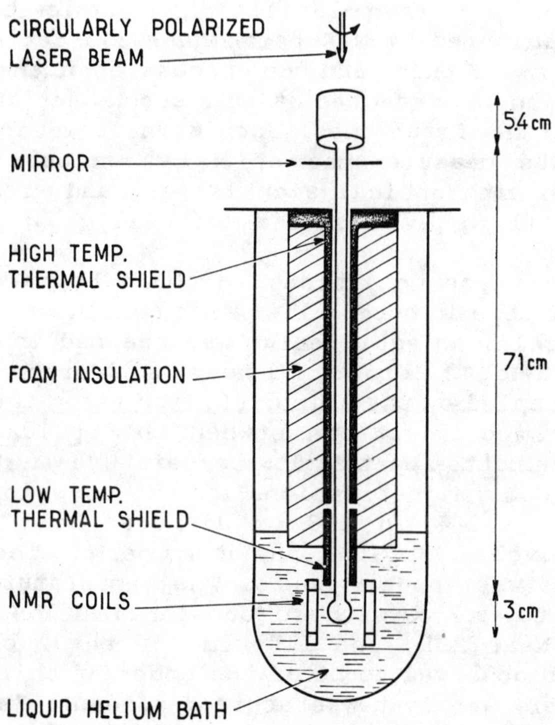

Figure 5 : Sketch of the double cell experiment for polarizing ^{3}He at low temperature. The upper cell at 300 K is submitted to optical pumping. The nuclear polarization is transferred by spin exchange to the low cell immersed in a liquid helium bath.

rature gradients were controlled along the tube so as to localize the temperature transition region where the hydrogen coating stops. Two copper thermal shields were used for this purpose, separated by a distance of the order of 1 cm (figure 5). The entire double cell was placed inside a vertical magnetic field of $\simeq 4$ Gauss provided by a set of three circular coils. Currents in the coils were adjusted so as to reduce the field gradient along the coil axis to a few 10^{-4}. The polarization of the low temperature cell was monitored by NMR techniques : a set of 2 coils provided a $\pi/2$ RF pulse and the precession of the ^{3}He magnetization was detected using a second set of coils orthogonal to the first ones. Such signals were calibrated for absolute measurements of M by various techniques (comparison with optical signals [29] and with an auxiliary RF coil in place of the ^{3}He cell).

Under these conditions, M could be measured as a function of time in both cells. With continuous pumping of the upper cell, an equilibrium was reached in both cells after a delay of the order of 1 hour. Both orientations are strongly coupled by the spin diffusion along the tube. The transfer of polarization between the optically pumped cell and the bottom part at low temperature was found to be close to 100%. Magnetizations of the order of 50% were recorded at 4.2 K in a 3 cm diameter cell containing approximatively 1.5×10^{18} ^{3}He atoms per cc. The T_1 relaxation time was about 6 hours. The temperature could be decreased to 1.7 K without substantial loss of polarization. Near 2 K, the lifetime of the nuclear polarization is observed to be of the order of thirty minutes and only very weakly temperature dependent. Further cooling of the atoms below 1.7 K should be thus possible with an adequate cryostat. It is thus expected that the study of quantum effects in the transport properties of ^{3}He should be possible in the 1.2 – 2 K temperature range, where they are strongest. Furthermore, polarizing ^{3}He below 1 K should be possible by using superfluid ^{4}He coatings for preventing too fast relaxation of ^{3}He on the cell walls [34].

ACKNOWLEDGMENTS

The authors want to thank L.F. Mollenauer (Bell Laboratories) for his tremendous help at the early stage of this experiment, as well as H. Eisele and H. Paus

(Stuttgart University) for their help in preparing the colour centre laser crystals.

REFERENCES

[1]   J. Physique Coll., C-7 (1980); "Spin polarized quantum systems" and references contained.

[2]   C. Lhuillier et F. Laloë, J. Physique, 43, 197 and 225 (1982).
       C. Lhuillier, J. Physique, 44, 1 (1983).

[3]   H.H. Mc Adams and G.K. Walters, Phys. Rev. Lett. 18, 436 (1967).
       H.H. Mc Adams, Phys. Rev. 170, 276 (1968).

[4]   B. Castaing and P. Nozières, J. Physique, 40, 257 (1979).

[5]   G. Schumacher, D. Thoulouze, B. Castaing, Y. Chabre, P. Segranson and J. Joffrin, J. Physique Lett. 40, L143 (1979).

[6]   M. Chapellier, G. Frossati and F.B. Rasmussen, Phys. Rev. Lett. 42, 904 (1979).

[7]   L.J. Friedman, P.J. Millet, R.C. Richardson, Phys. Rev. Lett. 47, 1078 (1981).

[8]   M. Chapellier, J. Physique Lett. 43, L609 (1982).

[9]   F.D. Colegrove, L.D. Schearer and G.K. Walters, Phys. Rev. 132, 2561 (1963).

[10]   L.F. Mollenauer, Opt. Lett. 5, 188 (1980).

[11]   P.J. Nacher, M. Leduc, G. Trénec and F. Laloë, J. Physique Lett. 43, L525 (1982).

[12]   G.C. Phillips, R.R. Perry, P.M. Windham, G.K. Walters, L.D. Schearer and F.D. Colegrove, Phys. Rev. Lett. 9, 502 (1962).

[13]   S.D. Baker, G. Roy, G.C. Phillips and G.K. Walters, Phys. Rev. Lett. 15, 115 (1965).

[14]   L. Passell and R.I. Schermer, Phys. Rev. 150, 146 (1966) and L. Passell, private communication.

[15]   M.V.Mc Cusker, L.L. Hatfield and G.K. Walters, Phys. Rev. Lett. 22, 817 (1969); Phys. Rev. A5, 177 (1972).

[16]   P.J. Keliher, R.E. Gleason and G.K. Walters, Phys. Rev. A11, 1279 (1975).

[17]   L.A. Hodge, F.B. Dunning and G.K. Walters, Rev. Sci. Instr. 50, 1 (1979).
       L.B. Gray, K.W. Giberson, Chu Cheng, R.S. Keiffer, F.B. Dunning and G.K. Walters, Rev. Sci. Instr. 54, 271 (1983).

[18] S.D. Baker, E.B. Carter, D.O. Findley, L.L. Hatfield, G.C. Phillips, N.D. Stockwell and G.K. Walters, Phys. Rev. Lett. $\underline{20}$, 738 (1968).

[19] T.W. Riddle, M. Onellion, F.B. Dunning and G.K. Walters, Rev. Sci. Instr. $\underline{52}$, 797 (1981).

[20] D.E. Murnick, Appl. Phys. Lett. $\underline{42}$,, 544 (1983).

[21] H.G. Robinson and Than Myint, Appl. Phys. Lett. $\underline{5}$, 116 (1964).

[22] L.D. Schearer, F.D. Colegrove and G.K. Walters, Rev. Sci. Instr. $\underline{34}$, 1363 (1963).

[23] M.A. Taber, J. Physique Coll. $\underline{39}$, C6, 192 and Thesis, Stanford Univ. (1978).

[24] P.J. Nacher and M. Leduc, in preparation for J. Physique.

[25] J. Dupont-Roc, M. Leduc and F. Laloë, J. Physique, $\underline{34}$, 961 and 977 (1973).

[26] J.M. Daniels and R.S. Timsit, Can. J. Phys. $\underline{49}$, 525 and 545 (1971).

[27] W. Happer, Rev. Mod. Phys. $\underline{44}$, 169 (1972).

[28] G. Trénec, P.J. Nacher and M. Leduc, Opt. Commun. $\underline{43}$, 37 (1982).

[29] M. Pavlovic et F. Laloë, J. Physique $\underline{31}$, 173 (1970). M. Pinard and J. Van der Linde, Can. J. Phys. $\underline{52}$, 1615 (1974).

[30] R. Barbé, F. Laloë and J. Brossel, Phys. Rev. Lett. $\underline{34}$, 1488 (1975).

[31] V. Lefèvre-Seguin, P.J. Nacher, J. Brossel, W.N. Hardy and F. Laloë, to be published in J. Physique.

[32] F.D. Colegrove, L.D. Schearer and G.K. Walters, Phys. Rev. $\underline{135}$, A353 (1964).

[33] S.D. Rosner and F.M. Pipkin, Phys. Rev. $\underline{A5}$, 1909 (1972).

[34] M. Himbert, V. Lefèvre - Seguin, P.J. Nacher, J. Dupont-Roc, M. Leduc and F. Laloë, to be published in J. Physique Lettres· (July 1983).

ELECTRON PARAMAGNETIC RESONANCE, DYNAMIC POLARIZATION AND LOW TEMPERATURE HIGH FIELD EXPERIMENTS ON ^{3}He AND FLOUROCARBON SPHERES

A. Schuhl, L. Sniadower, H. Alloul and M. Chapellier

Orme/Orsay, Laboratoire de Physique des Solides, Bat. 510
Universite de Paris-Sud, 91 405 ORSAY (FRANCE)

We report the presence of very stable paramagnetic centers, which seems to be peroxy radicals (both mid-chain and end-chain polymer radical). These radicals give an EPR line of around 12 Gauss width at room temperature which broaden to 60 Gauss at helium temperature (EPR at 10 GHz). The concentration is around 10^{18} centers/cc. When the sample is deoxygenated the T_1 of the fluorine is long and allows, in spite of the large linewdith at He temperature, a dynamic polarization of fluorine with an enchancement factor up to 30 (at 3.8K). This enchancement factor is also observed on a single layer of ^{3}He, which is the first observance of a dynamic polarization of ^{3}He in dense form.

If the coupling between F and ^{3}He is attributed to the presence of the electronic center and if these centers are fixed in space we should observe a $1-P_e^2/H^2$ dependance of the coupling (P_e is the electronic polarization P_e = th $\mu H/2kT$ and H is the applied field).

This was not observed by Hammel et al(*) at 1 kG and T down to 1 mK. We have made experiments at field up to 3.7 T and temperature down to 30 mK. The coupling is still present, and give as measured on fluorine relaxation and directly on He^3 a relaxation of ^{3}He governed by $T_1 T/H$ ct.

This may indicate a relaxation in the first monolayer governed by some defects (fundamental vacancies?).

The presence of a coupling between ^{3}He and F nuclei still active at very high polarization of these electronic center means either that the coupling has nothing to do with these electrons or that some molecular movement still persist in this glassy material even at very low temperature. More experiments on spheres without electronic centers are needed.

*R.C. Richardson, private communication June 82, and Hammel et al, this conference.

HEAT CAPACITY MEASUREMENTS AND A SEARCH FOR SUPERFLUIDITY
IN ^{3}He-^{4}He MIXTURES AT VERY LOW TEMPERATURES

R.M. Mueller, H. Chocholacs, Ch. Buchal, M. Kubota,
J.R. Owers-Bradley and F. Pobell
Institut für Festkörperforschung, Kernforschungsanlage Jülich
5170 Jülich, W. Germany

Measurements of the heat capacity of dilute solutions of ^{3}He in ^{4}He give useful information about the interactions of the ^{3}He quasiparticles; however, they should also be a reliable indicator for any phase transition which occurs in the solutions. The most recently published calculation, that of Owen[1], predicts a transition of the quasiparticles into a superfluid phase at mixture temperatures which are now becoming available for experiments.

Above the transition temperature, the quasiparticles constitute a Fermi liquid with a heat capacity which is expected to be linearly proportional to temperature, if the temperature is sufficiently below the Fermi temperature. As the concentration of a mixture is varied, the effective mass, m_3^*, of the ^{3}He quasiparticles varies because of interactions between quasiparticles. The relationship of the effective mass to the specific heat capacity C of a solution is

$$\frac{m_3^*}{m_3} = \frac{\hbar^2 \, (3\pi^2 \, N_3/V)^{2/3} \, C}{\pi^2 m_3 \, k_B \, x \, R \, T}$$

where m_3 is the mass of an isolated ^{3}He atom, N_3/V is the number of particles per unit volume of solution and $x = N_3/(N_3+N_4)$ is the concentration of the ^{3}He in a solution consisting of N_3 atoms of ^{3}He and N_4 atoms of ^{4}He.

The heat capacity of solutions containing 1.3%, 5% and 9.2% of ^{3}He were measured at various pressures and for temperatures in the range from 320 μK to 25 mK, an order of magnitude colder than for any previous heat capacity measurements. The values of m_3^*/m_3 calculated from the heat capacities are given in the table.

Table I Values of m_3^*/m_3 derived from the measured heat capacities

x(%)	1.3	1.3	5	9.2	9.2
P(bar)	0	10	0	7.3	14.1
m_3^*/m_3	2.40±0.05	2.63±0.04	2.47±0.02	2.51±0.05[#]	2.99±0.05[#]

Preliminary Values

Though it was possible to cool the 1 mole sample to 215 μK, no evidence of a superfluid transition of the ^{3}He component was observed

REFERENCE

1. J.C. Owen, Phys.Rev.Lett. <u>47</u>, 586 (1981)

HEAT TRANSFER IN DILUTE ^{3}He-^{4}He MIXTURES

F. Guillon, J.P. Harrison and A. Sachrajda
Queen's University, Kingston, Ontario, Canada

Heat transfer between ^{3}He quasiparticles and heat exchangers is invariably much better than expected upon the basis of the large acoustic mismatch between the quasiparticles and phonon in the solid. An experiment was designed to study the two steps in the heat transfer (quasiparticles to helium phonon and helium phonon to solid phonons) as a function of temperature ^{3}He concentration in the ^{3}He-^{4}He mixture and whether or not sintered submicron copper powder was in the cell. For the bulk helium experiment, the heat transfer coefficient between ^{3}He quasiparticles and helium phonons was in reasonable agreement with theory for the one mixture with known viscosity (0.1%). With sinter with pore size $\sim$0.25 μm occupying one third of the cell there was a substantial enhancement in this heat transfer coefficient for the 0.03% mixture at 20 mK with the enhancement decreasing the increase of temperature and ^{3}He concentration. A temperature time analysis showed that the enhancement was in the <u>bulk</u> component of mixture in the cell and therefore it might have arisen from a process occurring at the bulk mixture/sinter interface.

NMR MEASUREMENT ON TEXTURES IN ROTATING SUPERFLUID ^{3}He-B

Yu. M. Bunkov*, P. J. Hakonen, and M. Krusius
Low Temperature Laboratory, Helsinki University of Technology,
SF-02150 Espoo 15, Finland

ABSTRACT

Results from cw NMR measurements on superfluid ^{3}He-B in rotation
are discussed. The measurements have been performed with the magnetic field oriented at different angles with respect to the rotation
axis. The results illustrate the effect of rotation on different
order parameter textures in a cylindrical container. In particular,
they allow the textural vortex free energy to be determined. The influence of boundary conditions on different textures has been investigated.

INTRODUCTION

Recent experiments[1] with a rotating cryostat on the ^{3}He superfluids have uncovered a wealth of new phenomena associated with vorticity in anisotropic superfluids. The measurements made so far have
focused on the NMR properties of the rotating ^{3}He superfluids. The
distinct changes which are observed in the NMR absorption spectra as
a function of rotation speed allow one to infer a detailed understanding of the new and unexpected features of B-phase vorticity. The
imprint of rotation on the order parameter texture has proved to be
more prominent than originally expected. Furthermore, the NMR results have been readily amenable to theoretical interpretation and
thus a consistent picture of the B-phase vortex phenomena has started
to emerge. Here we report on recent results from NMR measurements in
the B-phase which amply exemplify the rich new features displayed by
the vortex textures.

The measurements have been performed on a liquid ^{3}He sample in
a cylindrical container with its axis parallel with the rotation
axis. When the magnetic field $\vec{H}$ is also oriented along this common
axis, a highly symmetric "flare-out texture" is favoured, corresponding to a well-defined free energy minimum, and rigidly stabilized by the orienting effect of the walls of the cylinder.[2] In a
flare-out texture the order parameter, i.e. the $\hat{n}$-vector, is closely
aligned along the axial direction in the central region of the
cylindrical cross section. A large fraction of the total NMR absorption is therefore concentrated close to the Larmor frequency region.
Nevertheless, the orienting effect of the walls, which in the axial
case results in a tilt of 63.4° of $\hat{n}$ with respect to $\vec{H}$, causes the
angle β between $\hat{n}$ and $\vec{H}$ to open up linearly with increasing radial
distance in the central region.[3,4,5,6] The linear dependence of the
$\hat{n}$-vector deflection creates a harmonic potential well for localized
spin waves, which are responsible for the most striking NMR signature

* Institute for Physical Problems of the Academy of Sciences of the
  USSR, Moscow 117334, USSR.

of the flare-out texture, namely a set of sharp NMR absorption
maxima centered on evenly shifted harmonic oscillator frequencies.
During rotation a lattice of linear and singular vortices is formed
such that their separation is less than the characteristic scale of
bending of the B-phase texture. Therefore, the textural effect of
the vortices is to introduce an additional average tilt of $\hat{n}$ towards the transverse plane, as can be understood to result e.g. from
the presence of the local superflow circulating around the cores.
In the rigid flare-out texture the effect of rotation is borne out
as a general reduction in the NMR absorption in the Larmor region
and as a noticeable increase of the spin wave frequency splitting.[7]

If the magnetic field is tilted at an angle $\mu > 14.5^o$ from the
axial direction, a texture with a considerably reduced symmetry is
formed. One severe impediment of this texture is the apperance of
linear textural defects localized along the wall, imposed by the discontinuous nature of the boundary conditions as enforced by the minimum energy configuration of the wall orientation.[8] The remarkable
charateristic feature of the "tilted field texture" is the fact that
its NMR response from the central region closely adheres to the bulk
liquid behaviour during rotation.[9] For this reason it has turned
out to provide a useful quantitative measure of the textural vortex
free energy.

In the course of the present NMR measurements it has became clear
that in most cases the B-phase textures are inflicted by persistent
textural defects if the texture is rapidly precipitated after substantial supercooling of the A-liquid, as is often the case in practice. Rotation, in particular rapid acceleration or deceleration,
provides the first really effective external means of annealing and
homogenizing textures.[1,9] For instance, the spin wave absorption
modes usually become clearly distinguishable only after rotation.[7]
On this basis it is understandable why in the past various kinds of
experiments have notoriously suffered from difficulties and inconsistencies if the outcome of the measurement has depended crucially
on the particular texture involved. Only the simplest geometry,
the 1-dimensional parallel plate configuration, appears to be less
susceptible to defects.[10] On the other hand, once the different
possible equilibrium textures and their corresponding NMR spectra
have been classified, one might also study the structure of textural defects which have been deliberately introduced in a controlled fashion. This can be accomplished, for instance, by starting from a well-defined equilibrium texture and by rotating the
magnetic field to a new orientation. At low temperatures the boundary conditions are rigidly frozen on the walls and thus the original
wall orientations will persist largely unaltered. As an example
of such metastable textures one might rotate the field from a tilted
orientation into the axial position and study the implications from
the wall defects on the texture in an axially symmetric environment.

Undoubtedly the most significant phenomenon associated with B-
phase vorticity is the discovery of a phase transition, apparently
involving a restructuring of the vortex core. The structural change
of the core has a substantial quantitative effect on the contribution
of the vortices to the texture. Moreover, it turns out that an even

more dramatic change occurs at the transition in the magnitude of a
small frequency shift which depends on the relative direction of the
rotation with respect to the magnetic field.* The textural orienting effect from the vortices is composed of two components, namely
of the local superflow[11] and the boundary conditions imposed on the
order parameter at the exterior of the core.[12] The relative magnitude of these contributions is still a matter of dispute,[4,6] although most authors seem to stress the importance of the superflow
effect.[3,11,13] In this respect the experimental observations presented in this work indicate that at low temperatures and high pressures the core transition controls a prominent part of the orienting
effect. These features are poorly understood and insufficiently
covered by experiment at present.

In this report we shall confine ourselves to the experimental
characteristics of the different vortex textures on which detailed
information exists. After a brief summary of the experimental
essentials we shall first discuss the axial field texture, next the
tilted field case and finally we shall present some examples of non-
equilibrium textures. All of the measurements described in this
report have been performed either in the stationary state or using
a rotation speed of 1.40 rad/s at the liquid ^{3}He pressures of 20.5
and 29.3 bar in a magnetic field of 28.4 mT. With respect to the
angular velocity dependence we here refer to other texts.[1,9]

EXPERIMENTAL DETAILS

The rotating nuclear demagnetization cryostat employed in the
present measurements is in most respects of standard current design
and has been described in detail elsewhere.[14] It incorporates a ^{3}He/
^{4}He dilution refrigerator with large liquid volumes such that it can
be operated in the single cycle mode for 14 consecutive hours. During this time the still is pumped with a charcoal cryoadsorption
pump which is an integral part of the cryostat and is located above
the vacuum jacket in the liquid He bath. Thus the cryostat can be
completely disconnected from the external pumping manifold and can
be brought into rotational mode of operation without a need for any
vacuum tight rotating seals to external pumping lines. The whole
cryostat including its liquid He dewar and the pumping head rotates
on compressed air bearings driven by an electric feed back controlled motor by means of a rubber belt. The analog electronics is
stacked on a rack rotating with the cryostat. The communication
between the laboratory and the rotating frame is managed via a
digital data bus.

The static polarization field for the NMR is produced with
superconducting magnets located in the He bath. They include an
end compensated solenoid with a winding diameter of 150 mm inside
a pair of saddle-shaped coils. The fringing field of the large 8 T
superconducting solenoid for adiabatic nuclear demagnetization has

-----

* In this work the relative orientations of $\vec{\Omega}$ and $\vec{H}$ have been chosen
such that $0 < \arccos (\vec{\Omega}, \vec{H}) < \pi/2$. This choice produces the largest
frequency shifts.

been carefully compensated in the region of the NMR pick-up coil. The final tuning of the axial field homogeneity is accomplished with two independent axial superconducting coils by adjusting their currents for minimum linewidth. The field homogeneities as determined by the ^{3}He NMR linewidths are 4×10^{-5} in the axial direction and 8×10^{-5} in the horizontal direction. Two pairs of large saddle-shaped copper coils, located on a diameter of 1.5 m around the cryostat, are employed for cancelling the horizontal component of the earth's magnetic field in order to avoid a modulation of the horizontal NMR field during rotation.

The transverse NMR signal is measured in the cw mode by sweeping the polarization field with a digitally controlled sweep. A frequency synthesizer with a short term frequency stability of 10^{-8} is used as an rf oscillator and a lock-in amplifier followed by a chart recorder are employed for signal detection. The liquid ^{3}He sample is contained in an epoxy tower with an inside diameter of 5 mm and a length of 30 mm. The rf pick-up coil consists of a pair of saddle-shaped coils with a length of 12 mm, wound and epoxied directly on the inside surface of the tower. In the present measurement the primary quantity of interest is the frequency shift. In order to ensure reliable control of all the measurement system on the NMR frequency axis a parallel measurement of the Fermi liquid signal is recorded during each field sweep. For this end a second rf coil is used which is thermally anchored to the mixing chamber and is monitored with an independent signal detection system. Using this technique the overall precision in the frequency shift measurement is estimated to be better than ± 25 Hz. The temperature is determined from the integrated nuclear free precession signal from platinum powder immersed in liquid ^{3}He. The temperature scale is fixed to the superfluid transition temperature T_c by identifying T_c from the NMR response on warming slowly through T_c after each cooldown into the superfluid state.

AXIAL FIELD TEXTURE

The NMR absorption response from an annealed flare-out texture is shown in the insert of Fig. 1 both in the stationary and rotating ($\Omega = 1.40$ rad/s) states at the temperature $T = 0.51\ T_c$ and the pressure 29.3 bar. The signals display a series of spin wave resonances with amplitudes rapidly decreasing towards larger frequency shifts. Obviously the effect of rotation is to enforce a larger deflection of $\hat{n}$ from the axial orientation since the spin wave peaks, and much of the signal intensity, shift to higher frequencies. In fact, while the spin wave frequency splitting provides a sensitive probe of the local slope of $\beta(r)$, on the other hand, the averaged resonance absorption maps the global distribution of $\beta(r)$ by virtue of a simple local oscillator type of behaviour.

The temperature dependence of the three lowest spin wave eigenfrequencies has been plotted in Fig. 1 for $\Omega = 0$ and 1.40 rad/s. The frequencies have been expressed as normalized shifts, i.e. the plotted quantity is $\sin^2 \beta \cong 2\nu_o(\nu_t - \nu_o)/(\nu_L^B)^2$ where ν_t is the measured eigenfrequency and ν_L^B the longitudinal B-phase frequency.[15]

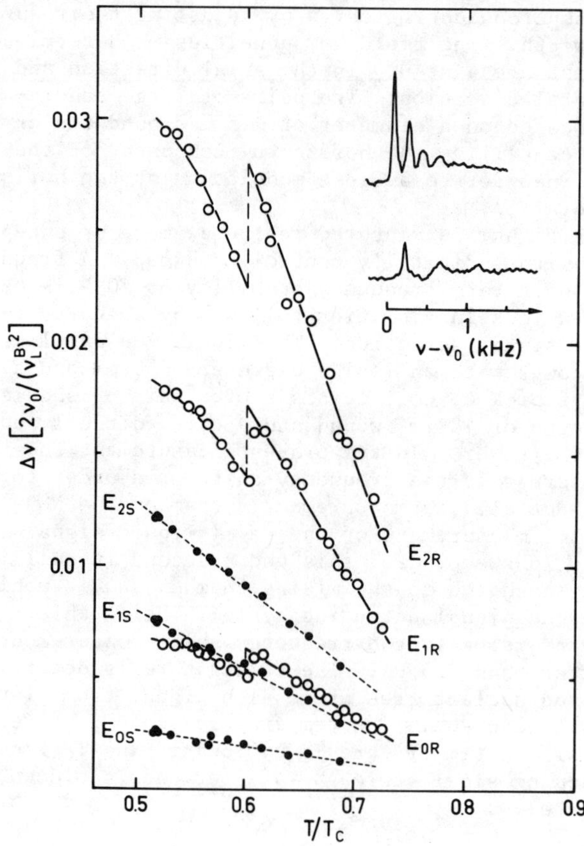

Fig. 1. Normalized frequencies of spin wave resonances in the axial field at 29.3 bar in the stationary ($\bullet$, $\Omega = 0$) and rotating (o, $\Omega = 1.40$ rad/s) states. The insert shows the NMR signals at $T = 0.51$ T_C.

A prominent feature of the spin wave frequencies is the discontinuity at $T = 0.60$ T_C, which is observed only in the rotating state. It represents a phase transition between two different vortex core structures for the following reasons: The temperature of this transition is independent of Ω but does depend on pressure. Tentative measurements show that the transition moves closer to T_C when the pressure is reduced, such that at 20.5 bar it occurs at $T = 0.71$ T_C. The size of the discontinuity scales with Ω, as it should if it is composed of the contributions from individual vortices, but it decreases towards lower pressures becoming increasingly more difficult to distinguish. The frequency shifts of the spin wave resonances exhibit remarkable metastable behaviour on warming or cooling through the vortex transition while the cryostat is simultaneously continuously rotated.[1,7] In order to avoid such effects and also to ensure an equilibrium vortex density, the data in Fig. 1 has been measured during a slow warm-up at a rate of roughly 60 nK/s by simultaneously slowing down the rotation to a stop every 10 - 20 minutes.[1,9]

The even eigenvalues corresponding to the axially symmetric eigenfunctions of an isotropic 2-dimensional harmonic oscillator follow the sequence $E_n = (2n + 1)E_o$, with n = 0, 1, 2, ... The spin wave frequencies in Fig. 1 closely fit this pattern. The most accurately measured quantity is the difference $\Delta E_{10} = E_1 - E_o$. Comparing it with the frequency spacing on either side one finds that e.g. in the stationary state at T = 0.65 T_c $2E_o/\Delta E_{10}$ = 1.25 $\pm$ 0.2 and $\Delta E_{21}/\Delta E_{10}$ = 1.16 $\pm$ 0.2. Similarly in the rotating state the corresponding values are 1.15 $\pm$ 0.2 and 0.98 $\pm$ 0.2, respectively. With increasing temperature one would expect anharmonic distortion to influence the spin wave spectrum.[5] The magnetic bending length, which at 29 bar roughly equals $\xi_H \cong 40(1 - T/T_c)/H$ mm x mT, increases towards low temperatures and makes the texture more resistant to bending. As a result, the large β value at the walls is transmitted more effectively into the central regions of the cylinder. This acts to increase the slope of $\beta(r)$ in the center and tends to reduce the curvature in the spin wave potential $\propto \sin^2\beta(r)$ in the region where the first few spin wave functions have more weight. Correspondingly the anharmonicity should become apparent towards high temperatures as a relative increase in the frequency spacing between the higher modes which sample more of the outer regions of the potential. However, this feature is not very prominently illustrated by the results in Fig. 1. The same situation prevails at lower pressures where, within the experimental precision, the above ratios apply for the spacing between the spin wave resonances at 20.5 bar. However, on reducing the pressure ξ_H increases and the normalized frequency shifts become larger. In the temperature region T > 0.6 T_c one would have to scale the data in Fig. 1 with a factor of roughly 2.0 in order to obtain the corresponding spin wave frequencies at 20.5 bar both in the stationary and rotating states.

A quantitative extraction of the orientational effect of rotation on the flare-out texture requires a numerical calculation of the complete texture in order to fix $\beta(r)$ in the center of the cylinder. This determines the spin wave potential in the spin wave equation. A detailed comparison of the calculated and measured spin wave eigenfrequencies involves as parameters the bending length ξ_H, the dipolar healing length ξ_D, and the vortex parameter λ, which in this context is the object of interest. For details we refer the reader to Ref. 6.

TILTED FIELD TEXTURE

The stationary state NMR response in the tilted magnetic field is depicted in the insert of Fig. 2. This signal has a rather different appearance from that originating from the axial flare-out

200

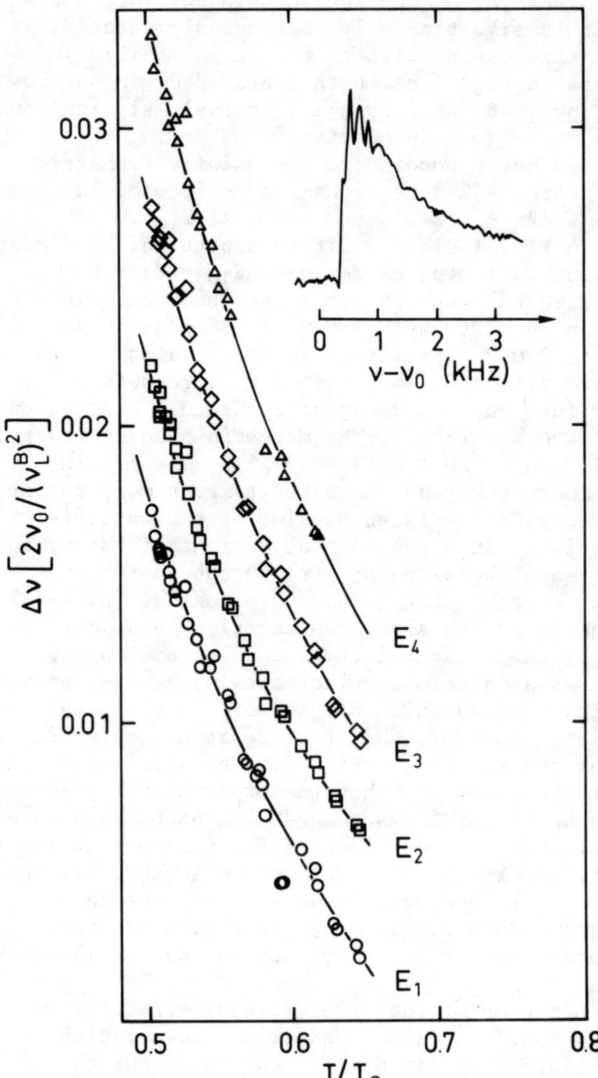

Fig. 2. Normalized frequencies of spin wave resonances in the stationary state at 29.3 bar in a magnetic field tilted by 25° from the axial direction. The insert shows the NMR signal at $T = 0.53\ T_C$.

texture shown in Fig. 1. It features a sharp cut-off towards the Larmor frequency ν_0 and a massive total absorption intensity concentrated in the Larmor region. A series of spin wave resonances lies crowded close to the cut-off. Their frequencies have been plotted in Fig. 2 as functions of temperature for the case when $\vec{H}$ is tilted at an angle of $\mu = 25°$ with respect to the axial direction. The resonances are again rather accurately equidistant, for example, at $T = 0.60\ T_C$ $\Delta E_{32}/\Delta E_{21} = 1.1$ and $\Delta E_{43}/\Delta E_{21} = 1.2$. This spacing is comparable to that of the flare-out texture since, for instance, at $T = 0.60\ T_C$ $\Delta E_{21}(\mu = 25°)\,/\,\Delta E_{10}(\mu = 0) = 1.1$ with, however, the clear distinction that the whole spin wave spectrum is now shifted to much higher frequencies. In this case the lowest curve E_1 in

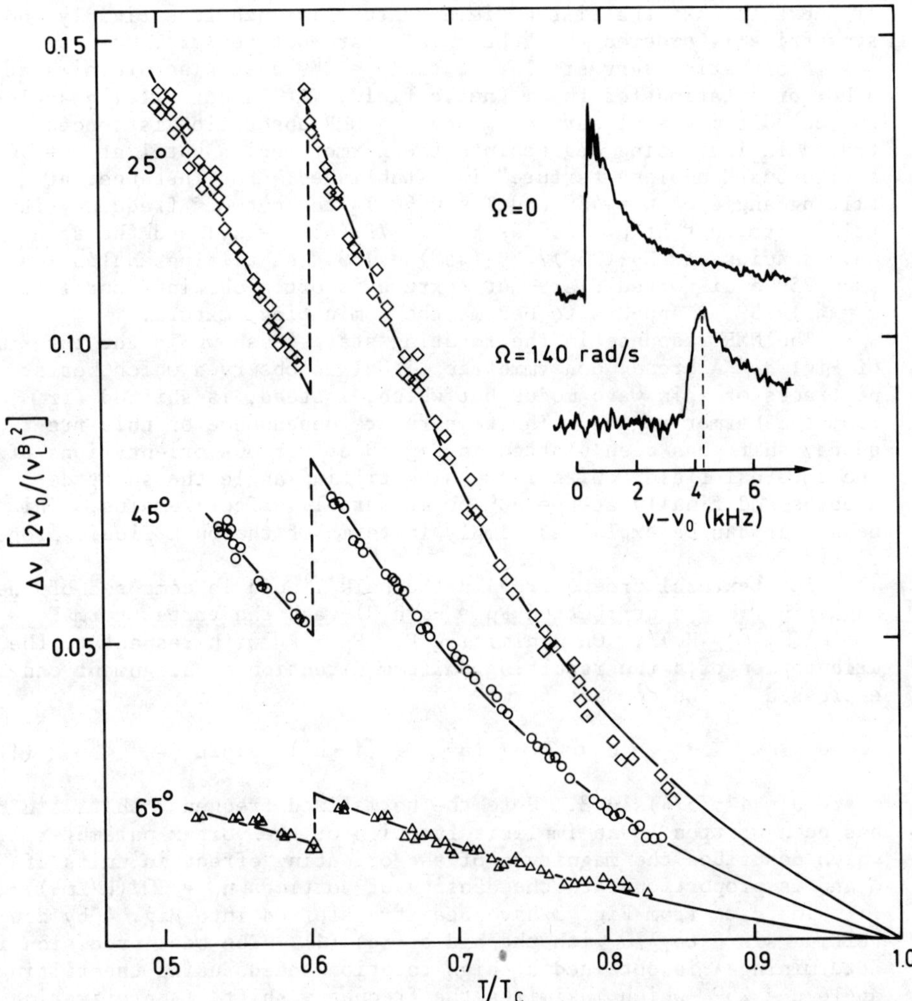

Fig. 3.   Normalized frequency shifts at different tilted field orientations at 29.3 bar and 1.40 rad/s: $\Diamond$, $\mu = 25°$; $\bigcirc$, $\mu = 45°$; and $\triangle$, $\mu = 65°$. The NMR signals have been measured at $T = 0.58\,T_c$.

Fig. 2 corresponds closely to the cut-off frequency where the first inadequately traced resonance peak is located. Obviously, at low temperatures little if any absorption takes place in the region between the cut-off and ν_0. This feature clearly reflects the influence from the boundaries. However, in contrast to the axial case now much of the total resonance absorption is piled against the cut-off close to the Larmor region. Consequently, in a large portion of the total cross section of the cylinder the texture is able to re-

202

duce its magnetic field energy by adjusting its average orientation
closer to the external field direction. This property is a result
of the fact that the tilted field texture is much less rigidly and
symmetrically ordered than the axial flare-out texture.

A situation very similar to the $\mu = 25^{o}$ case also prevails at
other orientations of the magnetic field. At larger tilting angles
the cut-off moves closer to ν_{o} and the NMR absorption is concentrated in increasing degree into the Larmor region until at $\mu = 90^{o}$
a rigorous "in-plane texture" is established. For instance, at a
tilting angle of $\mu = 45^{o}$ and $T = 0.60\ T_{c}$ the cut-off frequency is
related to that at $\mu = 25^{o}$ as $E_{1}(25^{o})/E_{1}(45^{o}) = 2.1$ and the spin
wave spacing as $E_{21}(25^{o})/E_{21}(45^{o}) = 1.9$. At tilting angles less
than 25° a distorted flare-out texture is often obtained until at
$\mu = \lesssim 14.5^{o}$ it appears to become the dominating texture.

The NMR response in the rotating state is shown in the insert
of Fig. 3. A broad, nonsymmetric signal is observed which bears
no traces of spin wave modes but which, instead, is shifted far
from the Larmor region. The temperature dependence of this frequency shift has been plotted in Fig. 3 at various orientations of
the external field. With increasing tilting angle the shift decreases and finally at $\mu = 90^{o}$ no measurable effect remains. This
behaviour can be explained simply in terms of the bulk liquid properties.[9]

The textural free energy in the bulk liquid is composed of the
magnetic field contribution $F_{H} = -a(\hat{n}\cdot\vec{H})^{2}$ and the vortex term $F_{V} =
(2/5)a\lambda(\hat{\Omega}_{i}\cdot\overleftrightarrow{R}_{ik}\cdot\vec{H}_{k})^{2}$. On minimizing $F = F_{H} + F_{V}$ with respect to the
orientation of $\hat{n}$ the resulting uniform direction of alignment can be
expressed in the compact form [8]

$$\frac{1}{\lambda} = u\cos 2\mu + (u^{2} - \frac{1}{2})(1 - u^{2})^{-\frac{1}{2}}\sin 2\mu \tag{1}$$

where $u = 1 - (5/4)\sin^{2}\beta$. Here the normalized frequency shift $\sin^{2}\beta$
has been written as an implicit function of the vortex parameter λ,
which describes the magnitude of the orienting effect in units of
a and is proportional to the density of vortices $n_{V} = 2\Omega/(h/2m_{3})$.

The data from Fig. 3 have been transferred into Fig. 4 by con-
verting $\sin^{2}\beta$ to λ/Ω with the aid of Eq. (1). The best precision in
determining λ is obtained at high rotation speeds using the tilting
angle $\mu = 25^{o}$, which maximizes the frequency shift. Approximating
λ with a linear temperature dependence in the high temperature
region yields $\lambda/\Omega = 2.0(1 - T/T_{c})$ at 29.3 bar. The data from 20.5
bar at one orientation of the magnetic field have also been included in Fig. 4, yielding $\lambda/\Omega = 3.5(1 - T/T_{c})$. However, the most
significant feature displayed in this figure is the excellent fit to
Eq. (1), such that all the data obtained at different field orientations fall on a single curve. The same conclusion is derived if the
frequency shifts are measured at fixed μ with Ω as a variable parameter.[9] It is therefore to be concluded that the texture in a cylinder of 5 mm diameter is controlled to a smaller extent by boundary
effects in a tilted magnetic field than in the axial case, and that
on the average it is able to conform to the bulk liquid behaviour at
high rotation speeds. Finally, it should be noted that rotation at

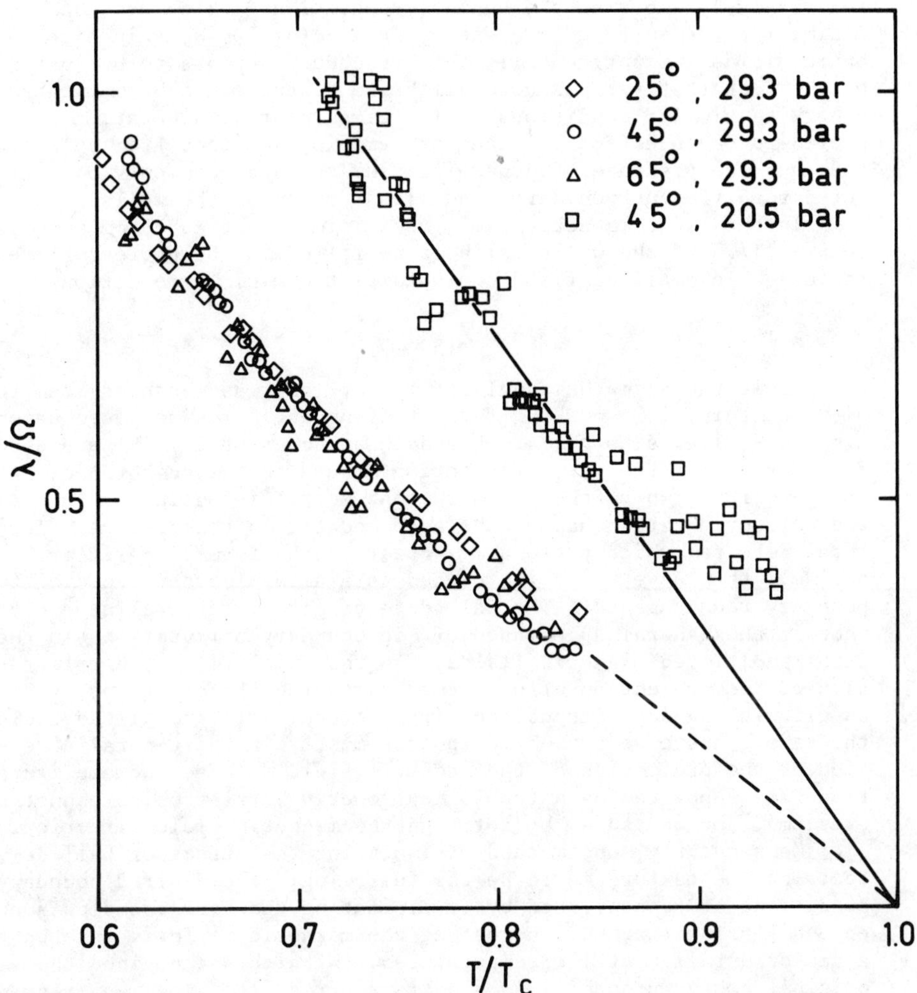

Fig. 4. Vortex parameter λ/Ω measured at 1.40 rad/s for two different pressures.

angular velocities of order 1 rad/s is comparable to the magnetic field as a textural orienting effect beyond the Ginzburg-Landau regime. An intuitive approach to the tilted field texture with its complicated boundary conditions easily leads to misconceptions. A detailed numerical analysis of the tilted field textures in the presence of different possible boundary conditions is presently under way (see Ref. 6).

NONEQUILIBRIUM TEXTURES

The orienting effect of the walls can apparently be accounted for safely by imposing the boundary conditions[3] which are obtained by minimizing the surface free energy $F_S = -d(\hat{s}_i \cdot \overleftrightarrow{R}_{ik} \cdot \vec{H}_k)^2$ in high magnetic fields. Experimentally this procedure appears to be justified by the fact that various nonequilibrium textures can be maintained in otherwise similar conditions at low temperatures.[9] Rotation has proved to be an effective means of removing solitons from bulk liquid[1] and thus the presence of nonequilibrium textures can only be attributed to different boundary conditions frozen on the walls.

In the axial magnetic field the minimum configuration requires $\cos\beta = \pm 1/\sqrt{5}$ at the walls which is fulfilled in the perfect flare-out texture. In a tilted field one obtains the minimum condition[8]

$$\sin^2\beta = \frac{4}{5}(1 \pm \sin\mu\cos\phi) \ , \tag{2}$$

where ϕ is the azimuthal angle of the cylinder as measured from the plane containing $\vec{\Omega}$ and $\vec{H}$. If the tilting angle is increased beyond $\mu = 14.5^\circ$, i.e. $\sin\mu > 1/4$, then both branches of Eq. (2) are needed in order to fix the wall orientation around the circumference. At the junctions, where the two solutions meet, discontinuous jumps in the wall orientation have to be accommodated in the texture. Textural defects, which presumably appear in the form of a linear boojum,[8] thus have to be introduced as a natural consequence of the boundary requirements. The wall defects give rise to a large difference in the general appearance of the boundary orientations in the axial and tilted magnetic fields. In the axial field the orientation of the projection of $\hat{n}$ in the transverse plane rotates by 2π on circulating once around the circumference. In the tilted field the same transverse projection points mostly in the general direction of the projection of the magnetic field. These two configurations are separated by a topological energy barrier which cannot be overcome, for instance, by rotating the magnetic field direction.

Consequently, one method of isolating the effect of wall defects on the texture could be the following: The desired boundary conditions are established by precipitating the corresponding annealed equilibrium texture, whereafter the magnetic field is rotated to a new orientation at low temperatures. With this technique the original boundary conditions are transferred into a new environment and a metastable nonequilibrium state is created. Obviously the boundary orientation will attempt to adjust to the change in the magnetic field direction as far as possible in a continuous manner. However, the surface defects will resist additional distortion and thus small changes in the field direction will not produce major reordering in the wall orientations. On warming up the metastable texture it will at some point relax to the equilibrium configuration, usually at temperatures in the region of $T \cong (0.7...0.8)T_c$.

In Fig. 5 the results are given on a simple test of this approach which clearly demonstrate the metastable properties and the importance of the boundary conditions both in the stationary and rotating states. In this case the cooldown was performed with the

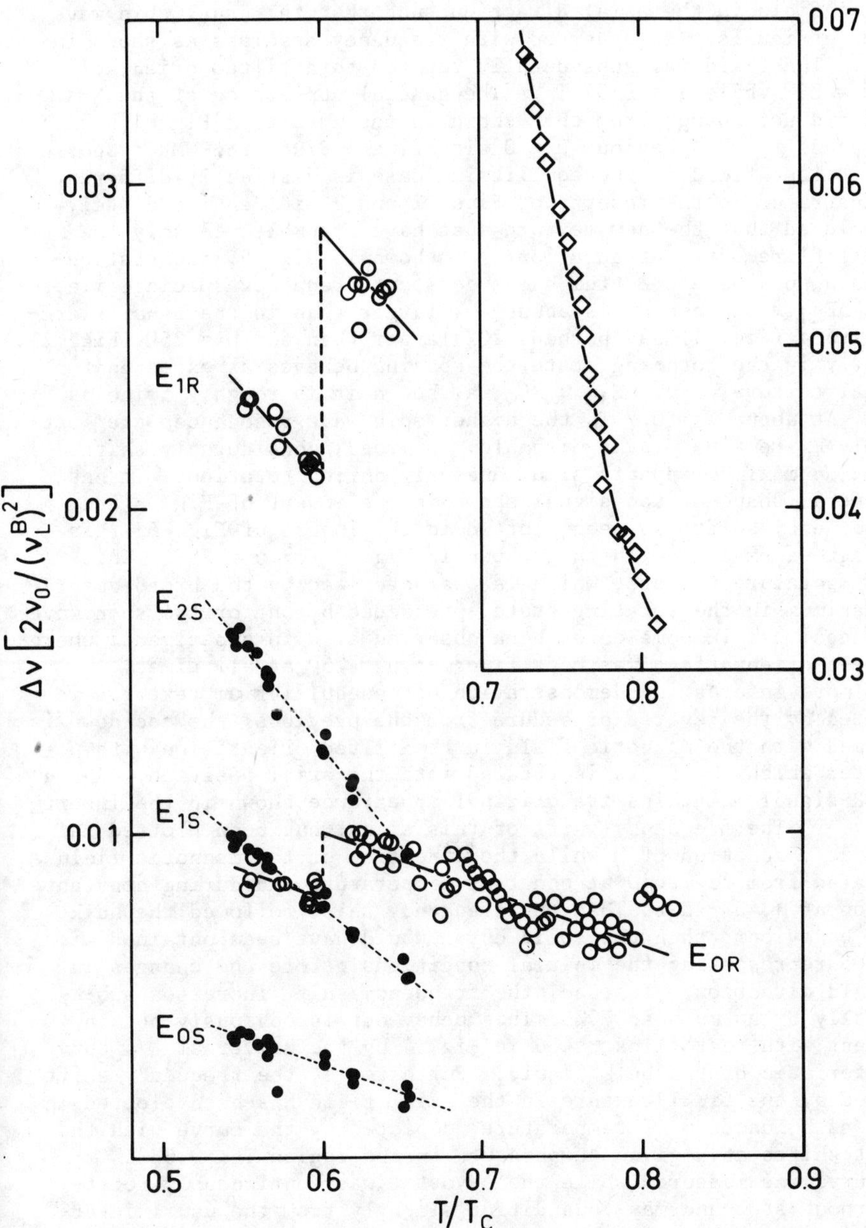

Fig. 5. Normalized frequencies of spin wave resonances at 29.3 bar after rotating the magnetic field direction from 0 to 25°: •, $\Omega = 0$ and o, $\Omega = 1.40$ rad/s. The frequency shift of the broad absorption anomaly appearing at high temperatures T > 0.7 T_C during rotation is shown in the insert (◇, $\Omega = 1.40$ rad/s).

206

magnetic field in the axial direction such that the usual spin wave
resonance signals were observed with frequency spacings as shown in
Fig. 1. The field was subsequently rotated to a tilted orientation
with μ = 25° while T < 0.54 T_c. The general appearance of the NMR
signal did not change from that shown in the insert of Fig. 1, which
corresponds to the previous μ = 0 situation. Since the NMR response
in the tilted field in the equilibrium case is distinctly different,
as illustrated in the inserts of Figs. 2 and 3, it can immediately
be concluded that the new texture must have resembled closely the
original flare-out configuration. As shown in Fig. 5, the stationary state spin wave spectrum has a constant frequency spacing, i.e.
$\Delta E_{10} \cong \Delta E_{21} \cong 2E_0$, which is about 40% larger than in the usual flare-
out texture (Fig. 1) and perhaps 20% larger than for μ = 25° (Fig. 2).
Similarly in the rotating state the spacing behaves as expected for
an axial texture, i.e. $\Delta E_{10} \cong 2E_0$, although it is roughly twice as
large. At about T $\sim$ 0.7 T_c the higher spin wave resonances are lost
and only E_0 remains distinguishable. A broad, considerably shifted
absorption maximum appears simultaneously during rotation. It has
the general shape of the signal shown in the insert of Fig. 3 and
its frequency shift has been plotted in the insert of Fig. 5; this
shift agrees exactly with that shown in Fig. 3 for μ = 25°. This
high temperature texture, which is characterized by the broad absorption maximum in the rotating state, preceeded by one or two spin wave
resonances, has in some cases been observed also in experiments where
the field orientations was kept fixed at μ = 25° at all times.

A more interesting demonstration of nonequilibrium textures is
generated by the reverse procedure from the previous: the cooldown is
performed with the magnetic field in the tilted orientation with μ =
25° after which the field is rotated into the axial position. Again
the NMR signal maintains its original appearance shown in the insert
of Fig. 3. The frequency shift of this signal has been plotted in
Fig. 6 as a function of μ while the direction of the magnetic field
is rotated from 25° to 0 at constant temperature and during constant
rotation at 1.40 rad/s. Had the frequency shift followed the bulk
liquid behaviour, then the solid curve would have been obtained with
λ = 1.05 representing the initial conditions before the changes in
the field direction. Instead, the frequency shift increases substantially by as much as 70%. This behaviour is obviously not in
agreement with that illustrated in Fig. 4 by the universal λ/Ω curve
characteristic of the bulk liquid. Furthermore, the frequency shift
produced by the final texture in the axial field has been plotted in
Fig. 7 as a function of temperature, depicted by the curve with the
largest shifts at a given temperature in the region T < 0.68 T_c.
This curve was measured while the cryostat was continuously rotating without stop and may thus differ slightly from the usual inter-
mittently rotated result.

Two more frequency shift curves have been included in Fig. 7
for comparison, corresponding to closely related textures with similar
signal behaviour. One of them represents a regular equilibrium case
from Fig. 3, namely the frequency shift measured at μ = 25° characterizing the initial state before rotation of the field. The second
curve reproduces the frequency shift from a nonequilibrium texture in

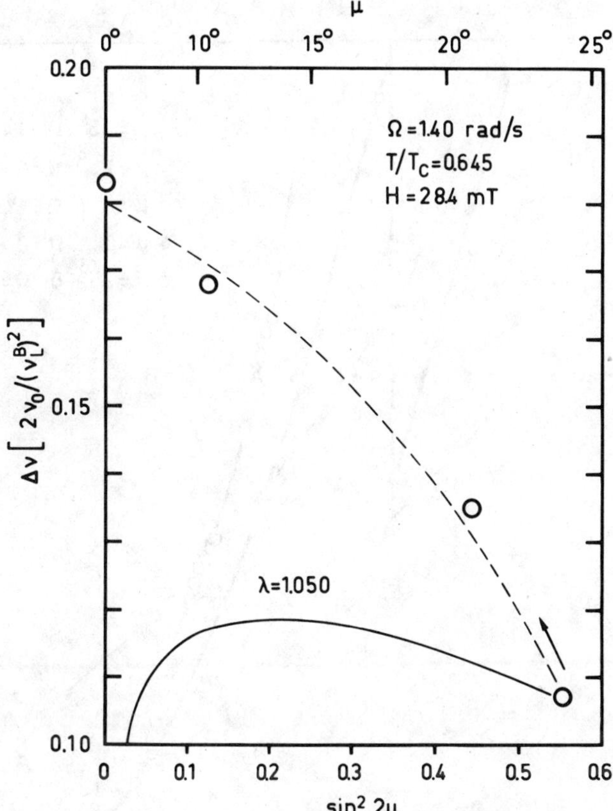

Fig. 6. Normalized frequency shift during a change of the
field orientation from $\mu = 25°$ to $\mu = 0$ at constant temperature and in continuous rotation at 1.40 rad/s. The solid
curve represents the behaviour expected for bulk liquid
according to Eq. (1).

the axial field. This texture is sometimes observed in the axial
field at high pressures if rapid acceleration and deceleration of
the rotational motion is carefully avoided ($|\dot{\Omega}| < 0.01$ rad/s^2). The
frequency shift has been measured following the usual routine of in-
terrupting the rotation at regular intervals. No spin wave resonances have been detected in this case, not even in the stationary
state, which signals the nonuniform structure of this texture. At
$T \cong 0.74\ T_C$ the texture was observed to transform into the flare-
out configuration. If one were to convert the measured frequency
shifts to λ/Ω values by the use of Eq. (1), one would find that λ/Ω
remains Ω-dependent and lies well above the 29.3 bar data of Fig. 4.
In short, we believe that this texture corresponds to a case where
surface defects prevent the formation of the regular flare-out
texture but where the tilted field frequency shift behaviour accord-

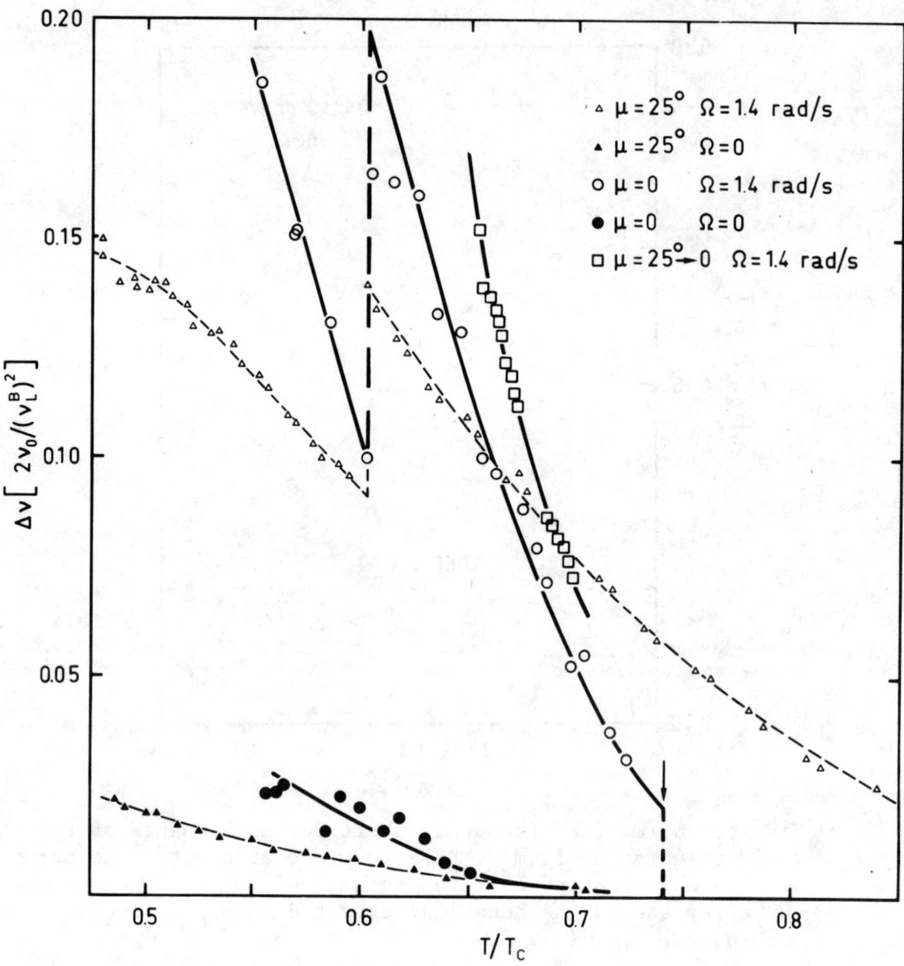

Fig. 7. Normalized frequency shifts at 29.3 bar for two non-equilibrium textures: □, after a change of the field direction from 25° to 0 during continuous rotation at 1.40 rad/s; o, Ω = 1.40 rad/s and ●, Ω = 0 describe a nonequilibrium texture in the axial field. The data from Fig. 3 for μ = 25° has been included for comparison: Δ, Ω = 1.40 rad/s; and ▲, Ω = 0.

ing to Eq. (1) is similarly inhibited.

Finally, we conclude by observing that textural defects and singularities localized on the walls appear to remain partially "frozen-in" and, in particular, will not be erased by a small change in the direction of the magnetic field as is proved by the general behaviour of the NMR response in the above examples. This provides a

means of investigating nonequilibrium textures. These observations
can be put on a solid basis when comparisons between numerical calculation and NMR experiment will be tested quantitatively for different
arrangements of the boundary conditions.

ACKNOWLEDGEMENTS

This work has benefitted from a close interaction with our
theoretical support, Drs. A.D. Gongadze, M.M. Salomaa, and G.E. Volovik. It has been conducted under the auspices of the Academy of
Sciences of the USSR and the Academy of Finland.

REFERENCES

1. P.J. Hakonen, O.T. Ikkala, S.T. Islander, O.V. Lounasmaa, and
    G.E. Volovik, J. Low Temp. Phys., to be published; O.T. Ikkala,
    Ph.D. Thesis, Helsinki University of Technology, 1982.

2. H. Smith, W.F. Brinkman, and S. Engelsberg, Phys. Rev. $\underline{B15}$, 119
    (1977).

3. K. Maki and M. Nakahara, to be published.

4. P.J. Hakonen and G.E. Volovik, J. Phys. C: Solid St. Phys. $\underline{15}$,
    L1277 (1982).

5. K.W. Jacobsen and H. Smith, J. Low Temp. Phys., to be published.

6. M.M. Salomaa and G.E. Volovik, these proceedings.

7. O.T. Ikkala, G.E. Volovik, P.J. Hakonen, Yu.M. Bunkov, S.T. Is-
    lander, and G.A. Kharadze, Pis'ma ZhETF $\underline{35}$, 338 (1982); JETP
    Lett. $\underline{35}$, 416 (1982).

8. G.E. Volovik, A.D. Gongadze, G.E. Gurgenishvili, M.M. Salomaa,
    and G.A. Kharadze, Pis'ma ZhETF $\underline{36}$, 404 (1982).

9. Yu.M. Bunkov, M. Krusius, and P.J. Hakonen, Pis'ma ZhETF, to be
    published.

10. D.D. Osheroff, Physica $\underline{90B}$, 20 (1977).

11. A.D. Gongadze, G.E. Gurgenishvili, and G.A. Kharadze, Fiz. Nizk.
    Temp. $\underline{7}$, 821 (1981); Sov. J. Low Temp. Phys. $\underline{7}$, 397 (1981).

12. T. Passvogel, N. Schopohl, and L. Tewordt, J. Low Temp. Phys.
    $\underline{50}$, 509 (1983).

13. S. Theodorakis and A.L. Fetter, preprint.

14. P.J. Hakonen, O.T. Ikkala, S.T. Islander, T.K. Markkula, P.M.
    Roubeau, K.M. Saloheimo, D.I. Garibashvili, and J.S. Tsakadze,
    Cryogenics, to be published.

15. J.C. Wheatley, Prog. Low Temp. Phys., Vol. VII A, North-
    Holland Publ. Co., 1978, p. 38.

ON THE THEORY OF TEXTURES IN ROTATING SUPERFLUID ^{3}He

M. M. Salomaa
Low Temperature Laboratory, Helsinki University of Technology
SF-02150 Espoo 15, Finland

and

G. E. Volovik
Landau Institute for Theoretical Physics, Academy of Sciences
of the USSR, 117334 Moscow, USSR

ABSTRACT

Theoretical principles underlying the unusual signatures of vortices in textural NMR on rotating superfluid ^{3}He are summarized here. We emphasize the evidence for continuous vorticity in the A-phase and a complicated structure of the vortex core in the B-phase, manifested through the observed first-order phase transition of the core of vortices. Textures in a rotating cylinder of ^{3}He-B which result under various combined orientating effects of surfaces, vortices, and the normal fluid-superfluid counterflow are discussed.

INTRODUCTION

Presently it is firmly established that a rotating bucket containing superfluid ^{4}He is perforated with quantized vortices. The superflow velocity around vortices, $\vec{v}_s = (\kappa_4/2\pi)\vec{\nabla}\phi$ is proportional to the gradient of the phase of the order parameter; in the coefficient of proportionality $\kappa_4 = h/m_4 \cong 0.997 \times 10^{-3}$ cm^2/s is the quantum of circulation in ^{4}He. Assuming p quanta of circulation in the rotating liquid, energy considerations yield the expression $E_v(p) \sim p^2 E_v(1)$ for the energy of a single vortex line, assumed to be quantized p times. Clearly for rotating He-II the energy attains a minimum for singly quantized vortices. Under conditions of rotating equilibrium, the vortices form a hexagonal lattice. In superconductors, on the other hand, magnetic field plays a role analogous to that of circulation in ^{4}He. Magnetic flux penetrates type II superconductors in single units of the flux quantum $\phi_0 = e/2h \cong 2.07 \times 10^{-7}$ Gcm^2. Again, these elementary flux lines tend to form a regular lattice structure, the so-called Abrikosov vortex array.

It was expected that superflow in rotating ^{3}He would resemble that in He-II, at least in some respects. Even more important was thought to be the analogy with superconductors, since in ^{3}He there occurs generalized BCS-type pairing. However, we now recognize that vortices in rotating superfluid ^{3}He possess quite unique and unusual properties, which were not anticipated at all before the discovery of superfluid ^{3}He. Two most striking features are (i) the continuously distributed vorticity in the A-phase, and (ii) the complicated structure of the core of the vortex in the B-phase, and phase transition(s) inside the vortex core. The first aspect had been conceived on theoretical grounds prior to the first experiments[1-4] on rotating

^{3}He-A, while the latter one was first discovered in the course of the ^{3}He-B experiments.[3-5]

Both liquids, ^{3}He-A and ^{3}He-B, are nuclear antiferromagnets, due to the peculiar breakdown of spin-orbit symmetry in superfluid ^{3}He. Therefore, NMR serves as a particularly useful and versatile tool for investigating both local and global properties of these superfluids, including the effects due to rotation, as we shall discuss in the following.

ROTATING SUPERFLUID ^{3}He-A

The A-phase is quite an unusual superfluid as it shares certain properties with both antiferromagnets and liquid crystals. Combined effects of all these features result in extraordinary behaviour. The varied properties are consequences of the specific structure of the order parameter characterizing ^{3}He-A; the A-phase order parameter is

$$A_{ik} = \sqrt{\tfrac{3}{2}}\Delta(T)(\hat{d})_i(\vec{\Delta}' + i\vec{\Delta}'')_k \quad , \tag{1}$$

where the part $\vec{\Delta}' + i\vec{\Delta}''$ describes a ferromagnetic symmetry break in the orbital space. The cross product

$$\vec{\ell} = \vec{\Delta}' \times \vec{\Delta}'' \tag{2}$$

defines the $\vec{\ell}$-vector. Cooper pairing in ^{3}He-A takes place in the L = 1 state, with the projection of L on the axis of quantization (the $\vec{\ell}$-vector): $L_z = 1$.

The $\vec{\ell}$-vector has many important meanings: (i) it defines the direction of the orbital momentum of the pairs, (ii) it serves as an anisotropy axis in the liquid, as in liquid crystals, such that all superfluid properties are anisotropic with respect to this axis. This is because (iii) there are nodes in the excitation spectrum in the directions $\pm \vec{\ell}$, which leads to the vanishing of the Landau critical velocity in these directions. A remarkable property of the A-phase is that (iv) textures of the $\vec{\ell}$-vector give rise to continuous vorticity (Mermin and Ho[6]), because the $\vec{\ell}$-texture is inherently connected with the superfluid part of the order parameter, such that

$$v_s^i = \frac{\hbar}{2m_3}\Delta_\alpha'\nabla^i\Delta_\alpha'' \quad . \tag{3}$$

The spin part of the A-phase order parameter corresponds to S = 1; $\vec{S}$ projected onto an axis $\vec{d}$ is zero: $\vec{S}\cdot\vec{d} = 0$. Thus $\vec{d}$ is the anisotropy axis for the spin properties, like an axis in antiferromagnets. There exists a weak spin-orbital interaction, which couples $\vec{\ell}$ and $\vec{d}$. Due to this effect, there is a possibility to probe the $\vec{\ell}$-texture with the use of NMR, and hence with it, the vortex texture.

Note in particular that vorticity, defined as the path integral

$$\kappa(r) = \oint d\vec{\sigma}\cdot\vec{v}_s \tag{4}$$

may, according to Eq. (3), be continuously distributed. Here we want to discuss briefly two examples of structures with continuous vorti-

212

city. (i) The singular vortex disgyration has the following axially
symmetric distribution of the order parameter, superflow, and vorticity:

$$\hat{\Delta}' + i\hat{\Delta}'' = -\hat{r}\sin\eta(r) + i\hat{\phi} + \hat{z}\cos\eta(r)$$

$$\hat{\ell} = -\hat{r}\cos\eta(r) - \hat{z}\sin\eta(r)$$

$$\vec{v}_s = \frac{\hbar}{2m_3 r}\sin\eta(r)\hat{\phi}$$

$$\kappa(r) = (h/2m_3)\sin\eta(r) \quad , \tag{5}$$

where $\hat{r}$, $\hat{\phi}$, and $\hat{z}$ are unit cylindrical coordinates. The function
$\kappa(r)$ varies from 0 on the vortex axis at $r = 0$ to $\pi/2$ far from it.
The radial distribution of the $\hat{\ell}$-field, the superfluid circulation,
and v_s are illustrated in Fig. 1a. At the vortex center, $\kappa = 0$ and
it varies continuously to 1 at r larger than the dipole length ξ_D^A.
The $\hat{\ell}$-vector is perpendicular to the vortex axis at $r = 0$, producing
a singularity in the $\hat{\ell}$-field. (ii) In the Anderson-Toulouse texture
the fields are again axially symmetric, but $\hat{\ell}$ is here continuosly
distributed. The corresponding vortex structure is described by

$$\hat{\Delta}' + i\hat{\Delta}'' = [-\hat{r}\sin\eta(r) + i\hat{\phi} + z\cos\eta(r)]e^{i\phi}$$

$$\hat{\ell} = -\hat{r}\cos\eta(r) - \hat{z}\sin\eta(r)$$

$$\vec{v}_s = (\hbar/2m_3 r)[1 + \sin\eta(r)]\hat{\phi}$$

$$\kappa(r) = (h/2m_3)[1 + \sin\eta(r)] \quad , \tag{6}$$

with $\eta(r)$ varying from $-\pi/2$ at $r = 0$ to $\pi/2$ at large distances. This
vortex structure is shown in Fig. 1b.

In modified form, both vortex structures form a lattice, which
in a large enough magnetic field $\vec{H}$ has the form shown in Fig. 2. The
$\vec{d}$-vector is constrained into the plane perpendicular to $\vec{H}$ due to the
large magnetic anisotropy

$$F_H = \frac{1}{2}\chi_a(\vec{d}\cdot\vec{H})^2 \quad , \tag{7}$$

where the susceptibility anisotropy χ_a is of the order χ. Note that
$\hat{\ell}$ also tends to be locked with $\vec{d}$ due to the dipole interaction

$$F_D = -g_D(\vec{d}\cdot\vec{\ell})^2 \quad , \tag{8}$$

where $g_D \sim \chi_a H_c^2$ with $H_c \sim 25$ Oe.

In each lattice, the circulation around the boundary of the unit
cell must be quantized in units of the circulation quantum in super-
fluid ^{3}He: $\kappa_3 = h/2m_3 \cong 0.662 \times 10^{-3}$ cm^2/s. In order to imitate
solid body like rotation of the superfluid component, the mean vor-
ticity should correspond to

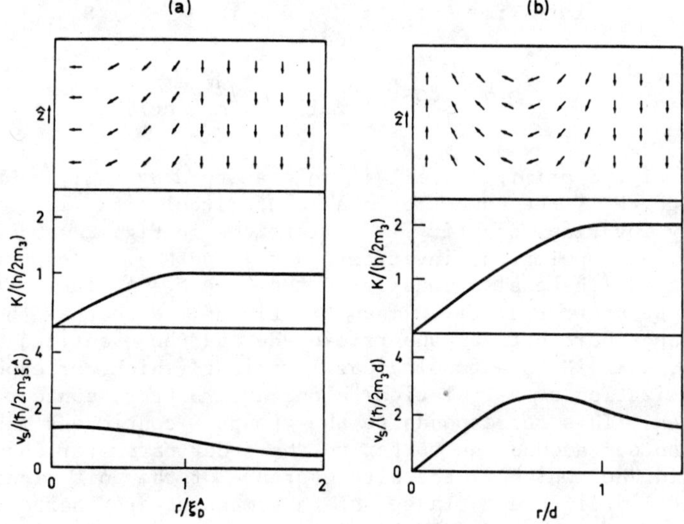

Fig. 1. Radial distributions of the $\hat{\ell}$-field, continuous vorticity, and superflow velocity for (a) the vortex-disgyration texture (Volovik and Mineev) and for (b) the Anderson-Toulouse texture, whose length scale is fixed by the vortex density: $d \cong n_\Omega^{-\frac{1}{2}}$.

Fig. 2. The fields for $\hat{\ell}$ (full arrows) and $\hat{d}$ (dashed lines) in a plane perpendicular to the magnetic field $H > H_c$, parallel to $\vec{\Omega}$. (a) The singular vortex texture where ξ denotes the superfluid coherence length. (b) The array of continuous vortices.

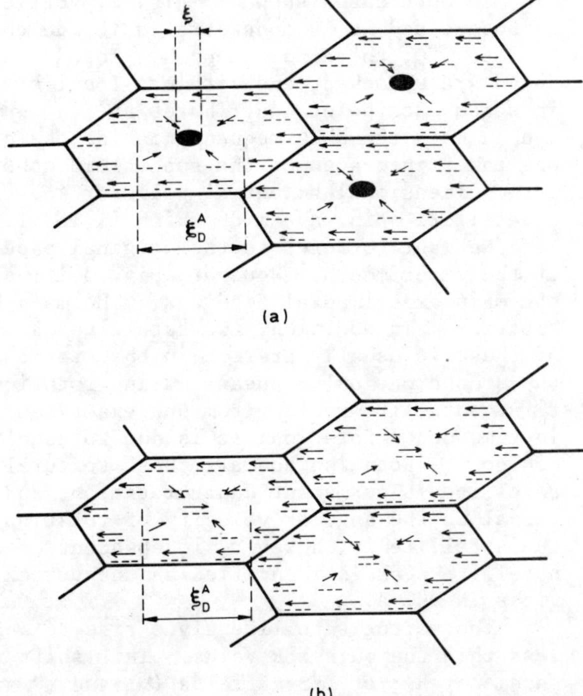

$$2\vec{\Omega} = \langle \operatorname{rot} \vec{v}_s \rangle = \hat{z} \int_{\text{cell}} d^2 r (\operatorname{rot} \vec{v}_s \cdot \vec{ds})/S_{\text{cell}}$$

$$= \hat{z} \oint_{\text{cell}} \vec{d\sigma} \cdot \vec{v}_s / S_{\text{cell}} = \hat{z}\frac{Nh}{2m_3}/S_{\text{cell}} \quad . \tag{9}$$

The area of the primitive cell is related to Ω by $S_{\text{cell}} = Nh/4m_3\Omega$. The integer N is the number of quanta of circulation, it is a topologically invariant quantity. The lattices in Fig. 2a and 2b correspond to the topological invariants N = 1 and N = 2, respectively. Fujita et al.[7] have also considered the case N = 4; their structure seems to be preferred at low magnetic fields. Note that there is a difference here between superfluid ^{3}He and the quantized vortices in He-II, where N is a topological invariant (which corresponds to the quantization of circulation) along an arbitrary contour around the vortex. This corresponds to the group of continuous mappings of the contour around the vortex to the order parameter space in superfluid ^{4}He, which is the circumference of the unit circle. The topologically different classes of this mapping are characterized by the integer number N. In ^{3}He-A, on the other hand, circulation depends on the contour, due to the continuously distributed vorticity, and is quantized only along the boundary of the elementary vortex lattice cell. In this case, the integer N corresponds to the mapping of a two-dimensional torus (the elementary cell in a two-dimensional lattice is topologically equivalent with the torus) to the unit sphere of the $\vec{\ell}$-vector.

In both cases shown in Fig. 2, vorticity is continuously distributed, and it is concentrated inside the soft vortex core of radius $\sim \xi_D^A$, the dipole length. Inside this soft core radius, $\vec{\ell}$ and $\vec{d}$ are unlocked, and within a local oscillator approximation, in which each point contributes at its own NMR-frequency, there is a change in the NMR frequency inside the core region ($< \xi_D^A$). If one takes into account the superfluid coherence, then this area is a two-dimensional potential well for the spin waves, and one expects a satellite spin wave mode which is localized in this potential well.

We want to refer to the original papers[1,2,4] for full details of the experiments. However, Fig. 3 illustrates for convenience the main experimental findings. The main NMR peak broadens during rotation. In addition, two satellites are observed in general: one of these is usually present in the stationary situation $\Omega = 0$, while the second one only appears during rotation. The intensity of the first satellite varies from one experiment to another, and it decays in time under rotation; it is due to usual solitons[8] created during cooldown. Rotation anneals these textural defects away. The second satellite is permanent under rotation, and its intensity is proportional to the angular velocity of rotation. The frequency shift of the satellite is, however, independent of rotation. The second satellite peak is identified as the vortex peak which is due to vortices in ^{3}He-A.

The vortex satellite gives rise to absorption at frequencies less than the bulk NMR value. This shift is usually described by a parameter R_t^2; at large fields (ν_o and ν_L^A are the Larmor and the A-

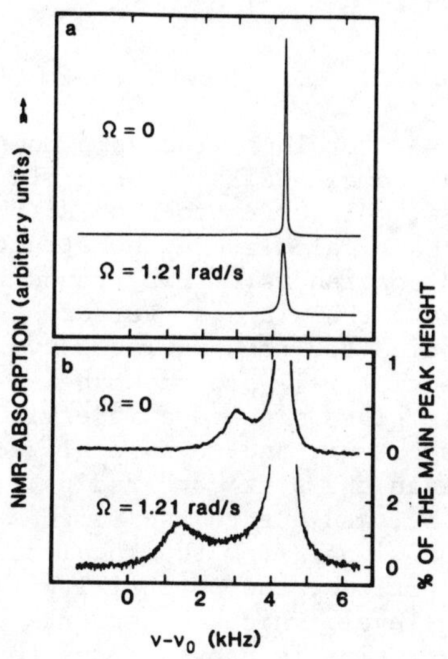

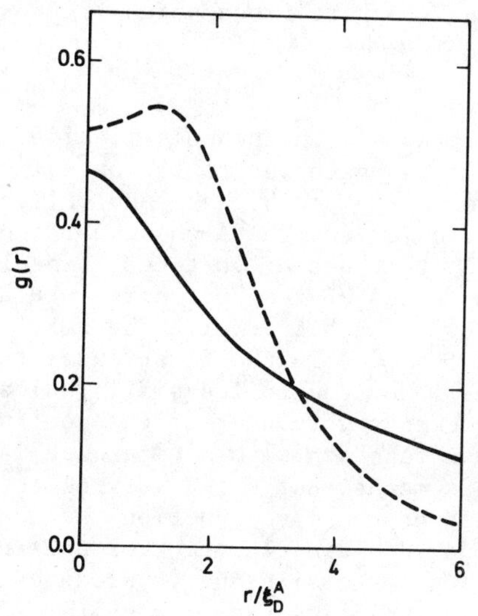

Fig. 3. (a) The broadening
of the main NMR peak in ^{3}He-A
under rotation. (b) The
soliton $(\Omega = 0)$ and vortex
$(\Omega = 1.21$ rad/s) satellites.

Fig. 4. The normalized wave
functions g(r) of a localized
state in the potential well
formed by singular (solid line)
and continuous (dashed line)
vortices (Ref. 9).

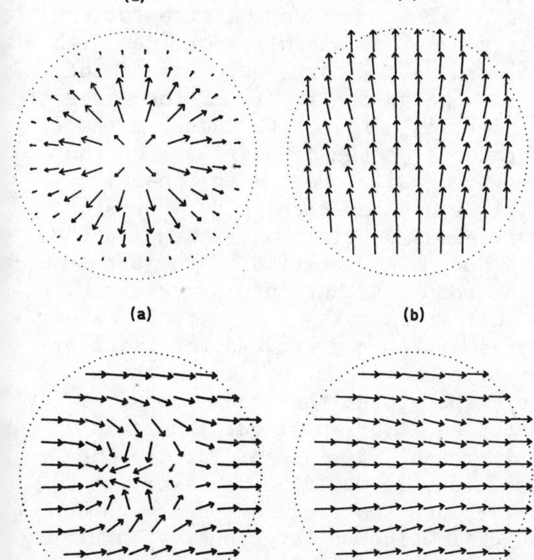

Fig. 5. The continuous z-
vortex (for $\vec{H} \parallel \hat{z}$). (a) The
distribution of ℓ_x and ℓ_y
and (b) the components of
the spin anisotropy axis,
d_x and d_z. The radius r =
$5\xi_D^A$ has been used.

Fig. 6. As in Fig. 5, but
for the continuous x-vortex.
(a) The components ℓ_x and
ℓ_y and (b) the components d_x
and d_z.

216

phase Leggett frequencies):

$$\nu = \nu_o + \frac{1}{2}R_t^2(\nu_L^A)^2/\nu_o \; , \tag{10}$$

with $R_t^2 = 1$ in the bulk liquid, and $R_t^2 < 1$ for localized spin modes.
This is analogous to the one-dimensional domain walls in superfluid
^{3}He, discussed by Maki and collaborators.[8] In order to find R_t^2 theoretically for the A-phase vortices, we have calculated[9] the structure
of both types of vortices, singular and continuous: it was found that
the sizes of the soft cores are different: the singular vortex has a
radius $\sim 2.5\ \xi_D^A$, while the soft core radius for the continuous vortex
is $\sim 5\ \xi_D^A$. In the first case, the potential well proved to be
shallower due to its small radius. This gives rise to a mode with
frequency very near to the continuous spectrum, and to a localized
wave function with a size much larger than the soft core radius.
This means that R_t^2 is very close to unity, and due to the large extent of the wave function the absorption signal from the localized
mode should be large. For continuous vortices, the potential well
has a large size and it produces a deep level, which corresponds to a
R_t^2-value being far from unity. The absorption is proportional to the
area of the soft core because the wave function is concentrated in
the potential well. Fig. 4 displays the calculated radial wave
functions for the local spin-wave mode in these two cases.

The experimental values for R_t^2 and also for the absorption intensity were quite consistent with the results obtained for the continuous vortices. This gives strong support for the conclusion that
continuous vorticity was actually observed in the experiments.

Energy considerations, however, show that at moderate angular
velocities, with $\Omega < 3 - 4$ rad/s, which are used in the experiments,
the singular vortex structure is energetically more advantageous.
This disagreement may be understood if we compare the time for the
creation of both types of vortices. In the singular vortices with
superfluidity broken inside the hard vortex core, which is of the
order of the coherence length, the energy density is of the order of
the condensation energy. It is difficult to create these in the
bulk liquid; they may only be produced on the boundaries. On the
contrary, continuous vortices may be easily created in the bulk[10]
and one may expect that they will be created first after the start
of rotation. As soon as the continuous vortices have occupied the
bulk liquid, producing the solid body like rotation of the superfluid component, there remains no room for the singular vortices.
The difference in the energy of both structures is, in fact, rather
small when compared with the total difference between the rotating
state without vortices and the solid body like rotation with an
equilibrium density of vortices. Therefore, the transition from
the continuous vortex array to the equilibrium singular vortex
array seems to have a very small probability indeed.

In order to have this transition, it would be necessary that
each continuous vortex would split into two singular ones with one
quantum of circulation each, which involves a large energy barrier.
Another argument in favour of continuous vortices is the existence
of a critical velocity in the experiment; it has the same order of

magnitude as predicted by Anderson and Toulouse for the creation of continuous vorticity.

To conclude the discussion of the rotating A-phase, we note that more detailed calculations of the vortex structure are needed, where all parameters are included, in particular the spin part. Such calculations are presently pursued, see Figs. 5 and 6, which illustrate the calculated $\hat{l}$- and $\hat{d}$-fields for the continuous z-vortex and the continuous x-vortex. These calculations will allow one to obtain more detailed agreement between experiment and theory for the intensity and the position of the vortex satellite mode. Calculations on the broadening of the main NMR line in the A-phase[11] should be carried out more quantitatively using the calculated vortex structure.

In a tilted magnetic field the $\vec{l}$-vector, which tends to align itself with the $\vec{d}$-vector, lies in a plane perpendicular to $\vec{H}$. On the other hand, $\vec{l}$ is oriented by the vortex lattice due to local superflow in the plane perpendicular to the rotation axis $\hat{\Omega}$. These effects serve to lock $\vec{l}$ in the direction $\vec{H} \times \vec{\Omega}$. This anisotropy may be probed by ion mobility or ultrasound experiments combined with NMR.

ROTATING SUPERFLUID ^{3}He-B

For the rotating A-phase, most of the important theoretical concepts had already been introduced prior to the experiments. The situation was quite opposite in the case of the B-phase, where the theory evolved only after unexpectedly interesting experimental results had been obtained. Just as for He-II, the superfluid velocity $\vec{v}_s$ in ^{3}He-B is the gradient of the phase of the order parameter $\vec{v}_s = (\hbar/2m_3)\vec{\nabla}\phi$. In addition, vortices carrying one quantum of circulation were expected.[12] Due to the isotropy of the B-phase order parameter, the spin-orbital part of the order parameter R_{ik} and the superfluid part $e^{i\phi}$ in the total order parameter $A_{ik} = \Delta(T)e^{i\phi}R_{ik}$, are practically not coupled with each other. (Above $R(n,\theta_o)_{ik} = \delta_{ik}\cos\theta_o + (1 - \cos\theta_o)n_in_k + \varepsilon_{ikj}n_j\sin\theta_o$ is the matrix of rotation from the state ^3P_o, corresponding to $J = 0$, via a relative rotation of the spin and orbital spaces by the "magic angle" $\theta_o \cong 104^o$ about the $\hat{n}$-vector.) This made one suspect that it would be impossible to find any experimental evidence for vortices in ^{3}He-B using NMR techniques. However, textural spin waves[13] were expected.

The only exception was that there would be some transient process during which an insufficient density of vortices is present. This is possible since the rate of creation of vortices in the B-phase is much slower than that in the A-phase because B-phase vortices are singular, while the A-phase vortices can be continuous. During the transient situation, there is produced a large superfluid-normal fluid counterflow $\vec{v}_s - \vec{v}_n$, which was expected to give rise to $\sin^2\beta \neq 0$, where β is the angle between $\hat{n}$ and $\vec{H}$. This would give rise to the frequency shift $\nu - \nu_o = \sin^2\beta(\nu_L^B)^2/2\nu_o$ seen in transverse cw NMR, where ν_o is the Larmor frequency and ν_L^B is the Leggett frequency.

In the absence of other orientating effects, β is zero due to the magnetic energy

$$F_M = -a(\hat{n}\cdot\vec{H})^2 \quad , \qquad (11)$$

218

which aligns $\hat{n}$ and $\vec{H}$. The small parameter $a \sim 10^{-5}\chi$ arises due to the tiny spin-orbit interaction, due to which the B-phase order parameter slightly deviates from the rotation matrix, and the B-phase becomes somewhat anisotropic.

The orientating effect of the counterflow is described by the term[14]

$$F_f = -\frac{2}{5}a[(v_i^s - v_i^n)R_{ik}H_k]^2/v_c^2 \ , \tag{12}$$

where $v_c = h/2m_3\xi_D^B \sim 1$ mm/s. Minimization of the terms $F_M + F_f$ yields $\sin^2\beta \neq 0$. In equilibrium the counterflow is practically compensated. The small uncompensated counterflow is produced by the local superflow around the vortices. However, their averaged orientating effect[15]

$$F_v^{(1)} = \frac{2}{5}a\,\lambda^{(1)}(\hat{z}_i\overleftrightarrow{R}_{ik}\vec{H}_k)^2 \tag{13}$$

was estimated to be quite small in comparison with F_f, i.e. $\lambda^{(1)} \ll 1$ at the available angular velocities such that only a very small[15] change in the NMR spectrum was expected.

However, the experiments[3-5] not only showed the existence of the transient effect, which of course confirmed the existence of vortices, but also showed noticeable nontransient rotation-dependent stationary phenomena in the rotating equilibrium state, when the counterflow is practically absent. Namely, the rotation-dependent change in the splitting of the standing spin-wave modes,[5] which arises due to the texture, was of the same order of magnitude as the splitting itself. This is illustrated in Fig. 7a. It means that λ is more effective than expected on the basis of the local superflow alone and is, in fact, of order 1. The spin-wave modes arise due to the texture, which becomes important with the large magnetic healing length ξ_H^B; which is comparable with the radius $R = 2.5$ mm of the vessel. This gives rise to a nonuniform distribution of the order parameter, and a texture with $\sin^2\beta(r) \neq 0$, in general. Taking into account the superfluid coherence, this texture gives rise to standing spin-wave modes localized on the texture. Moreover, the experiment showed an essential dependence of λ on temperature, in disagreement with the theory of Gongadze et al.[15] However, most striking is the abrupt change at the temperature $T = 0.6\,T_c$, which is independent of the angular velocity of rotation, see Fig. 7b.

The discovered phase transition cannot be attributed to a change in the vortex array structure, because in this case it would depend on the angular velocity of rotation in an essential way. Therefore, this transition is related with a change in the individual properties of a vortex, namely the change of the vortex core structure on the distance of the order of the coherence length ξ. This structure is very rigid and does not depend on the angular velocity of rotation. Thus the nontransient effect should be related with the vortex core which changes at the transition point. The mechanism of the orientating effect of the vortex core on the order parameter is the large magnetic susceptibility anisotropy[16]

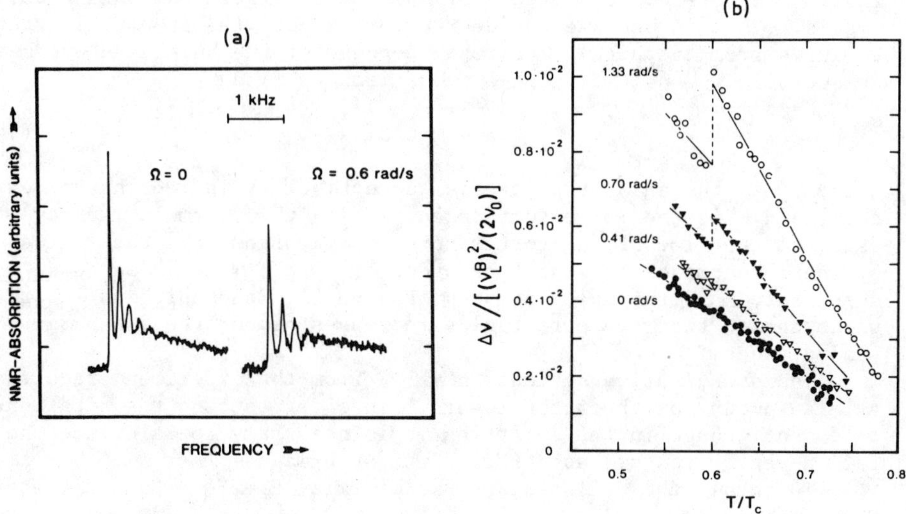

Fig. 7. (a) Typical measured spin wave spectrum in an axial field in the stationary situation and during rotation. Note the increase in the spin wave splitting with Ω. (b) The spin wave splitting between the two lowest spin eigenmodes. Note that the position of the discontinuity signalling the transition of the vortex core is inpdependent of Ω.

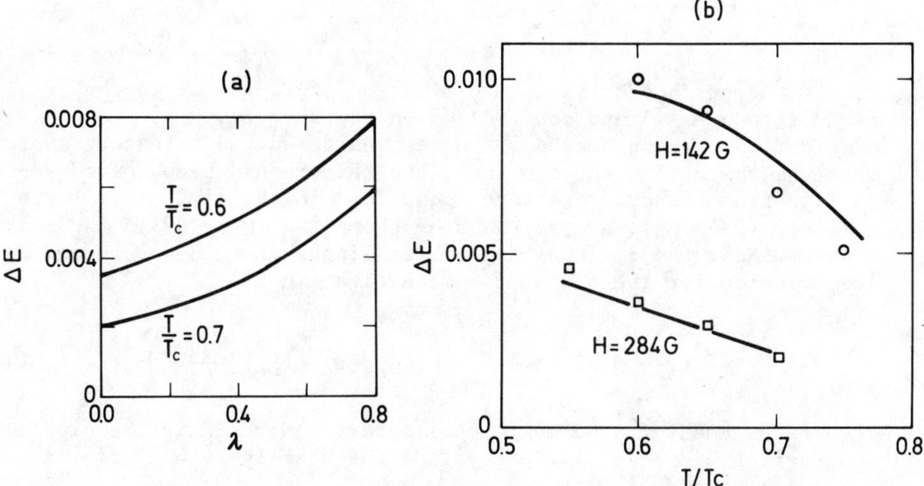

Fig. 8. (a) The calculated spin wave splitting as a function of the vortex parameter λ at two temperatures and (b) a comparison with the experimental splitting in the stationary situation at the two field values used in the experiments (from Ref. 19).

220

inside the core, 5 to 6 orders of magnitude larger than in the bulk liquid. This is because inside the vortex core the superfluid state deviates from the highly isotropic B-phase of the bulk, and becomes anisotropic. Considering this mechanism, we find

$$\lambda = \chi_a n_\Omega S/2a \quad , \tag{14}$$

where χ_a is the anisotropy of the susceptibility inside the vortex core, and S is the vortex core area. Using this formula, we may estimate the size of the vortex core; for λ of order 1 the radius of the core proves to be on the order of 10 ξ. This is in good agreement with the calculation by Passvogel, Schopohl, and Tewordt,[17] which showed that the core radius increases beyond the Ginzburg-Landau range.

The quantitative estimation of λ from the experiment requires a large amount of theoretical work. In order to find the rotation-dependent change in the splitting it is necessary to calculate the full $\vec{n}$-vector texture and to find the spin-waves.

To obtain the full texture, we minimize $F = F_M + F_V + F_g$, where F_g is the gradient energy

$$F_g = \frac{16}{13} a H^2 \left(\xi_H^B\right)^2 [\, (\nabla_i \cdot \vec{n})^2 - \frac{1}{2}\left(\sqrt{\frac{5}{8}} \vec{n} \cdot \vec{\nabla} \times \vec{n} + \sqrt{\frac{3}{5}} \vec{\nabla} \cdot \vec{n}\right)^2 \,] \quad , \tag{15}$$

F_M and F_V are as in Eqs. (11) and (13), where λ is now given by Eq. (14). It is convenient to parametrize the $\hat{n}$-vector in cylindrical coordinates as follows $\hat{n} = - \sin\beta\cos\alpha \hat{r} + \sin\beta\sin\alpha \hat{\phi} + \cos\beta \hat{z}$, where α and β denote the azimuthal and polar angles of $\hat{n}$. On the boundary $\hat{n}$ is determined by minimization of the surface energy[14]

$$F_s = - d(\hat{s} \cdot \vec{R} \cdot \vec{H})^2 \quad , \tag{16}$$

with $\hat{s}$ the surface normal. Minimization of F_s yields in the axial field case $\alpha = \pi/3$ and $\cos\beta = 1/\sqrt{5}$ on the surface.

While Maki and Nakahara[18] have considered the $\vec{n}$-texture approximately assuming α to be constant, the texture has been solved exactly[16] from the Euler-Lagrange equations in the limit of a large cylinder $(R \gg \xi_H^B)$, and it has been shown that $\alpha = 38^{\circ}$ in the bulk liquid. Near the cylinder axis, β is linear in r, $\beta \sim \beta_1 r$, and the equation for the transverse magnetization S^+

$$(\nu - \nu_o)S^+ = \frac{(\nu_L^B)^2}{2\nu_o} \left(\sin^2\beta(r) - (\xi_D^B)^2 \frac{48}{65}\nabla^2\right)S^+ \tag{17}$$

reduces to that of a harmonic oscillator. (Here ξ_D^B is the dipole length in the B-phase.) This yields the equally spaced spin wave eigenmode frequencies

$$\frac{\nu_n}{(\nu_L^B)^2/2\nu_o} = 2(n + 1)\left(\frac{48}{65}\right)^{\frac{1}{2}}\xi_D^B\beta_1 \quad , \tag{18}$$

where only the even harmonics can be excited in a uniform magnetic

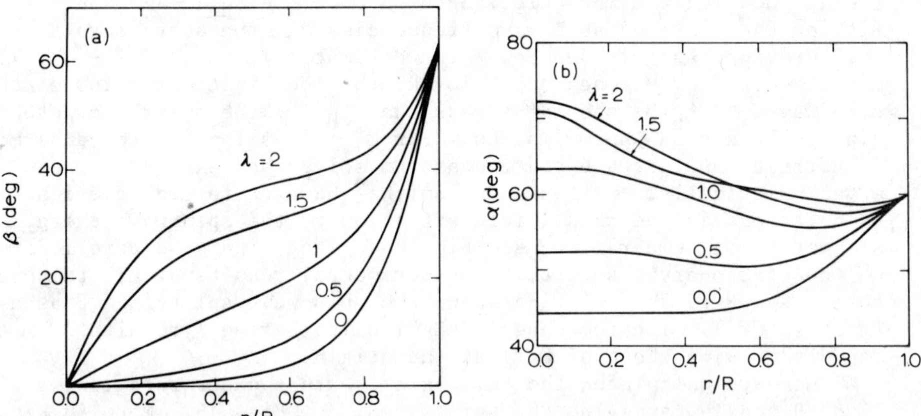

Fig. 9. The computed radial distribution of (a) the polar angle β between $\hat{n}$ and $\vec{H}$ and (b) the azimuthal angle α of $\hat{n}$ as functions of the vortex parameter λ, calculated for parameters appropriate at the temperature $T = 0.8\ T_C$. Note the formation of a plateau in $\beta(r)$ for $\lambda \gtrsim 1$, and the associated increase in $\alpha(r)$ with λ.

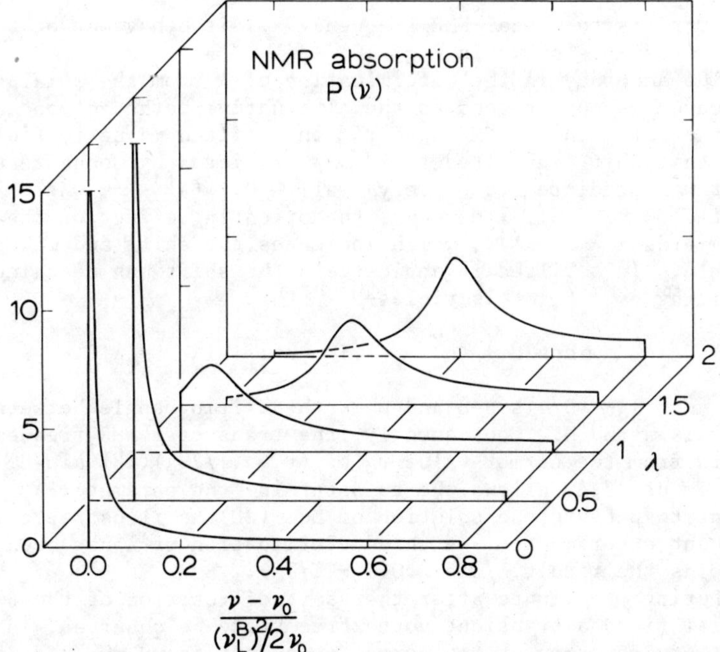

Fig. 10. The overall spectral distribution of the NMR absorption as a function of $\lambda(\Omega,T)$ for the texture in Fig. 9, using the local oscillator approximation. Note the shift of the main NMR peak to higher frequencies.

field. Scaling the magnetic length into $\xi_H^{eff} = \xi_H(1 - \lambda)^{-\frac{1}{2}}$ and fitting the measured spin-wave frequencies to the above result in the rotating case yielded for λ the estimate $\lambda/\Omega \cong 2.8(0.9 - T/T_c)$.

As noted by Hakonen and Volovik, the localization radius of the spin waves $\cong \sqrt{\xi_D^B/\beta_1}$ should be less than ξ_H^B for the quadratic potential to be a good approximation. For $\beta_1 \to 0$ as $T \to T_c$, the anharmonic corrections increase and one has to solve the texture and spin waves numerically. Jacobsen and Smith[19] have performed numerical calculations in the axial field situation on the spin-wave energy as a function of temperature, see Fig. 8a. They find good agreement in the stationary case with the experimental results at all temperatures, see Fig. 8b. Fig. 8a shows the spin-wave splitting ΔE as a function of λ. The increase in ΔE with increasing λ results from the rapidly growing slope of $\beta(r)$ at the origin.

We have calculated the full axial texture under rotation, see Figs. 9 and 14, and also the spin waves, see Fig. 15. Note in Fig. 9a the textural change, i.e. β develops a broad plateau for λ increasing beyond 1. Also α increases for small r. The plateau in β corresponds to a new maximum in the NMR spectrum, as calculated within the local oscillator approximation

$$P(\nu) = \frac{2}{R^2} \int_o^R dr \, r\delta \left[\nu - \nu(r)\right] , \qquad (19)$$

where $\nu(r)$ is the local NMR frequency. This behaviour of $P(\nu)$ as a function of λ is shown in Fig. 10.

The accuracy in the determination of λ from the axial field spin waves is not as good as the alternative method of obtaining λ from the shift in the NMR spectrum in a tilted magnetic field. First this shift in tilted field was predicted by Gongadze et al.,[15] but it was predicted to be very small $\sim 0.1 \, (\lambda^{(1)})^2$, where $\lambda^{(1)}$ was expected to be small. However, the orienting effect of the vortex core yields a $\lambda \gg \lambda^{(1)}$, which increases the shift and makes it observable. In a tilted magnetic field the shift may be calculated by minimizing $F_M + F_V$; this yields[20]

$$u\cos2\mu + (u^2 - \frac{1}{2})(1 - u^2)^{-\frac{1}{2}}\sin2\mu = \lambda^{-1} , \qquad (20)$$

where $u = 1 - (5/4)\sin^2\beta$ and μ is the tilting angle between the rotation axis $\hat{\Omega}$ and $\vec{H}$. Consequently, the transverse NMR frequency is shifted from the Larmor value ν_o by $\Delta\nu = (1/2\nu_o)(\nu_L^B)^2\sin^2\beta$, and the measurement of $\Delta\nu$ allows one to determine the parameter λ. It is convenient to picture the solution of Eq. (20) as illustrated in Fig. 11. In recent experiments[21,22] this determination of λ has been realized; it yields the result $\lambda/\Omega = 2.0(1 - T/T_c)$.

During one minute after the start of rotation of the B-phase in an axial field a transient vortexfree state is observed, in which solid body rotation of the normal component is $\vec{v}^n = \hat{\Omega} \times \vec{r}$, and the superfluid component remains at rest. The $\hat{n}$-texture which occurs in transient state[20] is obtained by minimization of $F_M + F_f + F_g$; the results are shown in Fig. 12 for different values of Ω at the temperature $T = 0.7 \, T_c$. The NMR power spectrum $P(\nu)$ calculated within the

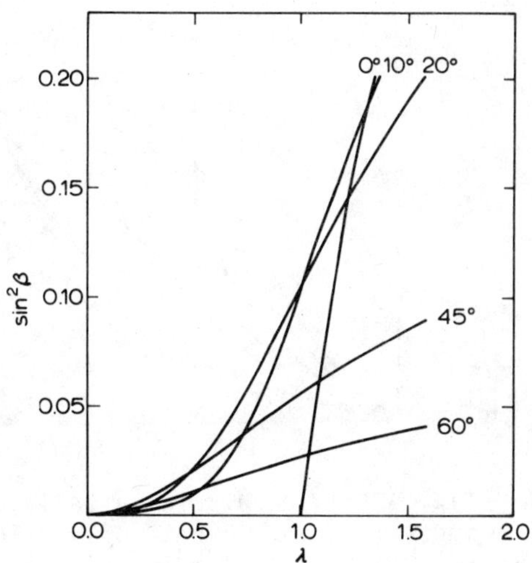

Fig. 11. The solution of Eq. (20) for the frequency shift $\propto \sin^2\beta$ of the NMR line in bulk ^{3}He-B in a tilted magnetic field as a function of the vortex parameter λ and for different tilting angles μ. The measured shift in a tilted field is easily converted to a λ-value using such curves.

local oscillator approximation is displayed in Fig. 13. For small Ω there is a sharp absorption maximum at the Larmor frequency ν_0. With increasing Ω the center of mass of the spectrum shift to higher frequencies and at $\Omega \cong 0.8$ rad/s a new peak appears. The occurrence of this peak is in rough agreement with the one observed in the experiments, except that it is already seen at $\Omega = 0.6$ rad/s. An explanation for the difference is the deviation of parameters, such as the critical velocity v_c, from their respective Ginzburg-Landau values. On further increasing Ω, singularities appear in $P(\nu)$ which result from the extrema of $\beta(r)$ in Fig. 12a.

Further interesting textural changes occur in tilted magnetic fields. In an axial field there are four different orientations of $\hat{n}$ which all correspond to the same surface energy. In a tilted magnetic field, however, it is possible to create line singularities[20] on the surface. These separate regions with different orientations of $\hat{n}$, and they are called boojums. For a tilted field, $\sin\beta$ possesses two branches on the boundary as a function of the azimuthal angle ϕ from the plane $(\hat{z},\hat{H})$:

$$\sin\beta(\phi) = \frac{2}{\sqrt{5}}(1 \pm \sin\mu\cos\phi)^{\frac{1}{2}} \ , \tag{21}$$

which correspond to the same surface energy. For $\sin\mu > 1/4$, i.e. $\mu > 14.5°$, both branches have a gap, which gives rise to two boojums, separating the regions with different branches of $\sin\beta$.

As the boojums modify the $\hat{n}$-texture in the bulk, one obtains a change in the NMR spectrum for tilting angles $\mu > 14.5°$. In the transverse field case with $\mu = 90°$, for example, assuming that boojums occur on the lines $\phi = \pi/2$ and $\phi = -\pi/2$, there exists a line on which β vanishes. To estimate the NMR power spectrum consider a small difference $\tilde{\nu} = \nu - \nu_0$ from the Larmor frequency. The power spectrum is obtained from $P(\tilde{\nu}) \propto \int \delta[\tilde{\nu} - \tilde{\nu}(r)]d^2r$, where $\tilde{\nu}(r) \cong$

224

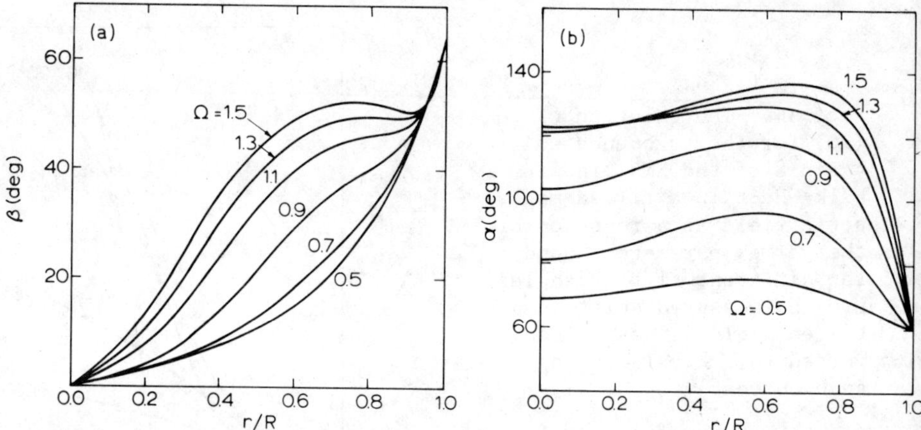

Fig. 12. The calculated radial distribution of (a) the polar angle β and (b) the azimuthal angle α in the transient texture in ^{3}He-B at the start of rotation under the orientating effect due to normal fluid-superfluid counterflow. The calculation was done at T = 0.7 T_c and H = 284 G, and Ω is in units of rad/s.

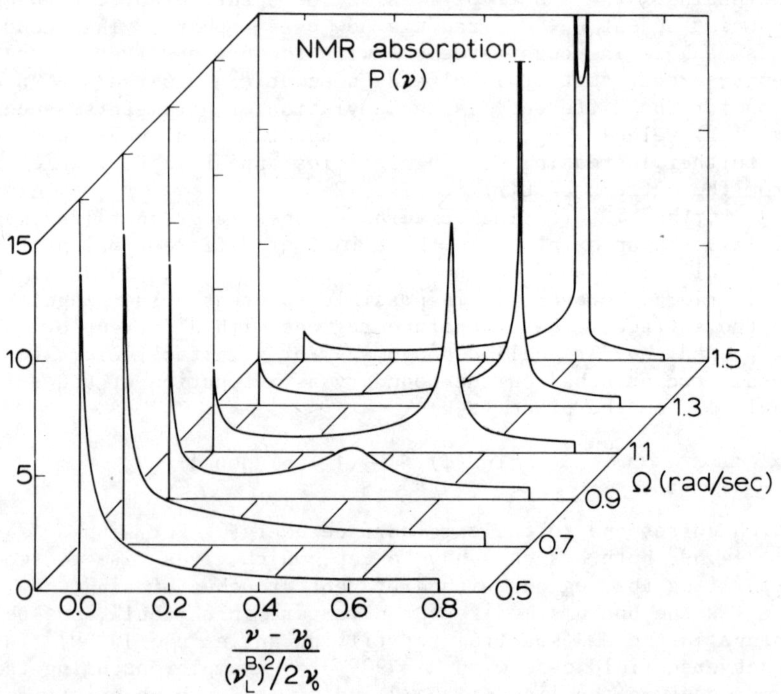

Fig. 13. The spectral distribution of the NMR absorption P(ν) for the transient texture in Fig. 12, calculated using the local oscillator approximation.

$x^2 f(y)$. Upon integrating over the cross-sectional area of the cylinder, this yields

$$P(\tilde{\nu}) \cong \frac{1}{\sqrt{\tilde{\nu}}} \int \frac{dy}{\sqrt{f(y)}} \quad , \tag{22}$$

where the integral term is a constant. This result implies a square root singularity in $P(\tilde{\nu})$ at the threshold Larmor frequency. Therefore, the change in the topology of the spin-wave potential due to boojums displays observable effects on the measurable NMR response.

Moreover, there exist intriguing metastability effects due to the boojums. This arises because it is difficult to create these linear defects. Hence the topology of the axial field texture tends to be maintained even if the field is tilted to an angle $\mu > 14.5°$ after cooldown to B-phase. Just as the boojums are difficult to be created, they are difficult to be annihilated as well. Hence it is possible to rotate the field from an angle $\mu > 14.5°$ to the axial direction and yet maintain the topology of the tilted field texture.[22] Hence the boojums tend to be "frozen" into the texture once one has cooled sufficiently far below T_c. Numerical calculations of these tilted field textures are in progress and will be reported elsewhere. Figs. 14 and 15 show first results of this general program in the special case of an axial field, which can be checked against previous calculations.[19] Detailed comparisons of the calculated textures with the experiments are possible through computed spin wave spectra and overall form of the envelope of the NMR lineshape.

DISCUSSION

It is necessary to calculate the vortex core structure for ^{3}He-B accurately, and to determine the configurations which correspond to the two (or more) observed[22] phases for the core. Theodorakis and Fetter[23] have recently presented a variational model for the structure of the B-phase vortex. We want to point out here that a spontaneously broken rotational symmetry of vortices in ^{3}He-B is possible, such that one would have a transition from an axially symmetric vortex core to one which is axially nonsymmetric.

Let us consider the orientating effect of an axially nonsymmetric vortex in the B-phase. In this case the vortex free energy term [Eq. (13)] is modified into

$$F_V = \frac{2}{5} a \lambda \{ (\hat{z}_{vortex} \cdot \overleftrightarrow{R} \cdot \vec{H})^2 + \delta (\hat{\nu}_{vortex} \cdot \overleftrightarrow{R} \cdot \vec{H})^2 \} \quad . \tag{23}$$

Here δ denotes the anisotropy of the vortex, and $\hat{\nu}$ is the direction of the anisotropy in the plane perpendicular to the axis ($\hat{z}$) of the vortex, see Fig. 16a. The anisotropy axis $\hat{\nu}$ will adjust itself in such a manner as to minimize F_V, i.e. (i) $\hat{\nu} \parallel (\overleftrightarrow{R} \cdot \vec{H})_\perp$ for $\delta < 0$, and (ii) $\hat{\nu} \perp (\overleftrightarrow{R} \cdot \vec{H})_\perp$ for $\delta > 0$. Above $(\overleftrightarrow{R} \cdot \vec{H})_\perp = \overleftrightarrow{R} \cdot \vec{H} - \hat{z}(\hat{z} \cdot \overleftrightarrow{R} \cdot \vec{H})$ denotes the transverse component of $\overleftrightarrow{R} \cdot \vec{H}$.

In the second case we find that $(\hat{\nu} \cdot \overleftrightarrow{R} \cdot \vec{H})^2 = 0$, thus F_V does not change from Eq. (13). However, in the first case we obtain $(\hat{\nu} \cdot \overleftrightarrow{R} \cdot \vec{H})^2 =$

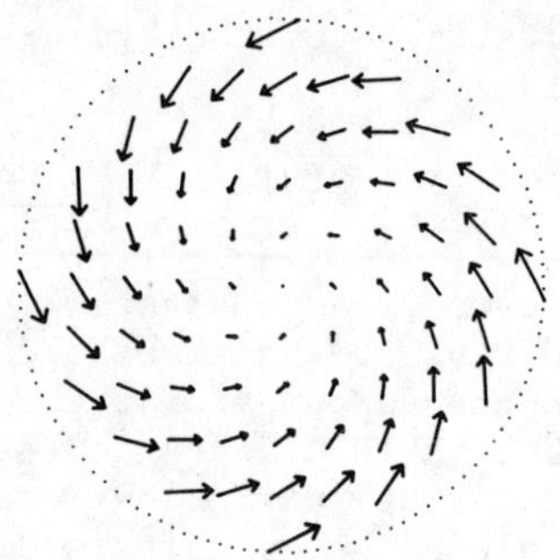

Fig. 14. The components n_x and n_y in a flare-out $\hat{n}$-texture in the plane perpendicular to $\vec{\Omega}$ at $T = 0.7\ T_c$ and rotation speed corresponding to $\lambda = 1$.

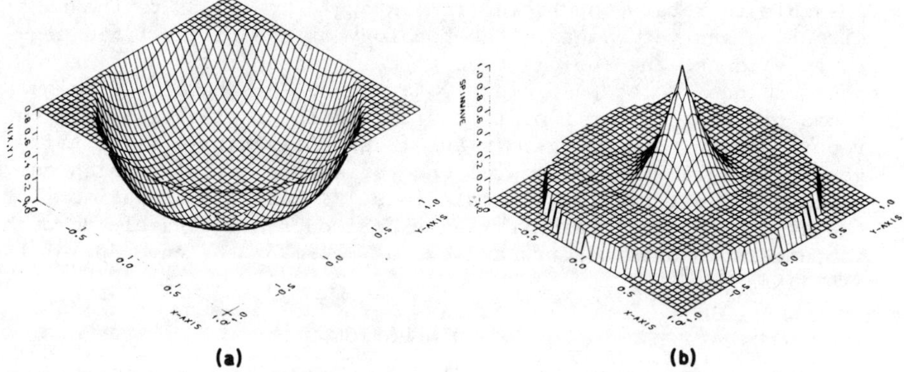

(a)   (b)

Fig. 15. (a) The spin wave potential for the texture in Fig. 14 and (b) the first spin wave function $S^+(x,y)$ in this potential. Wiggles in S^+ are due to the numerical (finite element) method employed.

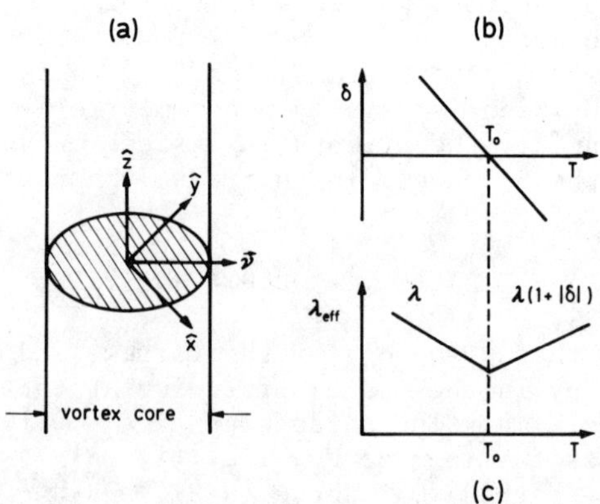

Fig. 16. The phenomenological model for a nonaxial vortex in ^{3}He-B. (a) The orientation of the anisotropy axis $\vec{\nu}$. (b) An assumed temperature dependence of δ and (c) the resulting effective λ-value.

$$[(\overleftrightarrow{R}\cdot\vec{H})_\perp]^2 = H^2 - (\overleftrightarrow{R}\cdot\vec{H})_\parallel^2 , \quad \text{such that}$$

$$F_V\Big|_{min} = \frac{2}{5}a\lambda(1-\delta)(\hat{z}\cdot\overleftrightarrow{R}\cdot\vec{H})^2 . \tag{24}$$

Now assume that $\delta(T)$ would change sign at $T = T_o$, as in Fig. 16b. We thus find

$$F_V\Big|_{min} = \begin{cases} \frac{2}{5}a\lambda(\hat{z}\cdot\overleftrightarrow{R}\cdot\vec{H})^2 & T < T_o \\[2ex] \frac{2}{5}a\lambda(1+|\delta|)(\hat{z}\cdot\overleftrightarrow{R}\cdot\vec{H})^2 & T > T_o . \end{cases} \tag{25}$$

Hence the effective λ-value, λ_{eff}, would display a discontinuity in slope, as depicted in Fig. 16c. This would be observed in the NMR properties.

ACKNOWLEDGEMENTS

We are grateful to Yu.M. Bunkov, P.J. Hakonen, O.T. Ikkala, S.T. Islander, M. Krusius, O.V. Lounasmaa, and H.K. Seppälä for co-operation and I.A. Fomin, A.D. Gongadze, G.A. Kharadze, and V.P. Mineev for useful discussions. This work has been supported by the Finnish-Soviet Commission for Scientific and Technical Cooperation, by the Academy of Finland, and by the Academy of Sciences of the USSR.

REFERENCES

1. P.J. Hakonen, O.T. Ikkala, S.T. Islander, O.V. Lounasmaa. T.K. Markkula, P. Roubeau, K.M. Saloheimo, G.E. Volovik, E.L. Andronikashvili, D.I. Garibashvili, and J.S. Tsakadze, Phys. Rev. Lett. __48__, 1838 (1982).

2. P.J. Hakonen, O.T. Ikkala, and S.T. Islander, Phys. Rev. Lett. __49__, 1258 (1982).

3. O.T. Ikkala, Ph.D. Thesis, Helsinki University of Technology, 1982.

4. P.J. Hakonen, O.T. Ikkala, S.T. Islander, O.V. Lounasmaa, and G.E. Volovik, J. Low Temp. Phys., to be published.

5. O.T. Ikkala, G.E. Volovik, P.J. Hakonen, Yu.M. Bunkov, S.T. Islander, and G.A. Kharadze, Pis'ma ZhETF __35__, 338 (1982) [transl.: JETP Lett. __35__, 416 (1982)].

6. N.D. Mermin and T.-L. Ho, Phys. Rev. Lett. __36__, 594 (1976).

7. T. Fujita, M. Nakahara, T. Ohmi, and T. Tsuneto, Prog. Theor. Phys. __60__, 671 (1978).

8. R. Bruinsma and K. Maki, Phys. Rev. B __18__, 1101 (1978).

228

9. H.K. Seppälä and G.E. Volovik, J. Low Temp. Phys. $\underline{51}$, 279 (1983).

10. T.-L. Ho, Phys. Rev. Lett. $\underline{41}$, 1473 (1978).

11. I.A. Fomin and V.G. Kamenskij, Pis'ma ZhETF $\underline{35}$, 241 (1982) [transl.: JETP Lett. $\underline{35}$, 302 (1982)].

12. G.E. Volovik and V.P. Mineev, Pis'ma ZhETF $\underline{24}$, 605 (1976) [transl.: JETP Lett. $\underline{24}$, 561 (1976)].

13. D.D. Osheroff, Physica $\underline{90B}$, 20 (1977).

14. W.F. Brinkman and M.C. Cross, in Progress in Low Temperature Physics, ed. by D.F. Brewer (North-Holland, Amsterdam, 1978), Vol. VII A, p. 105.

15. A.D. Gongadze, G.E. Gurgenishvili, and G.A. Kharadze, Fiz. Nizk. Temp. 7, 821 (1981) [transl.: Sov. J. Low Temp. Phys. $\underline{7}$, 397 (1981)].

16. P.J. Hakonen and G.E. Volovik, J. Phys. C: Solid St. Phys. $\underline{15}$, L1277 (1982).

17. T. Passvogel, N. Schopohl, and L. Tewordt, J. Low Temp. Phys. $\underline{50}$, 509 (1983).

18. K. Maki and M. Nakahara, preprint.

19. K.W. Jacobsen and H. Smith, J. Low Temp. Phys., to be published.

20. G.E. Volovik, A.D. Gongadze, G.E. Gurgenishvili, M.M. Salomaa, and G.A. Kharadze, Pis'ma ZhETF $\underline{36}$, 404 (1982).

21. Yu.M. Bunkov, M. Krusius, and P.J. Hakonen, Pis'ma ZhETF, to be published.

22. Yu.M. Bunkov, P.J. Hakonen, and M. Krusius, these proceedings.

23. S. Theodorakis and A.L. Fetter, preprint.

WHAT CAN NMR SAY ABOUT TEXTURES IN ^{3}He-A?

Alexander L. Fetter
Institute of Theoretical Physics, Physics Department
Stanford University, Stanford, CA 94305

ABSTRACT

Longitudinal and transverse NMR in ^{3}He-A can be reduced to equivalent Schrödinger equations, with attractive potentials localized in the regions where $\hat{\ell}$ differs from $\pm\hat{d}$. These potentials typically have depth and range of order unity, producing only one bound state. The associated eigenvalue characterizes the shift in the NMR frequency from the bulk value, and the range of the wave function determines the relative absorption strength. The theory is applied to narrow channels and to vortices in rotating superfluid ^{3}He-A.

I. INTRODUCTION

In the absence of external perturbations, the order parameter in a condensed medium typically assumes a uniform configuration to minimize the gradient terms in the free-energy density. This bulk state is generally well understood. In ^{3}He-A for example, it represents the state with $\hat{d}$ and $\hat{\ell}$ aligned, perpendicular to the applied magnetic field, producing the well-known shift in transverse NMR and the longitudinal NMR. When additional forces act on the medium, however, the competition between various terms in the free energy can produce new and unusual nonuniform static configurations, in which the order parameter either varies continuously in space or acquires one or more singularities. These textures change with the applied stress, and there can be sharp transitions between distinct textural states. Various methods have been used to study textures in superfluid ^{3}He, the most common being the altered NMR spectrum relative to the bulk phase. It must be noted, however, that the external magnetic field itself affects the texture. Consequently, some interesting textures (for example, certain nonsingular vortices in rotating ^{3}He-A) are inaccessible to high-field NMR. In this case, ultrasonic attenuation or ions could be very useful.[1]

The present work studies the NMR in the presence of textures-- in particular, how the NMR spectrum differs from that in bulk and what specific features of the texture are probed. It aims to develop an intuitive picture that relies on experience with quantum mechanics and Schrödinger wave functions. For definiteness, the discussion focuses on the A phase, but similar techniques apply to ^{3}He-B. Section II formulates the general problem and reduces it to an effective Schrödinger equation, one each for longitudinal and transverse applied fields. As a simple example, Sec. III applies this formalism to a slab, where a textural transition occurs at a characteristic width of order 20 µm. Section IV considers the more complicated and interesting case of bulk rotating ^{3}He-A in a large magnetic field parallel to the rotation axis. Various vortex configurations have been proposed, and the detailed form of the NMR spectrum should help identify which structure actually occurs.

II. GENERAL FORMALISM

The fundamental quantities are the order parameter and the free-energy density, which depends on the order parameter and its gradients. We shall use the hydrodynamic approximation,[2] which applies throughout the whole A-phase region of the phase diagram but holds only over distance scales large compared to the zero-temperature coherence length $\xi_0 \approx 200$ Å. As a result, it can describe fully the relevant spatial variations occurring on the scale of the dipole interactions ($L_D \approx 10$ μm), but it cannot treat the deviations from the A-phase that are believed to occur in the center of certain singular vortices. Since the NMR is insensitive to such short-distance phenomena, this defect is not serious.

The A-phase order parameter is characterized by a real unit vector $\hat{d}$ that couples to the spin coordinates and by a complex orbital vector $\hat{\Delta}_1 + i\hat{\Delta}_2$, where $\hat{\Delta}_1$ and $\hat{\Delta}_2$ are real orthonormal vectors. If we define $\hat{\ell} = \hat{\Delta}_1 \times \hat{\Delta}_2$, then the orbital part can be characterized by the equivalent orthonormal oriented triad $(\hat{\Delta}_1, \hat{\Delta}_2, \hat{\ell})$. The free-energy density will be measured in units of the dipole coupling constant λ_D (see Table I) and has the corresponding dimensionless form

$$f = f_{orb} + f_{gd} + f_D + f_M, \tag{1}$$

where f_{orb} involves only the orbital part of the order parameter,

$$f_{gd} = \frac{1}{2} \, K_1 (\hat{\ell}\cdot\vec{\nabla}) d_\mu (\hat{\ell}\cdot\vec{\nabla}) d_\mu + \frac{1}{2} \, K_2 (\hat{\ell}\times\vec{\nabla})_i \, d_\mu (\hat{\ell}\times\vec{\nabla})_i d_\mu \tag{2}$$

is the gradient energy for the $\hat{d}$ vector, $f_D = -\frac{1}{2}(\hat{d}\cdot\hat{\ell})^2$ is the dipole contribution, and $f_M = \frac{1}{2}(\hat{d}\cdot\vec{h})^2$ is the magnetic term, with

$h = H/H_D$, and $H_D \approx 30$ Oe. In addition, the order parameter obeys certain boundary conditions, with $\hat{\ell}$ normal to the surface and $\hat{d}$ having zero normal derivative.

In bulk, $\hat{d}$ and $\hat{\ell}$ tend to be uniform, aligned by the dipole energy and oriented perpendicular to the magnetic field by f_M. More generally, for any given geometry, there is assumed to be some equilibrium configuration of $\hat{d}$ and $\hat{\ell}$ that minimizes the total free energy, obtained by integrating f over the region in question. In some cases, this static equilibrium can be found easily, (for example, Sec. III), whereas its determination can be very difficult in others (particularly those involving bulk flow, such as helical textures or a rotating system, as in Sec. IV). We shall let $\hat{\ell}$ denote the equilibrium orbital texture, although further variables are required in the presence of hydrodynamic flow. This orbital

texture remains stationary when an rf magnetic field is applied.
Let $\hat{d}_0$ denote the corresponding equilibrium "spin" part of the
order-parameter texture; in contrast to the orbital part, however,
$\hat{d}$ deviates from $\hat{d}_0$ under the influence of an rf field, and we study
the linearized equations for the small deviation $\hat{d}'$. Since $\hat{d}$, is,
by definition, a unit vector, $\hat{d}'$ is perpendicular to $\hat{d}_0$, leaving
only two degrees of freedom associated with the spin dynamics of the
order parameter.

We assume that the magnitude of the static magnetic field $H_0\hat{z}$
far exceeds H_D, so that $\hat{d}_0$ lies in the xy plane (neglecting corrections of order H_D^2/H_0^2). In addition, $\hat{d}_0$ and $\hat{\ell}$ will be generally
parallel (or antiparallel), to minimize the bulk dipole energy, but
deviations from this aligned state can occur over distances of order
of the dipole length L_D (see Table I). This situation can produce
a boundary layer near a wall. It can also lead to a local bending
of $\hat{\ell}$ within a region of radius $\approx L_D$ around the center of a vortex.
Outside this "core" region, the superfluid velocity has the usual
form: approximately circular streamline, $v_s = \hbar/2m_3r$ (assuming a
singly quantized vortex), and $\hat{d}_0$ and $\hat{\ell}$ locally uniform and aligned.
It will be convenient to choose a coordinate system with $\hat{x} = \hat{d}_0$ and
$\hat{z}$ along the static magnetic field $\vec{H}_0$, but the inclusion of small
local deviations in $\hat{d}_0$ from $\hat{x}$ is not difficult.

Table I. Characteristic scales and unlocked elastic hydrodynamic
parameters for ^{3}He-A near the melting pressure. (from
M. R. Williams, Ph.D. Thesis, Stanford University, 1979,
unpublished)

T/T_c	λ_D $(10^{-5}$ erg/cm$^3)$	H_D (Oe)	L_D (μm)	K_1	K_2	K_s	K_b
.99	0.60	29.1	5.96	1.00	2.00	0.50	1.50
0.9	4.89	31.2	6.95	0.89	1.67	0.43	1.44
0.8	8.22	33.0	8.10	0.79	1.36	0.37	1.39
0.7	11.06	34.9	9.21	0.68	1.10	0.31	1.33

A weak rf magnetic field induces small oscillations in the
spin density $\vec{S}$ and the spin part $\hat{d}$ of the order parameter. A transverse field H_y' induces small rotations θ_y about $\hat{y}$, whereas a longitudinal field H_z' induces rotations θ_z about $\hat{z}$ (see Fig. 1). These
two independent quantities form a more convenient parameterization
of $\hat{d}'$ than the equivalent set $d_y' = -\theta_z$ and $d_z' = \theta_y$. Thus the problem is to obtain and solve the coupled differential equations
for the space and time variations in the induced spin density $\vec{S}'$
and the two local rotations θ_y and θ_z. In the linearized theory,
the induced response occurs at the applied frequency ω, with the
common time dependence $e^{-i\omega t}$.

The basic equations for the spin dynamics follow from Leggett's
theory and will not be given here. In the present case, it is most
convenient to eliminate $\vec{S}'$ explicitly, leading to two coupled equations for the local rotation angles (this can be done either
directly from the linearized equations of motion,[2] or, more efficiently, with the local Euler-angle formalism[2-4] since θ_z and θ_y are

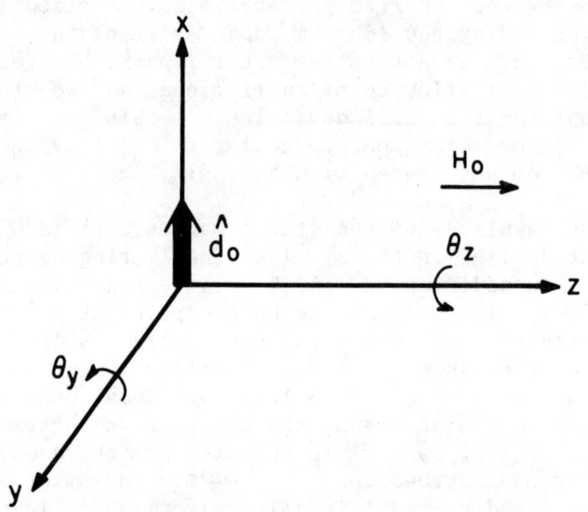

Fig. 1. Rotations of $\hat{d}$ induced by longitudinal (θ_z) and transverse (θ_y) rf fields

just the azimuthal and polar angles of $\hat{d}$). The resulting equations take the form of an inhomogeneous Schrödinger equation for a two-component wavefunction

$$\mathcal{D}\Psi + \mathcal{V}\Psi + \begin{pmatrix} 1-\omega^2/\Omega_A^2 & 0 \\ 0 & 1+(\omega_L^2-\omega^2)/\Omega_A^2 \end{pmatrix} \Psi = \frac{i\omega\gamma_0}{\Omega_A^2} \mathcal{H}' \tag{3}$$

where $\Psi(\vec{x}) = \begin{pmatrix} \theta_z(\vec{x}) \\ \theta_y(\vec{x}) \end{pmatrix}$ characterizes the spatial variation of the induced oscillations in the order parameter, and $\mathcal{H}' = \begin{pmatrix} H_z' \\ H_y' \end{pmatrix}$. Here

$$\mathcal{D} = -L_D^2 K_2 \nabla^2 - L_D^2 (K_1-K_2)\vec{\nabla}\cdot[\hat{\ell}(\hat{\ell}\cdot\vec{\nabla})] \tag{4a}$$

is a diagonal differential operator (multiplied by an implicit unit matrix) analogous to the kinetic energy operator, and

$$\mathcal{V} = \begin{pmatrix} -\ell_z^2-2\ell_y^2 & \ell_y\ell_z \\ \ell_y\ell_z & -2\ell_z^2-\ell_y^2 \end{pmatrix} \equiv \begin{pmatrix} V_1 & V_{12} \\ V_{12} & V_2 \end{pmatrix} \tag{4b}$$

is a localized attractive potential that depends only on the deviation of $\hat{\ell}$ from the local direction of $\hat{d}_0$. The explicit 2x2 matrix in (3) that depends on the frequency plays the role of the eigenvalue, and the right-hand side is an inhomogeneous driving term, proportional to the rf fields. In addition, γ_0 is the gyromagnetic ratio, $\omega_L = \gamma_0 H_0$ is the Larmor frequency, and Ω_A is the Leggett frequency that characterizes the bulk NMR shifts.

The present approach concentrates on the order parameter, but the induced oscillating magnetization is readily found with the relations

$$\gamma_0 M_x' = \chi\omega_L\theta_y$$

$$\gamma_0 M_y' = \chi(-i\omega\theta_y + \gamma_0 H_y') \tag{5}$$

$$\gamma_0 M_z' = \chi(-i\omega\theta_z + \gamma_0 H_z')$$

Furthermore, the integrated time-averaged energy absorption is

$$\overline{\frac{dU}{dt}} = \frac{1}{2}\omega \int dV \ \mathrm{Im}\ \vec{H}'\cdot\vec{M}'$$

$$= (\omega\chi/2\gamma_0)\,\mathrm{Im} \int dV[-i\omega(\theta_z H_z' + \theta_y H_y')]. \tag{6}$$

In the simple case of a uniform bulk medium, with $\hat{\ell} = \pm\hat{d}_0$, $\mathbf{\mathcal{V}}$ vanishes and, Ψ is a spatial constant found directly from Eq. (3). The usual causal prescription ($\omega \to \omega+i\varepsilon$) then yields the mean energy absorption per unit volume for (say) longitudinal excitation

$$\frac{1}{\mathrm{vol}}\,\overline{\frac{dU}{dt}} = \frac{\pi}{2}\,|\omega|^3\chi\,H_z'^2\delta(\omega^2-\omega_\ell^2), \tag{7}$$

which occurs only at the longitudinal frequency $\omega_\ell = \Omega_A$. The similar formula for transverse absorption differs only in the replacements H_y' and $\omega_t^2 = \omega_L^2 + \Omega_A^2$ for H_z' and ω_ℓ^2.

In general, the off-diagonal elements of $\mathbf{\mathcal{V}}$ couple the two components of the order-parameter oscillations. As a result, a pure transverse applied field will excite θ_z as well as θ_y, and the response will contain some component near the longitudinal frequency ω_ℓ as well as the dominant part near ω_t. For high fields, however, this coupling can be neglected apart from corrections of order $(\Omega_A/\omega_L)^2 \ll 1$. To prove this assertion, it is convenient to introduce a diagonal matrix Green's function $\mathbf{\mathcal{G}} = \begin{pmatrix} G_1 & 0 \\ 0 & G_2 \end{pmatrix}$ whose elements satisfy the simpler inhomogeneous equations

$$(\mathbf{\mathcal{D}}+V_1)G_1(\vec{x},\vec{y})+(1-\omega^2/\Omega_A^2)G_1(\vec{x},\vec{y}) = \delta(\vec{x}-\vec{y})$$

$$(\mathbf{\mathcal{D}}+V_2)G_2(\vec{x},\vec{y})+[1+(\omega_L^2-\omega^2)/\Omega_A^2]G_2(\vec{x},\vec{y}) = \delta(\vec{x}-\vec{y}) \tag{8}$$

They can be constructed from normalized eigenfunctions ψ_j and eigenvalues λ_j of homogeneous equations of the form

$$(\mathbf{\mathcal{D}}+V_1)\psi_j = \lambda_j\psi_j \quad \text{or} \quad (\mathbf{\mathcal{D}}+V_2)\psi_j = \lambda_j\psi_j \tag{9}$$

Since V_i has the typical range L_D, it is clear from (4a) that L_D also characterizes the spatial variations of ψ_i, and that the eigenvalues are of order unity. A standard calculation yields the solutions

234

$$G_1(\vec{x},\vec{y}) = \Sigma_j \ \frac{\psi_j(\vec{x})\psi_j(\vec{y})^*}{\lambda_j + 1 - \omega^2/\Omega_A^2} \ , \quad G_2(\vec{x},\vec{y}) = \Sigma_j \ \frac{\psi_j(\vec{x})\psi_j(\vec{y})^*}{\lambda_j + 1 + (\omega_L^2 - \omega^2)/\Omega_A^2} \ ,$$

$$(10)$$

where the eigenfunctions and eigenvalues are those of Eq. (9) appropriate for the potentials V_1 or V_2, respectively.

If $\mathcal{V}$ were diagonal, the problem would be solved in principle, since Eq. (3) would then separate into two uncoupled equations. Let $\mathcal{V}'$ denote the off-diagonal part of $\mathcal{V}$ in (4b). The Green's function allows us to recast Eq. (3) as an integral equation

$$\Psi(\vec{x}) = \int dV_y \ \mathcal{G}(\vec{x},\vec{y}) \ \frac{i\omega\gamma_0}{\Omega_A^2} \ \mathcal{H}' - \int dV_y \mathcal{G}(\vec{x},\vec{y}) \mathcal{V}'(\vec{y}) \Psi(\vec{y}) \qquad (11)$$

analogous to the usual formulation of scattering theory. The first term on the right-hand side is the solution neglecting the coupling between the longitudinal and transverse response ($V_{12} = 0$) and the second term contains all the remaining corrections.[12] If Eq. (11) is iterated in ascending powers of V_{12}, every term but the first contains both G_1 and G_2 at least once. As seen below, the relevant physical quantity is the solution Ψ evaluated at the resonance frequencies, which are of order Ω_A for longitudinal excitation and ω_L for transverse excitation. In either case, the structure of Eq. (10) shows that the resulting product $G_1 G_2$ is smaller than the leading term in Eq. (11) by a factor of $(\Omega_A/\omega_L)^2 \ll 1$. Thus the coupling V_{12} indeed induces small corrections in the resonance frequencies and absorption strengths, but these will be negligible compared to the dominant absorption in the present high-field limit. Evidently, the situation would be much more complicated in the low-field limit, when H_0 is comparable with the dipole field H_D.

This approximation reduces the problem to two separate eigenvalue equations (9). For the longitudinal response (for example), the leading term of Eq. (11) gives

$$\theta_z(\vec{x}) = \frac{i\omega\gamma_0}{\Omega_A^2} \ \int dV_y G_1(\vec{x},\vec{y}) H_z' = i\omega\gamma_0 \ \Sigma_j \ \frac{\psi_j(\vec{x}) \int dV_y \psi_j(\vec{y}) H_z'}{\Omega_A^2(\lambda_j+1) - \omega^2}$$

$$(12)$$

showing that the important quantity is the overlap integral of the eigenfunction $\psi_j(\vec{y})$ with the uniform rf field H_z'. Substitution into Eq. (6) yields the corresponding integrated mean energy absorption[5]

$$\frac{\overline{dU_\ell}}{dt} = \frac{\pi}{2} \ |\omega|^3 \chi \ H_z'^2 \ \Sigma_j \ \delta(\omega^2 - \omega_j^2) \ \left| \int dV_x \ \psi_j(\vec{x}) \right|^2 \qquad (13)$$

where the set of squared frequencies

$$\omega_j^2 = \Omega_A^2(1 + \lambda_j) \qquad (14a)$$

characterizes the resonant absorption for longitudinal excitation, with a strength proportional to the squared overlap integral $|\int dV_x \psi_j(\vec{x})|^2$. Only modes with a significant uniform component will be observed in an ordinary NMR experiment. Note that Eq. (13) is very similar to (7) that describes the bulk absorption; the ratio of coefficients of the delta functions can be taken as a measure of the relative absorption strength. The transverse excitation has the same form as (13), with H_y' replacing H_z' and the resonant frequencies given by

$$\omega_j^2 = \omega_L^2 + \Omega_A^2(1+\lambda_j).\qquad(14b)$$

In addition, the eigenvalues λ_j and eigenfunctions ψ_j are computed from the eigenvalue equation (9) for the transverse potential V_2 instead of the longitudinal one V_1.

In this way, the problem of NMR for ^{3}He-A in the presence of textures becomes one of finding the wave functions and eigenvalues of an attractive potential V_1 or V_2 that depends solely on the static orbital texture $\ell(\vec{x})$. Furthermore, an experiment will detect only those modes that have appreciable overlap with the applied rf field. Strictly speaking, the equation is slightly more complicated than the usual Schrödinger equation because the differential operator (4a) also depends on $\hat{\ell}$, but the qualitative situation is the same. In particular, a variational estimate for the lowest eigenvalue follows from the functional[6]

$$I_i[\psi] = \frac{\int dV[K_2|\vec{\nabla}\psi|^2+(K_1-K_2)|\hat{\ell}\cdot\vec{\nabla}\psi|^2+V_i|\psi|^2]}{\int dV|\psi|^2} \qquad i=1,2 \qquad(15)$$

whose minimum value is an upper bound on λ_0. In principle, higher states can be determined from (15) by constructing suitably orthogonalized trial states, but numerical solution of Eq. (9) is probably more direct. Note that λ_j is the quantity of most direct physical interest, for it is just the shift in the squared frequency (in units of Ω_A^2) from the appropriate bulk value. Evidently, the longitudinal and transverse resonances measure somewhat different combinations of the $\hat{\ell}$ texture, and thus complement each other.

III. NMR IN A SLAB GEOMETRY

The application of these ideas is most easily seen in examples, and we start with the A-phase confined to a slab of width W, in a strong perpendicular magnetic field. If there is no hydrodynamic flow, the $\hat{\ell}$ texture lies in a single plane (taken as the xz plane), with $\hat{d}_0 = \hat{x}$ and $\hat{\ell} = \hat{x}\sin\beta + \hat{z}\cos\beta$. All quantities depend only on the single coordinate z, which will be dimensionless, measured in units of L_D. Thus z runs from 0 to $w \equiv W/L_D$. The great simplicity of this problem is that the $\hat{\ell}$ texture can be found analytically, and has the following properties.[7] (a) for w less than a critical value $w_c = \pi K_b^{1/2}$, which depends weakly on temperature through the dimensionless constant K_b (see Table I), the walls predominate and $\hat{\ell}$ is along the z axis (parallel to the external field), with $\beta = 0$. (b) For w just greater than the critical value, the polar angle β is small, with the form $\beta(z) = \beta_m \sin(\pi z/w)$, and the maximum deviation is given by $\beta_m^2 = (2/\pi^2 K_s)(w^2 - w_c^2)$, (c) for wide channels ($w/w_c \gtrsim 2$, say), the solution has the form of a boundary layer, with $\sin\beta(z) \approx \tanh(z/K_s^{1/2})$ near $z = 0$ and a symmetrical extension near $z = w$. Note that the characteristic width of the boundary layer is comparable with the dipole length L_D, which itself is temperature dependent (see Table I). Furthermore, the critical width $W_c = \pi L_D K_b^{1/2}$ increases by nearly 50% as the temperature decreases from T_c to 0.7 T_c. Since the onset of deformation of $\hat{\ell}$ affects the NMR through the potentials V_1 and V_2, the temperature dependence might be directly measurable.

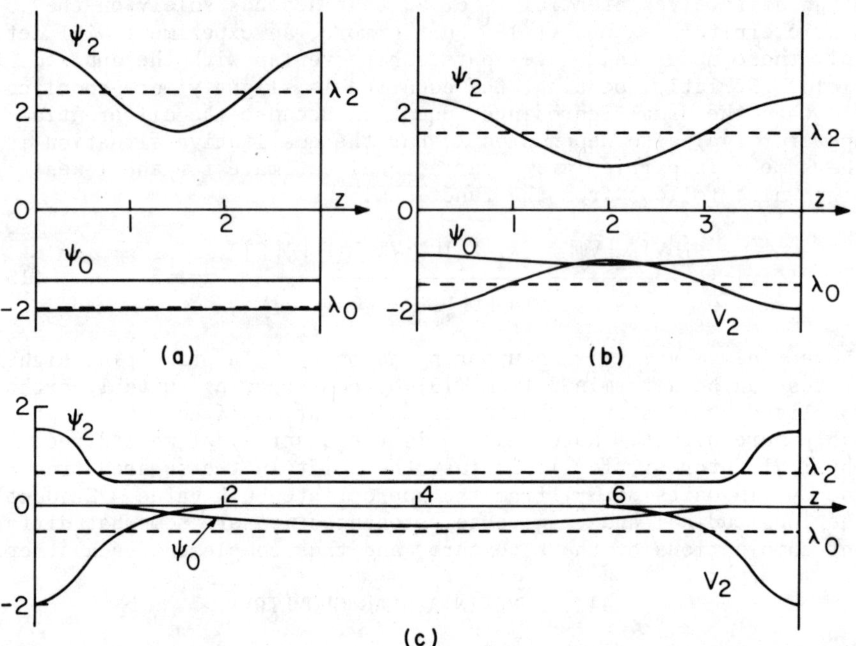

Fig. 2. NMR in a slab geometry, showing the attractive potential V_2 for transverse excitation for widths (a) w = 3 (b) w = 4 and (c) w = 8 (here we use the critical width w_c = 3.85). Also shown are the lowest state (ψ_0) and first even excited state (ψ_2) with the corresponding eigenvalues (λ_0 and λ_2) indicated as dashed lines.

In the present situation, the $\hat{\ell}$ texture is planar, and the potential matrix in Eq. (4b) is automatically diagonal, with $V_1 = -\cos^2\beta$, and $V_2 = 2V_1 = -2\cos^2\beta$. Thus the "Schrödinger" equation (9) becomes one-dimensional, and Fig. 2 illustrates the transverse potential V_2 for the three cases listed above. The only significant difference from the usual Schrödinger equation is the altered boundary condition on the wave function, which must have vanishing derivative at $z = 0$ and w, instead of the usual vanishing value. In addition, only even solutions can be excited by a uniform rf field. In case (a) ($w \lesssim w_c$), the potentials are square wells, and the solutions are proportional to $\cos(2j\pi z/w)$, with $j = 0,1,2,\ldots$. Only the mode with $j = 0$ is excited by a uniform rf field, with the corresponding eigenvalue $\lambda_0 = -1$ for longitudinal and -2 for transverse excitation. Equations (14a) and (14b) then give the frequencies $\omega_\ell^2 = 0$ and $\omega_t^2 = \omega_L^2 - \Omega_A^2$, which also follow from the general expressions for uniform textures.[8] Thus there is no longitudinal excitation in a narrow channel, and the transverse frequency is shifted below the Larmor frequency. The first two even eigenfunctions and eigenvalues for the transverse situation are shown in Fig. 2a.

The behavior in case (b) (the width just exceeds w_c and the maximum deviation β_m is small) can be examined with perturbation theory, as in the usual Schrödinger problem. The potentials V_1 and V_2 acquire small humps near the center of the channel with magnitudes of order β_m^2 (see Fig. 2b). There is a weak longitudinal absorption at a frequency $\Omega_A\beta_m/\sqrt{2}$, with an intensity proportional to β_m^3, but this effect is probably unobservable. The deformation of $\hat{\ell}$ also shifts the transverse frequency upward to $\omega_t^2 = \omega_0^2 \equiv \omega_L^2 - \Omega_A^2 + \Omega_A^2\beta_m^2$, which could be used to study the onset of the deformation as the temperature is reduced. In addition (see Fig. 2b), the state ψ_2 acquires a small uniform component, producing a much smaller transverse absorption near $\omega_L^2 = \omega_0^2 + 4K_1\Omega_A^2/K_b$.

For wide channels, the situation is quite different, because the potentials are attractive only near the walls, with a range of order 1 in the present dimensionless units. It is obvious by inspection that there are two classes of solutions. Those with negative eigenvalues (analogous to bound states) have NMR frequencies shifted down from the bulk, whereas those with positive eigenvalues (analogous to standing-wave continuum states) have NMR frequencies above the bulk values. To the extent that the two sides of the channel are independent, the bound states (those with $\lambda_j < 0$) are equivalent to even bound states in the potential $V_1 = -\mathrm{sech}^2(z/K_s^{1/2})$ or $V_2 = 2 V_1$. Unfortunately, the $\hat{\ell}$ dependence of the differential operator (4a) precludes an analytic solution, but the trial function $\mathrm{sech}^\mu(z/K_s^{1/2})$ easily gives a variational estimate from Eq. (15). These solutions are very similar to the spin eigenmodes for solitons.[3,9] As in that case, it is convenient to characterize the shifted frequencies by ratios R_ℓ and R_t, defined by

$$R_\ell^2 = 1+\lambda_0 \text{ for } V_1, \qquad R_t^2 = 1+\lambda_0 \text{ for } V_2. \tag{16}$$

(Note, however, that R_t^2 can be negative in certain cases). In addition, the absorption strength may be represented by Q, defined as the ratio of the coefficients of the delta functions in Eqs. (13) and (9). For the bound states in a wide channel, Q is proportional to w^{-1}, since the bound-state wave functions have a width of order 1. Table II gives the variational estimates of these parameters for various temperatures, and Fig. 2c shows the potential V_2 and lowest even solution ψ_0 for the transverse case.

Table II. Variational estimate of lowest NMR frequencies $\omega_\ell^2 = R_\ell^2 \Omega_A^2$ and $\omega_t^2 = \omega_L^2 + R_t^2 \Omega_A^2$, and absorption strengths Q_ℓ and Q_t (relative to bulk) for a wide channel ($w \gg 1$).

T/T_c	R_ℓ	wQ_ℓ	R_t	wQ_t
1.0	0.92	17.2	0.69	10.9
0.9	0.92	14.7	0.68	9.8
0.8	0.91	13.3	0.67	8.7
0.7	0.91	11.3	0.66	7.7

More generally, Fig. 2 illustrates several features that are expected to occur in any situation. Since the absorption is proportional to the squared overlap integral of the eigenfunction with the uniform rf field, only states with the appropriate (here even) symmetry need be considered. In this one-dimensional configuration the jth even mode has 2 j nodes (j = 0,1,2,...). The "potential" is always attractive at the walls with dimensionless negative value of order unity. As the width of the channel increases, the potential remains flat until a critical width w_c of order 4 (in units of L_D) beyond which the potential starts to rise toward zero near the center. For $w/w_c \gg 1$, the potential is essentially zero apart from a boundary region near the walls (of order 1 in range and depth). Since an attractive potential in one dimension always has at least one bound state,[10] the lowest eigenvalue λ_0 is negative, implying a downward shift in the NMR frequency relative to bulk. It is clear that λ_0 is most negative for a narrow channel and approaches a finite negative value as $w \to \infty$. If ψ_0 is normalized, the overlap with 1 is of order unity for all w.

The first excited even state ψ_2 has two nodes; the corresponding eigenvalue λ_2 appears to be positive in all cases, which is not surprising since a potential with range and depth of order unity typically has only one bound state in one dimension. For a narrow channel, ψ_2 is proportional to $\cos(2\pi z/w)$ and has zero overlap integral with 1; thus the only observable resonance is that associated with ψ_0. As the width of the channel increases, ψ_2 becomes increasingly distorted, with the nodes remaining near the walls. Thus its overlap integral with 1 increases and becomes of order $\sqrt{w}$ for $w \to \infty$. In this limit, λ_2 approaches 0 from above, and nearly all the NMR absorption is associated with this state (apart from a fraction of order w^{-1} associated with the bound state ψ_0).

The eigenfunction ψ_2 is now analogous to the lowest continuum state of a one-dimensional Schrödinger equation with box normalization (and a modified boundary condition). This latter correspondence explains why none of the higher even states contributes appreciably to the NMR. In a narrow channel, they are obviously orthogonal to 1; in a wide channel, they are analogous to excited continuum states with finite wavenumbers and are therefore again orthogonal to 1 (apart from small corrections from the region near the walls). Note that the NMR strength shifts continuously from the state ψ_0 to ψ_2 as w increases from 1 toward ∞, with λ_0 remaining negative and λ_2 positive. For dimensionless widths of order 5 (say), both states should contribute significantly. Since the critical width $\pi L_D K_b^{1/2}$ increases from $\approx 23\mu m$ near T_c to $\approx 33\mu m$ at $0.7\ T_c$, the effect might be detected in a single layered cell with spacing of order 40-50 μm. An experimental search for this behavior would be most interesting.

IV.   NMR IN ROTATING ^{3}He-A

Consider a long cylinder of radius R, placed in a large axial static magnetic field $H_0\hat{z}$ and rotating about its axis with an angular velocity Ω. We follow Volovik and Hakonen[5] in assuming that $\hat{d}_0$ is essentially uniform and perpendicular to the field ($\hat{d}_0$ along $\hat{x}$, say). The fundamental unit of circulation is $\kappa = h/2m_3$, where the factor 2 reflects the Cooper pairing. Exactly the same argument as for rotating superfluid ^{4}He implies that an array of p-fold quantized vortices will have a density $n_v = 2\Omega/p\kappa$ per unit area. We ignore the detailed lattice structure of the vortices and instead consider a circular Wigner-Seitz cell of area n_v^{-1} and radius

$$a = (\pi n_v)^{-1/2} = \left(\frac{p\kappa}{2\pi\Omega}\right)^{1/2} = \left(\frac{p\hbar}{2m_3\Omega}\right)^{1/2} . \tag{17}$$

If $a \ll R$, then the cylinder contains many vortices; in addition, currently available rotation speeds[11] imply $a \gg L_D$ ($a\approx100\mu m$ for $p = 1$ and $\Omega = 1$ rad/sec). Hence the vortices are widely separated, and the dipole energy f_D predominates over most of the sample. As noted previously, this coupling tends to align $\hat{\ell}$ and $\hat{d}_0$, but $\hat{\ell}$ can deviate from the vector $\hat{d}_0$ over regions of characteristic size L_D. In the present case, we therefore can think of $\hat{\ell}$ as uniform and parallel to $\hat{d}_0 = \hat{x}$ throughout the cylinder, apart from essentially circular "core" regions of radius $\approx L_D$ about the center of each vortex (this picture ignores the bending of $\hat{\ell}$ near the cylinder's boundary, which becomes negligible as the number of vortices becomes large). Within this "outer core" region of radius $\approx L_D$ around the vortex, the order parameter retains its A-phase form, except perhaps for a much narrower "inner core" of radius $\approx \xi(T)$, where Cooper pairing may be suppressed (note that $\xi \ll L_D$ for all practical situations).

For further progress, it is essential to consider the detailed form of the order parameter in ^{3}He-A. In the present situation, it is convenient to characterize the ⁻ ⁻⁻tal part by the orthonormal triad $\hat{\Delta}_1$, $\hat{\Delta}_2$ and $\hat{\ell}$. The orientation of this triad can be specified by the usual Euler angles (α, β, γ) needed to reach the given orientation from the standard set $\hat{x}$, $\hat{y}$, $\hat{z}$. Thus α is the azimuthal angle of $\hat{\ell}$, and β is its polar angle relative to the axes xyz. The third angle γ then represents an additional rotation about the direction $\hat{\ell}$. In general, the order parameter must be continuous and single-valued throughout the sample, but it may acquire discrete line singularities at the center of the vortices. If $\hat{\ell} \neq \pm\hat{z}$, the rotations α and γ are about distinct axes. In this case, the conditions on the order parameter require that β be continuous and single-valued, lying between 0 and π; in contrast, α and γ need not be single valued, but can instead increase by an integral multiple of 2π in following some closed circuit in a region where $\sin\beta$ never vanishes. Consequently, both α and γ are analogous to the phase Φ of the order parameter for superfluid ^{4}He; it is the occurence of two such functions that endows ^{3}He-A with a rich variety of possible vortex structures.[1]

The superfluid velocity in the A phase is given in terms of the Euler angles by

$$\vec{v}_s = -(\hbar/2m_3)(\vec{\nabla}\gamma + \cos\beta\vec{\nabla}\alpha). \tag{18}$$

This expression is more complicated than the analogous $(\hbar/m_4)\vec{\nabla}\Phi$ for ^{4}He II. In particular, curl $\vec{v}_s$ does not, in general, vanish in the A phase; instead it depends on $\hat{\ell}$ and its derivatives. As a corollary, the circulation of $\vec{v}_s$ need not be quantized about any contour along with the order parameter retains its A-phase form. (This differs from ^{4}He II, where curl $\vec{v}_s = 0$ except for core regions of radius $\approx \xi_4 = $ a few Å.)

With our Wigner-Seitz model, the picture becomes much simpler in the region outside the vortex cores, where $\hat{\ell} = \hat{x}$, $\alpha = 0$, and $\beta = \pi/2$. As a result, Eq. (18) reduces to $-(\hbar/2m_3)\vec{\nabla}\gamma$ beyond the outer core, and the cylindrical symmetry around each vortex requires that γ be $\pm$p times the local azimuthal angle about the center of the vortex. Thus the exterior superfluid velocity is locally that of a p-fold quantized vortex. As for ^{4}He II, an array of such vortices induces a net overall rotation, which is cancelled by the transformation to the frame rotating with the normal fluid and the container.[1,12] There is no difficulty in constructing an appropriate quasiperiodic function $\gamma(\vec{r})$ that has all the necessary properties throughout the region exterior to the vortex cores. Furthermore, $\hat{\ell}$ is uniform in the same region, so that continuity and single-valuedness are assured if $\hat{\ell}$ approaches $\hat{x}$ at the edge of each outer core. Thus the problem of constructing an order parameter for rotating ^{3}He-A in a large axial magnetic field reduces to that of finding local vortex textures that can join smoothly to $\hat{\ell} = \hat{x}$ and $\vec{v}_s = \pm(\hbar/2m_3)(p/r)\hat{\theta}$ on a circle of radius $\approx L_D$.

Which specific state actually occurs can be determined only by comparing the free energy per unit of circulation. For p = 2, say, the texture need not have a singularity on the axis,[13,14] eliminating a logarithmic contribution of order $\ln(L_D/\xi)$. On the other hand, the kinetic energy increases like p^2, favoring a singular vortex with $p^2 = 1$. Variational calculations with various trial functions are currently being performed; these should lead to a prediction of the sequence of vortex types that occurs in rotating ^{3}He-A as the angular velocity is increased (keeping a large magnetic field). Since the various hydrodynamic parameters depend on temperature (see Table I), it is not obvious that the same sequence of vortices will occur at all temperatures; an exploration of the whole A-phase region of the phase diagram is of interest.

Current experiments have studied these vortices with transverse NMR.[11] The formalism developed in Sec. II applies directly, and we again reduce the problem to effective Schrodinger equations with attractive potentials[5,6]

$$V_1 = -\ell_z^2 - 2\ell_y^2, \qquad\qquad V_2 = -2\ell_z^2 - \ell_y^2, \qquad\qquad (19)$$

for the longitudinal and transverse excitations, respectively. Owing to the boundary conditions at the outer-core radius, the texture and the resulting potentials will not be axisymmetric. Thus a given eigenfunction of Eq. (9) will not have a simple azimuthal dependence; instead, it must be expanded as a Fourier series in all the different angular components. Nevertheless, the final absorption will be determined by the axisymmetric component, and, as a first approximation, we may assume that the relevant eigenfunctions depend only on the radial distance from the center of the vortex. It follows from Eq. (15) that these two-dimensional analogs of s-wave solutions satisfy simple radial equations with the potentials in Eq. (19) replaced by their angular averages $\overline{V}_1$ and $\overline{V}_2$.

In this simplified picture, the longitudinal and transverse resonances differ principally in their sensitivity to the different components of ℓ. If the core texture involves a rotation of ℓ about $\hat{y}$, then $\hat{\ell}_z^2$ will predominate in the core, and the transverse potential should be the more attractive; such a texture would have a bigger transverse NMR shift from the bulk than a longitudinal one. Conversely, Seppälä and Volovik[6] have proposed vortex cores in which $\overline{V}_1$ and $\overline{V}_2$ are essentially identical, which would imply similar transverse and longitudinal NMR shifts. As in the discussion at the end of Sec. III, the potentials have a range and depth of order unity, so that only one bound state is likely to occur[10] (namely only one state will have the NMR frequency shifted down from that in bulk). Of the remaining infinite set of "continuum" states, only one will have large overlap with 1, and its frequency will be essentially that of the bulk. Detailed study of the wave functions for these two cases indicates that the absorption strength for the bound state relative to bulk is comparable with the fractional area of the vortex cores. Since their radii are of order L_D, this absorption is proportional to $(L_D/a)^2$, and hence to the number of vortices [see Eq. (17)]. In addition,

242

the overall numerical coefficient depends strongly on the range of the bound-state wave function, and hence on the binding energy.

The result of this analysis is that NMR can probe only a few rather gross features of the core texture in an A-phase vortex. In particular, it is sensitive only to the $\hat{\ell}$ configuration, and not the superfluid velocity. Furthermore, the problem is somewhat analogous to determining the internucleon potential from the binding energy of the deuteron. Since only one bound state is expected, quite different detailed $\hat{\ell}$ textures can yield the same bound-state energy (and hence NMR shift). Fortunately, the relative absorption also depends on the range of the wave function, and this additional bit of information can help in determining the specific form of the texture. It is clear that simultaneous study of longitudinal and transverse NMR doubles the available information and could thus prove invaluable in choosing among the various possible textures that have been proposed. Calculations of the resulting eigenfunctions and eigenvalues for both $\overline{V}_1$ and $\overline{V}_2$ are under way and will be reported separately.

The preceding picture applies whenever the vortices are widely separated (a $\gg$ radius of outer core $\approx L_D$). As the angular velocity increases, however, the Wigner-Seitz cell boundary a shrinks, and the exponential tails of the bound-state (spin-wave) eigenfunctions start to overlap. The resulting tunneling between different vortices in the lattice will broaden the degenerate bound-state eigenvalue into a band. Such an effect might be observable at higher angular velocity (say ≈ 10 - 20 rad/sec), but it depends sensitively on the range of the eigenfunctions. At still higher angular velocities ($\gtrsim 30$ rad/sec), the outer core radii themselves start to overlap, leading to a deformation of the $\hat{\ell}$ texture within each vortex. There may also be discrete textural transitions between different core structures, since the free energy of each different vortex depends on the ratio L_D/a.

It is illuminating to compare the NMR in the A and B phases. For both, the length L_D characterizes the typical dimensions of the spin-wave eigenfunctions. In the A phase, as shown here, L_D also determines the range of the deviation between $\hat{d}$ and $\hat{\ell}$, and hence the range of the attractive potential for NMR. Thus only one (or conceivably two) NMR absorption line with frequency below the bulk should be seen. Since the "potential" never varies slowly relative to the "wavelength" L_D, local-oscillator (quasi-classical) approximations will not work for the A phase. The B phase, in contrast, has textures that vary on a much larger scale[2] (which is also field dependent). As a result, the equivalent potentials in the B phase can have many bound states, whose relative spacing and intensities provide important detailed information on the textures. Since NMR cannot yield comparable information for ^{3}He-A, other probes (sound attenuation, ion mobility) are likely to be especially important in the A phase.

<u>ACKNOWLEDGMENTS</u>

This work is supported in part by the NSF, Grant No. DMR 81-18386.

It has benefited greatly from discussions with J. A. Sauls, D. L. Stein, and V. Vulovik.

REFERENCES

1. A. L. Fetter, J. A. Sauls, and D. L. Stein, to be published.
2. W. F. Brinkman and M. C. Cross, Prog. Low. Temp. Phys. VIIA, 105, (1978).
3. See, for example, K. Maki and P. Kumar, Phys. Rev. B17, 1088 (1978).
4. D. J. Bromley, Phys. Rev. B21, 2754 (1980); see especially the Appendix.
5. G. E. Volovik and P. J. Hakonen, J. Low Temp. Phys. 42, 503 (1981).
6. H. K. Seppälä and G. E. Volovik, to be published.
7. A. L. Fetter, Phys. Rev. B15, 1350 (1977).
8. S. Takagi, J. Phys. C8, 1507 (1975).
9. L. J. Buchholtz, Phys. Rev. B18, 1107 (1978).
10. L. D. Landau and E. M. Lifshitz, Quantum Mechanics (Pergamon, London, 1965), 2nd ed., Ch. VI.
11. P. J. Hakonen, O. T. Ikkala, and S. T. Islander, Phys. Rev. Lett. 49, 1258 (1982); and to be published.
12. V. K. Tkachenko, Zh. Eksp. Teor. Fiz. 49, 1875 (1965) [Sov. Phys.— JETP 22, 1282 (1966)].
13. G. E. Volovik and V. P. Mineev, Zh. Eksp. Teor. Fiz. 72, 2256 (1977) [Sov. Phys.— JETP 45, 1186 (1978)].
14. N. D. Mermin, Rev. Mod. Phys. 51, 591 (1979).

ANALYTIC VORTICES IN ROTATING SUPERFLUID ^{3}He-A

Kazumi Maki
Department of Physics
University of Southern California
Los Angeles, CA 90089-0484

ABSTRACT

Recent works on analytic vortices in rotating ^{3}He-A are reviewed. It is shown that the circular-hyperbolic vortex lattice has the lowest free energy in an axial magnetic field. The associated nuclear magnetic resonance satellites account for not only the observed satellite frequencies in a rotating ^{3}He-A experiment but also the observed intensity of the satellite resonances.

INTRODUCTION

Perhaps analytic vortices[1-3] in superfluid ^{3}He-A are one of the most curious objects in nature. However, for a long time these objects have remained as elegant but useless theoretist's constructions. In the last year this deplorable state of affairs is completely changed due to a beautiful series of experiments performed in rotating superfluid ^{3}He by Hakonen et al.[4,5] Indeed we believe now that the transverse NMR satellites generated in rotating superfluid ^{3}He-A in an axial magnetic field are due to analytic vortices. In general vorticity of superfluid in rotating ^{3}He-A is carried by quantized vortices, which can be ordinary singular vortices or analytic vortices. We shall, however, limit ourselves to 2π-analytic vortices, since the vortex lattice formed with 2π-analytic vortices[6] appears to have the lowest free energy. Fujita et al[6] have shown that it is possible to form a regular two dimensional lattice with the circular and the hyperbolic vortices with the lowest free energy in the absence of magnetic field in the Ginzburg Landau regime. We extend their result in the presence of an axial magnetic field.[7] Furthermore their stability is valid beyond the Ginzburg-Landau regime.[7] Only in the extreme low temperatures the radial-hyperbolic pairs become more favorable than the circular-hyperbolic pairs.[7] These analytic vortices give rise to the NMR satellites. Contrary to the earlier assertion,[7] the observed transverse NMR satellite frequencies[4,5] lie right in between those associated with the circular vortices and the hyperbolic vorteces,[8] suggesting that the rotating ^{3}He-A experiment[5] indeed observes the circular-hyperbolic vortex lattice. Furthermore the observed Ω-dependence of the satellite intensity is also consistent with the present interpretation.

THREE TYPES OF ANALYTIC VORTICES

We shall first study the free energy of analytic vortices in the presence of a magnetic field along the z direction, which is parallel to the rotation axis. In this situation the spatial configuration of $\hat{d}$ in rotating ^{3}He-A is mapped to the spatial distribution of disgyrations. Indeed we can fill the two dimensional space with a regular array of disgyrations with n=1 (radial or circular) and n=-1 (hyperbolic), so that the sum of the indices n's over all disgyrations are zero. Far away from the center of each vortex, $\hat{\ell}$ the orbital vector is parallel to $\hat{d}$ to minimize the dipole energy. Therefore in the presence of a strong magnetic field we have three types of 2π analytic vortices; the circular, the hyperbolic and the radial vortex as shown in Fig. 1. From these three types of vortices we can form a two dimensional regular vortex lattice as shown in Fig. 2. As easily seen these are the square lattices. Note that unlike the case of the singular vortices, where we can form the hexagonal lattice, it is impossible to form the hexagonal lattice with analytic vortices. For example the configuration proposed by Volovik and Kopnin[3] requires singular sheets or solitons. Therefore their configuration cannot be the low free energy state with the given vorticity. We shall present here our analysis of the vortex free energy in the Ginzburg Landau regime for simplicity. Furthermore we consider single vortices as the intervortex distance r_0 in the Helsinki experiment is much larger than $\xi_\perp$ ($=10\mu$m) the dipole coherence distance. For this purpose we start with the texture free energy of ^{3}He-A in a magnetic field given by

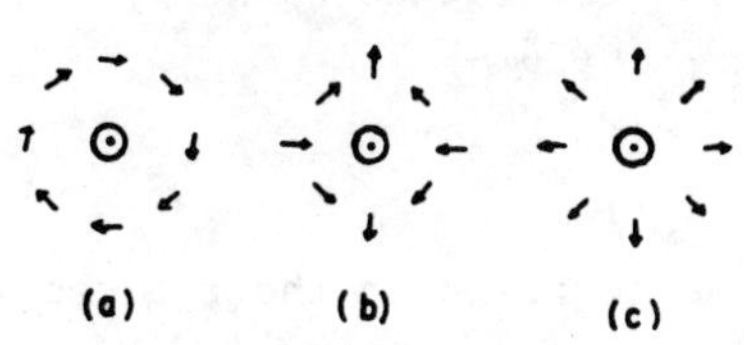

Fig. 1. Three types of analytic vortices are shown, where arrows indicate the direction of the $\hat{\ell}$ vector. a) circular, b) hyperbolic and c) radial.

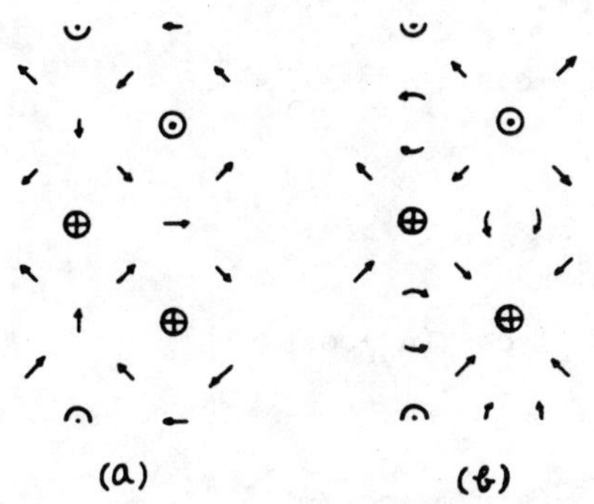

Fig. 2. Two types of the vortex lattice a) the circular-hyperbolic and b) the radial-hyperbolic type.

246

$$F = F_{kin} + E_D + E_H \tag{1}$$

where

$$F_{kin} = \frac{1}{2} K \int d^3r \left[3|\vec{\nabla}\cdot\vec{\Delta}|^2 + |\vec{\nabla}\times\vec{\Delta}|^2 + 2|(\vec{\Delta}\cdot\vec{\nabla})\hat{d}|^2 \right.$$
$$\left. + |\vec{\Delta}|^2 \, (|\vec{\nabla}\cdot\hat{d}|^2 + |\vec{\nabla}\times\hat{d}|^2) \right] \tag{2}$$

and

$$K = \frac{6}{5} \frac{N}{8m^*} \frac{7\zeta(3)}{(2\pi T_c)^2} \tag{3}$$

and E_D and E_H are the dipole energy and the magnetic energy. Here we have written the order parameter of ^{3}He-A as

$$A\mu\nu = (\Delta_o/\sqrt{2})\vec{\Delta}_\mu\hat{d}_\nu \tag{4}$$

In the presence of a strong magnetic field in the z direction, we can parameterize $\hat{d}$ and $\vec{\Delta}$ as,

$$\hat{d} = (-\cos\psi\hat{x}+\sin\psi\hat{y})\sin\chi+\cos\chi\hat{z}$$

$$\vec{\Delta} = e^{i\alpha}[\cos\beta(\cos\gamma\hat{x}-\sin\gamma\hat{y})+\sin\beta\hat{z}$$
$$+i(\sin\gamma\hat{x}+\cos\gamma\hat{y})] \tag{5}$$

and assuming that $\hat{d}$ and $\vec{\Delta}$ depend only on x and y, we find the free energy for a single vortex per unit length f;[7,10]

$$f = \frac{1}{2}A\int d^2r \{4[(\vec{\nabla}\alpha)+\cos\beta(\vec{\nabla}\gamma)]^2+3\sin^2\beta(\vec{\nabla}\gamma)^2$$

$$-2\sin^2\beta[(\cos\gamma\alpha_x-\sin\gamma\alpha_y)^2+(\sin\gamma\alpha_x+\cos\gamma\alpha_y)^2]$$

$$+|\vec{\nabla}\beta|^2+2\sin^2\beta(\cos\gamma\beta_x-\sin\gamma\beta_y)^2$$

$$+2\sin\beta[\beta_x(\alpha_y+\cos\beta\gamma_y)-\beta_y(\alpha_x+\cos\beta\gamma_x)$$

$$+2(\cos\gamma\beta_x-\sin\gamma\beta_y)(\sin\gamma(\alpha_x+\cos\beta\gamma_x)+\cos\gamma(\alpha_y+\cos\beta\gamma_y))]$$

$$+2[(1+\cos^2\beta)(|\vec{\nabla}\chi|^2+\sin^2\chi|\vec{\nabla}\psi|^2)$$

$$+\sin^2\beta((\sin\gamma\chi_x+\cos\gamma\chi_y)^2=\sin^2\chi(\sin\gamma\psi_x+\cos\gamma\psi_y)^2)]$$

$$+4\xi_\perp^{-2}[1-(\cos\chi\cos\beta+\sin\chi\sin\beta\cos(\gamma-\psi))^2]$$

$$+4\xi_H^{-2}\cos^2\chi\} \tag{6}$$

where $\xi_\perp$ and ξ_H are the dipolar coherence length ($\xi_\perp = C/\Omega-10\mu m$)

and the magnetic coherence length $[\xi_H=(H_o/H)\xi_\perp$ where $H_o{\sim}27{\,}Oe$ at the melting pressure]. In the experiment of Hakonen et al[4,5] with $H=284$ Oe, we obtain $\xi_H{\sim}10^{-1}\xi_\perp$.

First let us consider a family of radial-circular vortices, which is given by[6,7]

$$\alpha = \phi$$

$$\psi = \gamma = -\phi+\gamma_o \qquad (7)$$

where ϕ is the azimuthal angle and γ_o is a constant; $\gamma=0$ corresponds to the radial vortex, while $\gamma_o = \frac{\pi}{2}$ to the circular vortex. Then it is easily seen that β and χ depend only on r the radial distance from the center of the vortex. In the case of the radial vortex, more accurate numerical solutions are now available.[11] However, we shall determine here β and χ variationally, which will be adequate for our purpose. We assume that β and χ are given by

$$\cos\beta = e^{-(\eta r)^2} \qquad\qquad \cos\chi = e^{-(\zeta r)^2} \qquad (8)$$

where η and ζ are variational parameters.

Substituting Eqs. (7) and (8) into Eq. (6), we obtain;

$$f = \pi A\{(5-2\sin^2\gamma_o)\ell n(\sqrt{\gamma}^*\eta r_o)+4\ell n(\sqrt{\gamma}^*\zeta r_o)$$

$$+(\frac{1}{2}-\sin^2\gamma_o)\ell n2+(\sin^2\gamma_o-2\zeta(3)\cos^2\gamma_o)(\eta/\zeta)^2$$

$$+\frac{5}{12}\pi^2+\cos^2\gamma_o+\xi_\perp^{-2}(\eta^{-2}-\zeta^{-2})+\xi_H^{-2}\zeta^{-2}\} \qquad (9)$$

where we kept up to the lowest order terms in $(\eta/\zeta)^2$ as $\eta/\zeta{\sim}10^{-1}$. Here we have cut off the r integral at r_o the intervortex distance. Minimizing f in η and ζ, we obtain

$$\eta^{-2} = \frac{1}{2}\xi_\perp^2[5-2\sin^2\gamma_o+(\sin^2\gamma_o-2\zeta(3)\cos^2\gamma_o)(\eta/\zeta)^2]$$

$$\zeta^{-2} = 2\xi_H^2 \qquad (10)$$

The size of the d-core (ζ^{-1}) is independent of γ_o, while that of the ℓ-core of the analytic vortex depends on γ_o. The η's for the radial and the circular vortices are given by

$$\eta_r^{-2} = \frac{1}{2}\xi_\perp^2(5-\frac{8}{5}\zeta(3)P)^{-1} \qquad (11)$$

$$\eta_c^{-2} = \frac{1}{2}\xi_\perp^2(3+\frac{4}{3}P)^{-1} \qquad (12)$$

with $p=(H_o/H)^2$, respectively. It is easy to see that the circular vortex ($\gamma_o=\pi/2$) has the lowest energy, while the radial vortex ($\gamma_o=0$) has the highest energy. Also it is clear from Eqs. (11) and (12) that the circular vortex is the smallest in size in the radial-circular family.

To go beyond the Ginzburg Landau regime, we make use of the generalized Ginzburg-Landau free energy due to Cross.[12] A similar analysis is carried out in this more general case. The lowest order corrections in $\varepsilon(=1-T/T_c)$ are easily calculated. In this approximation η^{-2}'s for the radial and the circular vortices are given by[7]

$$\eta_r^{-2} = \frac{1}{2} \xi_\perp^2 [5 + \frac{2}{3}(4A_1 - 5B_1)\varepsilon] \tag{13}$$

$$\eta_c^{-2} = \frac{1}{2} \xi_\perp^2 [3 + 4(\frac{1}{3}A_1 - B_1)\varepsilon] \tag{14}$$

where

$$A_1 = F_1/(1 + \frac{1}{3}F_1)$$

$$B_1 = F_1^a/(1 + \frac{1}{3}F_1^a) \tag{15}$$

and F_1 and F_1^a are the Fermi-liquid coefficients. At the melting pressure making use of $F_1 = 15.66$ as tabulated by Wheatley[13] and $F_1^a = -1.33$ deduced from the NMR satellite frequencies associated with the composite soliton,[14,15] it is easily seen that $\eta_r > \eta_c$, implying that the circular vortex has the lower free energy than that of the radial vortex in this temperature region.[7] Therefore we believe that the two dimensional vortex lattice formed by the circular-hyperbolic vortex pairs is the one with the lowest free energy in the temperature region of experimental interest. $(0 < \varepsilon < 0.4)$ On the other hand near $T = 0K$, it is shown that the radial vortex has the lower free energy than that of the circular vortex. The hyperbolic vortex is constructed as[7]

$$\alpha = \phi$$

$$\psi = \gamma = \phi + \gamma_0 \tag{16}$$

where γ_0 is again a constant. In this case the vortex energy does not depend on γ_0, since γ_0 determines the orientation of the hyperbolic vortex in the x-y plane. Contrary to a family of the circular-radial vortices, the hyperbolic vortex does not possess the axial symmetry. However, for simplicity, we assume that β and χ are still given by Eq. (8) but η depends now on ϕ as well. Then we obtain

$$f_h = \frac{1}{2}A\int_0^{2\pi}d\phi\{(4+\cos(4\phi))\ln(\sqrt{\gamma}*\eta\pi_0) + 4\ln(\sqrt{\gamma}*\zeta r_0)$$

$$+ \sin^2(2\phi)(1+\zeta(3))(\eta/\zeta)^2 + \frac{\pi^2}{12}(1+(\eta_\phi/\eta)^2)$$

$$+[\cos(2\phi) - \sin(2\phi)\eta_\phi/\eta]^2 + \frac{\pi^2}{6} + \xi_\perp^{-2}(\eta^{-2}-\zeta^{-2})$$

$$+\xi_H^{-2}\zeta^{-2}\} \tag{17}$$

Minimizing Eq. (17), we obtain

$$\zeta^{-2} = 2\xi_H^2$$
$$\eta_h^{-2} = \frac{1}{2}\xi_\perp^2 (4+\cos(4\phi) + (1+\zeta(3))P) \tag{18}$$

From this we conclude that

$$\eta_r^{-1} > \eta_h^{-1} > \eta_c^{-1}$$

and

$$f_r > f_h > f_c \tag{19}$$

where f_r, f_h and f_c are the free energy of the radial, the hyperbolic and the circular vortex. The ε corrections to η_h is obtained similarly as

$$\eta_h^{-2} = \frac{1}{2}\xi_\perp^2[4+2(A_1 - \frac{11}{6}B_1)\varepsilon] \tag{20}$$

we conclude that the two dimensional vortex lattice formed by the circular-hyperbolic vortex pairs is the most stable one in the presence of a magnetic field as well in the low temperature region. This generalizes the result due to Fujita et al.[6]

NUCLEAR MAGNETIC RESONANCES

The nuclear magnetic resonances (NMR) is the most sensitive method to unravel the textural defects with localized deficit in the dipole energy.[14] In the presence of such a defect the local deficit in the dipole energy gives rise to the localized potential for the spin oscillations. In particular the bound spin wave states can be detected as satellites in the NMR experiments. In the case of analytic vortices in an axial magnetic energy, the vortices give rise to the two-dimensional potential well with the diameter of the order 2ξ. The corresponding satellite frequencies are expressed in terms of eigenvalues λ_g and λ_f as $\omega_t = (\omega_o^2 + \lambda_g\Omega_A^2)^{\frac{1}{2}}$ and $\omega_\ell =(\lambda_f)^{\frac{1}{2}}\Omega_A$ for the transverse and the longitudinal satellites respectivelr, where $\omega_o =\gamma H$ is the Larmor frequency and Ω_A is the Leggett frequency. Since the eigenvalue equations are slightly different for the circular, hyperbolic and radial vortices, we shall consideer the three cases separately
a. Circular Vortex

250

The circular vortex is an analytic vortex with the lowest free energy among three analytic vortices so far considered. The corresponding potential is the smallest of the three's. The eigen value equations are[7,8]

$$\lambda_g g = -\frac{1}{r}\frac{d}{dr}(rg_r) + (1 - 2\cos^2\beta)g \tag{21}$$

$$\lambda_f f = -\frac{1}{r}\frac{d}{dr}(rf_r) + (1 - \cos^2\beta)f \tag{22}$$

where the length r is measured in the unit of $\xi_\perp$ and $\cos\beta$ is given by

$$\cos\beta = e^{-(\eta_c r)^2} \tag{23}$$

With η_c defined in Eq(14). We have solved Eqs(21) and (22) variationally and find;[8]

$$\lambda_g = 0.874 - .192(\tfrac{1}{3} A_1 - B_1)\epsilon \tag{24}$$

$$\lambda_f = 0.995 - .022(\tfrac{1}{3} A_1 - B_1)\epsilon \tag{25}$$

b. Hyperbolic Vortex

The corresponding eigen equations are given by;[7,8]

$$\lambda_g g = -\frac{1}{4r}\frac{d}{dr}(r(3+\cos^2\beta)g_r) + (1 - 2\cos^2\beta)g \tag{26}$$

$$\lambda_f f = -\frac{1}{4r}\frac{d}{dr}(r(3+\cos^2\beta)f_r) + (1 - \cos^2\beta)f \tag{27}$$

where $\cos\beta$ is still given by Eq(24) but η_c replaced by η_h given in Eq(18). A variational calculation yields

$$\lambda_g = .720 - .218(A_1 - \tfrac{11}{6} B_1)\epsilon \tag{28}$$

$$\lambda_f = .958 - .052(A_1 - \tfrac{11}{6} B_1)\epsilon \tag{29}$$

c. Radial Vortex

The corresponding eigen equations are given by

$$\lambda_g g = -\frac{1}{2r}\frac{d}{dr}(r(1+\cos^2\beta)g_r) + (1 - 2\cos^2\beta)g \tag{30}$$

$$\lambda_f f = -\frac{1}{2r}\frac{d}{dr}(r(1+\cos^2\beta)f_r) + (1 - \cos^2\beta)f \tag{31}$$

where $\cos\beta$ is given by Eq(24) but η_c replaced by η_r in Eq(13). The eigenvalues are

$$\lambda_g = 0.4928 - .223(A_1 - \frac{5}{4} B_1)\epsilon \tag{32}$$

$$\lambda_g = 0.889 - .0852(A_1 - \frac{5}{4} B_1)\epsilon \tag{33}$$

So far we limit ourselves to the resonance frequencies. However, it is easy to calculate the intensity of the corresponding satellite resonances. The intensity of the satellite resonances from a single vortex is expressed in terms of the eigen functions g for the case of the transverse resonance (and f for the case of the longitudinal resonance) as

$$I = 2\pi \left| \int_0^\infty r \, dr \, g \right|^2 / \int_0^\infty r \, dr \, g^2 \tag{34}$$

Substituting the corresponding eigen functions we obtain;

$$I_c = 14.8(1 + \frac{8}{3} (\frac{1}{3} A_1 - B_1)\epsilon)(2\pi\xi_\perp^2) \tag{35}$$

$$I_h = 6.4(1 + (A_1 - \frac{11}{6} B_1)\epsilon)(2\pi\xi_\perp^2) \tag{36}$$

$$I_r = 2.9 (1 + \frac{4}{15} (4A_1 - 5B_1)\epsilon)(2\pi\xi_\perp^2) \tag{37}$$

for the transverse satellite associated with the circular, the hyperbolic and the radial vortex respectively. Since two types of the vortex shares the angular momentum of the superfluid the relative weight of the satellite resonance due to the circular vortex is then given by

$$I_{sat} = I_c \Omega / 2K \tag{38}$$

where Ω is the rotation speed and $K = h/2m_3 = 0.661 \times 10^{-3} cm^2/sec$.

COMPARISON WITH EXPERIMENTS

The transverse satellite resonance frequencies observed by Hakonen et at[4,5] are conveniently fitted by

$$(\lambda_g^{exp})^{1/2} = 0.895 - 1.24 \, \epsilon \tag{39}$$

for $0.15 < \epsilon < 0.4$, although there is a small fluctuations (of the order of 5%) in the experimental data. Extrapolating this result to $T = T_c$ (i.e. $\epsilon = 0$), we find that the experimental data lie right in between the satellite frequencies due to the circular vortex and those of the hyperbolic vortex.[8] Therefore it is quite natural to suppose that the observed satellites corresponding to the two separate satellites, one due to the circular vortex and the other due to the hyperbolic vortex. This interpretation can account for the reason why the observed satellite peaks are so broad. Furthermore, assuming that the satellite intensity is the sum of the intensity due to the circular vortex and that due to the hyperbolic vortex, this is again consistent with the observed intensity. The temperature dependence of the satellite position (i.e. $R_t = (\lambda_g)^{\frac{1}{2}}$) can be fitted by choosing $F_1^a = -1.82$. Such a fit is plotted in Fig.3. We see that the experimental data lie just in the middle of two satellites predicted theoretically.[8] In this fitting we made use of the value of $F_1 = 15.66$ tabulated by Wheatley.[13] The value used for F_1^a is somewhat larger in the absolute magnitude than that deduced from the satellite frequency of the composite soliton,[14] but is not inconsistent. We believe that, if the present interpretation is correct, the observed satellites will split into two distinct satellites at low enough temperatures. In any case it appears that the two dimensional lattice formed by the circular and hyperbolic vortices as predicted by Fujita et al[6] is consistent with the observed satellites in rotating ^{3}He-A. However more works are clearly desirable.

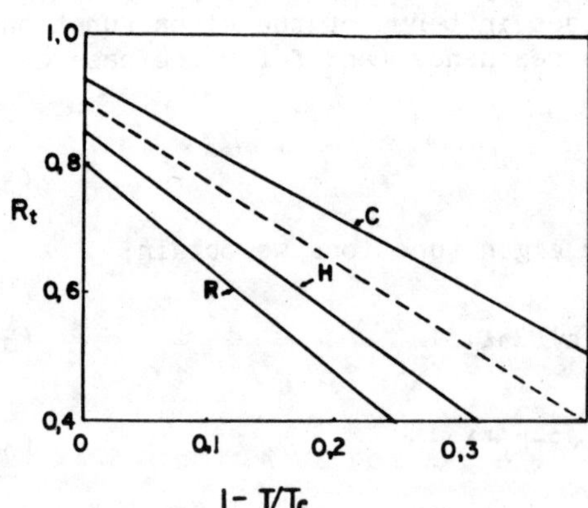

Fig. 3. The R_t's associated with three types of aanalytic vortices are shown as functions of $\varepsilon = 1-T/T_c$. The letter c, h, and r indicate the circular, hyperbolic and radial vortices. The experimental data are shown by a broken line.

ACKNOWLEDGEMENT

I would like to thank Dr. Hakonen et al. who kept me informed about their experimental results prior to publication. This work is supported by National Science Foundation under grant number DMR-8214525.

REFERENCES

1. N. D. Mermin and T. L. Ho, Phys. Rev. Lett. <u>36</u>, 594 (1976).

2. P. W. Anderson and G. Toulouse, Phys. Rev. Lett $\underline{38}$, 508 (1977).

3. G. E. Volovik and N. B. Kopnin, Pis'ma Zh. Eksp. Teor. Fiz. $\underline{25}$, 26 (1977) [JETP Lett. $\underline{25}$, 22 (1977)].

4. P. J. Hakonen et al., Phys. Rev. Lett. $\underline{48}$, 1838 (1982).

5. P. J. Hakonen, O. T. Ikkala, and S. T. Islander, Phys. Rev. Lett $\underline{49}$, 1258 (1982); J. Low Temp. Phys. (in press).

6. T. Fujita, M. Nakahara, T. Ohmi, and T. Tsuneto, Prog. Theor. Phys. $\underline{60}$, 671 (1978).

7. K. Maki, Phys. Rev. $\underline{B27}$, 4173 (1983).

8. K. Maki, to be published.

9. K. Maki, J. Low Temp. Phys. $\underline{32}$, 1 (1978).

10. We have corrected here a few misprints in Eq(4) appeared in ref. 7.

11. T. Passvogel, N. Schopohl, M. Warnke, and L. Tewordt, J. Low Temp. Phys. $\underline{46}$, 161 (1982); T. Passvogel, N. Schopohl and L. Tewordt, J. Low Temp. Phys. (in press).

12. M. C. Cross, J. Low Temp. Phys. $\underline{21}$, 525 (1975).

13. J. C. Wheatley, Rev. Mod. Phys. $\underline{47}$, 415 (1975).

14. K. Maki and P. Kumar, Phys. Rev. $\underline{B16}$, 182 (1977); $\underline{B17}$, 1088 (1978).

15. Similar value of F_1^a is deduced in D. D. Osheroff, W. van Roosbroeck, H. Smith and W. F. Brinkman, Phys. Rev. Lett. $\underline{38}$, 134 (1977).

QUASIPARTICLE BALLISTICS IN ^{3}He-A*

N. A. Greaves[§]
School of Mathematical and Physical Sciences
University of Sussex, Falmer, Brighton BN1 9QH, Sussex, UK

A. J. Leggett[†]
Laboratory of Atomic and Solid State Physics, and
Materials Science Center, Cornell University, Ithaca, NY 14853

ABSTRACT

We consider the propagation of quasiparticles in superfluid ^{3}He-A cooled (in a magnetic field $\gtrsim$ 6 kG) to temperatures low enough that the quasiparticle mean free path is substantially larger than the characteristic scale of a texture. Under these conditions quasiparticles propagate ballistically and can be reflected from textural variations, a phenomenon similar to, but rather simpler than, the Andreev reflection well known in superconductors. It is shown that because of this phenomenon the transport of heat and momentum in a flat-slab geometry should be extremely sensitive to the orientation of the magnetic field. Some of these effects should probably be observable with current cryogenics.

As is well known, the superfluid A phase of liquid ^{3}He displays anisotropic properties similar to those occurring in a liquid crystal. As regards the orbital properties, the anisotropy is defined by a characteristic unit vector $\hat{\ell}$, which is odd under both space reflection and time reversal and may be intuitively thought of as the "direction of the angular momentum of the Cooper pairs" (see e.g. ref. (1)). Since the gross energies which are responsible for the formation of the superfluid state are invariant under rotation of the spatial coordinates, the existence of such a characteristic vector indicates the occurrence of a broken symmetry, and the direction of $\hat{\ell}$ is not fixed by these gross energies but must be determined by much weaker effects such

---

*Work supported in part by a studentship from the Science and Engineering Research Council.

§Present address: Audit Division, Esso Italiano S.p.A., Palazzo Esso, Viale Castello della Magliana, 25, 00148 Roma, Italy. Please send reprint requests to second author.

†On leave of absence from the University of Sussex, U.K. and the University of Illinois. Address from August 1983: Department of Physics, University of Illinois, Urbana, IL 61801.

as the nuclear dipole energy, interactions with walls and so on.
Because of this, situations often arise in which, while the
general structure of the A phase is homogeneous, its orientation
(and in particular the direction of $\hat{\ell}$) varies in space: such a
configuration is known as a <u>texture</u>. In particular, in a flat-
slab geometry (i.e., with the liquid contained between two effectively infinite flat walls) the interaction with the boundary tends
to force $\hat{\ell}$, in the region near the walls, to lie perpendicular to
them, while the combination of the nuclear dipole interaction and
the polarizing effect of a magnetic field will tend to orient
$\hat{\ell}$ in the bulk, in a direction perpendicular to the field.[1] As a
result one finds that for a magnetic field in the plane of the
walls the liquid will adopt texture I of Fig. 1, or less probably
texture II: while if the field is turned into the direction perpendicular to the walls, these textures will deform into textures
Ia and IIa, respectively. The characteristic "scale" of textures
Ia and IIa, that is the distance D_C (Fig. 1) one has to go from
the wall before the $\hat{\ell}$-vector has turned through nearly $\pi/2$, is of
the order of the so-called dipole healing length,[1] which is of the
order 10^{-4}-10^{-2} cm depending on pressure and temperature. We
note also, for future reference, that if the magnetic field is
tilted slightly out of the plane of the walls we will get a
variant of texture Ia in which the $\hat{\ell}$-vector rotates only a small
angle away from perpendicular, rather than through $\pi/2$. We call
this the "slightly tilted" texture.

Consider now the motion of a normal excitation (quasiparticle
or quasihole) in such a texture. Over most of the pressure and
temperature range investigated so far, the excitation mean free
path is much shorter than the scale of the texture, so that it is
an adequate approximation to treat its dynamics as if $\hat{\ell}$ were completely uniform: most, if not all, considerations of the transport properties of ^{3}He-A in the literature to date have explicitly
or implicitly made this assumption. Under these conditions one
can define "local" transport coefficients for each individual
region of the liquid, which may of course depend on the direction
of $\hat{\ell}$ in this region, and then obtain the overall transport proper
ties by putting them in series. Such considerations lead to relatively unspectacular dependences of the transport properties on
texture. In particular, if the magnetic field is tilted through
an angle θ out of the plane of the walls, we expect the coefficients to have the general form $A + B \cos^2\theta$, with A and B in
general temperature-dependent constants: thus the properties of
the "slightly tilted" texture are only slightly different from
those of texture I.

However, by applying a magnetic field of a few kilogauss it
is possible to stabilize ^{3}He-A down to very low temperatures:
values of T/T_C as low as 0.33 have already been achieved,[2] and
there seems no obvious reason why considerably lower values could
not be reached were there an incentive to do so. At these low
temperatures excitations are few, collisions between them are rare
and the excitation mean free path therefore becomes very long.
Thus we may reasonably ask the question: If an excitation is

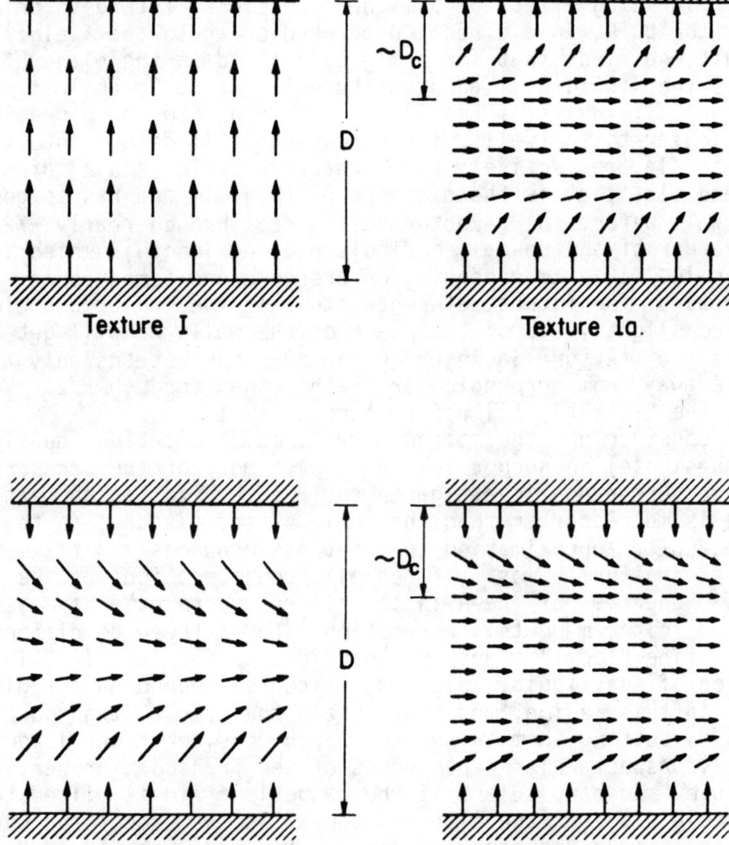

Fig. 1. Possible textures in a slab geometry.

emitted into a texture whose scale is shorter than the mean free path, how does it behave? The reason this is a nontrivial question is that, for an excitation with wave vector $\underset{\sim}{k}$, the energy depends on the orientation of the $\underset{\sim}{\ell}$-vector:

$$E(\underset{\sim}{k}:\hat{\underset{\sim}{\ell}}) = [\varepsilon_k^2 + \Delta_0^2(\hat{\underset{\sim}{k}} \times \hat{\underset{\sim}{\ell}})^2] \tag{1}$$

where the notation is that of ref. (1) except that Δ_0 denotes the <u>maximum</u> value of the energy gap rather than the root-mean-square value. Thus it is not immediately clear what will happen if $\hat{\underset{\sim}{\ell}}$ varies in space.

An answer to the question just posed is developed in ref. (3), to which we refer for details of the work reported here (see also ref. (4)). There we find that, provided the temperature is low enough that the superfluid component can be treated as an inert matrix for the propagation of excitations, then the motion of an excitation in the extreme ballistic limit (infinite mean free path) is described by the simple and intuitively plausible equations

$$\frac{d\underset{\sim}{r}}{dt} = \frac{1}{\hbar} \frac{\partial E}{\partial \underset{\sim}{k}} \tag{2a}$$

$$\frac{d\underset{\sim}{k}}{dt} = - \frac{1}{\hbar} \frac{\partial E}{\partial \underset{\sim}{r}} \tag{2b}$$

where the quasiclassical energy function $E(\underset{\sim}{k},\underset{\sim}{r})$ is the obvious generalization of eqn. (1):

$$E(\underset{\sim}{k},\underset{\sim}{r}) \equiv [\varepsilon_k^2 + \Delta_0^2 k_F^{-2}(\underset{\sim}{k} \times \hat{\underset{\sim}{\ell}}(\underset{\sim}{r}))^2] \tag{3}$$

(For the slight difference between (1) and (3) in the form of the second term in brackets, see ref. (3).) Bearing in mind that the ratio Δ_0/ε_F (where ε_F is the Fermi energy) is small ($\sim 10^{-2}$ - 10^{-3}) we can deduce from eqns. (2) that under almost all conditions (see below for the exception) the trajectory of an excitation through a texture will simply be (nearly) a straight line, but will be traced out with varying velocity $\underset{\sim}{v}(\underset{\sim}{r})$:

$$\underset{\sim}{v}(\underset{\sim}{r}) \cong \frac{\hbar\underset{\sim}{k}}{m^\star} \left(\frac{\varepsilon_k}{E_k}\right) = \frac{\hbar\underset{\sim}{k}}{m^\star} \frac{(E^2 - \Delta_0^2 k_F^{-2}(\underset{\sim}{k} \times \hat{\underset{\sim}{\ell}}(\underset{\sim}{r}))^2)}{E} \tag{4}$$

where $\underset{\sim}{k}$ is nearly constant and E is the quasiparticle energy, which as a consequence of eqns. (2) is rigorously conserved. Note that the velocity $\underline{v}$ is parallel to the wave vector $\underline{k}$ for a quasi-particle ($\varepsilon_k > 0$) but antiparallel to it for a quasihole ($\varepsilon_k < 0$).

However, there is one situation which clearly needs special consideration. What happens if the energy E is less than Δ_0 and, at some point on the trajectory which the excitation would otherwise have followed, the quantity $\Delta_0^2 k_F^{-2} (\underline{k} \times \hat{\underline{\ell}}(r))^2$ becomes greater than E^2? Such a situation would arise, for example, for a quasiparticle emitted into texture Ia (Fig. 1) with $\underline{k}$ normal to the wall and $E_0 < \Delta$. In such a case we see, by integrating the equations of motion (2) carefully, that at the point where the right-hand side of eqn. (4) tends to zero the quantity ε_k changes sign, i.e., the quasiparticle changes into a quasihole and begins to propagate <u>backwards</u> with respect to its wave vector $\underline{k}$. In the process the excitation energy E is changed not at all, and the momentum $h\underline{k}$ very little (by an amount $\sim\Delta_0/\varepsilon_F$), but since the velocity is reversed, the <u>transport</u> of energy and (all components of) momentum by the excitation changes sign: Thus the texture effectively reflects energy and momentum. This effect is closely analogous to the phenomenon of Andreev reflection[5] well known in the intermediate state of superconductors: the principal difference in the present case (which actually makes the theory considerably simpler) is that the characteristic scale of variation, which is of the orientation rather than the magnitude of the order parameter, is always long compared to the dimension of a Cooper pair (the zero-temperature coherence length ξ_0). We will refer to this phenonemon as "textural Andreev reflection."

Let us now consider the transport of heat and (transverse) momentum across a flat slab of ^{3}He-A at low temperatures ($T \ll T_c$). At such temperatures almost all the excitations which carry the heat or momentum will have energies small compared to the maximum gap Δ_0: since for all textures $\hat{\underline{\ell}}$ is perpendicular to the walls in the region near them, this means that, irrespective of the texture, these excitations will all have wave vector $\underline{k}$ nearly perpendicular to the walls. For texture I these excitations will all simply propagate ballistically to the opposite wall,* and the transport of heat and momentum will be limited only by the effective number of carriers available (which varies as a power of T). For texture Ia, on the other hand, most of the excitations leaving the wall will be reflected by the texture: the only ones which get acro<u>ss</u> to the opposite wall will be those with energy greater than $\Delta_0/\sqrt{2}$ and traveling approximately at an angle of 45° to the right (it is easy to convince oneself that this is the "easiest" direction, i.e., the maximum value of $(\underline{k} \times \hat{\underline{\ell}})^2$ experienced by

---

*We assume for simplicity that the mean free path is much longer than the slab width as well as much longer than the distance D_c. It is easy to treat also the opposite case (see Ref. (3)).

excitations traveling in this direction ($\Delta_0/\sqrt{2}$) is lower than that of any other). Thus we expect the transport of heat and momentum to be proportional to $\exp(-\Delta_0/\sqrt{2}k_BT)$.[†]

It is possible to make these considerations quantitative by making suitable assumptions about the distribution of quasiparticles emitted by a heated or moving wall: see ref. (3) for details. We find that in the case of heat flow the effect of the texture is to generate a "pseudo-Kapitza resistance" R_{PK} which is in series with the true Kapitza resistances due to the solid-liquid interfaces: the order of magnitude of R_{PK} is given by the expressions (valid for $T \ll T_c$)

$$R_{PK} \sim 0.1(T_c/T)^3 \text{ cm}^2 \text{ K/W} \qquad \text{(texture I)} \qquad \text{(5a)}$$

$$R_{PK} \sim 0.3(T_c/T)^{\frac{1}{2}}\exp(\Delta_0/\sqrt{2}k_BT) \text{ cm}^2 \text{ K/W} \qquad \text{(texture Ia)} \qquad \text{(5b)}$$

Similarly, in the case of momentum transfer (e.g., in an oscillating-disc experiment) we can define a "momentum transfer coefficient" $K \equiv \underline{P}/\underline{v}$ where $\underline{P}$ is the momentum transfer per unit area per unit time across the texture and $\underline{v}$ is the velocity discontinuity across it (see ref. (3) for details). For texture Ia (but not for II or IIa) K is actually a tensor since wall velocities parallel and perpendicular to the plane of the texture are not equivalent. The order of magnitude of K is given by[3]

$$K \sim 200 \, (T/T_c)^4 \text{ gm cm}^{-2} \text{ sec}^{-1} \qquad \text{(texture I)} \qquad \text{(6a)}$$

$$K \sim 200 \, (T/T_c)^n \, e^{-\Delta_0/\sqrt{2}k_BT} \text{ gm cm}^{-2} \text{ sec}^{-1} \qquad \text{(texture Ia)} \qquad \text{(6b)}$$

where n is 3/2 or 5/2 depending on whether the walls move parallel or perpendicular to the plane of the texture.

It is fairly clear that the pseudo-Kapitza resistance of texture I (eqn. (5a)) will always be small compared to the true Kapitza boundary resistance and hence is unlikely to be observable. However, crude estimates suggest[3] that for texture Ia (eqn. (5b)) the effect may just be observable at the lowest temperatures currently achievable. A much more promising candidate is the effect on momentum transfer, as observed for example in the damping of an oscillating disc. If the disc is made of sinter

---

[†] Note for texture IIa (in which the $\hat{\underline{\ell}}$-vector turns through a full π as we go across the slab) the corresponding factor is $\exp(-\Delta_0/k_BT)$: and that for the "slightly tilted" texture with tilt angle $\theta \ll 1$ it is $\exp(-\theta\Delta_0/2k_BT)$.

then the analogue of the Kapitza resistance is essentially zero (i.e., the (normal component of the) liquid very close to the wall moves with the wall), and the textural impedance K^{-1} is effectively in series only with the bulk viscosity of the liquid. A simple calculation then shows that the effective complex damping coefficient Γ of the disc at frequency ω (i.e., the ratio of the viscous drag to velocity in sinusoidal oscillation) is given by

$$\Gamma = \frac{A}{K^{-1} + \dfrac{\sqrt{2}\ \delta(\omega)}{(1+i)\eta_n}} \tag{7}$$

where A is the (effective[3]) area, η_n the viscosity of the normal component and $\delta(\omega) \equiv (\eta_n/\rho_n\omega)$ is the usual viscous penetration depth (ρ_n is the normal density). To see the effect of textural Andreev reflection we evidently need to go to temperatures low enough that K^{-1} becomes comparable to $\delta(\omega)/\eta_n$: using the quantitative formulae for K given in ref. (3) and the fact that $p_F^2(dn/d\varepsilon)(T/\Delta_0)^4 \sim \rho_n$, we find that for texture I the condition is approximately $\rho_n/\rho \lesssim \omega\tau_n$, where τ_n is the normal-state excitation lifetime at T_c ($\approx 10^{-7}$ secs). For $\omega/2\pi \sim 1$ kHz, a typical disc oscillation frequency, this is not too easy. However, for texture Ia we need only the (order-of-magnitude) condition $\rho_n/\rho \lesssim \omega\tau_n \exp(\Delta_0/\sqrt{2}k_B T)$, which should be relatively easy to obtain with current cryogenics. Indeed, it may not be impossible to see the effect of tilting the field (and hence the bulk value of $\hat{\ell}$) through angles small compared to $\pi/2$ (see ref. (3) for details): this would be particularly satisfying confirmation of the above ideas since at sufficiently low T the effect of even a small tilt is quite dramatic and quite different from the behavior in the hydrodynamic regime.

References

(1) A. J. Leggett, Revs. Mod. Phys. **47**, 331 (1975).
(2) H. N. Scholz, Ph.D. Thesis, Ohio State University, 1981 (unpublished).
(3) N. A. Greaves and A. J. Leggett, J. Phys. C., in press.
(4) N. A. Greaves, D. Phil. Thesis, University of Sussex, 1981 (unpublished).
(5) A. F. Andreev, Zh. Eksp. Teor. Fiz. **46**, 1823 (1964) (translation: Soviet Physics JETP **19**, 1228 (1964)).

SUPERFLUID CURRENTS IN ^{3}He A AT T = 0

R. Combescot and T. Dombre
Groupe de Physique des Solides de l'Ecole Normale Supérieure
24 rue Lhomond, 75231 Paris Cédex 05, France

ABSTRACT

We consider the problems raised by the hydrodynamics of the A phase at T = 0. We discuss Volovik and Mineev's proposal of a non-zero normal density at T = 0. Then, on the basis of our exact solution of the linearized problem, we obtain the superfluid current which displays a number of non analytical second order terms. We present the excitation spectrum and obtain the superfluid density.

INTRODUCTION

The A phase of superfluid ^{3}He seems to behave in a rather strange way at very low temperature. The source of the problems is the existence of two nodes for the gap on the Fermi surface. At finite temperature, these nodes are so to speak hidden, but they become naked at T = 0.

A major problem with the A phase is that it is apparently not possible to write a well behaved non linear hydrodynamics at T = 0 while with the B phase there is no difficulty with the hydrodynamics [1] because the gap is finite everywhere.

Indeed, if we take the Josephson equation [2] :

$$- \frac{\hbar}{2m} \frac{\partial \phi}{\partial t} = \mu + \vec{V}_n . \vec{V}_s + \frac{\hbar}{4m} \hat{\ell} . \vec{\nabla} \times \vec{V}_n \quad , \qquad (1)$$

and calculate the chemical potential μ, taking into account the generally admitted expression for the current :

$$\vec{g} = \rho \vec{V}_s + \frac{\hbar}{4m} \vec{\nabla} \times (\rho \hat{\ell}) - \frac{\hbar \rho}{2m} \hat{\ell} (\hat{\ell} . \vec{\nabla} \times \hat{\ell}) \quad , \qquad (2)$$

we are left in Eq.(1) with a term $(\hbar/2m)\vec{V}_n . \hat{\ell}(\hat{\ell} . \vec{\nabla} \times \hat{\ell})$ which comes from the last term in the current. This is the only term containing $\vec{V}_n$ and it makes a trouble : physically, we would not expect $\vec{V}_n$ to stay in the equations at T = 0 and mathematically, we have no equation of motion for $\vec{V}_n$ since the term $\rho_n \partial \vec{V}_n / \partial t$ in the momentum conservation law disappears because $\rho_n = 0$.

Meeting similar problems, Volovik and Mineev [3] have suggested that there is still some normal fluid left at T = 0, so that $\rho_n \neq 0$. They notice that in the presence of a texture the low energy excitations cannot move freely and their spectrum is modified. To obtain this change, they use a "generalized gauge transform" : for an order parameter $\Delta = \Delta_0 e^{i\phi}$, the excitation energy becomes $E_k + \vec{k} . \vec{V}_s$ where $\vec{V}_s = (\hbar/2m)\vec{\nabla}\phi$. Now, for the A phase, $\Delta_k = \delta \sin\theta \exp(i\phi_k)$ with $\phi_k = \text{Arctg}(\hat{k}_y / \hat{k}_x)$ which leads to a spectrum $E_k + (\hbar/2m)\vec{k} . \vec{\nabla}\phi_k$ with $\vec{k} . \vec{\nabla}\phi_k = (\vec{k} . \hat{\ell})(\hat{k} . \vec{\nabla})\hat{\ell}| . |\vec{k} \times \hat{\ell}|/\hat{k}^2$. Therefore, there is a finite density of states for zero energy $N(0) = \sum_k \delta(E_k + (\hbar/2m)\vec{k} . \vec{\nabla}\phi_k)$ and a normal density

$(\rho_n)_{ij} = \Sigma \; k_i \kappa_j \; \delta(E_k + (\hbar/2m)\vec{k}.\vec{\nabla}\phi_k)$ which leads to $(\rho_n)_{zz} = k_F^2 \, N(0) = (3/4)\rho(V_F|(\hat{\ell}.\vec{\nabla}).\hat{\ell}|/\delta)$. This implies that in Eq.(2) $\rho\vec{V}_s$ should be replaced by $\rho\vec{V}_s - (\rho_n)_{zz} \, \hat{\ell}(\hat{\ell}\vec{V}_s)$ in order to satisfy Galilean invariance.

CURRENT AT T = 0

In the regime we are interested in (T = 0, finite gradients), Eq.(2) is not justified since it is obtained in the regime $qv_F \ll kT$. We have studied the opposite situation $qv_F \gg kT$ in order to see if the expression for the current is not deeply modified. Muzikar and Rainer[4] have made a similar study in considering the validity of Volovik and Mineev's approach. Looking at the regime $qV_F \gg kT$ implies that one cannot perform a gradient expansion anymore. One has to solve exactly Gorkov's equations :

$$\begin{pmatrix} i\omega_n - \xi_k & \Delta_k(r) \\ -\Delta_k^*(r) & -i\omega_n - \xi_k \end{pmatrix} g + i\vec{V}_k.\vec{\nabla}g = 1 \quad . \tag{3}$$

This does not sound possible for general $\Delta_k(\vec{r})$. But after linearizing the spatial dependence of $\Delta_k(\vec{r})$, we have been able to solve these equations exactly. If we expand $\Delta_k(\vec{r})$ around the origin as :

$$\Delta_k(\vec{r}) = \Delta_k^o + \vec{r}.\vec{\nabla}\Delta \quad , \tag{4}$$

and set $\hat{k}.\vec{\nabla}\Delta = \alpha$ which can always be taken real positive for fixed $\hat{k}$, we obtain for the ξ-integrated Green's function at the origin :

$$\int g^{11}d\xi = -\, i\sqrt{\pi/2} \left[\lambda F(|\lambda|^2+1, x_0) + \lambda^* F(|\lambda|^2, x_0)\right] \quad , \tag{5}$$

where $\lambda = (\omega_n + i \, \mathrm{Im} \, \Delta^\circ)/\sqrt{2\alpha V_F}$, $x_0 = \sqrt{2/\alpha V_F} \, \mathrm{Re} \, \Delta^\circ$ and :

$$F(|\lambda|^2, x_0) = \Gamma(|\lambda|^2) \, U(|\lambda|^2-1/2, x_0) \, U(|\lambda|^2-1/2, -x_0) \tag{6}$$

where U is the parabolic cylinder function. The current is then obtained by :

$$\vec{g} = N_0 T \sum_n \int d\xi (d\Omega/4\pi)\vec{k} \, g^{11} \quad . \tag{7}$$

At T = 0, we obtain, in addition to the regular part of the current Eq.(1), a singular part coming from the vicinity of the nodes of the gap. For the part of the current perpendicular to $\hat{\ell}$, we find up to second order in gradients :

$$(\vec{g}_{sing.})_\perp \sim \rho\frac{\hbar}{m} \, (\mathrm{curl} \, \hat{\ell})_\perp \frac{V_F|(\hat{\ell}.\vec{\nabla})\hat{\ell}|}{\delta} \quad , \tag{8}$$

which has also been obtained by Muzikar and Rainer. For the current parallel to $\hat{\ell}$, the result is more complicated :

$$(\vec{g}_{\text{sing.}})_{/\!/} \sim \rho \frac{V_F |(\hat{\ell}.\vec{\nabla})\hat{\ell}|}{\delta} \left[A(V^S_{/\!/} - \frac{\hbar}{4m}\hat{\ell}.\text{curl}\hat{\ell}) + B\frac{\hbar}{4m}(\partial_x\hat{\ell}_y + \partial_y\hat{\ell}_x) \right] \quad (9)$$

where the x and y axes are along $(\text{curl}\hat{\ell})$ and $\hat{\ell}\times\text{curl}\hat{\ell}$ respectively, and A and B are numerical coefficients. Therefore, we find various non analytical second order contributions, which does not improve at all the situation for the current.

EXCITATION SPECTRUM

Actually, we have not done our best in order to obtain the $\vec{V}_s$ contribution to the current. Instead of linearizing, it is much better to perform a gauge transformation in order to get rid of $\vec{V}_s$. This amounts to $i\omega_n \rightarrow i\omega_n - k.V_s$ and use again our solution to find the current. Before doing this, it is interesting to look at the excitation spectrum. This is done easely by continuing analytically g^{11} to the real frequency axis. For fixed $\vec{k}$, one obtains a set of poles at $\omega = \sqrt{2\alpha p V_F + (\text{Im}\,\Delta^0)^2}$ where p is a positive or zero integer. From the spectral density, one can see that the low energy excitations (which implies a small $\text{Im}\,\Delta_0$) are built from free excitations with small ξ_k and $\text{Re}\,\Delta_0$, that is free excitations near the nodes of the gap. One can also see that these excitations are spatially localized. Therefore the qualitative picture is very similar to the one proposed by Volovik and Mineev, but the excitation spectrum is very different from what they suggest.

For low energies, we have $\text{Re}\,\Delta^° \simeq \delta k_y$, $\text{Im}\,\Delta^° \simeq -\delta k_x$ and $\alpha = \delta|(\hat{\ell}.\vec{\nabla})\hat{\ell}|$ and the density of excitations is :

$$N(\omega) = N_0(\omega) + \sum_{p=1}^{\infty} N_p(\omega) \quad , \quad (10)$$

with

$$N_0(\omega) = \frac{N_0}{2}\frac{\alpha V_F}{\delta^2} \qquad N_p(\omega) = N_0\frac{\alpha V_F}{\delta^2}\frac{\omega}{\sqrt{\omega^2 - 2\alpha p V_F}} \quad . \quad (11)$$

It can be checked that for $\omega \gg \sqrt{\alpha V_F}$, $N(\omega)$ gives the ordinary density of states $N_0(\omega/\delta)^2$ for the low lying excitations in the homogeneous system. The whole structure for the density of states Eq.(10) is somewhat reminiscent of the effect of the Landau levels in a three dimensional electron gas. Now Eq.(10) gives a finite density of states for $\omega = 0$:

$$N(0) = \frac{N_0}{2}\frac{V_F}{\delta}|(\hat{\ell}.\vec{\nabla})\hat{\ell}| \quad , \quad (12)$$

which is twice Volovik and Mineev's result (as corrected by Muzikar and Rainer).

Therefore, we expect also a normal density at $T = 0$ linked to $N(0)$ and a corresponding correction to ρ_s in order to insure Galilean invariance. Indeed, if we calculate the current at $T = 0$ from the gauge transformation, we have to integrate over ω a function of $\omega + i\vec{k}.\vec{V}_s$. When we shift the integration path to the real ω axis, we obtain contributions from the poles corresponding to low frequency excitations. This gives finally :

$$(\rho_s)_{ij} = \rho(\delta_{ij} - \frac{3}{2} \frac{V_F |(\hat{\ell}.\vec{\nabla})\hat{\ell}|}{\delta} \hat{\ell}_i \hat{\ell}_j) \tag{13}$$

In conclusion, we find from our exact solution localized excitations with a finite density of states at zero energy, which gives rise to a nonzero normal density at $T = 0$. However because of the presence of non analytical contributions to the current, hydrodynamics does not seem in good shape.

REFERENCES

1. R. Combescot and T. Dombre, Phys. Lett. 76A, 293 (1980).
2. C.R. Hu and W.M. Saslow, Phys. Rev. Lett. 38, 605 (1977).
3. G.E. Volovik and V.P. Mineev, Sov. Phys. JETP 54, 524 (1981).
4. P. Muzikar and D. Rainer, to be published.

PROBING ^{3}He EXCITATION KINETICS WITH A VIBRATING WIRE

H.E. Hall
Schuster Laboratory, University of Manchester, Manchester M13 9PL
and Laboratory of Atomic and Solid State Physics
Cornell University, Ithaca, NY 14853

INTRODUCTION

Most of the experiments I want to discuss in this talk have very recently been published[1], and I shall therefore not give a detailed account of either the experimental method or the results. Rather, I shall emphasize the special features of the vibrating wire viscometer that have enabled this rather prosaic piece of apparatus to yield information about previously unexplored aspects of the dynamics of excitations in superfluid ^{3}He, and indicate possible directions of future work.

There are three features of the transverse motion of a small cylinder through a liquid that we have been able to exploit:

(1) There are two characteristic length scales associated with the motion, the viscous penetration depth $\delta = (2\eta/\rho_n\omega)^{1/2}$ and the radius a of the wire. In ^{3}He the regime $\delta \ggg a$ is readily attainable, and it is then feasible to have the mean free path ℓ of excitations such that $\ell \gg a$ but $\ell \ll \delta$. Since $(\ell/\delta) \sim (\omega\tau)^{1/2}$ this has the consequence that we can study long mean free path effects with $\omega\tau \ll 1$, so that ordinary hydrodynamics is nevertheless applicable over most of the liquid. We shall see that this makes it possible to determine both the true viscosity and the mean free path in the same experiment.

(2) The geometry is such that the moving wire pushes the liquid as well as slides past it. This is an incidental benefit in mean free path experiments, because it means that momentum exchange between the wire and the liquid is far less dependent on the specularity of excitation scattering than in the case of a plane boundary moving parallel to itself. More importantly, this pushing tests the compressibility of the excitation gas, and our experiments provided the first evidence of the second viscosity ζ_3 associated with this compressibility. This effect has since been invoked by Brand and Cross[2] to explain the large dissipation observed in U-tube flow[3]; there the 'pushing' is done by the free surface of the liquid rather than by a solid object moving through it.

(3) The third feature is rather obvious; the wire senses the present condition of the liquid within a small region of size δ. It can therefore be used as an instantaneous local thermometer, and has been used to study the propagation of heat pulses[4].

MOTION OF THE WIRE

The hydrodynamic force per unit length on a wire vibrating on the axis of a cylinder of radius b with transverse velocity $\vec{v}$ at

266

angular frequency ω in an isotropic superfluid may be written as

$$\vec{F}_H = i\pi a^2 \rho\omega\vec{v}\left(1 - \frac{\rho_n/\rho}{C+iC'}\right) \tag{1}$$

where C and C' are calculable functions of (δ/a), (b/a) and (δ_2/a). Here δ_2 is a second penetration depth associated with compressible flow of the excitation gas and is given by

$$\delta_2^2 = \frac{2\rho_s}{\rho\rho_n\omega}\left(\frac{4}{3}\eta + \zeta_3\rho^2\right) = 2D_\eta'/\omega \tag{2}$$

in which ζ_3 is the second viscosity associated with irreversible compression of the excitation gas at constant fluid density. When a driving force F_D is applied to the wire, by passing A.C. current through it in a transverse magnetic field, the equation of motion for the displacement y is

$$\pi a^2 \rho_w \ddot{y} + \pi a^2 \rho_w \omega_v^2 y = (F_D + F_H)e^{-i\omega\tau}, \tag{3}$$

where ρ_w is the density of the wire and ω_v the vacuum resonance frequency; nuisance damping was negligible in our experiments. A particularly convenient mode of operation is to drive the wire at the temperature independent 'magic' frequency given by $\omega_m^2 = \omega_v^2/(1 + \rho/\rho_w)$, since the motion then has an amplitude y_0 given by

$$y_0 = \frac{F_D(C+iC')}{\pi a^2 \rho_n \omega_m^2} \tag{4}$$

This has the effect of removing the irrotational first term from (1), and the reactive and resistive voltages across the wire and then simply proportional to C and C'.

Even in the absence of mean free path effects our two measured quantities (C/ρ_n) and (C'/ρ_n) are functions of three quantities, η, ζ_3 and ρ_n. We have, therefore, taken ρ_n as known from accurate torsion pendulum experiments[5] in order to analyze our data. A graph illustrating the functional dependence of C, C' on η, ζ_3 is shown in Figure 1. The most noteworthy feature is that the main effect of compressibility (small ζ_3) is to increase the value of C, restoring it towards the value for an unbounded medium. Compressibility allows the long range normal flow around the wire to be replaced by short range superflow, so that the effect of surrounding solid boundaries is diminished.

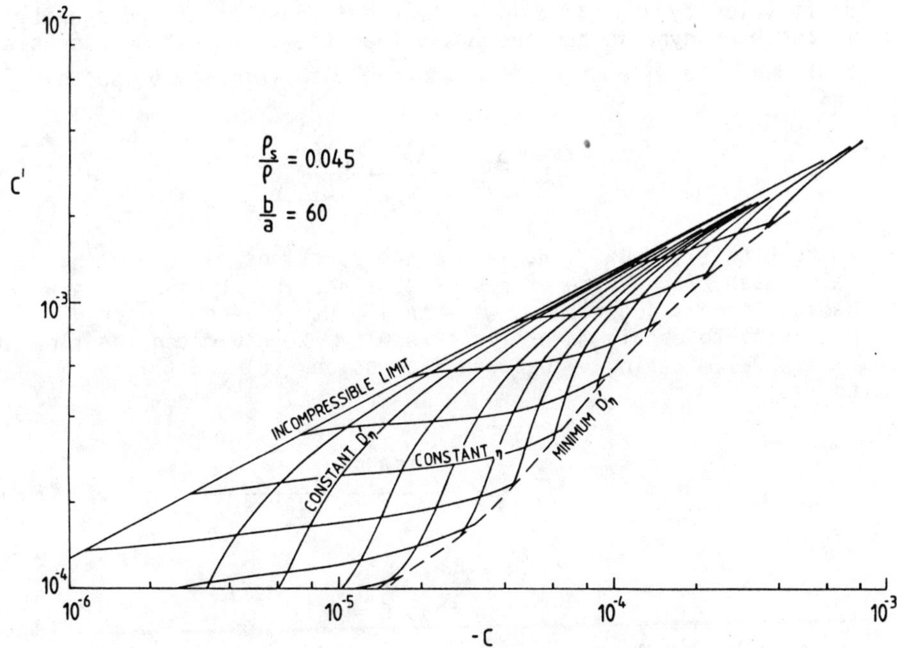

Fig. 1. Calculated compressibility effects; contours are at 2dB intervals of η and D'_η.

MEAN FREE PATH EFFECTS

As indicated in the introduction, our analysis of mean free path effects hinges on the fact that the flow can be divided into a Knudsen region $r \lesssim \ell$ and a hydrodynamic region $r \gtrsim \ell$. We first divide the hydrodynamic force into an irrotational and a viscous part

$$F_H = F_{irr} + F_v \tag{5}$$

and assume that the former contribution is always present, since liquid has to get out of the way whatever the mean free path (F_{irr} is essentially the superfluid contribution in low temperature ^{3}He-B). In the Knudsen region simple kinetic theory arguments give

$$\vec{F}_v = (1/\alpha)\pi a\rho_n v_g(\vec{v}_D - \vec{v}) \tag{6}$$

where v_g is the excitation group velocity and $\vec{v}_D$ the excitation drift velocity at $r \sim \ell$; α is a numerical constant of order unity. In the hydrodynamic region we say that the flow must be essentially that due to a wire of radius ℓ moving with velocity $\vec{v}_D$ so that

$$\vec{F}_v = - \frac{i\pi a^2 \rho_n \vec{v}_D}{C_o + iC'_o} \tag{7}$$

where $C_o(\eta,\zeta_3)$ and $C'_o(\eta,\zeta_3)$ are the functions already calculated, for a wire of radius ℓ; since the dependence on wire radius is only logarithmic we actually use C_o and C_o' for radius a, to obtain an explicit result. We now match the flow at $r \sim \ell$ by eliminating v_D between expressions (6) and (7) for F_v to yield

$$\vec{F}_v = - \frac{i\pi a^2 \rho_n \vec{v}}{C_o + iC_o' + \alpha(i\omega a/v_g)} \tag{8}$$

$$\text{so that } C = C_o \qquad\qquad\qquad \tag{9}$$
$$C' = C'_o + \alpha(\omega a/v_g)$$

We expect the result (9) to be qualitatively valid for any value of (ℓ/a), though the numerical value of α may change somewhat between the small (ℓ/a) and large (ℓ/a) limits. The small (ℓ/a) limit of α has been calculated by Jensen et al[6] (it is weakly temperature dependent in the superfluid state) but the long mean free path limit has not yet been calculated accurately; we therefore adopt an empirical approach.

We now find that our measured C and C' again depend on three parameters η, ζ_3 and α - though the dependence on α is mercifully simple (equation 9). Our data analysis was done[1] by a mixture of assuming a theoretical ζ_3 and assuming α, in different regions. Here I illustrate only the former procedure. The important feature of the result (9) is that given ζ_3, η can be calculated from C independently of any mean free path effects. Such values are shown in figure 2, together with values calculated ignoring compressibility of the excitation gas ($\zeta_3 \to \infty$); it is clear that the agreement with other workers is much better when the theoretical ζ_3 is used. When η has been obtained in this way ($C_o'(\eta,\zeta_3)$ is known, so an experimental value of $\Delta C' = (C' - C_o')$ can be found. This is illustrated in figure 3, where the full curve shows

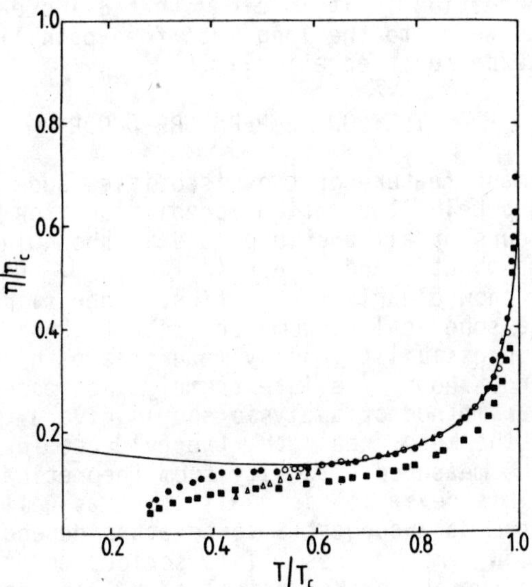

Fig. 2. Viscosity at 20 bar. Solid circles assume theoretical ζ_3, solid squares assume $\zeta_3 \to \infty$. Open circles and triangles are the Cornell spherical viscometer and first order slip corrected torsion pendulum, respectively.

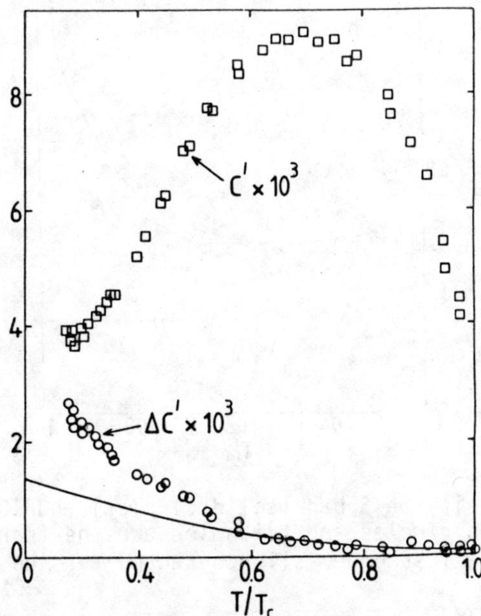

Fig. 3. Experimental mean free path correction; the solid curve is the first order slip correction.

the value of $\Delta C'$ for α as calculated [6] for the short mean free path limit (slip correction). It is clear that α increases by a factor of two or so as we go to the long mean free path limit, a result that is not unexpected theoretically.[1]

THE LOW TEMPERATURE DROOP

A significant feature of the viscosities shown in figure 3 is the marked droop below theoretical predictions for $(T/T_c) \lesssim 0.5$. This effect occurs at all pressures and is shown in more detail for measurements at about 5 and 10 bar in figure 4. While the torsion pendulum data[7] show clearly that a first order slip correction is inadequate, the spherical viscometer[7] results, which are not affected by compressibility, or by mean free path in the temperature range shown, are in extremely good agreement with ours. Since our method of analysis should give a true viscosity, independent of the mean free path, the evidence for a serious deviation of the measured viscosity from theoretical prediction is very strong. This deviation is qualitative as well as quantitative; the data suggest a temperature dependence at least as fast as T^2 at low temperatures. This serious and surprising discrepancy deserves both theoretical study and experimental investigation at lower temperatures.

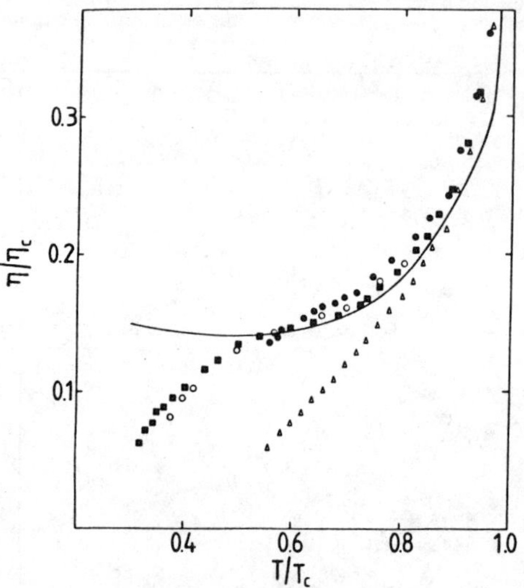

Fig. 4. Viscosity at 5 bar (solid circles) and 10 bar (solid squares). Open circles and triangles are the Cornell spherical viscometer and first order slip corrected torsion pendulum, respectively.

VIBRATING WIRE THERMOMETRY

By using two vibrating wires at different distances from a source of heat the propagation of heat pulses in normal and superfluid ^{3}He has been studied[4]. Since this work is still in progress and will be the subject of other contributions to this conference I shall not describe it in detail, but simply draw attention to two significant results that have emerged so far.[8] An order of magnitude calculation shows that the counterflow mechanism can make a significant contribution to heat flow only close to T_c, essentially because of the large normal viscosity and low entropy. We find that near T_c our observed pulse propagation speeds are intermediate between those expected for normal thermal conduction and those expected for thermal conduction plus counterflow. They can be modelled by the reasonable assumption of counterflow limited by a critical velocity of the order of mm/sec.

For $(T/T_c)<0.4$, where counterflow is certainly negligible, the ordinary thermal conductivity may be deduced. It is not significantly different from the value at T_c, in contrast to the expected variation roughly as $(1/T)$.[9]

More recently, the Lancaster group[10] has used a vibrating wire to monitor the cooling of both saturated ^{3}He - ^{4}He solutions and pure ^{3}He in their nuclear refrigeration cryostat. In these experiments they reach the extreme long mean free path regime, where the force on the wire is given approximately by equation (6) with $\nu_D = 0$. In this regime the vibrating wire essentially measures the density of excitations.

FUTURE PROSPECTS

Because of its insensitivity to specular scattering and ability to work in a hydrodynamic regime even when the mean free path is long the vibrating wire has some merit as an instrument for investigating absolute viscosity at very low temperatures, providing the theoretical foundation of equation (8) can be made more secure, in particular the separation of the irrotational force according to equation (5).

While the parameters of a wire can be optimised for the measurement of ζ_3, it is also likely that equally good or better geometries for the measurement of ζ_3, can be devised, for example diaphragm driven flow experiments (the free surface flow considered by Brand and Cross[2], though sensitive to ζ_3, is hard to analyze accurately).

The most interesting prospect for the applications of methods like the vibrating wire is as an ultra-low temperature thermometer for liquid ^{3}He[10]. The important feature of equation (6) is that α is only weakly dependent on scattering specularity whereas ρ_n is exponentially dependent on temperature. Therefore, with a modicum of faith in BCS theory, the absolute temperature of the liquid is determined with moderate precision and great reliability; in the ultra-low temperature region, where any thermal contact is dubious, the reliability obtained by using the liquid itself as a thermometric substance is of paramount importance. At temperatures of 100μK or less nuisance damping will be important; a likely development, therefore, is the fabrication of oscillating objects from silicon single crystals,[11] which can have an extremely high Q.

REFERENCES

1. D.C. Carless, H.E. Hall and J.R. Hook, J. Low Temp. Phys. __50__, 583, 605 (1983).
2. H. Brand and M.C. Cross, Phys. Rev. Lett. __49__, 1959 (1982).
3. J.P.Eisenstein and R.E. Packard, Phys. Rev. Lett. __49__, 564 (1982).
4. N.V. Wellard, P.W. Alexander, H.E. Hall and J.R. Hook, Physica __109__ and __110B__, 2096 (1982).
5. J. Saunders, D.G. Wildes, J. Parpia, J.D. Reppy and R.C. Richardson, Physica __108B__, 791 (1981).
6. H.H. Jensen, H. Smith, P. Wölfle, N. Nagai and T.M. Bisgaard, J. Low Temp. Phys. __41__, 473 (1980).
7. Y.A. Ono, J. Hara and K. Nagai, J. Low Temp. Phys. __48__, 167 (1982).
8. N.V. Wellard, Ph.D. Thesis, University of Manchester (1981)
9. P. Wölfle, Prog. Low temp. Phys. Vol VII (a) (ed. D.F. Brewer: North Holland, 1978) p. 193.
10. A.M. Guénault, V. Keith, C.J. Kennedy and G.R. Pickett Phys. Rev. Lett. __50__, 522 (1983) and paper by G.R. Pickett at this conference.
11. J.B. Angell, S.C. Terry and P.W. Barth, Scientific American __248__ (4), 44 (1983).

THE MECHANICAL PROPERTIES OF THE CONDENSATE
IN ^{3}He-B IN THE QUASIPARTICLE-FREE REGIME

A. M. Guénault, V. Keith, C. J. Kennedy and G. R. Pickett
Department of Physics, University of Lancaster, U. K.

ABSTRACT

We have made measurements of the damping and resonant frequency
of a vibrating wire in superfluid ^{3}He-B at temperatures down to
about $0.12T_c$ at 0.0 bar and at 7.3 bar pressure. At the lowest
temperatures the wire is effectively in a quasiparticle vacuum,
the contribution to the damping from the quasiparticles having
disappeared. In this regime the behaviour of the wire becomes
strongly non-linear reflecting the mechanical response of the superfluid condensate.

INTRODUCTION

The superfluid phases of ^{3}He are of particular interest owing
to the unique directional properties of the liquid arising from the
vector nature of the order parameter. Of the range of phenomena
embracing this directionality, the texture, it would be fair to say
that theory has far outstripped experimental evidence since
textural experiments are by no means simple to devise or execute.
Most information has to date been obtained via the medium of NMR,
and the majority of experimental and theoretical work has concentrated on the Ginzburg-Landau regime near T_c since that is the most
accessible experimentally. One area which has not received much
attention is that of the purely mechanical nature of the texture.

We begin by considering a container filled with B-phase (we are
limited by experimental necessity to measurement in the B-phase)
which is stationary, in equilibrium and in the absence of a
magnetic field. A static deformation of the container will lead to
a rearrangement of the $\underline{n}$ vector field and give rise to an energy
change from the bending of $\underline{n}$ which may be either positive or negative according to the particular geometry. If the deformation is
carried out rapidly then a further range of textural energies
governed by the superflow come into play. In consequence the deformation gives rise to both a static and a dynamic textural energy
which we may regard in the light of potential and kinetic energies
and the liquid will behave under deformation as an elastic solid,
albeit a rather unusual one. Unfortunately these energies are
small; for example the bulk flow energy term is on a scale several
orders of magnitude less than the simple kinetic energy $\frac{1}{2}\rho_s v_s^2$
associated with the superflow. One could contemplate setting up a
mechanical oscillator to study these elastic properties but would
find that the damping effect of the quasiparticles would completely
mask any textural effects. The only conditions under which such a
measurement could be made is in the region approaching $T=0$ where the
quasiparticle density has fallen essentially to zero and only the

274

bare condensate remains.

Using a vibrating wire resonator we have been able to observe
these textural effects. With our experimental configuration the
texture begins to dominate the behaviour once the normal fluid
fraction ρ_n/ρ has fallen to around 10^{-5}. In ^{3}He-B at zero pressure
(which is the easiest pressure for us to work with) this condition
holds for T < 180 μK. In practice[1] we can cool the B-phase at
this pressure to ~125μK at which point we begin to lose thermal
contact with the thermometer owing to the rapidly diminishing quasiparticle density. In the range 125 - 180 μK we can clearly see the
mechanical response of the condensate which, as one might expect, is
rather unusual and has a very clear signature.

EXPERIMENTAL CONFIGURATION

The ^{3}He is cooled by the nuclear demagnetization of copper. The
specimen is contained in a 0.25 mm wall plastic inner cell which
also holds nine 1 mm slabs of copper coated with sintered silver.
The cell has a 1.5 mm diameter tower which holds the wires of a Pt
NMR thermometer which is heat-sunk to the liquid via a sintered
silver pad. The vibrating wire resonator for probing the
properties of the liquid consists of a 0.124 mm diameter Ta wire
bowed into an approximate semicircle of 4 mm radius. The resonator
is surrounded by a free volume of liquid such that all walls are at
least 1 mm from the moving parts of the wire[2]. The cell is precooled by a high purity silver link via a monolithic single crystal
Aℓ heat switch[3]. The filling tube of 1 m of 0.3 mm i.d. Cu
capillary isolates the inner cell from the outer cell in which it is
immersed. The outer cell also contains ^{3}He liquid and copper
refrigerant in flake form. The inner cell is thermally hung completely from the outer cell. The outer cell also contains the
excitation coil for the inner cell NMR thermometer. The outer cell
is isolated from the mixing chamber of the precooling dilution
refrigerator by a further two metres of Cu capillary. The inner
cell contains about 1.1 moles of Cu and about 0.1 moles of ^{3}He. The
complete double cell contains about 75 cm^3 of ^{3}He in total. The
experiment is precooled to ~10 mK in an rms field of 6.4 T. The
base temperature of the refrigerator[4] is 3 mK so that considerably
lower starting temperatures could be achieved by more patience if
necessary. The precise demagnetization profile has virtually no
effect on the final temperature of the liquid which is governed only
by the heat leak and therefore we have made no serious effort to
optimise the demagnetization procedure. After demagnetization the
liquid temperature continues to fall for a number of days by a few
μK per day, which presumably reflects the cooling of plastic parts
of the inner cell since the thermal relaxation time of the liquid
is less than a minute at the lowest temperatures. Other than this
there is no evidence of any time-dependent heat leak into the inner
cell, the heat leak being largely determined by the viscometer and
NMR measurements and is a fairly reproducible 50 pW/mole of
refrigerant.

It is fortunate that the temperature is so steady at the lowest temperatures since the frequency widths measured by the vibrating wire are only a few mHz and require sweeps of several hours to be fully resolved. The wire is driven by an HP 3325A frequency synthesizer controlled by an HP 85 desktop computer. The output is phase sensitively detected by a Brookdeal(EG&G) 5206 lock-in amplifier also controlled by the HP 85. After sweeping the line in 100 discrete steps the computer executes a simple fitting program and returns the central frequency, f_o, the frequency width at half height, Δf_2, and the height, independently for both the in phase and quadrature components.

RESPONSE OF THE QUASIPARTICLES

The response of a vibrating wire in a liquid can be briefly summarized. In the hydrodynamic regime with a vanishingly small viscosity the movement of the wire implies the backflow of a similar volume of liquid. The resonator thus has a slightly lower resonant frequency than in vacuum since the backflow contributes to the effective mass of the wire. As the viscosity increases, which corresponds in normal ^{3}He and dilute solutions to a reduction in temperature, the wire moves an increasing mass of liquid as the viscous penetration depth increases and thus f_o falls further. At the same time the width, Δf_2, increases owing to the increasing viscous damping. This trend continues until the mean free path of the quasiparticles becomes comparable to the radius of the wire at which point the system gradually goes over to the collisionless quasiparticle gas regime, where the concept of viscosity no longer has any meaning. As the mean free path increases further the liquid moving with the wire finally falls back to its simple backflow value and hence f_o returns to the zero viscosity value while the damping saturates. The damping simply measures the momentum transferred from the wire to the liquid[2]. In the collisionless regime that is simply proportional to the number of quasiparticles encountered by the wire as it moves and so is a direct measure of the quasiparticle density. At low pressures in ^{3}He the mean free path has not reached a value comparable to the wire radius before the liquid becomes superfluid. However, below the superfluid transition ρ_n drops very rapidly and the extreme collisionless limit is reached (for P=0) at around 0.25 mK. Below this temperature Δf_2 simply reflects the falling normal fluid fraction and the frequency width becomes linear in $\exp(-\Delta(0)/kT)$. A plot of $\ln(\Delta f_2)$ against $1/T$ therefore yields the gap directly.

A series of measurements of Δf_2 at 0.0 bar and 7.3 bar are shown in Fig.1. The data show more scatter than would a professional vibrating wire measurement but it should be remembered that the signals are very small since the resonator itself is short, the field is relatively low (13 mT) and the drive velocity may not be too high otherwise dissipation is observed. The lowest temperatures reached by the liquid at both 0 and 7.3 bar (we only had enough ^{3}He to fill our 75 cm^3 cell to around 7 bar) are in approximate proportion to the respective gaps and correspond to similar frequency widths and hence to approximately the same normal fluid densities. This is an indication that in this experiment the Kapitza resistance has been effectively overcome and we are instead running into problems with the low thermal conductance of the bulk liquid as the quasiparticle density falls (or that the Kapitza conductance is a function of the normal fluid density). The zero bar data implies a width of ~480 Hz at T_c (the vacuum central frequency is 896 Hz) which falls to ~0.003 Hz at the lowest temperatures, a difference of more than five orders of magnitude. A computer fit to the P=0 data yields a gap $\Delta(0)$ of the BCS value (1.76T$_c$) enhanced by 10%, which seems in line with current thinking.

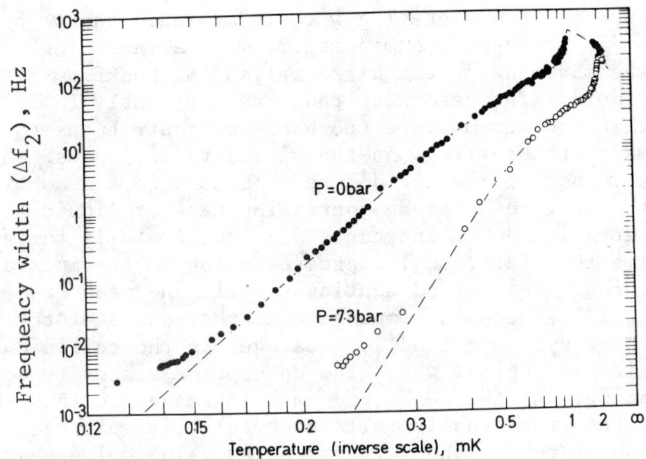

Fig.1. Frequency width, Δf_2, in Hz plotted against temperature. The dashed lines are computer calculations using a value for the gap, $\Delta(0)$, of 1.76T$_c$ enhanced by 10%.

RESPONSE OF THE TEXTURE

It is in the low temperature regime where Δf_2 begins to deviate from the linear dependence of Fig.1 that the textural effects begin to be visible. The first noticeable feature of the behaviour is that the central frequency f_o becomes dependent on the drive level. For zero pressure this is shown in Fig. 2 where we have plotted $\ln(\Delta f_2)$ against f_o for a number of drive levels. If there were no influence other than that of the quasiparticles one would expect f_o to approach asymptotically a fixed value as Δf_2 goes to zero. As can be seen the central frequency increases by approximately the same frequency interval for each factor of two reduction in drive level. At the highest temperatures in the figure the widths Δf_2 are much wider than the drive dependent shifts and the effect becomes lost. Within the accuracy of the data the results for each drive

level lie on parallel curves. The doubling back of the curves is an
artifact of the lineshape.

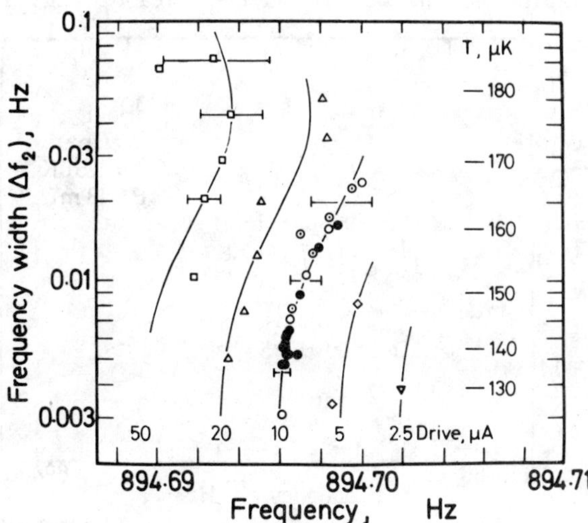

Fig. 2. The frequency width of the resonance plotted against the
central frequency, f_o, for a number of different drive levels.

As the damping from the quasiparticles falls the second
singular feature becomes apparent as the lineshape of the resonance
becomes heavily skewed with a sharp rise at low frequencies and a
long tail towards the higher frequencies. This lineshape indicates
a non-linearity becoming apparent in the resonance. We have made
very slow sweeps through the line at low temperature to try to resolve the rise on the low side and also to determine if there is any
hysteresis. A series of such measurements is shown in Fig. 3 where
the slowest sweeps span the rise region alone and were taken over 5
hours. There is hysteresis in the sense that the rise occurs at a
different frequency for up sweeps and for down sweeps but does not
appear to be a result of a re-entrant amplitude-frequency relation,
but is more complex. In a "simple" non-linear system where, for
example, the potential is not strictly harmonic but droops a little
at large distances, then to first order we can think of the resonance as harmonic but with an average force constant which depends on
amplitude[5]. Consequently the amplitude-frequency curve becomes
asymmetrical and shifts to lower frequencies at higher amplitudes
and may give rise to triple values of amplitude on the low
frequency side. This would lead to straightforward hysteresis. Our
results do not follow this pattern but rather seem to indicate a
self-regulation of the oscillation. In Fig. 3 the high frequency
tail follows, within the accuracy of the data, an exponential curve.
The low frequency rise seems to be a switching point between a state
of no movement and a state on the exponential curve. The point at
which the switching occurs depends on both sweep rate and on whether

the sweep is up or down. (It is the gradual appearance of this
lineshape from the Lorentzian shape from the normal fluid damping
which gives rise to the wobble in the data of Fig. 2).

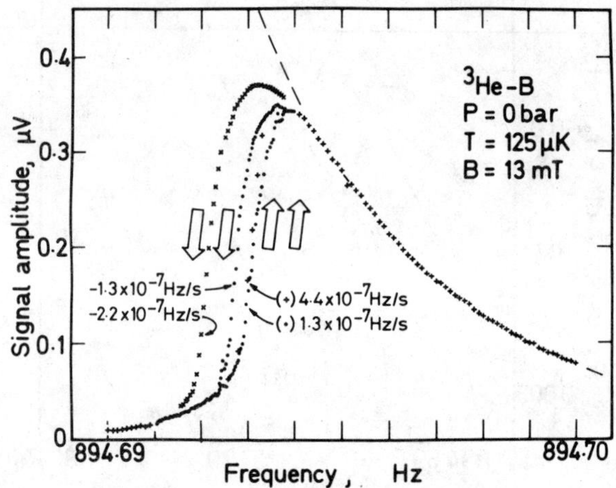

Fig. 3.   The lineshape of the resonance in the absence of quasiparticle damping for a number of different sweep rates. The dashed
line is an exponential curve which fits the data on the high frequency side.

However, when we superimpose the data from a number of drive
levels the true importance of the exponential region is seen. A
series of measurements at 6 mT are shown in Fig. 4 for a number of
different drive levels. All are taken sweeping down, the low side
rise being slightly smeared by the relatively fast sweeps of only
one hour per scan at each level. Leaving for a moment the data from
the highest drive level the exponential region of the curve is
common to all drive levels. The only difference between them is
that the switching point is drive-dependent. This means that if a
fixed frequency is chosen somewhere in the middle of the figure and
the drive level is gradually increased then at some point the
oscillation suddenly starts and then holds a fixed amplitude independently of the still increasing drive. This seems a rather unusual
self-regulation mechanism. The curve for the 50 μA drive level does
not conform to this pattern being cosiderably broader and clearly
shows the effects of dissipation. The signal at maximum corresponds to an rms wire velocity of about 0.5 cm/s, a velocity at
which dissipation is seen in a number of other experiments.
In Fig. 5 we have plotted the frequency at maximum against
drive level for four different magnetic fields at around 135 μK and
for P=0. Within the experimental error all the points lie on
similar curves which are linear except at the highest drives levels
where there is a small curvature which seems to be caused by the
onset of dissipation. The displacement of the curves along the

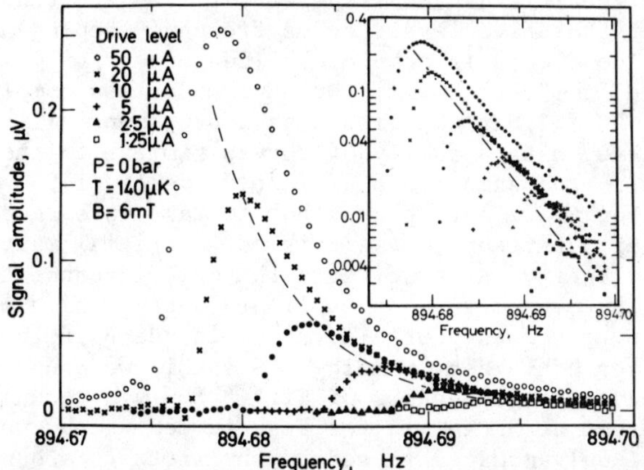

Fig. 4.   The lineshapes for several drives showing that the exponential part of the curve is common to all drive levels.   The inset
is the same data plotted as the logarithm of signal amplitude to
make this clearer.   The 50 μA curve is broader and shows signs of
dissipation.

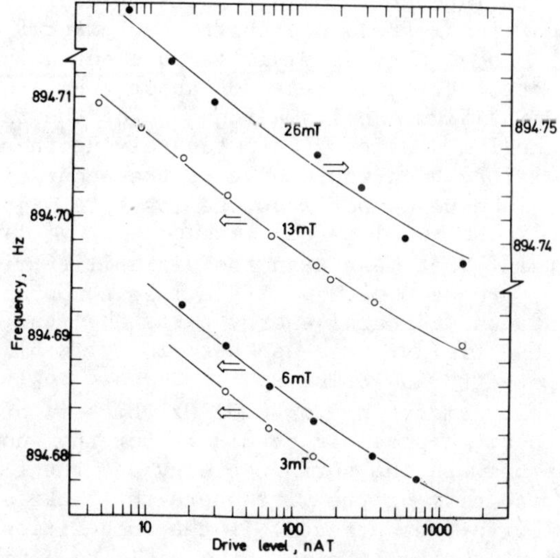

Fig. 5.   The dependence of the central frequency, f_o, on drive level
(here field times current) for a number of magnetic fields.

frequency axis is not a function of the ^{3}He.   The use of a superconducting wire as a mechanical resonator in a steady field provided
by a superconducting solenoid brings in a restoring force on the
wire from magnetic flux compression as the wire moves which leads to
an increase in the resonant frequency proportional to B^2.   The displacements in Fig. 5 are consistent with values for this effect

measured at higher fields and temperatures. We are thus fairly
confident that the drive level versus frequency dependence of the
resonance in the liquid is completely independent of magnetic field
over the range from 3 to 26 mT. We have made less complete measurements also at 7.3 bar and find behaviour essentially the same as
at zero pressure, with a family of curves similar to those of Fig. 2
for P=0 and with the same slope $df_o/d\ell n$(drive level).

One difficulty we have had is how to take into account the
vacuum width of our resonator. It is not a trivial matter to
measure this quantity since firstly the cell is connected to the
outside world by many metres of capillary tubing which means we
reach a much better quasiparticle vacuum in ^{3}He-B at low temperature
than we can ever achieve by pumping. Secondly we cannot cool the
cell evacuated. The lineshapes of Fig. 3 imply a vacuum width of
<0.001 Hz whereas at 3K and in the best pumped vacuum we could
attain we have only managed to reduce the width to around 0.01 Hz
which is clearly not the residual value. However, at these widths
in vacuum there was no drive dependence of f_o and the lines were
completely Lorentzian.

CONCLUSION

To be able to interpret these results it would be an advantage
to know what additional frequency shift the textural effects contribute over and above that from simple pure potential backflow.
The backflow contributes a decrease of about 2 Hz from the vacuum
value, the precise figure being a function of the ratio of the wire
density to the helium density. Unfortunately neither of these is
known to a precision which would give us the accuracies of a few mHz
that we need. Since we cannot know the absolute magnitude of the
frequency shift contributed by the texture, all we can assume is
that it is certainly not less than the variability with drive which
we observe, i.e. around 0.01 Hz. If we insert appropriate values
for B and v_s into the textural energy terms then the bulk flow term
in $(\underline{n}\cdot\underline{v}_s)^2$ and the field-flow cross term in $(\underline{v}_s\cdot R\cdot\underline{B})^2$, using the coefficients of Leggett[6] and Brinkman and Cross[7], together yield a
scale for the shift amounting to $\sim$0.02 Hz which is of the right
order of magnitude. A possible candidate for the non-linearity is
the competition between the surface energy and the flow energy in
the immediate viscinity of the wire where the backflow velocity is
high and parallel to the surface. If the competition forces the
distribution of superflow at high drive rates to extend further out
into the liquid than pure potential flow would suggest then more
liquid will move with the wire and f_o will fall. However, this does
not explain why the drive level dependence should be logarithmic,
which if maintained to the lowest drives suggests that the texture
is infinitely stiff to infinitesimal forces. To advance much
further than order of magnitude calculations we need a clearer
insight into what processes are occurring. One is quickly led to
the conclusion that an orthodox vibrating wire is not the most
convenient probe for a number of reasons. Firstly even at a given

instant in time the wire does not move with uniform velocity but has a range of velocities from zero at the anchorage points to a maximum at the furthest radial distance which makes analysis difficult if say critical velocity phenomena are involved. Secondly the influence of the texture is inconveniently low since it only yields widths of the order of mHz; we need a somewhat larger width to allow measurements to be made over a timescale faster than 20,000 seconds.

Perhaps the most important conclusion to be drawn from these measurements is simply that the texture does have mechanical properties and that the temperature range in which they can be readily observed is experimentally accessible. Hopefully some attention to this problem by the theoretical community will lead to the design of more specific and more readily interpretable experiments in this area in the near future.

ACKNOWLEGEMENTS

This work is supported by SERC (grant GR/C/05960). We are also grateful to H. E. Hall and J. R. Hook for many useful discussions and to I. E. Miller and N. S. Lawson and others at Lancaster for much assistance.

REFERENCES

1. A. M. Guénault, V. Keith, C. J. Kennedy, I. E. Miller, G. R. Pickett, Nature (1983) (in press).
2. A. M. Guénault, V. Keith, C. J. Kennedy, G. R. Pickett, Phys. Rev. Lett. $\underline{50}$, 522 (1983).
3. N. S. Lawson, Cryogenics $\underline{22}$, 667 (1982).
4. D. I. Bradley, T. W. Bradshaw, A. M. Guénault, V. Keith, B. G. Locke-Scobie, I. E. Miller, G. R. Pickett, W. P. Pratt, Cryogenics $\underline{22}$, 296 (1982).
5. e.g. N. N. Bogoliubov, Y. A. Mitropolsky, Asymptotic Methods in the Theory of Non-linear Oscillations (trans. Hindustan Publ. Corp., Delhi 1961) pp 254-267.
6. A. J. Leggett, Rev. Mod. Phys. $\underline{47}$, 331 (1975).
7. W. F. Brinkman, M. C. Cross, Prog. Low Temp. Phys. $\underline{VIIA}$, 105 (1978).

TEXTURE TRANSITIONS IN FLOWING ^{3}He[*]

D. M. Bates, C. M. Gould[†], and H. M. Bozler
Physics Department, University of Southern California
Los Angeles, CA 90089-0484

ABSTRACT

A system for the study of flowing superfluid ^{3}He has been developed. The cell, which is mounted on a nuclear demagnetization cryostat, utilizes a bellows drive system that can produce a flow at a well defined velocity. This cell is being used to look for texture changes caused by flow using the anisotropy of the attenuation of zero sound in the A phase. Preliminary results indicate that we can observe reproducible texture changes and that these transitions may be consistent with the onset of helical textures.

For several years there have been predictions[1] for the onset of helical textures and other flow induced textural changes in ^{3}He-A. In this paper we present the design of an apparatus we are using to look for these effects. We also give preliminary results for a helical texture phase diagram.

Superfluid flow can be used to study the competition between the several terms in the free energy of the A phase of liquid ^{3}He. These energies can be written in terms of the components of the orbital part of the order parameter $\hat{\Delta}_1, \hat{\Delta}_2, \hat{\ell}=\hat{\Delta}_1\times\hat{\Delta}_2$; the spin part of the order parameter $\hat{d}$; an applied magnetic field $\vec{H}$; and the superfluid velocity $\vec{v}_s$. The relevant energies are the dipole energy E_D, the susceptibility anisotropy energy E_H, flow energies E_v, and bending energies E_{grad}. These energies are given explicitly below[2].

$$F = E_D + E_H + E_v + E_{grad}$$

$$E_D \propto -\Omega_A^2(\hat{\ell}\cdot\hat{d})^2$$

$$E_H \propto (\hat{d}\cdot\vec{H})^2$$

$$E_v = {}^1\!/_2\rho_s v_s^2 - {}^1\!/_2\rho_0(\hat{\ell}\cdot\vec{v}_s)^2 + c(\vec{v}_s\cdot\vec{\nabla}\times\hat{\ell}) - c_0(\vec{v}_s\cdot\hat{\ell})(\hat{\ell}\cdot\vec{\nabla}\times\hat{\ell})$$

---

[*] Work supported by the National Science Foundation grant numbers DMR82-00661 (HMB,DMB), DMR82-08760 (CMG), and Research Corporation (CMG).

[†] Alfred P. Sloan Research Fellow

$$E_{grad} = \tfrac{1}{2}K_s(\vec{\nabla}\cdot\hat{\ell})^2 + \tfrac{1}{2}K_t(\hat{\ell}\cdot\vec{\nabla}\times\hat{\ell})^2 + \tfrac{1}{2}K_b[\hat{\ell}\times(\vec{\nabla}\times\hat{\ell})]^2$$

$$+ \tfrac{1}{2}(K_1-K_2)[(\hat{\ell}\cdot\vec{\nabla})\hat{d}]^2 + \tfrac{1}{2}K_2\partial_i d_\alpha \partial_i d_\alpha$$

In most of our experiments, we apply a magnetic field parallel to the direction of flow. In the case where we have a small v_s but significant H the stable texture is expected to be uniform with $\hat{d}||\hat{\ell}|\vec{v}_s$. On the other hand, with small H, the flow should line up $\hat{\ell}$ so that $\hat{d}||\hat{\ell}||\vec{v}_s$. The theoretical predictions for the breakdown of the uniform textures are in qualitative agreement[1], and based on these we can draw a proposed phase diagram in v_s vs. H space in figure 1. The regions of stability for the helical textures depend on both the velocity and the magnetic field. The behavior in the A phase is expected to be hysteretic and the arrows indicate the direction that we should move to reproducibly go through these boundaries.

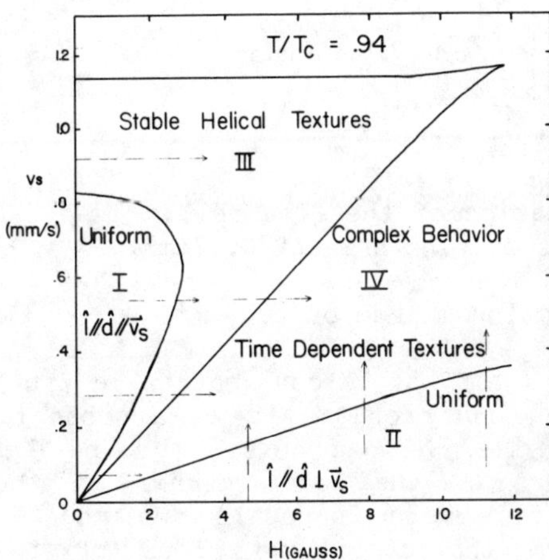

Fig. 1. Proposed phase diagram for $\vec{v}_s || \vec{H}$.

The boundary of region I depends crucially on the value of the terms in the flow energy E_v, in particular on c_o. We should note that while time independent helical textures would not produce dissipation, transients and time dependent textures will.

We have developed a tower that is directly mounted on our nuclear demagnetization apparatus to study the H vs. v_s phase diagram. This cell, shown in figure 2, uses bellows to drive ³He through the experimental region at a well defined velocity. Most of the metal parts including the bellows are made from Berylco-25. However the "piston" is silver with a sintered silver heat exchanger on both sides. The superleak is an array of ~830,000 3mm long glass capillaries with 2μm diameters. The ⁴He driving bellows carries a superconducting lead strip whose position can be measured with 0.2μm resolution using a pair of mutual inductors mounted on the frame of the tower. No moving electrical leads are required. The signal from the position sensor is compared with a linear ramp signal and fed back to a ⁴He pressure bomb located in the helium dewar outside of the exchange gas can. This servo-loop controls the bellows motion to produce the desired v_s.

284

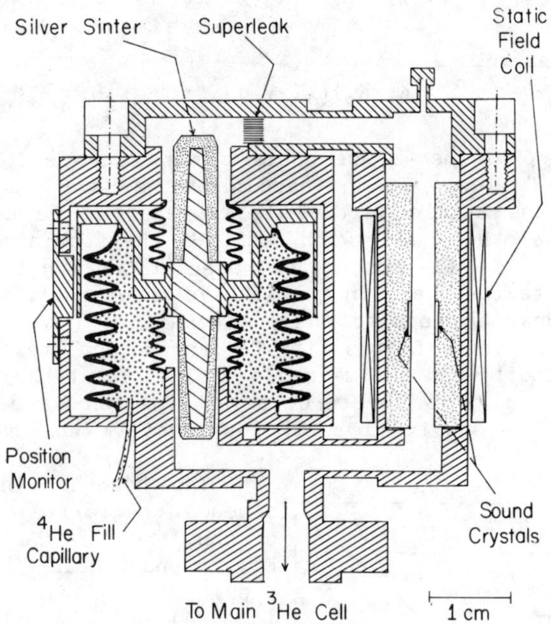

Fig. 2. Schematic representation of the flow cell. The
³He bellows have an effective driving area of 0.77cm². A
removable epoxy insert holds the crystals and forms the
flow channel with a cross sectional area of .054cm².

We are currently using a pair of X-cut quartz crystal
transducers with a fundamental frequency of 10 MHz as a probe of
textural changes in the flowing ³He. Two precision ground 3mm
thick quartz bars form the sides of the flow channel in the
experimental region. The 0.40cm diameter crystals are mounted
flush in epoxy walls that form the other two sides of the
rectangular channel (3mm x 1.75mm). In this cell $\vec{q}$ will be
perpendicular to $\vec{v}_s$.

Textural changes are detected by the anisotropy of zero sound
attenuation[3]. The attenuation depends on the angle between $\hat{\ell}$ and
the sound propagation wavevector $\vec{q}$. To directly measure the
anisotropy we keep v_s equal to zero and slowly rotate a small
magnetic field (~10 gauss) in the plane perpendicular to the flow
channel axis. This causes $\hat{\ell}$ to be perpendicular to $\vec{q}$ twice every
cycle. Figure 3 shows a time trace of the received amplitude of 30
MHz sound and one component of the rotating magnetic field. This
technique can be used to find the parallel and perpendicular
components of the sound attenuation vs. temperature. At 30 MHz our
observations are limited to $T < .96T_c$ because of the very high
attenuation near T_c. All of the flow experiments reported here
were done at 28.8 bars and $T/T_c = .94$.

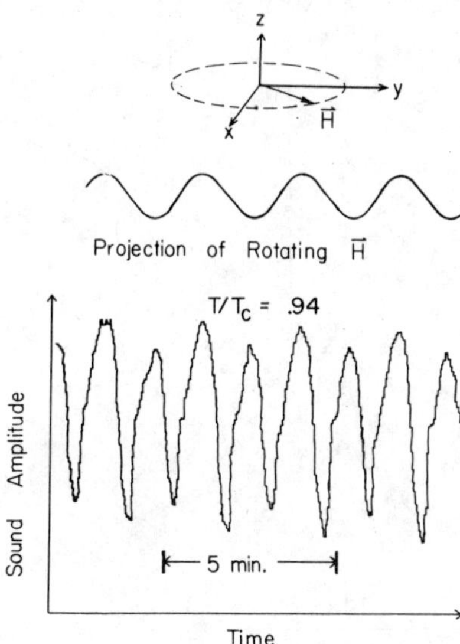

Fig. 3. Received signal caused by rotating the magnetic field in a plane perpendicular to the channel. The upper trace shows the projection of the rotating field H into the direction of q. The warming rate during this trace was ~ 1μK per minute.

In our first experiments using the flow tower, we have established a constant flow velocity and then ramped the magnetic field with $\vec{H}||\vec{v}_s$. Typical tracings of the received amplitude versus time are shown in figure 4. In trace (a) the sound starts at a low level and typically fluctuates, indicative of $\hat{\ell}$ staying nearly aligned to the sound direction. When the bellows start moving the sound level immediately increases to a value that is close to the maximum level found from the rotating field experiment--indicating that ℓ is perpendicular to $\hat{q}$ and is presumably aligned with the flow. The magnetic field is maintained at zero in this trace. After the bellows motion stops, the sound level slowly drifts back towards its original state. In trace (b) we start out with the same sequence. However, after the constant flow velocity has been established, a negative field ramp is started. (Both positive and negative field ramps have been used in these experiments.) After the magnetic field reaches a magnitude of ~ 1.5 gauss the sound level starts to diminish.

We can also see that in this event, when the field magnitude began to decrease, the sound level appeared to partially reverse until the flow stopped. In other traces for which the field is brought completely back to zero during the bellows motion the sound level does return to its original high value. This indicates that the transition is reversible. When we ramp the field to high values, however, the transition is not reversible. All of our data utilizes the down stroke of the bellows with the flow directed towards the superleak, since the reverse motion fails to hold ℓ aligned perpendicular to the sound transducers.

We should note that after creating a texture we cannot always get back to the uniform state. The symptom is that the sound level fluctuates and does not go to as high a level as we associate with alignment of $\hat{\ell}$ along the flow channel. Sometimes the temporary addition of a large (50 Oe) magnetic field seems to iron out the texture, but on other occasions we have had to wait several hours to get a uniform starting texture.

286

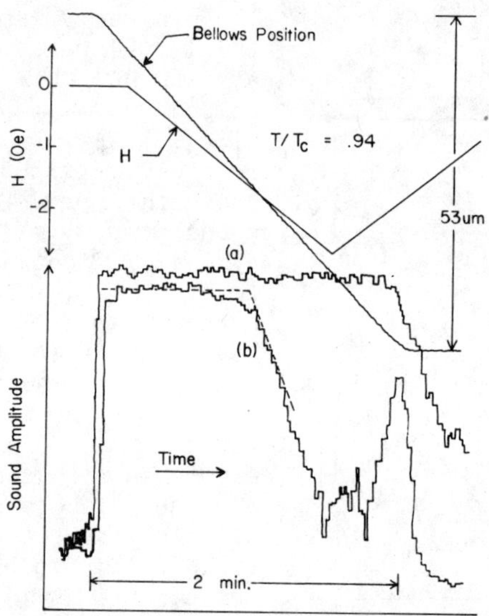

Fig. 4. Reproductions of representative chart recordings of the bellows position, magnetic field, and received sound signal. Sound traces (a) and (b) were taken at the same temperature with identical velocities. H=0 during trace (a). Trace (b) shows the effect of increasing the field magnitude during the motion.

From sound traces as shown in figure 4 we can estimate the field values that correspond to texture transitions. In our preliminary study we have used straight lines to obtain the values of field for the break in the sound level which we will designate H_c. Uncertainties in the values of H_c are due primarily to our present inability to pick the break point accurately. The superfluid velocity is calculated from the known volume displacement of the bellows and the measured[4] values of the superfluid density ρ_s. We assume that prior to the break, we can use the parallel component of ρ_s. Figure 5 shows our preliminary phase diagram. At low velocities the values of H_c appear to have the same qualititive behavior as predicted for the transition to helical textures. In addition, the maximum values of H_c agree reasonably well. However, we do not as yet see any upper boundary for the uniform texture region (region I in figure 1). Further studies of this phase diagram are continuing.

Previous experiments have looked at some aspects of flow related phenomena. Kleinberg[5] has performed similar measurements using heat flow instead of mass flow. Reproducibility was a problem in these experiments; however, similar signatures were observed. Paalanen and Osheroff[6] have measured dissipation and NMR frequency shifts in flowing ^{3}He-A, but in fields much greater than 30 gauss. Their results may indicate the high field limit of

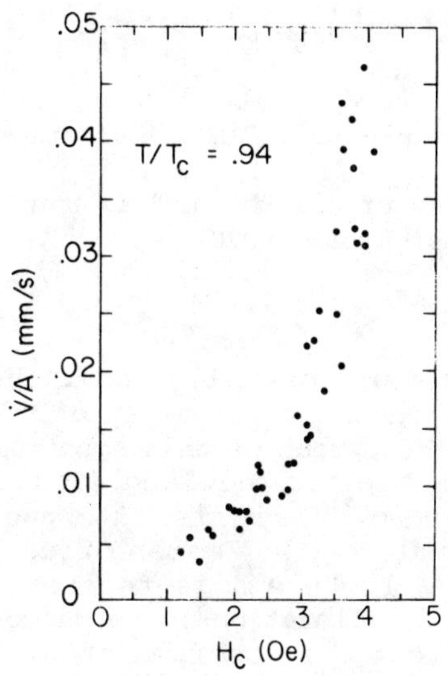

region II, but they get a critical velocity 40% smaller than predicted. Flint et al.[7] had previously performed experiments in high fields, but their work using a Pomeranchuk cell was subject to uncertainties in the normal fluid motion. As our preliminary results indicate, considerably more work will be needed to pin down the v_s vs. H phase diagram. We wish to thank K. Maki for useful discussions.

Fig. 5. Results of the field sweep measurements at constant velocity. $\dot{V}$/A is the volume flow rate divided by the flow channel area.

REFERENCES

1. D. Vollhardt, Y. R. Lin-Liu, and K. Maki, J. Low Temp. Phys. 37, 627 (1979); A. L. Fetter and M. R. Williams, Phys. Rev. B 23, 2186 (1981); and numerous references cited therein.
2. A. L. Fetter, Phys. Rev. B 20, 303 (1979), and M. C. Cross, J. Low Temp. Phys. 21, 525 (1975).
3. See for example D. N. Paulson, M. Krusius, and J. C. Wheatley, Phys. Rev. Lett. 37, 599 (1976), and D. N. Paulson, M. Krusius, and J. C. Wheatley, Phys. Rev. Lett. 37, 1322 (1978).
4. J. E. Berthold, R. W. Giannetta, E. N. Smith, and J. D. Reppy, Phys. Rev. Lett. 37, 1138 (1976).
5. R. L. Kleinberg, J. Low Temp. Phys. 35, 489 (1979).
6. M. A. Paalanen and D. D. Osheroff, Phys. Rev. Lett. 45, 362 (1980).
7. E. B. Flint, R. M. Mueller, and E. D. Adams, J. Low Temp. Phys. 33, 43 (1978).

PROBING COLLECTIVE MODES OF SUPERFLUID
^{3}He-B WITH ZERO SOUND

J. B. Ketterson, B. S. Shivaram, M. W. Meisel, Bimal K. Sarma
and W. P. Halperin
Department of Physics and Astronomy and Materials Research Center
Northwestern University, Evanston, IL 60201

ABSTRACT

We review what is currently known experimentally about the collisionless collective mode spectrum of the order parameter of the B phase of superfluid ^{3}He. The most powerful probe of this spectrum has proved to be longitudinal zero sound and our knowledge of the remaining modes comes about through mode coupling effects. A magnetic field splits some states into multiplets (the Zeeman effect) with the number of states given by 2J + 1 where J is the total angular momentum assigned to the state; an alteration in the spectrum, analogous to the Paschen-Back effect, is observed at high fields. The mode coupling leads to high attenuations over much of the range of interest and negates the use of the standard (phase sensitive) pulse transmission technique; this necessitates the use of acoustic impedance methods. Our acoustic and cryogenic techniques are discussed in some detail.

INTRODUCTION

Very shortly after the discovery of superfluid ^{3}He, experiments on the propagation of ultrasound were performed which showed dramatic structure.[1] Early theoretical studies showed that condensate pair breaking and collective mode interaction effects were primarily responsible for this structure.[2,3] In the B phase the observed collective mode is called the (imaginary) squashing (sq) mode. For all but the zero sound and spin modes, the frequencies, ω_i, of the collective modes are proportional to the temperature dependent energy gap, $\Delta(T)$, i.e., $\omega_i = a_i\Delta(T)$. On the other hand, the sound frequencies, ω_s, are confined to the (fixed) odd harmonics of a quartz transducer, $\omega_s = (2n+1)\omega_f$ where n is an integer and ω_f is the transducer fundamental resonant frequency. Mode-mode interaction effects occur primarily where $\omega_s = \omega_i$ and this matching is achieved, in a given run, by varying the temperature; thus collective modes appear as a peak (resonance) in the temperature dependent attenuation.

In later experiments the existence of another B phase collective mode, the real squashing (rsq) mode[4] was uncovered. Although the existence of this mode was already known theoretically[5], its coupling to zero sound was overlooked due to the fact that it primarily involves the spin (real) degrees of freedom of the order parameter. In the presence of the experimental data, the theory was extended[6] to include effects arising from an energy dependence of the density of states (so called particle-hole asymmetry) which leads to a coupling with (the imaginary or density mode) zero sound, albeit a very weak one. As a basis for parametrizing the interaction of zero sound with collective modes, a phenomenological model was introduced and applied to both the zero sound-squashing mode, and zero sound-real squashing mode couplings.[7] It was shown that the very strong temperature and pressure dependencies of the group and phase velocities as well as the attenuation, were well accounted for by this model, and the respective coupling strengths were obtained and compared with theory. The application of a magnetic field was observed to have a dramatic effect on the real squashing mode:[8] it split into five equally separated resonances (recent refinements are discussed later). This behavior, which is analogous to the Zeeman effect of atomic physics, had also been predicted theoretically.[9] The five lines are associated with the $2J + 1$ levels of the $J = 2$ total angular momentum assignment of the real squashing mode.[5]

The coupling of collective modes, especially the squashing mode, leads to enormous attenuations of sound, particularly at higher frequencies, making pulsed transmission acoustic studies (the most commonly applied acoustic technique) ineffective. This is just the regime where the acoustic impedance technique, applied earlier in normal ^{3}He,[10] is most effective. Acoustic impedance studies using both pulsed[11] and c.w.[12] variants were performed on the A and B phases; each variant has its own special strengths and weaknesses. The most recent work has centered on high field studies, primarily below .2T. A number of new phenomena have been observed including: 1) a splitting of the $J_z = 0$ rsq line at high fields;[13]; 2) a three-fold zero field dispersion induced splitting of the $J_z = 0, \pm 1$ and $+2$ rsq lines[13]; 3) field dependent effects in the squashing mode (Paschen Back analogue)[15]; 4) a non-linear evolution of the rsq Zeeman multiplet with field (also a Paschen-Back analogue);[13] and 5) unexplained structure near 2Δ (which may be associated with new modes). These phenomena will be the central focus of this review.

THEORETICAL BACKGROUND

A variety of techniques have been used to describe collective modes in superfluid ^{3}He. For those not schooled in many body theory, the mean field Boltzmann equation approach of Wolfle[16] is the easiest to understand. The number of modes, in the presence of only a p-wave pairing potential, V_1, (i.e. V_3 equal zero) is identical to the number of degrees of freedom of the complex amplitude of the L = 1, S = 1 condensate: $(2S+1) \times (2L+1) \times 2 = 18$. The total number of modes expected with $V_3 \neq 0$ and $F_i^{a,s} \neq 0$, $i > 2$, does not appear to have been fully explored. In particular, the fate of higher order normal fluid distortions (transverse zero sound, higher order spin waves) may not have been adequately studied. The nature of the modes differs substantially between the A and B phases, and we will discuss only the latter.

The isotropic nature of the B phase order parameter leads to J being a good quantum number, and J, together with identifying a real

(r) or imaginary (i) character, provides a specification of the mode. The 18 B phase modes are listed in Table I. Included in the table are (i) the J, J_z, and class (real or imaginary), (ii) the frequency of the mode in the form $\omega^2 = \omega_o^2 + a^2 V_F^2 q^2$ where a^2 is a dispersion constant, (iii) the Fermi liquid correction to ω_o^2, and (iv) the coupling to zero sound which is strong, weak (i.e., through particle–hole asymmetry), or zero (no coupling). The modes are referred to by the following names (in order of increasing frequency): [J = 0,i] zero sound, [J=1,r] spin waves, [J=2,r] real squashing mode, [J=2,i] (imaginary) squashing mode, [J=0,r] and [J=1,i] real and imaginary "gap modes." Of these, only the zero sound and the $J_z = \pm 1$ spin waves exist in the normal phase, with the latter Landau damped at H = 0. The zero sound, squashing and real squashing modes have been relatively well studied, i.e, 1+5+5 = 11 of the modes; the 4 gap modes (which may be overdamped, since they lie near the continuum) and the (collisionless) spin waves have yet to be observed.

The B phase work described here focuses on probing the squashing and real squashing modes with zero sound. To date, a theoretical treatment which includes corrections from V_3 and $F_i^{a,s}$ through i = 2, to both ω_o^2 and a^2 has not appeared. Sauls and Serene[17] have carefully examined the V_3 and F_i (to i=2) corrections to ω_o^2; for the squashing and real squashing modes, Combescot[18] has calculated the F_1 corrections to a^2. For the case $V_3=0$ the Sauls–Serene squashing and real squashing mode results are

$$\omega_o^2 \ (2,i) = 12/5 \, \Delta^2 \ (1 + 2/25 \ F_2^a) \tag{1a}$$

$$\omega_o^2 \ (2,i) = 8/5 \, \Delta^2 \ (1 + 3/25 \, \lambda \, F_2^a) \tag{1b}$$

where λ is the function introduced by Tsuneto.[17,19] The appearance of $F_2^{a,s}$ in these equations suggests that one may use the measured mode frequencies to determine these parameters about which little is known reliably; however, this requires independent information on the strong coupling effects and on the effects of V_3. The large coupling constant of the squashing mode to zero sound makes the dispersion difficult to observe; however, the real squashing mode dispersion has been observed (see Sec. IV), and Combescot's theoretical prediction[18] is given in Table I. The results with $F_1^a=0$ had been obtained earlier by Brusov and Popov[20]. We note in passing that (for $F_1^a=0$) the added J_z dependent dispersion induced splitting has the same form as the Stark effect of atomic spectra (i.e., $J_z^2 a^2$ with q playing the role of the electric field).

In the presence of a magnetic field, splitting is introduced into some of the mode frequencies.[8,9] Again it is the effect on the real squashing mode which has been measured; the theoretical prediction of Schopohl and Tewordt[9] for the low field regime is

$$\omega_o(J=2,H) = \omega_o(J=2,0) + M_J g(T) \omega_L(T,H) \tag{2a}$$

where the effective g factor and Larmor frequency are given by

291

TABLE I
B Phase Collective Modes

Specification	Mode Frequencies (ref. 20) $\omega^2 = \omega_0^2 + a^2 v_p^2 q^2$		Some Fermi Liquid Corrections (T = 0)		Coupling to Zero Sound
$[J, J_z, i \text{ or } r]$	ω_0^2	a^2	ω_0^2	a^2 (Ref 5, 18, and 20)	
$[0, 0, i]$	0	$1/3$	0	$1/3(1 + F_0{}^s)(1 + \frac{F_1{}^s}{3})$	1
$[1, 0, i]$	$4\Delta^2/h^2$	$0\text{-}111\text{-}0.169\ i$	---	---	?
$[1, \pm 1, i]$	"	$0.353\text{-}0.335\ i$	---	---	?
$[2, 0, i]$	$\frac{12}{5}\Delta^2/h^2$	0.502	$\frac{12}{5}\frac{\Delta^2}{h^2}(1 + \frac{2}{25} F_2{}^s)$	---	strong
$[2, \pm 1, i]$	"	0.418	"	---	"
$[2, \pm 2, i]$	"	0.164	"	---	"
$[0, 0, r]$	$4\Delta^2/h^2$	$0.237\text{-}0.295\ i$	---	---	weak
$[1, 0, r]$	~ 0	$\frac{1}{5}$	~ 0	$\frac{1}{5}(1 + \frac{2}{3} F_0{}^a)$	0
$[1, \pm 1, r]$	0	$\frac{2}{5}$	0	$\frac{2}{5}(1 + \frac{2}{3} F_0{}^a)$	0
$[2, 0, r]$	$\frac{8}{5}\Delta^2/h^2$	0.442	$\frac{8}{5}\frac{\Delta^2}{h^2}(1 + \frac{3}{25} \lambda F_2{}^a)$	$0.442 + 0.056 F_1{}^a$	weak
$[2, \pm 1, r]$	"	0.388	"	$0.386 + 0.140 \dfrac{0.333 + .056 F_1{}^a}{1 + 0.187\ F_1{}^a}$	"
$[2, \pm 2, r]$	"	0.224	"	$0.224 + 0.019 \dfrac{F_1{}^a}{1 + .187 F_1{}^a}$	"

$$g = 1/12[1 + (1-Y(T))/\lambda(\omega_o)] \qquad (2b)$$

$$\text{and } \omega_L = \gamma H/[1 + F_o^a(2/3 + 1/3Y)] \qquad (2c)$$

where Y and λ are the Yoshida and Tsuneto[19] functions, respectively. The above expressions have been generalized by Sauls and Serene[21] to include the effects of V_3 and $F_2^{s,a}$ which they have applied to the data of the French workers.[8] The resulting expressions are lengthy and will not be reproduced here. We remind the reader that data are generally taken by sweeping the temperature, this results in a five fold (temperature splitting) of the real squashing mode resonance.

As the magnetic field is increased, a non-linear regime is entered where the splitting of the levels of the multiplet is no longer proportional to H. Quite generally we expect H^2 contributions (quadratic Zeeman effect) and, on phenomenological grounds, we can write the field splitting in the form.

$$\Delta\omega = \alpha J_z H + \beta J_z^2 H^2 - \Gamma H^2 ; \qquad (3)$$

$\alpha = g\omega_L$, and β and Γ are constants. This phenomenon was predicted theoretically by Schopohl, Warnke, and Tewordt[22] and independently discovered experimentally by Shivaram et al[15]. The theoretical expressions for and will not be listed here. As Schopohl et al have shown[22], this effect arises from a field induced anisotropy in the otherwise isotropic B phase energy gap which is then described by two gap functions, Δ_1 and Δ_2.[23] They have termed the resulting evolution of the levels as the Paschen-Back effect.

EXPERIMENTAL TECHNIQUE

A. Cryogenics

Figure 1 shows a schematic diagram of the experimental region of our cryostat. Cooling is accomplished by a bundle of Cu wires which was demagnetized using an 8T superconducting solenoid (not shown in Fig. 1). Contact is made via a welded joint to a sintered copper heat exchanger with a total surface area of 34 m^2 and containing 20 cm^3 of liquid. The bundle and ^{3}He were precooled by the dilution refrigerator mixing chamber through a shielded Sn heat switch and a silver rod; the rod was in turn welded to a silver ring which was bolted to the heat exchanger. The magnetic fields applied to the ^{3}He were provided by a specially designed fourth order 3T solenoid[24] located in the ^{4}He bath (shown crossed in Fig. 1). Our requirement here was for the field to have a relatively high homogeneity in the region of the sound cell (1 part in 10^4 over a cm^3) yet fall off rapidly to minimize the spill over field at the bundle and near the mixing chamber. The field in the vicinity of the Sn heat switch was 30mT at maximum field which is easily shielded by a Nb annulus.

Two towers, of different heights, are attached to the upper plate of the heat exchanger. The smaller of these (left in Fig. 1)

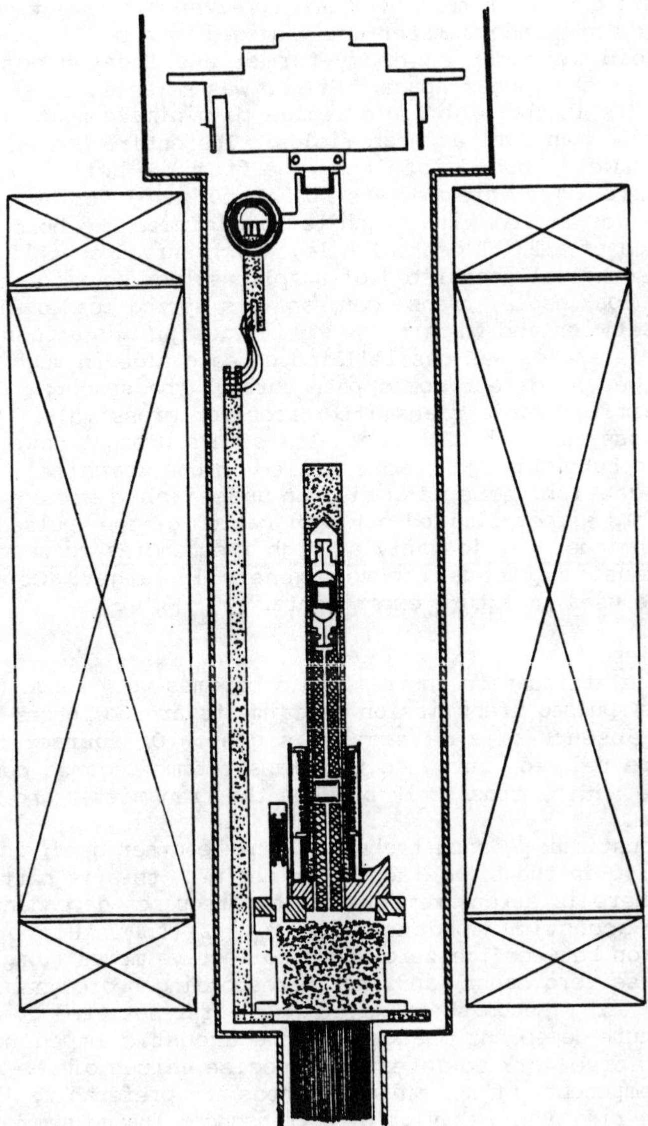

Fig. 1. Our experimental apparatus is shown,
and important features are discussed in the text.

is made from epoxy and contains an LCMN sample whose magnetization is measured using a.c. SQUID techniques. The larger column was the sonic tower (right in Fig. 1), which is made from silver, and contained the Pt NMR thermometer (sample and pickup/excitation coil) and the sonic cell near the lower and upper ends, respectively. The H_0 field for the Pt thermometer was provided by a small superconducting solenoid wound on an epoxy former and located outside the silver tube in the vacuum space. Silver was selected for the metallic components in the high field region to minimize heat loads from the nuclear Zeeman bath at high fields. The entire contents of the sonic tower were loaded into the tube from the bottom as a self-contained unit. The intermediate space, not used for sonic or NMR components, was filled with graphite to minimize ^{3}He hold up, with the exception of a 3/16" central hole, which served as a liquid heat pipe. The sound cell consisted of a split silver housing, each half involving a coax cable, transducer, and its spring loaded electrical contact. Between and within the split housing was a fused quartz annulus. This spacer was castlellated on each side in such a manner that there was no direct sound path through the spacer; this precaution minimizes sonic transmitter/receiver cross talk. To minimize r.f. cross talk, the CuNi coax lines were brought down opposite sides of the tube through tracks milled in the graphite. In spite of this precaution there was still an undesirable amount of cross talk, and this effect limited our application of the (pulsed) transmission technique, particularly at high frequencies where the coupling and acoustic attenuation were generally larger. Copper coax tubes may be used in future experiments.

B.   Acoustics

When the attenuation in the liquid becomes very large (typically 6 napiers) pulsed transmission experiments are no longer possible due to the absence of a detectable signal. Of course, the path length can be reduced, but when the transit time becomes comparable to the pulse width, cross talk between the transmitter and receiver is a problem.

The acoustic impedance technique, on the other hand, is usually most effective in the high attenuation regime; this is particularly true when there is strong temperature or frequency dependent component to the acoustic impedance, i.e. $Z_A = Z_A(T,\omega)$. Historically, the transition between the zero sound collective modes (longitudinal and transverse zero sound) and the corresponding hydrodynamic modes (first sound and viscous shear wave) were first detected by studying the temperature dependent changes in the acoustic impedance.[25,10] For quantitative work to determine precise values of the real and imaginary components of Z_A, pulsed methods are preferable; here one measures the ring down behavior of a transducer loaded symmetrically by the liquid. The quantitative superiority of pulsed methods stems from the fact that it is possible to time separate effects whose origin is primarily due to the acoustic properties of the liquid from those involving the electrical properties of the transducer, detection electronics, and interconnecting transmission lines; the ring down persists long after the primary pulse has disappeared.

In much of our recent work at Northwestern, we have employed a c.w. acoustic impedance technique. At this stage of the investigation, we are primarily interested in collective mode spectroscopy i.e., we are most interested in the temperature, magnetic field, and frequency dependence of the mode positions and are somewhat less concerned with the numerical strength of their coupling to zero sound, i.e., the magnitude of their effect on the acoustic impedance. There are two advantages of a c.w. method which help outweigh its lower quantitative accuracy: 1) the bandwidth is inherently much narrower; and 2) the amplitude of the acoustic wave induced in the liquid is much lower. The first of these is especially important when probing spectroscopically very sharp modes, e.g., the B phase real squashing mode. One obtains spurious line broadening if the spectral width of the sound wave is larger than that of the collective mode. For a pulsed technique this width is approximately the reciprocal of the pulse width; for an f.m. c.w. technique, it is on the order of the f.m. modulation frequency or frequency deviation, whichever is larger. The second factor (lower amplitude) helps to minimize amplitude dependent (non-linear) effects in the liquid.

Fig. 2 shows a block diagram of our f.m. c.w. spectrometer. A frequency modulated r.f. signal is generated by the HP 606B (low frequency) or 608F (high frequency) oscillator; the modulation signal may be applied either directly to the oscillator or to the synchronizer (when a higher frequency stability is desirable). The r.f. output is applied to the cryostat transmission line, and hence to the transducer, through a quadrature hybrid, the reflected signal appearing in the output arm of the hybrid. Since the transducer loaded line does not present a 50Ω load to the hybrid, there is a large unbalanced signal. This signal is nulled (using a phase shifter/attenuator combination) to minimize drift and noise components from the oscillator. Collective mode passage or other phenomena occurring in the liquid tend to upset the balance and cause d.c. signals to appear at the outputs of inphase and quadrature phase detectors, these signals being related to acoustic impedance changes in the liquid. Also appearing in the output are a.m signals at the f.m. modulation frequency (and also its harmonics). The a.m. components contain contributions arising from the frequency dependence of the resonant transducer and, near collective modes, the acoustic impedance, itself. The frequency dependence of the transducer impedance has dispersive and absorptive components characteristic of a resonant circuit; to the extent that these two contributions can be adjusted to be orthogonal, they produce a maximum and a zero in the a.c. output of the two phase detectors at the modulation frequency. These are the signals primarily studied in this work.

Concerning the f.m. technique, the following qualitative features should be noted. The transducer resonant frequency is rather sharp and is easily located; changes in the imaginary and real components of the acoustic impedance alter the resonant frequency and line width by amounts, $\delta\omega$ and $\Delta\omega$, which are given approximately[10]

296

by $\delta\omega = (\omega_o/\pi) \, X_A/R_Q$ and $\Delta\omega = (\omega_o/\pi) \, R_A/R_Q$, where X_A and R_A are the reactive and resistive components of the liquid impedance, ω_o is the resonant frequency, and R_Q is the quartz impedance. The width of the resonance is proportional to the dispersive component or the second harmonic signal at resonance.

The c.w. technique (acoustic impedance or transmission) is also useful in the low attenuation regime. Here the appearance of a reflected wave leads to a standing wave ratio in the cell. In the B phase the squashing mode causes the velocity to change significantly over a wide temperature range such that many standing wave resonances are swept out. From a node count we can deduce the temperature dependence of the phase velo-

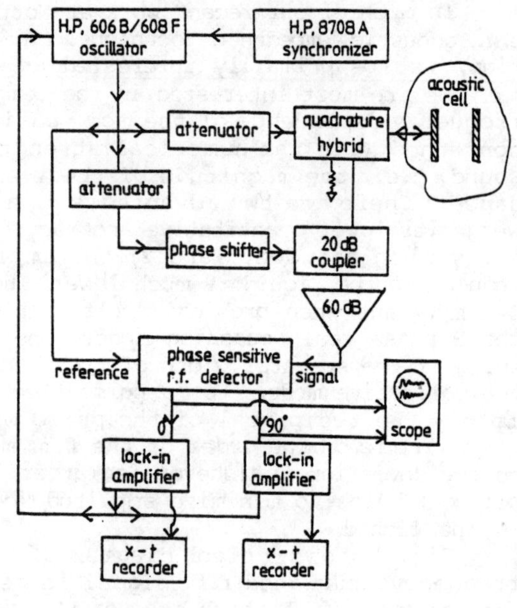

Fig. 2. A block diagram of our acoustic impedance technique is presented.

city, V_p, the phase of the received signal being given by $\phi = \omega d/V_p$. Non-parallel transducers and coupling to the transducers limits the standing wave ratio as the attenuation goes to zero and complicates analysis in this limit. However, as the attenuation increases, these effects can be neglected. When a weakly coupled mode such as the real squashing mode is superimposed on this oscillating background, the oscillations, in addition to becoming more rapid, locally reduce in amplitude, in many cases disappearing, and then grow again. From the envelope of the oscillations and the node count, accurate values of the attenuation and phase shift can be deduced.[13]

Finally, we note that there are actually <u>two</u> path lengths involved in our experiment. The primary path is defined by the 4mm fused quartz annulus. However, a 1 mm (relatively parallel) space exists on the other side of the transducer formed by the housing. Since this path is four times, shorter, standing wave resonances persisted to higher attenuation values, thus extending the range over which we could study this property.

C.   <u>Thermometry</u>

Temperatures in the [3]He were measured with a lanthanum diluted CMN thermometer calibrated with a platinum NMR thermometer. About 30-40 demagnetization runs could be achieved in a typical cool-down without warming the cell above mixing chamber temperatures. A single calibration was used for data taken for all the runs and was reestablished every time the cell was warmed up above 4 K. However, it was found that the shift in the calibration constants, on warming the cell to T > 20 K and cooling back again, was insignificant. The following is a brief description of the calibration procedure. The magnetization of LCMN was measured simultaneously along with the nuclear susceptibility of the platinum sample. The calibration of the LCMN thermometer is complete provided one knows the nuclear Curie constant of the platinum sample. This was obtained by comparing the acoustic signature at T_c against the phase-diagram established by Alvesalo <u>et al.</u>[26].

EXPERIMENTAL RESULTS

In the remainder of the paper, we will summarize the most recent observations of the Northwestern group. These have primarily used the acoustic impedance technique, but also include some pulsed transmission work.

<u>New Acoustic Impedance Phenomena in the B Phase</u>

Fig. 3a shows the output of the acoustic impedance spectrometer for the case H≅0. The oscillatory background arises from a combination of temperature induced dispersion (associated with the squashing and real squashing modes) and the presence of a non-unity standing wave ratio in the cell; each oscillation results from the addition or loss of one in the node number $n = 2d/\lambda$, where d is the spacer length. Three features marked A, B, and C are identified and are associated with the $J_z = 0, +1$ and $+2$ components of the dispersion split real squashing mode; the ratio of the $J_z = 0,1$ and $J_z = 0,2$ splitting is 1:3.8 which is to be compared with the $(F_1 = 0)$ 1:4 ratio quoted above. The coupling to the $J_z = 0$ mode (A feature) is significantly stronger than for $J_z = 1$ and 2 (B and C). The attenuation and change in phase velocity due to the $J_z = 0$ zero sound mode crossing (A feature) are shown in Fig. 3b. Less accurate (attenuation) information, due to the weak coupling, is also included for the $J_z = 1$ and 2 modes.

298

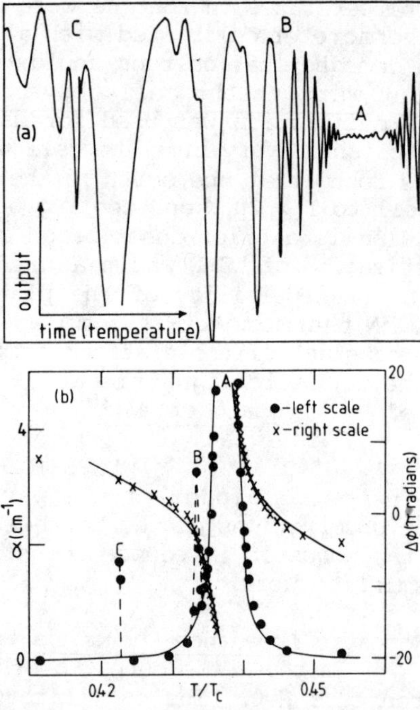

FIG. 3 The rsq mode for $H = 3$ mT at 63.8 MHz, 4.9 bars. In (a), the detected signal is given as a function of temperature. The attenuation and change in phase derived from the output in (a), as explained in the text, are plotted in (b) as functions of T/T_c. The solid curves are fits to the data from the phenomenological expressions of Calder *et al.* (Ref. 7). The broken curves are guides for the eye. The feature labeled A shows the region of large attenuation and dramatic phase velocity changes associated with the rsq-mode $J_z = 0$ state, superimposed upon the attenuation and phase velocity changes from the tail of the squashing mode. Features B and C are weaker anomalies associated with the $J_z = \pm 1$ and ± 2 states, respectively.

We now discuss the field evolution of the real squashing mode multiplet. Fig. 4 shows the "linear" (low T/T_c) regime while Fig. 5 shows the "non-linear" behavior (high T/T_c). In Fig. 4, we see that the spectrum evolves from the three line (dispersion split) low field spectrum into the high field six line spectrum involving the $J_z = 0$ "doublet". The field at which the doublet appears is history dependent. Experiments were performed where the liquid was first cooled into the superfluid with the field subsequently applied and vice versa. This behavior is shown in Fig. 4b. Note that above about 50mT the doublet splitting is history independent.

We now examine the non-linear Zeeman effect, depicted in Fig. 5. Experimentally, we can determine splittings more accurately than absolute values; hence, we fit the field dependence of the $J_z = 0$

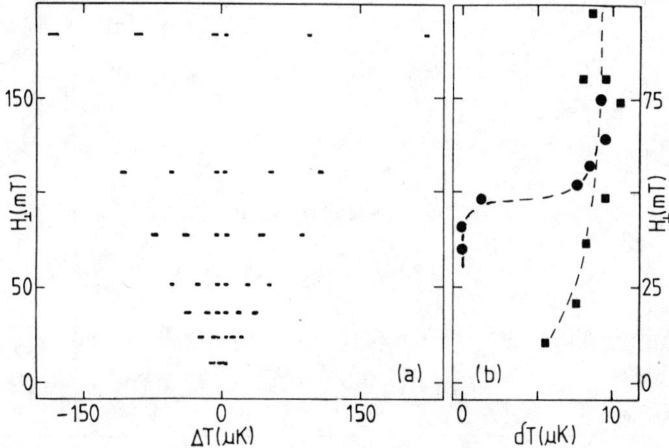

Fig. 4. (a) The magnetic field evoultion of the
rsq-mode into six states at 63.8 MHz, 5.3 bar and
with the liquid cooled in a magnetic field. The
bars indicate the "null" regions where the signal
drops below the noise; see Feature A of Fig.3 .
Clearly apparent is the new, additional structure
which appears to be a splitting of the center (J_z =
0) state. (b) The evolution of this new structure,
showing apparent dependence upon whether the liquid
was cooled into the superfluid phase in zero (circles) or non-zero (squares) magnetic field. The
positions of the states are taken to be the mid-
points of the T/T_c regions where a null in the signal occurs.

state to an H^2 dependence shown in 5b, and then measure the J_z =
+ 1 and + 2 splittings from the fitted J_z = 0 line in Fig. 5a. Note
that the extent of the non-linearity is such that the J_z = -1 and J_z
= -2 states actually cross around H = .13T. The data for all five
lines has been fit to Eq. 3, and the resulting values of α, β, and Γ
are plotted in Fig. 6. The values of $\alpha \equiv g\omega_L$ are in good agreement
with earlier measurements. Our Eq. 3 is phenomenological; however,
if we expand the expressions of Schopohl, Warnke, and Tewordt[22],
which involve field dependent gap functions Δ_1 and Δ_2, to second
order in H, we obtain the same form as Eq. 3 with $\Gamma = 2\beta$. The
rather good fit obtained to our data suggests that a quadratic form
is adequate for the field range investigated; however, we find Γ to
be somewhat greater than 2β (See Fig. 6).
    We now turn to a discussion of the squashing mode. Fig. 7
shows the spectrometer output in the vicinity of the resonance. The

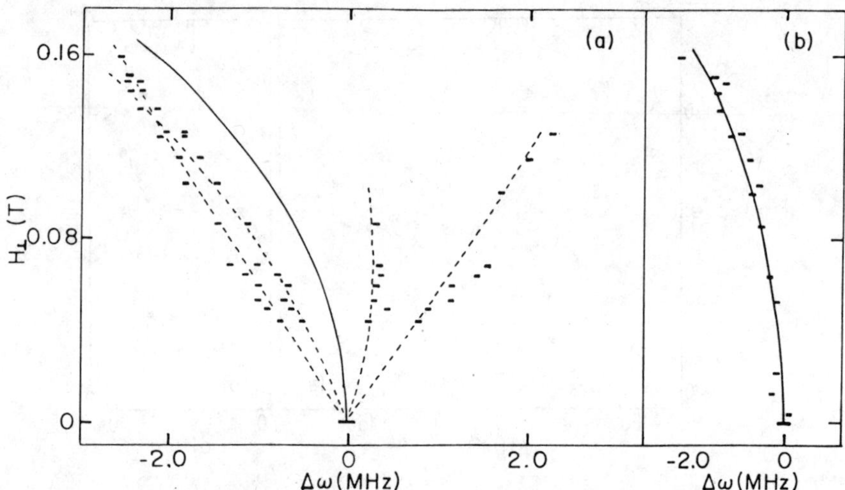

Fig. 5. (a) The magnetic field evolution of the rsq-mode
at 1.35 bar, 38.24 MHz, T/T_c = 0.767. The lines are a
result of the fit by Eq. 3 to the observed frequency shifts.
For the data shown α = 8.20 MHz T^{-2}. For fields greater
than 0.1 T, the J_z = +1 state is not observed. (b) The raw
data for the positions of the J_z = 0 state. The "doublet"
structure is not resolved at this temperature.

most prominent feature is the strong peak occurring at the mode
crossing.   As we go to lower temperatures, standing waves begin to
develop. Due to the large attenuation, they first appear as slow
oscillations associated with the d = 1 mm path on the driven side of
the transducer. As the temperature is lowered further and the
attenuation falls, a standing wave develops in the d = 4mm path
within the spacer, producing the faster oscillations; since the
relative round trip phase shift associated with the 4mm path is
larger than that of the 1mm path, the former produces a larger a.m.
component for a given (small) f.m. frequency deviation and thus
produces larger oscillations.

It would be especially interesting to study the Zeeman splitting associated with the squashing mode. Such experiments are
complicated by three factors: 1) the acoustic impedance resonances
are broad relative to the transmission "nulls" of the real squashing
mode, 2) the g factors, which are identical for both modes at T_c
become smaller and larger for the imaginary and real J = 2 modes
respectively (i.e., the splittings are smaller) and 3) for fields
and temperatures where the above two effects should not have de-
feated the observation of a splitting, only a single acoustic impedance signature is observed. From an analysis of the combined
effects of the magnetic field and boundary energies on the n̂ vector,
and the theoretical dependence of the mode coupling on n̂, Mast[27]

concluded that only the $J_z=0$ component of the multiplet should couple in an acoustic impedance experiment.

One can obtain indirect evidence for the Zeeman splitting of the squashing mode from the shift in the onset temperature of the standing wave oscillations shown in Fig. 7. An analysis of the data[14] indicates that at 1.35 bar and 38.24 MHZ, the squashing mode multiplet broadens by ~24μK on the application of 150mT.

The above problems do not interfere with a study of the non-linear behavior of the $J_z=0$ line which is governed by the H^2 term of Eq. 3. These results are reported in Ref. 14.

Pulsed Transmission Studies

In addition to our acoustic impedance experiments, pulsed zero sound measurements have been performed in the B phase for P < 5 bar at 12.697 MHz. We have completely resolved the pair-breaking attenuation peak for P < 2 bar. Historically, Paulson, Kleinberg and Wheatley[28] were the first to obtain data suggesting the separation of the pair-breaking and the sq-mode contributions to the attenuation. In later work, Giannetta et al.[5] were able to separate the pair-breaking and sq-mode contributions to the attenuation at 5.3 bar, 60 MHz. However, the high attenuation ($\alpha > 15$ cm^{-1}) of the pairbreaking and sq-mode features prevented complete resolution of either peak.

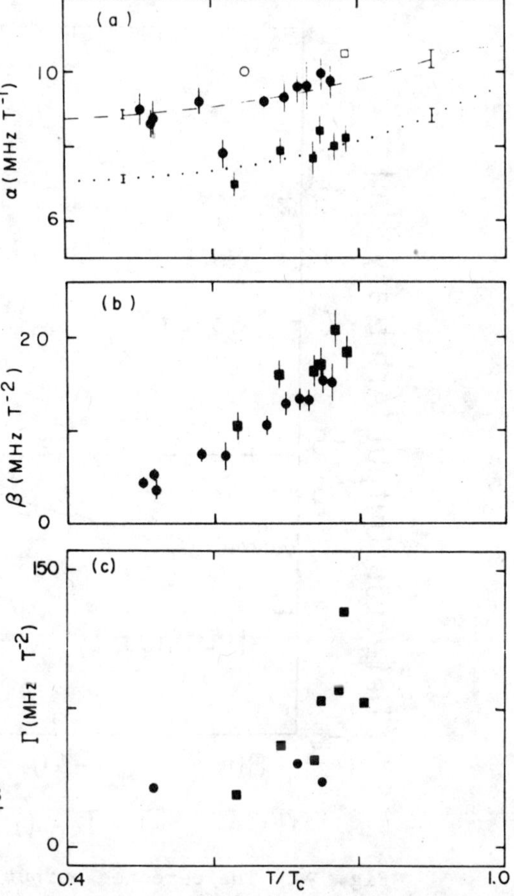

FIG. 6 The coefficients α, β, and Γ of Eq. 3 as functions of T/T_c. Closed circles (closed squares) represent data taken over a pressure range of 5.18 to 10.13 bars at 63.77 MHz (0.20 to 1.45 bars, 38.24 MHz). (a) The open circle (square) is the data from Ref. 8 at 11.0 bars, 74.5 MHz (3.5 bars, 44.7 MHz). The lines are the result of calculations using the expressions of Ref. 21 with $x_3^{-1} = 0$. $m*/m$ values of Ref. 29 have been used, and the dashed line (dotted line) corresponds to $P = 7.3$ bars, $F_0^a = -0.740$, $F_2^a = -1.13$ ($P = 0.80$ bar, $F_0^a = -0.709$, $F_2^a = -2.50$). The bars at $T/T_c = 0.9$ and 0.45 represent the shifts in the lines when F_0^a is changed over the pressure range of the data.

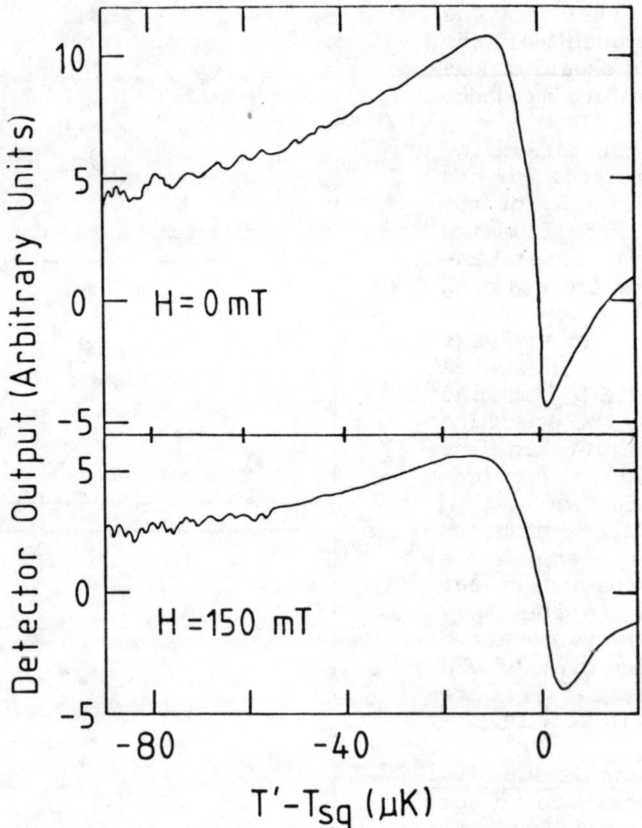

Fig. 7. The detected output for 1.35 bar, 38.24 MHz is shown as a function of temperature relative to the temperature of the $J_z = 0$ state, $T'-T_{sq}$, for two magnetic fields. The low temperature oscillations indicate the onset of interference effects in the low attenuation limit. The shift of these oscillations in a magnetic field to lower $T'-T_{sq}$ is clearly shown. In addition, the magnitude of the detected output around T_{sq} is shown to decrease in the presence of a field.

ACKNOWLEDGEMENTS

We acknowledge contributions from D. B. Mast, Jin Duo, J. R. Owers-Bradley and I. D. Calder in the early design and assembly of the Northwestern experiments. We thank N. Schopohl, M. Warnke and L. Tewordt; J. A. Sauls and J. W. Serene; V. Koch and P. Wolfle; R.

Combescot; M. E. Daniels et al.: and P. Bhattacharyya for communicating their work to us prior to publication. We are grateful to J. A. Sauls and J. W. Serene for sending us their computer programs. We appreciate the assistance of Pat R. Roach and the Argonne National Laboratory who made equipment available to us. Our measurements were performed in the Low Temperature Facility of the Northwestern University Materials Research Center supported in part by the NSF-MRL program Grant No. 79-23573. Support for this research was provided by the National Science Foundation through Grants No. DMR-8107385 and No. DMR-8108709.

REFERENCES

1. D. T. Lawson, W. J. Gully, S. Goldstein, R. C. Richardson and D. M. Lee, Phys. Rev. Lett. 30, 829 (1973); D. N. Paulson, R. T. Johnson, J. C. Wheatley, Phys. Rev. Lett. 30, 829 (1973); P. R. Roach, B. M. Abraham, M. Kuchnir, and J. B. ketterson, Phys. Rev. Lett. 34, 711 (1974); and P. R. Roach, B. M. Abraham, P. D. Roach and J. B. Ketterson, Phys. Rev. Lett. 34, 715 (1975).
2. P. Wolfle, Phys. Rev. Lett. 30, 1169 and 1437 (1973); H. Ebisawa and K. Maki, Prog. Theor. Phys. 51, 337 (1974); K. Maki, J. Low Temp. Phys. 16, 465 (1974).
3. J. W. Serene, Thesis, Cornell University (1974).
4. R. W. Giannetta, A. Ahonen, E. Polturak, J. Saunders, E. K. Zeise, R. C .Richardson and D. M. Lee, Phys. Rev. Lett. 45, 262 (1980); D. B. Mast, B. K. Sarma, J. R. Owers-Bradley, I. D. Calder, J. B. Ketterson and W. P. Halperin, Phys. Rev. Lett. 45, 266 (1980).
5. P. Wolfle, Physica 90B, 96 (1977).
6. V. E. Koch and P. Wolfle, Phys. Rev. Lett. 46, 486 (1981).
7. I. D. Calder, D. B Mast, B. K. Sarma, J. R. Owers-Bradley, J. B. Ketterson and W. P. Halperin, Phys. Rev. Lett. 45, 1866 (1980).
8. O. Avenel, E Varoquaux and H. Ebisawa, Phys. Rev. Lett. 45, 1952 (1980).
9. L. Tewordt and N. Schopohl, J. Low Temp. Phys. 37, 421 (1979); N. Schopohl and L. Tewordt, J. Low Temp. Phys. 45, 67 (1981).
10. P. R. Roach and J. B. Ketterson, Phys. Rev. Lett. 36, 736 (1976).
11. O. Avenel, P. Piche, W. M. Saslow, E. Varoquaux and R. Combescot, Phys. Rev. Lett. 47, 803 (1981).
12. D. B. Mast, J. R. Owers-Bradley, W. P. Halperin, I. D. Calder, B. K. Sarma and J. B. Ketterson, Physica 107B, 685 (1981).
13. B. S. Shivaram, M. W. Meisel, B. K. Sarma, D. B. Mast, W. P. Halperin and J. B. Ketterson, Phys. Rev. Lett. 49, 1646 (1982).
14. M. W. Meisel, B. S. Shivaram, B. K. Sarma, D. B. Mast, J. B. Ketterson and W. P. Halperin, to be published.
15. B. S. Shivaram, M. W. Meisel, B. K. Sarma, W. P. Halperin and J. B. Ketterson, Phys. Rev. Lett. 50, 1070 (1983).
16. P. Wolfle, Prog. Low Temp. Phys. 7a., ed. D. F. Brewer (North-

Holland, Amsterdam, 1978), p. 191.
17. J. A. Sauls and J. Serene, Phys. Rev. 23B, 4798 (1981).
18. R. Combescot, J. Low Temp. Phys. 49, 295 (1982).
19. T. Tsuneto, Phys. Rev. 118, 1029 (1960).
20. P. N. Brusov and V. N. Popov, Zh. Eksp. Teor. Fiz. 78, 2419
     (1980); Sov. Phys. JETP 51, 1217 (1980); see also K. Nagai,
     Prog. Theor. Phys. 54, 1 (1975).
21. J. A. Sauls and J. W. Serene, Phys. Rev. Lett. 49, 1183 (1982).
22. N. Schopohl, M. Warnke and L. Tewordt, Phys. Rev. Lett. 50,
     1066 (1983).
23. N. Schopohl, J. Low Temp. Phys. 49, 347 (1982); this anisotropy
     is expected on phenomenological grounds on the basis of the
     Landau-Ginzberg theory alone.
24. Darien Magnetics, Darien, Illinois.
25. B. E. Keen, P. W. Matthews and J. Wilks, Phys. Lett. 5, 5
     (1963); Proc. Roy. Soc. A284, 125 (1965).
26. T. A. Alvesalo, T. Haavasoja, M. T. Manninen and A. T. Soinne,
     Phys. Rev. Lett. 44, 1076 (1980).
27. D. B. Mast, Thesis, Northwestern University (1982).
28. D. N. Paulson, R. L. Kleinberg and J. C. Wheatley, J. Low
     Temp. Phys. 23, 725 (1976).

29. D.S. Greywall and P.A. Busch, Phys. Rev. Lett. 49,
     146 (1982).

ORDER PARAMETER MODES, ZERO SOUND, AND SYMMETRIES IN SUPERFLUID ^{3}He

J. W. Serene
Yale University, New Haven, Ct. 06520

ABSTRACT

In this paper I use general methods of linear response theory to discuss the consequences of gauge invariance, rotation invariance, reflection invariance, time reversal invariance, and approximate particle-hole symmetry for the coupling between order parameter modes and zero sound.

The utility of zero sound as a probe of superfluid ^{3}He results from its coupling to collective modes of the superfluid order parameter with dispersion relations of the form

$$\omega^2 = a\Delta(T)^2 + b(qv_F)^2, \tag{1}$$

where a and b are weakly temperature dependent functions of order one. The theoretical study of order parameter collective modes of this type is almost as old as the BCS theory itself. Their possible existence in superconductors was first suggested by Anderson and by Bogolyubov, Tolmachev, and Shirkov in 1958.[1,2] The following year Ginsberg, Richards, and Tinkham reported observing electromagnetic absorption at frequencies in the superconducting energy gap.[3] The idea that this anomalous absorption might be due to order parameter modes was explored in considerable detail by Tsuneto, by Bardasis and Schrieffer, and by Vaks, Galitskii, and Larkin.[4-6] Subsequent experimental work failed to confirm the results of Ginsberg et al., however, and the subject of order parameter modes in superconductors (other than the Goldstone mode of the phase required by gauge invariance) was generally forgotten. A likely additional cause for the lapse of interest in these modes was the theoretical result that their existence, frequency, coupling to radiation, etc., all depend critically on the strength of components of the pairing interaction with $\ell > 0$. In particular, if (as usual) one omits the $\ell > 0$ pairing interactions, the order parameter modes in the gap of a superconductor disappear altogether. In any case, when Paul Martin reviewed the subject for Parks' 1969 anthology, he wrote that "If the space in this volume were apportioned according to the experimental importance of the phenomena this chapter would be even shorter than it is ... The subject would be only slightly less academic if ^{3}He were to undergo a 'superconducting' transition."[7]

The study of order parameter collective modes revived in 1973 when Lawson, Gully, Goldstein, Richardson, and Lee observed large peaks in the zero sound attenuation in ^{3}He - A and Wölfle calculated a very similar structure for the attenuation using the $\ell = 1$ generalization of a high-frequency collisionless quantum kinetic equation for superconductors due to Betbeder-Matibet and Nozieres.[8-10] About another year passed before the general recognition that the major contribution to the attenuation calculated by Wölfle came not from pair-breaking but from coupling to $\ell = 1$ order parameter collective modes.[11] In contrast to the case of superconductors, these modes exist even when all components of the pairing interaction with $\ell \neq 1$ are omitted. In this respect $\ell = 0$ pairing is anomalous because the $\ell = 0$ order-parameter space is too small to contain anything other than the equilibrium order parameter itself. An inappropriate generalization from this special feature of $\ell = 0$ pairing apparently was responsible for much of the early confusion over the interpretation of Wölfle's calculation.[12]

Since 1973 a great deal of effort has been devoted to the experimental and theoretical study of these modes and of the analogous modes in ^{3}He-B. The theoretical work has been carried out in a number of superficially quite different but ultimately equivalent linear response formalisms.[13-15] Calculations of the mode frequencies and zero sound dispersion are straightforward but lengthy, and for this reason, together with the multiplicity of formal methods in use, it seems worthwhile to establish those features of the coupling between zero sound and order parameter modes which follow from symmetries alone, by a method independent of any particular calculational scheme.

To accomplish this goal we introduce the linear response functions which describe the response of the order parameter induced by a perturbation coupled to the density,[16]

$$\vec{\chi}''_{dn}(\vec{r}-\vec{r}',t-t';\hat{p};A,\vec{H}) =$$

$$= \frac{1}{2} \langle [\vec{d}(\hat{p},\vec{r},t),n(\vec{r}',t')] \rangle \quad , \tag{2}$$

$$\chi''_{d_0 n}(\vec{r}-\vec{r}',t-t';\hat{p};A,\vec{H}) =$$

$$= \frac{1}{2} \langle [d_0(\hat{p},\vec{r},t),n(\vec{r}',t')] \rangle \quad . \tag{3}$$

Here the triplet and singlet order-parameter operators are

$$\left.\begin{array}{c} \vec{d}(\hat{p},\vec{r},t) \\[2ex] d_0(\hat{p},\vec{r},t) \end{array}\right\} = \int \frac{d\Omega'}{4\pi} V(\hat{p}\cdot\hat{p}') \int_{-E_c}^{E_c} d\xi_{p'} \int d^3\rho \ e^{-i\vec{p}'\cdot\vec{\rho}} \times$$

$$\tag{4}$$

$$\frac{1}{2}(-i) \left\{\begin{array}{c} [\sigma_y \vec{\sigma}]_{\beta\alpha} \\[2ex] [\sigma_y]_{\beta\alpha} \end{array}\right\} \ \psi_\alpha(\vec{r} + \vec{\rho}/2,t)\psi_\beta(\vec{r} - \vec{\rho}/2,t),$$

where $V(\hat{p}\cdot\hat{p}')$ is the pairing pseudo-interaction and E_c is a low-energy cutoff chosen such that $k_B T_c \ll E_c \ll E_F$. The pointed brackets in eq's (2) and (3) denote an equilibrium average with density matrix $\exp[-\beta(H-\Omega)]$, and the same Hamiltonian H governs the time dependence of $\vec{d}(\hat{p},\vec{r},t), d_o(\hat{p},\vec{r},t)$ and $n(\vec{r},t)$. H contains the kinetic energy K, the bare interaction V_2 between ^{3}He atoms, the coupling $-\vec{M}\cdot\vec{H}$ between the total nuclear magnetic moment and an external magnetic field, and perhaps the nuclear dipole-dipole interaction H_{dipole}, together with an infinitesimal symmetry-breaking perturbation

$$H' = - \varepsilon N(E_F)\int d^3r\int\frac{d\Omega}{4\pi} \{\vec{d}(\hat{p},\vec{r})\cdot\vec{\Delta}(\hat{p})^* + \vec{d}^{\dagger}(\hat{p},\vec{r})\cdot\vec{\Delta}(\hat{p})\} , \tag{5}$$

whose only role is to select a particular equilibrium order parameter $\Delta_\mu(\hat{p}) = A_{\mu i}(\hat{p})_i$; one can take $\varepsilon \to 0^+$ after computing the expectation values in the thermodynamic limit. The equilibrium matrix order parameter A plays a formal role analogous to that of an external field, and thus appears together with $\vec{H}$ in the arguments of χ''. As usual we define the Fourier transforms of the response functions by

$$\chi''(\vec{q},\omega) = \int d^3r\int_{-\infty}^{\infty} dt\ e^{-i(\vec{q}\cdot\vec{r}-\omega t)}\chi''(\vec{r},t). \tag{6}$$

It is then straightforward to establish how the response functions transform under various symmetry operations. The simplest and best-known results follow immediately from the exchange antisymmetry of the two-Fermion order parameter,

$$\vec{\chi}''_{dn} (\vec{q},\omega;\hat{p};A,\vec{H}) = - \vec{\chi}''_{dn}(\vec{q},\omega;-\hat{p};A,\vec{H}) \tag{7a}$$

$$\chi''_{d_on} (\vec{q},\omega;\hat{p};A,\vec{H}) = \chi''_{d_on} (\vec{q},\omega;-\hat{p};A,\vec{H}). \tag{7b}$$

The invariance of $K + V_2 + H_{dipole}$ under global gauge transformations, rotations, reflections, and time reversal leads to the following transformation rules for $\vec{\chi}''_{dn}$ and χ''_{d_on}:

Global Gauge Transformations

$$\psi_\alpha(\vec{r}) \to e^{i\phi}\psi_\alpha(\vec{r}) \tag{8}$$

$$\vec{\chi}''_{dn} (\vec{q},\omega;\hat{p};A,\vec{H}) = e^{2i\phi}\vec{\chi}''_{dn} (\vec{q},\omega;\hat{p};e^{-2i\phi}A,\vec{H}) \tag{9a}$$

$$\chi''_{d_on} (\vec{q},\omega;\hat{p};A,\vec{H}) = e^{2i\phi}\chi''_{d_on} (\vec{q},\omega;\hat{p};e^{-2i\phi}A,\vec{H}). \tag{9b}$$

Rotations

Consider a rotation by the angle θ around the axis $\hat{\theta}$, and let R denote the corresponding three-dimensional rotation matrix. Then

$$\psi_\alpha(\vec{r}) \to [e^{\frac{i}{2}\vec{\theta}\cdot\vec{\sigma}}]_{\alpha\beta}\psi_\beta(R\vec{r}) \tag{10}$$

$$\vec{\chi}''_{dn}(\vec{q},\omega;\hat{p};A,\vec{H}) = R^{-1}\vec{\chi}''_{dn}(R\vec{q},\omega;R\hat{p};RAR^{-1},R\vec{H}) \tag{11a}$$

$$\chi''_{d_on}(\vec{q},\omega;\hat{p};A,\vec{H}) = \chi''_{d_on}(R\vec{q},\omega;R\hat{p};RAR^{-1},R\vec{H}) \tag{11b}$$

<u>Reflections</u> (Parity Transformations)

$$\psi_\alpha(\vec{r}) \to \psi_\alpha(-\vec{r}) \tag{12}$$

$$\vec{\chi}''_{dn}(\vec{q},\omega;\hat{p};A,\vec{H}) = -\,\vec{\chi}''_{dn}(-\vec{q},\omega;\hat{p};-A,\vec{H}) \tag{13a}$$

$$\chi''_{d_on}(\vec{q},\omega;\hat{p};A,\vec{H}) = \chi''_{d_on}(-\vec{q},\omega;\hat{p};-A,\vec{H}) \tag{13b}$$

<u>Time Reversal</u> (antiunitary)

$$\psi_\alpha(\vec{r}) \to [i\sigma_y]'_{\alpha\beta}\psi_\beta(\vec{r}) \tag{14a}$$

$$\vec{d}(\hat{p},\vec{r}) \to \vec{d}(\hat{p},\vec{r}) \tag{14b}$$

$$d_o(\hat{p},\vec{r}) \to d_o(\hat{p},\vec{r}) \tag{14c}$$

Because time reversal is represented by an antiunitary transformation, its consequences for the response functions are most transparent in terms of the Hermitian operators $\vec{d}'(\hat{p},\vec{r}) = (1/2)[\vec{d}(\hat{p},\vec{r}) + \vec{d}^\dagger(\hat{p},\vec{r})]$, $\vec{d}''(\hat{p},\vec{r}) = (1/2i)[\vec{d}(\hat{p},\vec{r}) - \vec{d}^\dagger(\hat{p},\vec{r})]$, $d'_o(\hat{p},\vec{r}) = (1/2)[d_o(\hat{p},\vec{r}) + d^\dagger_o(\hat{p},\vec{r})]$, and $d''_o(\hat{p},\vec{r}) = (1/2i)[d_o(\hat{p},\vec{r}) - d^\dagger_o(\hat{p},\vec{r})]$. The operators $\vec{d}'$ and d'_o are even under time reversal, while $\vec{d}''$ and d''_o are odd. Hence the response functions for these operators satisfy

$$\vec{\chi}''_{d'n}(\vec{q},\omega;\hat{p};A,\vec{H}) = -\,\vec{\chi}''_{d'n}(\vec{q},-\omega;\hat{p};A^*,-\vec{H}) \tag{15a}$$

$$\vec{\chi}''_{d''n}(\vec{q},\omega;\hat{p};A,\vec{H}) = \vec{\chi}''_{d''n}(\vec{q},-\omega;\hat{p};A^*,-\vec{H}) \tag{15b}$$

$$\chi_{d'_on}(\vec{q},\omega;\hat{p};A,\vec{H}) = -\,\chi''_{d'_on}(\vec{q},-\omega;\hat{p};A^*,-\vec{H}) \tag{15c}$$

$$\chi_{d''_on}(\vec{q},\omega;\hat{p};A,\vec{H}) = \chi_{d''_on}(\vec{q},-\omega;\hat{p};A^*,-\vec{H}). \tag{15d}$$

The Hamiltonian $K + V_2$ is separately invariant under orbital rotations and spin rotations, and hence in the absence of H_{dip} we have two further results for χ'':

Orbital Rotations

$$\vec{\chi}''_{dn}(\vec{q},\omega;\hat{p};A,\vec{H}) = \vec{\chi}''_{dn}(R\vec{q},\omega;R\hat{p};AR^{-1},\vec{H}) \tag{16a}$$

$$\chi''_{d_on}(\vec{q},\omega;\hat{p};A,\vec{H}) = \chi''_{d_on}(R\vec{q},\omega;R\hat{p};AR^{-1},\vec{H}) \tag{16b}$$

Spin Rotations

$$\vec{\chi}''_{dn}(\vec{q},\omega;\hat{p};A,\vec{H}) = R^{-1}\vec{\chi}''_{dn}(\vec{q},\omega;\hat{p};RA,R\vec{H}) \tag{17a}$$

$$\chi''_{d_on}(\vec{q},\omega;\hat{p};A,\vec{H}) = \chi''_{d_on}(\vec{q},\omega;\hat{p};RA,R\vec{H}) \tag{17b}$$

In addition to these exact symmetries, the nearly constant density of states for normal quasiparticles in the narrow low-energy band around the Fermi surface relevant for the superfluid properties leads to an approximate particle-hole symmetry. The particle-hole symmetry transformation interchanges a particle in the low-energy band above the Fermi surface with a hole an equal distance below the Fermi surface, and inverts the spins by a rotation through π around the y-axis. The associated unitary transformation of the field operators is

$$a_{\vec{p},\alpha} \rightarrow [i\sigma_y]_{\alpha\beta}a^\dagger_{\underset{\sim}{\vec{p}},\beta} \tag{18}$$

for $|\xi_{\underset{\sim}{p}}| < E_c$, where $\hat{\underset{\sim}{p}} = \hat{p}$ and $\xi_{\underset{\sim}{\vec{p}}} = -\xi_{\vec{p}}$. With a constant low-energy density of states, this transformation takes K into itself and leaves the low-energy part of $V_2 + H_{dip}$ invariant up to one-body correction terms which are negligibly small (of order $(E_c/E_F)^2$) in the thermodynamic limit. Under the unitary transformation (18) the order parameter operators and the low-energy part of the density operator transform according to

$$\vec{d}(\hat{p},\vec{r}) \rightarrow \vec{d}(\hat{p},\vec{r})^\dagger \tag{19a}$$

$$d_o(\hat{p},\vec{r}) \rightarrow d_o(\hat{p},\vec{r})^\dagger \tag{19b}$$

$$n(\vec{r}) \rightarrow -n(\vec{r}) + C \tag{19c}$$

where C is a constant; the total magnetization is invariant. Hence in the limit of exact particle-hole symmetry the response functions satisfy

$$\vec{\chi}''_{d'n}(\vec{q},\omega;\hat{p};A,\vec{H}) = -\vec{\chi}''_{d'n}(\vec{q},\omega;\hat{p};A^*,\vec{H}) \tag{20a}$$

$$\vec{\chi}''_{d''n}(\vec{q},\omega;\hat{p};A,\vec{H}) = \vec{\chi}''_{d''n}(\vec{q},\omega;\hat{p};A^*,\vec{H}) \tag{20b}$$

$$\chi''_{d'_on}(\vec{q},\omega;\hat{p};A,\vec{H}) = -\chi''_{d'_on}(\vec{q},\omega;\hat{p};A^*,\vec{H}) \tag{20c}$$

310

$$\chi''{}_{d''{}_o n}(\vec{q},\omega;\hat{p};A,\vec{H}) = \chi''{}_{d''{}_o n}(\vec{q},\omega;\hat{p};A^*,\vec{H}). \qquad (20d)$$

When $\vec{H} = 0$ these conditions are similar to, but stronger than, the conditions imposed by time reversal, since in the present case ω does not change sign between the left and right sides.

To illustrate the use of these symmetry relations, we will apply them to a familiar case, ^{3}He-B with $\vec{H} = 0$ and $H_{dipole} = 0$.[17] In this case $A_{\mu i}$ is proportional to a rotation matrix and can be taken to be real, $A_{\mu i}A_{\nu i} = \Delta(T)^2\delta_{\mu\nu}$ and $A_{\mu i} = A^*{}_{\mu i}$. From eq.'s (19) we immediately see that in the limit of exact particle-hole symmetry, only the imaginary order-parameter modes described by $\vec{d}''$ and d''_o can be excited by density fluctuations. The coupling to sound of the real modes $\vec{d}'$ and d'_o must be proportional to a particle-hole symmetry breaking parameter. The associated ultrasonic attenuation, which is proportional to the square of this coupling, must then be of second order in the particle-hole asymmetry. Eq.'s (9) tell us that $\vec{\chi}''_{dn}$ and $\chi''_{d_o n}$ must both contain one more factor of $A_{\mu i}$ than of $A^*{}_{\mu i}$. For $\chi''_{d_o n}$ this is inconsistent with spin rotation invariance, expressed by eq. (17b). Hence $\chi''_{d_o n} = 0$, and the even-ℓ spin-singlet order parameter modes are not excited by sound. As a special case of (9a) we have $\vec{\chi}''_{dn}(\vec{q},\omega;\hat{p};A,\vec{H}) = -\vec{\chi}''_{dn}(\vec{q},\omega;\hat{p};-A,\vec{H})$, which together with the parity rule, eq. (13a), implies that χ_{dn} is an even function of $\vec{q}$, $\chi''_{dn}(\vec{q},\omega;\hat{p};A,\vec{H}) = \vec{\chi}''_{dn}(-\vec{q},\omega;\hat{p};A,\vec{H})$, a result independent of any special assumptions.

If for simplicity we now restrict our attention to the $\ell = 1$ components of the order parameter, we easily see that consistency with eq.'s (7) – (17) requires $\vec{\chi}_{dn}$ to have the form

$$[\vec{\chi}''_{dn}(\vec{q},\omega;\hat{p};A)]_\mu = \chi''_0(q^2,\omega,\Delta^2)A_{\mu i}(\hat{p})_i +$$

$$+ \chi''_2(q^2,\omega,\Delta^2)[A_{\mu j}(\hat{q})_j(\hat{q})_i - \frac{1}{3}A_{\mu i}](\hat{p})_i. \qquad (21)$$

If we represent $\vec{\chi}''_{d'n}$ and $\vec{\chi}''_{d''n}$ in the form of eq. (21), then from time reversal symmetry the functions corresponding to χ''_0 and χ''_2 must be odd in ω in $\chi''_{d'n}$ and even in ω in $\chi''_{d''n}$. The subscripts of χ''_0 and χ''_2 were chosen because for the $J = 0$ equilibrium state, $A_{\mu i} = \Delta(T)\delta_{\mu i}$, the tensor multiplying χ''_0 has $J = 0$, and the tensor multiplying χ''_2 has $J = 2$ and $m_J = 0$ along $\hat{q}$. The $J = 1$, $m_J = 0$ mode corresponds to

$$A_{\mu k}\varepsilon_{kji}(\hat{q})_j(\hat{p})_i, \qquad (22)$$

whose presence in $\vec{\chi}''_{dn}$ is forbidden by parity together with spin rotation invariance and gauge symmetry. The coefficient of such a term in $\vec{\chi}''_{dn}$ would have to be a scalar function $\chi_1(\vec{q},\omega,A)$ which is odd in $\vec{q}$ and even in A. The simplest example with these properties is $A^*_{\mu\mu}(\hat{q})_i \varepsilon_{i\nu j} A_{\nu j}$, but this (and all other such functions) violates spin-rotation invariance.

We can easily understand why for a $J = 0$ equilibrium order parameter, only modes with $m_J = 0$ along $\hat{q}$ are excited by zero sound. Zero sound is by definition a spin scalar mode, so it carries no spin angular momentum along any axis. It is also a longitudinal density mode, which means that it carries no orbital angular momentum along its propagation direction. Hence the total angular momentum carried by zero sound along its propagation direction $\hat{q}$ is zero, and zero sound can only excite modes of a $J = 0$ equilibrium state with $m_J = 0$ along $\hat{q}$.

We can also draw some conclusions about the coupling strengths of the order-parameter modes based on the form of the distribution function for normal-state quasiparticles in the zero-sound mode.[18] This distribution function for a longitudinal sound mode has the general form

$$\delta n(\hat{p}) = \sum_{\ell} \delta n_{\ell} P_{\ell}(\hat{p} \cdot \hat{q}).$$ (23)

The continuity equation requires $\delta n_1 \sim s\, \delta n_0$, where $1/s = qv_F/\omega = v_F/C_0$ is a small expansion parameter in ^{3}He. In the hydrodynamic limit the components δn_{ℓ} with $\ell > 1$ are negligible. The hydrodynamic limit for normal ^{3}He requires $\omega\tau \ll 1$, while zero sound propagates in the limit $\omega\tau \gg 1$. In this case one finds $\delta n_{\ell+1}/\delta n_{\ell} \sim (1/s)$ for $\ell \geq 1$. For ^{3}He-B the hydrodynamic limit requires both $\omega\tau \ll 1$ and $\omega \ll \Delta$. Because the order parameter modes have frequencies of order Δ, except when $T \to T_c$ zero sound which excites these modes is never in the hydrodynamic limit, even at $T = 0$. Not surprisingly, one finds the same relative orders in $1/s$ for the δn_{ℓ} in the superfluid when $\omega \sim \Delta$ as in normal ^{3}He when $\omega\tau \gg 1$.

Maki pointed out long ago that with a $J = 0$ equilibrium order parameter, the order parameter collective modes at $q = 0$ can be classified by their total angular momentum J.[19] Sauls and Serene subsequently showed that the frequency of a mode with even J depends on T_c, on the transition temperatures $T_c^{(\ell)}$ with $\ell = J \pm 1$, and on F_J^s for an imaginary mode or on F_J^a for a real mode.[15] These conclusions also follow from symmetries alone. From eq. (1), the dispersion of the order parameter modes becomes appreciable when $q \sim \xi(T)^{-1} \sim \Delta(T)/\hbar v_F$. The zero sound dispersion relation crosses the $q = 0$ collective mode frequencies when $c_0 q \sim \Delta(T)$ or equivalently when $q\xi(T) \sim (1/s)$, and hence the total-J classification is an excellent approximation near the zero-sound crossing. The two modes which have been unambiguously identified to date are the real and imaginary modes with $J = 2$. All of the angular momentum of an excited order parameter mode must in this case ($J = 0$ equilibrium state) come from the orbital angular momentum of the zero sound mode. Thus, relative to the coupling of the $J = 2$ modes, the coupling of

the $J = 4$ modes is smaller by $\delta n_4/\delta n_2 \sim 1/s^2$, etc. (the odd-J modes
are all excluded by arguments analogous to that given above for
$J = 1$.) The ultrasonic attenuation goes like the square of this
coupling, and therefore the only mode with $J > 2$ which might be observable in the attenuation is the imaginary $J = 4$ mode. Whether
this mode falls below the quasi-particle continuum beginning at 2Δ
depends on the largely unknown parameters $F_4{}^s$, $T_c{}^{(3)}$, and $T_c{}^{(5)}$.
The precise dependence of the $J = 4$ frequency ω_{4_-} on these parameters of course cannot be determined on symmetry grounds alone; at
this point the hard work of microscopic theory must begin.[15]

In conclusion, I emphasize that while symmetry arguments of the
sort presented here cannot substitute for detailed microscopic calculations, they can serve two important purposes. The first is to
help us to understand why the existing microscopic calculations come
out the way they do. The second is to provide a check and a guide
for future microscopic calculations. Precisely because these calculations can be long and complicated, it is helpful to know in advance what results cannot be obtained, and what the functional form
of the answers must be.

ACKNOWLEDGMENTS

I have profitted greatly from conversations with J. A. Sauls.
This work was supported in part by the NSF through Grant No. DMR-
80000522.

REFERENCES

1. P. W. Anderson, Phys. Rev. 112, 1900 (1958).
2. N. N. Bogolyubov, V. V. Tolmachev, and D. V. Shirkov, "A New
    Method in the Theory of Superconductivity," (Academy of Science,
    Moscow, 1958; Consultants Bureau, New York, 1959).
3. D. M. Ginsberg, P. L. Richards, and M. Tinkham, Phys. Rev. Lett.
    3, 337 (1959).
4. T. Tsuneto, Phys. Rev. 118, 1029 (1960).
5. A. Bardasis and J. R. Schrieffer, Phys. Rev. 121, 1050 (1961).
6. V. G. Vaks, V. M. Galitskii, and A. I. Larkin, Zh. Eksp. Teor.
    Fiz. 41, 1655 (1961) [Sov. Phys. JETP 14, 1177 (1962)].
7. Paul C. Martin, in "Superconductivity," edited by R. D. Parks
    (Marcel Dekker, New York, 1969), vol. 1, chp. 7.
8. D. T. Lawson, W. J. Gully, S. Goldstein, R. C. Richardson, and
    D. M. Lee, Phys. Rev. Lett. 30, 541 (1973).
9. P. Wölfle, Phys. Rev. Lett. 30, 1169 (1973).
10. O. Betbeder-Matibet and P. Nozieres, Ann. Phys. (New York) 51,
    392 (1969).
11. P. Wölfle, in "Quantum Statistics and the Many-Body Problem,"
    edited by S. B. Trickey, W. P. Kirk, and J. W. Dufty (Plenum,
    New York, 1975).
12. P. Wölfle, private communication.
13. P. Wölfle, in "Progress in Low Temperature Physics," vol. VIIa,
    edited by D. F. Brewer (North-Holland, Amsterdam, 1978).

14. N. Schopohl and L. Tewordt, J. Low Temp. Phys. $\underline{45}$, 67 (1981).
15. J. A. Sauls and J. W. Serene, Phys. Rev. B$\underline{23}$, 4798 (1981).
16. D. Forster, "Hydrodynamic Fluctuations, Broken Symmetry, and Correlation Functions," (W. A. Benjamin, Reading, Mass., 1975), chp. 3.
17. W. P. Halperin, Physica $\underline{109}$ & $\underline{110}$ B, 1596 (1982).
18. G. Baym and C. Pethick, in "The Physics of Liquid and Solid Helium, Part II," edited by K. H. Bennemann and J. B. Ketterson (Wiley, New York, 1978), chp. 1.
19. K. Maki, J. Low Temp. Phys. $\underline{24}$, 755 (1976).

ATTENUATION OF ZERO SOUND IN ^{3}He-B;
OBSERVATIONS OF NEW STRUCTURE CLOSE TO THE PAIR BREAKING EDGE

J. Saunders
Physics Laboratory, University of Sussex, Brighton, BN1 9QH

and

M.E. Daniels, E.R. Dobbs, P.L. Ward
Department of Physics, Bedford College, University of London,
London NW1 4NS.

ABSTRACT

The attenuation of zero sound in superfluid ^{3}He-B has been measured for attenuations < 170 cm^{-1} at pressures < 3 bar. New structure
has been found in the temperature dependence of the acoustic attenuation in the vicinity of the pair breaking cut-off. The magnetic
field dependence of this structure has been investigated. Possible
origins of the structure are discussed including its possible identification with the predicted $J = 4^-$ mode, the existence of which
depends on significant $\ell > 1$ components of the pairing interaction.

INTRODUCTION

Acoustic order parameter mode spectroscopy has been reviewed
recently by Halperin (1) and Ketterson (2). Zero sound has been
found to couple to a number of collective modes of the order parameter and these may be visualised in terms of transitions of the
Cooper pair from its ground state to some excited state. The good
quantum number of the pair (neglecting the dipolar interaction) is its
total angular momentum J; for p-wave spin triplet pairing, believed
to describe ^{3}He-B, L = 1 and S = 1 so that the possible values of J
are 0,1,2. The ground state has J = 0 corresponding to an isotropic
energy gap.

More rigorously the collective modes consist of fluctuations of
either the real part or the imaginary part of the order parameter,
zero sound couples only weakly to the real modes due to the energy
dependence of the density of states at the Fermi surface (particle-
hole asymmetry). The modes which have so far been observed are
$J = 0^-$ (part of the zero sound mode), $J = 2^-$ (squashing mode) and
$J = 2^+$ (real squashing mode) where a +(-) superscript denotes participation in the mode of the real (imaginary) part of the order parameter. For zero sound of fixed frequency these modes will be observed
at a temperature such that $\hbar\omega_i = a_i\Delta(T)$ where the eigenvalues of the
modes a_i were originally predicted (3) as $\sqrt{12/5}$ for $J = 2^-$ and $\sqrt{8/5}$
for $J = 2^+$ in the absence of Fermi liquid corrections. It has been
found that the precise frequencies of these modes are modified by both
Fermi liquid corrections and the presence of higher angular momentum
components of the pairing interaction (in particular $\ell = 3$ components
(4). In addition to the excitation of these collective modes there
is a further contribution to the attenuation of zero sound in super-

fluid ^{3}He-B. This is for the zero sound to break up the Cooper pair, creating two quasiparticles and requires the condition $\hbar\omega > 2\Delta(T)$ to be satisfied.

The measurements described in this paper are of the acoustic attenuation close to the pair breaking edge $\hbar\omega = 2\Delta(T)$. They are of interest for two main reasons. Firstly, location of the pair breaking edge might seem to constitute a direct measure of the energy gap. In weak coupling BCS theory, the zero temperature gap is $\Delta(0) = 1.764\ k_B T_c$, where T_c is the superfluid transition temperature. $\Delta(T)/\Delta(0)$ has been tabulated by Mühlschlegel (5). If measurements are made at low pressures where strong coupling effects are presumed negligible and the specific heat jump at the superfluid transition has been measured to be in good agreement with the prediction of BCS theory (6), then identification of the pair breaking edge provides a direct measure of $\Delta(T)$ and hence T_c. The second motivation for this work is the theoretical prediction of various, and so far unobserved, collective modes in the vicinity of $\hbar\omega = 2\Delta(T)$. Of these perhaps the most interesting is a $J = 4^-$ mode, whose observation requires the non-vanishing of ≥ 3 components of the pairing interaction v_ℓ (4). These collective modes are discussed in greater detail later in this paper.

We now make some general qualitative remarks concerning the attenuation due to B phase collective modes. The maximum attenuation of zero sound of angular frequency ω at such a mode crossing is $\alpha_{MAX} \approx \dfrac{Z}{c}\dfrac{\omega^2 \lambda \eta^2}{\gamma}$. In this expression λ is a temperature and frequency dependent coupling strength and γ an inverse lifetime parameter; they have been calculated by Koch and Wölfle (7). η^2 is a number of order unity for the $J = 2^-$ mode, while for the $J = 2^+$ mode η^2 is the particle-hole symmetry violation parameter, of order 5.10^{-3} (7), Z is given by $(c_0-c_1)/c_1$, where c_0, c_1 are the velocities of zero sound and first sound respectively. Z has been measured by Abel et al (8) at low pressures, by Ketterson et al (9) at high pressures and by the Northwestern University group (10) at 13 bar. Z, the velocity of sound, and the quasiparticle relaxation time decrease with increasing pressure and all contribute to an associated substantial decrease in α_{MAX}. However the energy gap at a given reduced temperature increases with increasing pressure, somewhat faster than T_c due to strong coupling effects, and this requires measurements to be made at higher sound frequencies in order to obtain adequate temperature resolution of the sound attenuation structure of the superfluid phase. Such considerations impose important constraints on those regions of the phase diagram on which sound attenuation measurements at a given frequency, using a particular technique, can usefully be made.

The acoustic attenuation of superfluid ^{3}He has been measured using the transmission method by a number of groups. The size of the first received pulse v (relative to its amplitude v_c at T_c) is $v/v_c = e^{-(\alpha-\alpha_c)\ell}$ where α is the attenuation, α_c the attenuation at T_c and ℓ the path length. This corresponds to a signal loss of

316

- 8.68$(\alpha-\alpha_c)$dBcm^{-1}. Typical path lengths have been in the range 3-6 mm and this has limited the maximum observable attenuation relative to its value at T_c, to of order 20 cm^{-1}.

The earliest measurements by Ketterson et al (9) and Paulson et al (12) were confined to high pressures and low frequencies, less than 25 MHz. In this work the $J = 2^-$, squashing mode, and the pair breaking attenuation feature were unresolved. The first measurements to show the separation of these two features were those by Giannetta et al (13) at 60 MHz and 5.3 bar and are reproduced in Figure 1,

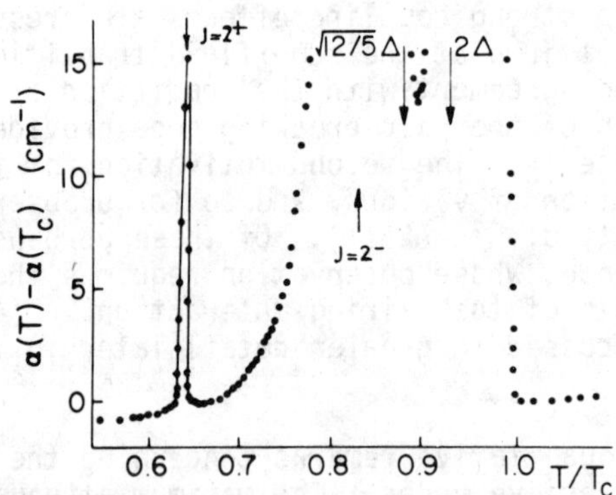

Fig. 1: Attenuation data of Giannetta et al (13) obtained at 60 MHz and 5.3 bar.

which also shows the $J = 2^+$, real squashing, mode crossing, simultaneously observed by Mast et al (14). The large acoustic attenuations associated with the $J = 2^-$ mode are best observed by means of the acoustic impedance technique, although there are problems in distinguishing between changes in the attenuation and phase velocity (2).

The approach taken in this work has been to attempt to push the transmission technique to its limit by reduction of the sound path length. With a 250 μm path length we have so far been able to measure attenuations as high as 170 cm^{-1}. A similar technique was used by Abel et al (8) to observe the transition from first sound to zero sound in the normal Fermi liquid. In general, coupling between the zero sound and the superfluid may be observed via the acoustic attenuation, phase velocity and group velocity. In this work we have made measurements of the acoustic attenuation only.

EXPERIMENTAL DETAILS

In our experimental cell, ultrasound is propagated through a plane parallel sample of ^{3}He, 250 μm in thickness. The acoustic experimental tower is illustrated in Figure 2. Longitudinal sound is generated in a 5 MHz fundamental X-cut quartz crystal, separated by a 250 μm thick fused quartz spacer from one end of a cylindrical Z-cut quartz rod. Thermal contact to the liquid ^{3}He in this space is provided by four slots 500 μm wide, 100 μm deep and 1 mm long cut

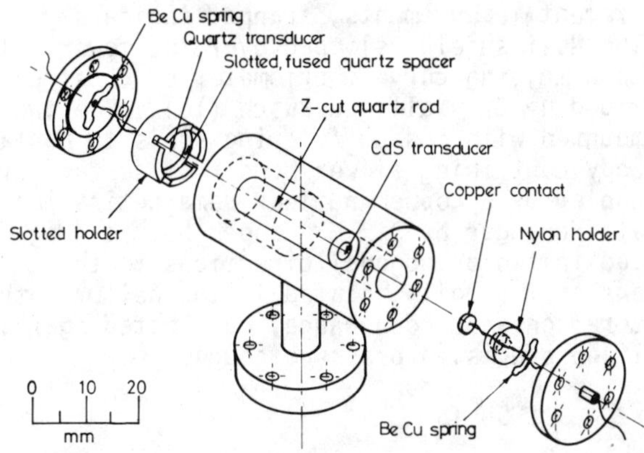

Fig. 2: Acoustic experimental tower. The cell body is made of coin silver.

in the spacer. The transmitted sound pulse is received by a transducer at the far end of the Z-cut quartz rod, which provides a delay of 3.14 µs and allows separation of the received sound pulse from crosstalk. The delay rod is coated with evaporated aluminium earthed to the cell body. The receiver transducer consists of a film of cadmium sulphide (15) evaporated on the surface, and has operated satisfactorily over the frequency range 40-100 MHz. Its insertion loss has been measured as -27 dB. The Q of the loaded quartz transducer is of order 10^3, measured at 94 MHz, a value somewhat larger than might be expected from a perfect bond between the crystal and the Stycast 1266 backing material. Tuning stubs were used to improve the electrical match to the quartz transducer, typical transmitter pulses were of order 10 dBm and 1.5 µs in length.

In these measurements the amplified received sound pulse was passed through a variable precision attenuator and then diode detected. The leading edge of the received pulse was averaged using a boxcar integrator. Because of the short ^{3}He path length care must be taken to sample signal corresponding to the first received pulse only. At a pressure of 3 bar, for example, the delay between the first received pulse and the first echo is 2.2 µs, shorter than the width of the envelope of the received pulse, so that interference between these two signals is a potential problem, especially at low values of the ^{3}He attenuation. The detector was found to behave accurately as a square law detector for the range of input signals to which it was subjected. This technique was insensitive to the phase of the received sound signal, however it is unsuitable for precise measurements in the vicinity of the squashing and real squashing collective modes, where dramatic changes in the group velocity are known to occur (10).

The thermometry is provided by measurement of the nuclear magnetic susceptibility of a sample of platinum powder, located in a separate experimental tower, by means of a PLM 3 spectrometer. The thermometer tower is constructed of Stycast 1266 and the NMR field of 14 mT is, in the most recent measurements, trapped inside a cylindrical superconducting NbTi shield, slipped over the tower. The cell is also equipped with a melting curve thermometer of construction similar to that described by Greywall and Busch (16). The two experimental towers are mounted with indium 'O' ring seals to a heat exchanger with a silver body containing silver powder of surface area ~ 100 m^2. The cell is cooled by a copper nuclear demagnetisation stage attached to the heat exchanger by six M3 copper bolts. Magnetic fields may be applied in two orthogonal directions to the entire sample cell by means of two coils located in the helium bath. The ^{3}He pressure is monitored on a Bourdon gauge, calibrated against a Texas Instruments Model 145 precision pressure gauge.

RESULTS

(i) Initial results in fixed magnetic field

The first series of experimental data (17) was taken in a fixed magnetic field of 14 mT orthogonal to the direction of propagation of the ultrasound pulse. The essential features of the attenuation of zero sound in ^{3}He-B are illustrated by the data of figure 3. The

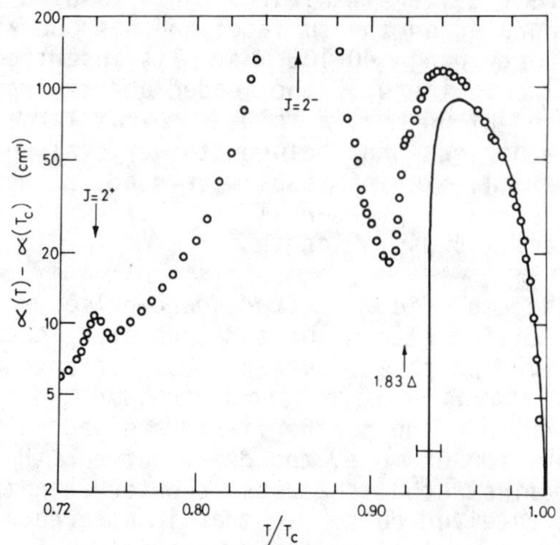

Fig. 3: Temperature dependence of the relative attenuation of 44.2 MHz sound at 2.44 bar. The curve is a calculated contribution due to pair breaking from equation (1). The (estimated) absolute accuracy of the pair breaking cut off is indicated by the error bar.

pair breaking contribution to the attenuation just below T_c is well resolved from the squashing mode (J = 2$^-$) whose maximum attenuation is is immeasurable due to loss of signal. The real squashing mode (J = 2$^+$) is also observed and the measured frequencies of these two modes $\hbar\omega_{2-} = 1.41 \, \Delta_{BCS}$ and $\hbar\omega_{2+} = 1.11 \, \Delta_{BCS}$ are in excellent agreement with the temperature dependent frequencies measured by the Northwestern University group (1,2). Due to the rather short pulse length employed in these measurements the two measured collective mode peaks suffer from a certain degree of instrumental broadening. A new feature which emerged from this data was that labelled at $\hbar\omega = 1.83 \, \Delta_{BCS}$. This feature is illustrated more clearly in Figure 4

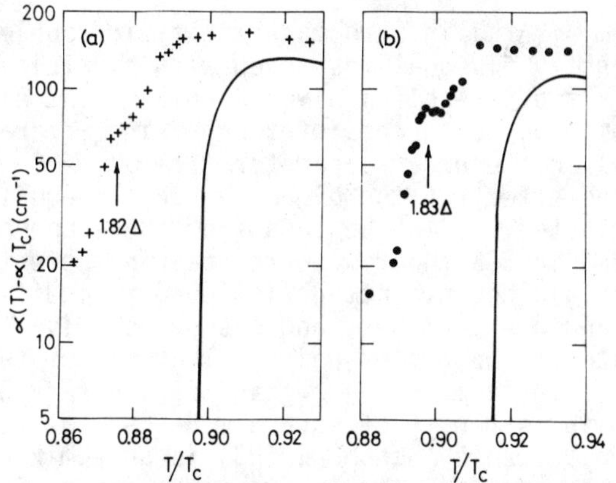

Fig. 4: Temperature dependence of the relative attenuation of sound in the vicinity of the pair breaking cut off. a) at 2.44 bar and 54.0 MHz; b) at 1.36 bar and 44.2 MHz. The curves are the calculated pair breaking contribution using equation (1)

which indicates its pressure and frequency dependence and a consistency in its observed position.

The main contribution to the ultrasonic attenuation near T_c is that due to pair breaking, α_{pb}, and occurs at temperatures such that $\hbar\omega > 2\Delta(T)$. This has been calculated by Wölfle (18) in the collisionless limit, with $F_2^s = 0$, for the Balian-Werthamer state as

$$\alpha_{pb} = \frac{qZ}{A} \left[\frac{6\pi}{5} \left(\frac{\Delta}{\hbar\omega}\right)^2 \tanh\left(\frac{\hbar\omega}{4k_BT}\right)\right] \left(1 - \frac{4\Delta^2}{\hbar^2\omega^2}\right)^{\frac{1}{2}} \theta(\hbar^2\omega^2 - 4\Delta^2) \qquad (1)$$

with $\quad A = 1 - \frac{12}{5}\left(\frac{\Delta}{\hbar\omega}\right)^2$

where ω, q are the acoustic angular frequency and wavenumber. The full expression for α_{ph} including the effects of collisions is given by Wölfle (18) and requires detailed numerical evaluation. The analytic expression above has been evaluated in figures 1,2 using the BCS weak coupling energy gap with the Helsinki T_c and a value of $Z = 30 \times 10^{-3}$ interpolated from references 8,9,10. In figure 1 the error bar indicates the uncertainty in the predicted reduced temperature of the pair breaking cut off due to the estimated ± 5% absolute accuracy of the temperature scale. Much of the apparent additional spectral weight could be explained by a reduction of these T_c values by approximately 7.5%. The maximum attenuation of the pair breaking feature calculated using equation (1) also increases if this adjustment is made.

However there is structure at the pair breaking edge which cannot be explained by the simple expedient of adjusting T_c. We now examine the predicted but so far unobserved collective modes of the B phase order parameter to see if any of them provides a possible explanation. The $J = 0^+$ mode has $\hbar\omega = 2\Delta(T)$ at $q = 0$ independent of Fermi liquid corrections (3), and would be expected to couple to zero sound due to particle-hole asymmetry. However Combescot (20) has shown that this mode should be destroyed by coupling to the particle-hole continuum.   The

$J = 1^-$ mode of frequency $\hbar\omega = 2\Delta(T)$ in zero magnetic field couples to spin density but not to density fluctuations, even with particle-hole asymmetry (21, 22).   Sauls and Serene (4) have calculated the effect of higher order angular momentum components of the pairing interaction (v_ℓ, $\ell > 1$) on fluctuations in the order parameter from a $J = 0$ equilibrium order parameter.   The effect of the higher v_ℓ on the equilibrium order parameter may be neglected, as may be seen from the exponential dependence of T_c, calculated in BCS theory, on the pairing potential. The $\ell = 3$ component of the pairing interaction is found to modify the frequencies of the $J = 2^-$ and $J = 2^+$ modes, and also gives rise to a possible $J = 4$ excited state of the Cooper pair.   They calculated the frequency of the $J = 4^-$ mode in terms of v_3, v_5 and F_4^s and find $\hbar\omega_4$- close to $2\Delta(T)$.   The coupling strength of such a mode has not been calculated.   A further conjecture by Schopohl (23) is an admixture of spin singlet pairing fluctuations, giving rise to a new set of $J = 2$ modes; this requires a significant tensor component of the pairing interaction.

The frequency of the $J = 4^-$ mode as a function of the strength of the components of the pairing interaction v_ℓ, at $T = 0$, is reproduced from Sauls and Serene's paper (4) in Figure 5. x_1 is defined as $1/v_1 - 1/v_\ell$.   As $T \rightarrow T_c$ the quantity $h\omega_4$-$/\Delta$ approaches 2 (24).   They have also calculated the effect of v_3 on the frequencies of the squashing and real squashing modes, which are also affected by the choice of F_2^s, F_2^a respectively.   Sauls and Serene (25) estimate v_3, F_2^a from $\hbar\omega_{2+}/\Delta$ and the Landé factor of the $J = 2^+$ mode to find a best estimate of $x_3^{-1} = 0.263$.   At reduced temperature $t = 0.9$ the predicted $h\omega_4$-$/\Delta$ = 1.993 (1.975 at $t = 0$) with $F_4^s = 0$.   In order to obtain agreement with our result of $\hbar\omega_4$-$/\Delta = 1.82$ all or any of the $\ell > 1$ interactions would be required to be considerably more attractive, which seems unreasonable in the light of the other acoustic attenuation data.

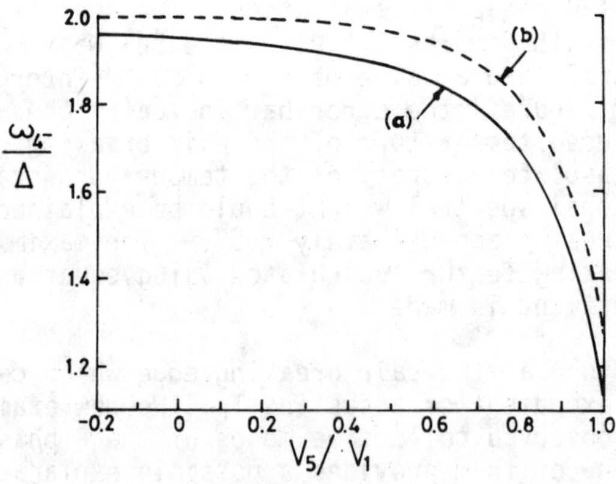

Fig. 5: Frequency of the $J = 4^-$ mode, $\hbar\omega_4$-$/\Delta$, at $T = 0$ with $F_4^s = 0$ for a) $x_3 = -2.31$; b) $x_3 = -15.0$ (a smaller v_3/v_1).

The resolution of this probably lies in the uncertainty as to the absolute magnitude of the energy gap, to which we have already referred. The interpretation of our data in terms of pair breaking processes and a single collective mode (here hypothetically identified as $J = 4^-$) necessarily implies a collective frequency close to $2\Delta(T)$. Estimating the energy gap in this way we obtain a result smaller by a factor $1.075 \pm .005$ than that expected on the basis of the Helsinki temperature scale and the assumption that the gap is of the BCS weak coupling form.   The $T = 0$ frequencies of the squashing and real squashing modes must then be rescaled to $\hbar\omega_{2-}/\Delta = 1.51$ and $\hbar\omega_{2+}/\Delta = 1.16$. This will then modify those values of F_2^s, F_2^a and v_3 which best fit this data and the measurements of the g-factor of the $J = 2^+$ mode (25).

(ii) Magnetic field dependence of new structure

A powerful technique that has been used in the study of the $J = 2^+$ and $J = 2^-$ modes is to apply an external magnetic field.   Avenel et al. (26) were the first to show that a field transverse to $\vec{q}$ lifts a five-fold degeneracy of the real squashing mode, and thereby established its $J = 2$ character. Subsequently the Northwestern University group have extended these studies of the Zeeman splitting of the $J = 2^+$ mode to higher magnetic fields (27) and observed large non-linearities. Their data has been explained by Schopohl et al. (28) in terms of the distortion of the isotropic B phase energy gap in an applied magnetic field.

The isotropic energy gap of the Balian-Werthamer phase becomes elliptical in an applied magnetic field, due to the suppression of the fraction of $S_z = 0$ pairs.   A longitudinal gap parameter Δ_2 gives the magnitude of the energy gap in the direction of the magnetic field and the transverse gap parameter Δ_1 is the energy gap perpendicular to the magnetic field. Δ_2 is less than Δ_1 and the maximum anisotropy of the gap occurs at the $A \leftrightarrow B$ transition. The gap distortion has been calculated in the Ginzburg-Landau regime by Fetter (29) and in a weak coupling calculation valid at all temperatures and including Fermi liquid factors via a renormalised Larmor frequency by Schopohl (30). This theory is able to fit measurements of the B phase magnetic susceptibility in various magnetic fields.

Schopohl et al (28) find a pair breaking cut off of $2\Delta_2 - \omega_L$ where ω_L is the renormalised Larmor frequency. However the full calculation of the field dependent pair breaking attenuation has yet to be performed. The gap distortion also causes the frequency of the $J_z = 0$ component of the $J = 2^+$ multiplet to shift downwards (27, 28).

The effect on the structure at the pair breaking edge observed in our experiment of applying a magnetic field is shown in Figure 6 for three different field strengths.   This data was obtained at a pressure of 2.90 bar at a frequency of 44.25 MHz. The increase of the sound attenuation at the lowest temperatures plotted corresponds to the onset of the squashing mode. Two clear features emerge from this data. If we describe the structure in terms of a new collective mode then the coupling constant of this mode increases rapidly with increasing magnetic field; secondly the width of the mode increases with increasing field. This last behaviour might be consistent with a mode of some non-zero J, but this interpretation is complicated by the possibility

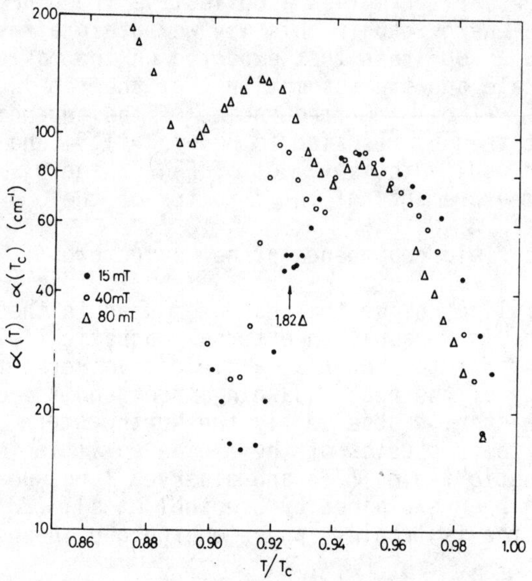

Fig. 6: Magnetic field dependence of the relative attenuation of 44.25 MHz sound at 2.90 bar.

of pair breaking processes damping the mode as a result of the effect of the gap distortion on the pair breaking cut off. Any apparent downward shift in the temperature of the peak is within the inaccuracies of our reduced temperature scale (due to residual uncompensated drifts in the NMR spectrometer).

Although the mode coupling strengths are temperature dependent (7) any possible small temperature shift in the peak position would not be sufficient to explain the apparent increase in coupling constant with applied field.   The data would appear to suggest, therefore, a mode of non zero J whose coupling strength is highly magnetic field dependent, reflecting perhaps a sensitivity to the induced distortions of the energy gap.

The frequency of the feature in 15 mT is $\hbar\omega = 1.82\ \Delta_{BCS}$ (assuming Helsinki T_c again) in excellent agreement with the earlier reported data in 14 mT. Not shown are measurements in zero magnetic field in which a separate feature is no longer resolvable. We have also applied a 40 mT field in a direction approximately parallel to the sound propagation direction.   A comparable increase in the peak height was observed and there was evidence of a narrower lineshape, but this data is somewhat preliminary in nature and suffered from a degraded performance of the NMR thermometer with this field direction.

Recently the Northwestern University group have measured the attenuation, group velocity and phase velocity at a pressure of 1.00

bar and a frequency of 12.7 MHz (32).   Under such conditions the pair
breaking cut off occurs at t = 0.993, so measurements of the attenuation do not clearly resolve the new mode from the pair breaking attenuation.   Interestingly, the group velocity data shows a dip at the
pair breaking edge consistent with the existence of a new collective
mode.

CONCLUDING REMARKS

At present no firm theoretical interpretation, beyond those specu-
lations outlined here, exists for the new structure in the B phase zero
sound attenuation in the vicinity of $\hbar\omega = 2\Delta(T)$.   Calculations of the
coupling constant to the $J = 4^-$ mode have yet to be published and it
would be interesting to know the effect of the field dependent gap distortion on this quantity.   The influence of v_3 on the gap distortion
might also be significant (33). Sauls and Serene point out that the coup-
ling constant to a mode of some J contains a term $\left|\dfrac{v_f}{c}\right|^{2J}$, which implies

that the $J = 4^-$ mode should tend to couple much more weakly at higher
pressures.   Also required is an evaluation of the contribution to the
acoustic attenuation of pair breaking processes in an applied magnetic
field.

Further measurements will attempt to uncover a pressure, frequency,
reduced temperature experimental window in which the new mode is accessible with this technique and sufficiently well resolved for its possible Zeeman splitting to be observed. If the interpretation as the
$J = 4^-$ mode is correct and it can be sufficiently well resolved from
the pair breaking edge, measurements of its frequency give important
information on v_3. If v_3 were known independently, the frequencies of
the $J = 2^-$ and $J = 2^+$ could be used as a sensitive measure of F_2^s, F_2^a.
Extensive measurements of the group velocity and the phase velocity
near 2Δ are clearly also desirable.

ACKNOWLEDGEMENTS

We thank D.C. Smith for invaluable technical assistance.   J.S.
would like to acknowledge the Royal Society for support in attending
this meeting. We would also especially thank J. Sauls, N. Schopohl,
M. Meisel and J. Owers-Bradley for useful correspondence and conversations. This work was supported by S.E.R.C. Grant No: GR/B.2700.5 .

REFERENCES

1.   W.P. Halperin, Physica 109 and 110B, 1596 (1982).
2.   J.B. Ketterson, these proceedings.
3.   See, for example, P. Wölfle, Physica 90B, 96 (1977) and references
      therein.
4.   J.A. Sauls and J.W. Serene, Phys. Rev. B 23 4798 (1981)
5.   B. Mühlschlegel, Z. Phys. 155, 313 (1959)
6.   T.A. Alvesalo, T. Haavasoja and M.T. Manninen, J.Low Temp. Phys.
      45, 373 (1981).

7. V.E. Koch and P. Wölfle, Phys. Rev. Lett. 46, 486 (1981).
8. W.R. Abel, A.C. Anderson and J.C. Wheatley, Phys. Rev. Lett. 17, 74 (1966).
9. J.B. Ketterson, P.R. Roach, B.M. Abraham and P.D. Roach in "Quantum Statistics and the Many Body Problem", ed. S.B. Trickey, W.P. Kirk and J.W. Dufty (Plenum, N.Y. 1975).
10. I.D. Calder, D.B. Mast, Bimal K. Sarma, J.R. Owers-Bradley, J.B. Ketterson, and W.P. Halperin, Phys. Rev. Lett. 45, 1866 (1980).
11. J.C. Wheatley, Rev. Mod. Phys. 47, 415 (1975).
12. D.N. Paulson, R.L. Kleinberg and J.C. Wheatley, J. Low Temp. Phys. 23, 725 (1976).
13. R.W. Giannetta, A. Ahonen, E. Polturak, J. Saunders, E.K. Zeise, R.C. Richardson and D.M. Lee, Phys. Rev. Lett. 45, 262 (1980).
14. D.B. Mast, Bimal K. Sarma, J.R. Owers-Bradley, I.D. Calder, J.B. Ketterson, and W.P. Halperin, Phys. Rev. Lett. 45, 266, (1980).
15. J. de Klerk in Physical Acoustics, vol. IVA (Academic Press, 1966).
16. D.S. Greywall and P.A. Busch, J. Low Temp. Phys. 46, 451 (1982).
17. M.E. Daniels, E.R. Dobbs, J. Saunders and P.L. Ward, Phys. Rev.B 27 (1983), to be published.
18. P. Wölfle, Phys. Rev. B 14, 89 (1976).
19. L. Tewordt and N. Schopohl, J. Low Temp. Phys. 37, 421 (1979).
20. R. Combescot, J. Low Temp. Phys. 49, 295 (1982).
21. N. Schopohl and L. Tewordt, J. Low Temp. Phys. 45, 67 (1981)
22. N. Schopohl, H. Streckwall, and L. Tewordt, J. Low Temp. Phys. 49, 319 (1982).
23. N. Schopohl, private communication.
24. J.A. Sauls, private communication.
25. J.A. Sauls and J.W. Serene, Phys. Rev. Lett. 49, 1183 (1982).
26. O. Avenel, E. Varoquaux and H. Ebisawa, Phys. Rev. Lett. 45, 1952 (1980).
27. B.S. Shivaram, M.W. Meisel, Bimal K. Sarma, W.P. Halperin and J.B. Ketterson. Phys. Rev. Lett. 50, 1070, (1983).
28. N. Schopohl, M. Warnke and L. Tewordt, Phys. Rev. Lett. 50, 1066, (1983).
29. A.L. Fetter, in "Quantum Statistics and the Many Body Problem", ibid.
30. N. Schopohl, J. Low Temp. Phys. 49, 347 (1982).
31. R.F. Hoyt, H.N. Scholz and D.O. Edwards, Physica 107B, 287 (1981).
32. M.W. Meisel, B.S. Shivaram, J.B. Ketterson and W.P. Halperin, preprint.
33. D. Rainer, private communication.

FLOW DISSIPATION IN THE ^{3}He SUPERFLUIDS

M. C. Cross

Bell Laboratories Murray Hill, New Jersey 07974

ABSTRACT

A review is given of the theoretical concepts relevant to experiments on the breakdown of superfluidity and flow dissipation in the ^{3}He superfluids.

INTRODUCTION

It is perhaps surprising that in the twelve years since the discovery of the two new superfluids - ^{3}HeA and ^{3}HeB - no clear understanding of their defining property has emerged. The microscopic nature of the phases is understood, and the broken gauge symmetry - the prerequisite for robust superfluidity - is well accepted. However the physics involved in the critical velocity, or even the existence of such a velocity of superflow below which there is no dissipation, and the nature of the dissipation above any critical velocity, are not well established. There are many elegant theoretical contributions, and a number of experiments on the subject, but the connection between theory and experiment is often tenuous.

These remarks are less surprising when one considers that after fifty years, the corresponding questions for superfluid ^{4}He are still receiving a great deal of attention, with significant new theoretical ideas[1] and experimental work[2] in the recent literature. The ^{3}He superfluids provide an even more difficult problem, with the complication of additional broken symmetries that may be involved in the flow, and the need for experiments at temperatures a factor one thousand lower. On the other hand better tools are available to investigate the flow (e.g., nuclear magnetic resonance), so that further progress may be easier. Similarities to, and differences from, superflow in ^{4}He should provide great insight into the general question of the stability of superflow. The purpose of this session is to review some of the recent experimental advances in this area. In the present article I will try to provide some background and theoretical concepts.

GENERAL REMARKS

The unifying feature, common to the various different superfluids, is the phase variable ϕ, which is the manifestation on the macroscopic scale of the phase of the wavefunction of the condensed particles. Spatial gradients of the

326

phase lead to a mass current

$$\vec{J}_s = \rho_s \vec{v}_s \,, \tag{1}$$

where

$$\vec{v}_s = (\hbar/M)\vec{\nabla}\phi \,, \tag{2}$$

with ρ_s the superfluid density and M the mass of the condensed object (twice the atomic mass m for BCS type superfluids). The dynamics of the phase variable is given by the Josephson relation

$$\hbar\dot{\phi} = - M\mu \,, \tag{3}$$

with μ the chemical potential per mass. The superflow, given by (1) and (2), will be dissipation free and will require no driving force ($\delta\mu = 0$ across the flow path) if there is no phase slip across the path:

$$\delta\mu = \int \vec{\nabla}\mu \cdot \vec{ds} = - \frac{\hbar}{M}\int \vec{\nabla}\left[\dot{\phi}\right] \cdot \vec{ds} \,. \tag{4}$$

Thus the question of critical velocities and flow dissipation reduces to a study of the macroscopic phase variable, and in particular of various phase slip processes defined by a non-zero value of the right hand side of (4). Note that the Landau critical velocity prescription

$$v_{cL} = \min(\epsilon(p)/p) \,, \tag{5}$$

with $\epsilon(p)$ the energy of an elementary excitation of momentum p, is simply an approximate estimate of an upper bound to the critical velocity: for $v \gtrsim v_{CL}$ one expects an instability to phase slip processes involving the creation of normal phase across the whole flow cross-section.

For orientation, it is first useful to consider ^{4}He. There the phase slip process is presumed to be the creation of quantized vortex lines, rings or tangles. The passage of such lines transverse to the flow path leads to a 2π phase slip per vortex. A typical estimate for the critical velocity is

$$v_c \sim \kappa/d \tag{6}$$

with $\kappa = h/M$ the quantum of circulation and d the (small) diameter of the flow channel. This expression can be rationalized in terms of the dynamics of single vortex lines spanning the flow channel, but is probably best thought of as dimensional analysis. More recently, the same quantity appears as the minimum velocity at which a turbulent vortex tangle is self sustaining.[1] The result seems to have some validity for a wide range of experiments,[3] at least at not too small d, and not to close to the transition temperature, although the fit to experiment involves a rather large prefactor, so that

$$v_c \simeq (100 \text{ microns/d}) \text{ cm/sec} . \qquad (7)$$

Close to the transition, thermal nucleation of vortex rings (with self induced velocity $< v_s$) in the bulk flow becomes the dominant process, leading to a size independent, but temperature dependent, critical velocity

$$v_c \sim \rho_s \kappa^3 / k_B T , \qquad (8)$$

at which dissipation dramatically increases.[4]

Presumably the same mechanisms would ultimately occur in ^{3}He, if not preempted by other processes. Thermal nucleation (8) will not be important in practice, since the thermal energy available is a thousand times smaller with the other energies comparable to those in ^{4}He. Equation (7) may be used to guess the critical velocity for singular vortex production in ^{3}He, although, since the physics involved is not clearly understood, this is not reliable (e.g., surface roughness on the scale of the coherence length may be relevant - a much longer length scale in ^{3}He than in ^{4}He). Observed critical velocities in ^{3}He are rather less than (7), and in fact work to date suggests that quite different mechanisms are involved.

In some sense, the two phases seem to occupy extreme positions. The A (axial) phase,[5] has been described as only "accidentally superfluid," rescued from bulk instabilities in arbitrary small flows only by numerical coincidence. On the other hand evidence is accumulating that the superfluid current in the B (pseudo-isotropic) phase[5] is limited simply by the kinetic energy of the flow competing with the condensation energy, leading to the ideal Landau-like critical velocity. I will discuss these ideas in turn.

A - PHASE FLOW DISSIPATION

The A phase is potentially more interesting in its superfluid properties, since the microscopic nature of the pairing requires a conceptual change in discussing the phase variable. In fact, since the pair wavefunction, proportional to the spherical harmonic Y_{11}, is complex on a microscopic scale, a macroscopic overall phase variable is not a good construction. Equations (1)-(4) remain schematically true - and with care can be made precise - but the phase variable is intimately tied up with the orbital anisotropy axis $\hat{\ell}$ (the axis of the Y_{11} wavefunction). This relationship is succinctly summarized by the Mermin-Ho equation[6]

$$\vec{\nabla} \times \vec{v}_s = (\hbar/2m)\, \frac{1}{2}\, \epsilon_{\alpha\beta\gamma} \ell_\alpha \vec{\nabla}\ell_\beta \times \vec{\nabla}\ell_\gamma \tag{9}$$

showing that circulation is no longer quantized if $\hat{\ell}$ is inhomogeneous; and the integrated Josephson equation for the chemical potential difference between two points connected by a path C,[7-9]

$$\delta\mu = -\int_C \vec{v}_s \cdot d\vec{s} + (\hbar/2m) \int_{\hat{\ell}(C)} (\hat{\ell}\times\dot{\hat{\ell}}) \cdot d\hat{\ell} . \tag{10}$$

The latter expression shows that, in complete contrast to conventional superfluids, even with no actual singularities crossing the path C (when the first term in (10) averages to zero) a chemical potential difference may occur, balanced simply by the motion of the $\hat{\ell}$ texture, with energy dissipated through the orbital viscosity. In fact the integral in the last term is simply the rate at which the contour $\hat{\ell}(C)$ sweeps out area on the unit sphere.[7-9] One immediate corollary is that $\hat{\ell}$ must pass through all directions (somewhere in space and time) to lead to continuous phase slip: motion in a plane, for example due to dipole locking in a magnetic field, is not sufficient. Examples of textural phase slipping motion are given by Hall and Hook[10] and Ho.[8] A consequence of (9) is the existence[11] of doubly quantized vortices, which possess a non-singular core in which $\hat{\ell}$ rotates over the unit sphere (e.g., rotating from up at large distances to down at the centre about the tangential direction). The passage of non-singular vortices across the path C is another way in which (10) may be satisfied without singular phase slip.

Thus superflow in the A phase is not protected by the need to form singular structures to relax a circulation quantization. It is in fact even more fragile than this! The state $\hat{\ell}$ parallel to $\vec{v}_s$ that minimizes the component of the flow energy $\frac{1}{2}\vec{v}_s\cdot\vec{\rho}_s\cdot\vec{v}_s$ is stable in large geometries only because the dipole energy locks deviations of $\hat{\ell}$ and the spin anisotropy axis $\hat{d}$ together, providing (at least near T_c where calculations are easy) sufficient restoring forces to inhomogeneous distortions. If this energy were absent, the flow state would be linearly unstable to a dynamic texture of $\hat{\ell}$ involving phase slip at arbitrarily small flows.[10,12] Practically, the stabilizing energy is reduced by going to lower temperatures[13] or by applying a magnetic field parallel to the flow. The initial instability then is to a time independent, helical $\hat{\ell}$ texture: subsequent transitions may lead to dissipative states.[14]

In the flow tubes of small cross-section of practical interest things become more complicated: since the $\hat{\ell}$ vector is anchored normal to surfaces, the phase variable becomes well defined at the edges of the flow tube. Circulation at the surface is then quantized as in conventional superfluids, and the usual phase slip arguments apply, provided there are no line singularities across the flow path at which the direction of $\hat{\ell}$ changes from up to down.[15] Dissipation of flow may then happen in two ways. Firstly, rotation of the $\hat{\ell}$ texture in the bulk may be associated with the motion of point singularities on the surface (e.g., "boojums" with their associated non-singular vortices in the bulk), or, in fact, with flow dissipation at line singularities of $\hat{\ell}$ on the surface. Ho[16] has constructed elegant examples showing that the latter motion may indeed relax the flow without leading to any "winding up" of the order parameter near the surface. An alternative mechanism,[16] that may be important if surface pinning is strong, is for superflow in the bulk to dissipate by the continuous creation of non-singular vortex structures, which migrate to the surface forming a screening layer of thickness κ/v_s maintaining the surface circulation. This idea has been implemented in phenomenological equations by Bagley et al.[7,17] to explain the dissipation observed in their torsional pendulum experiments.

Although all these arguments suggest that superflow in bulk ^{3}He-A is very fragile, this does not mean that critical velocities will necessarily be zero is practical situations. In fact, since textural transitions are probably involved, one would estimate a critical velocity

$$v_c \simeq (\kappa/d) = (6 \text{ microns}/d) \text{ cm/sec.} \tag{11}$$

by balancing flow and textural energies. This has the same form as the semi-empirical result for ^{4}He, but may be smaller in practice, since large numerical prefactors are not likely to be involved in (11). In situations where $\hat{\ell}$ is pinned perpendicular to a magnetic field through dipole coupling to $\hat{d}$ the length appearing in (11) should be the dipole coherence length $\xi_D \sim 10\mu$. Such an estimate agrees reasonably well with the temperature independent value of .52 mm/s measured by Paalanen and Osheroff,[18] although the remaining discrepancy is not understood. (It should be an easy transition to calculate!) A systematic test of (11) in the absence of magnetic fields remains to be done.

It is also interesting to investigate the nature of the dissipation due to textural motion for $v > v_c$. Naively one would estimate a frequency of rotation of the texture

$$\nu \sim \left[\frac{\rho_s}{\mu}\right] \left[\frac{\kappa}{\lambda}\right] v_s \tag{12}$$

with μ the orbital viscosity and λ the length scale of inhomogeneities in the texture: κ/λ is just v_s if the length scale is determined by the flow itself (c.f., Hall and Hook[10]). Eq. (12) leads to a pressure drop $\Delta P \sim \rho\kappa\nu$ per twist of $\hat{\ell}$, which is $(\rho_s\kappa/\mu) \times 10^{-3} (v_s/1 \text{ mm sec}^{-1})^2$ dynes cm^{-2}. The prefactor is

330

dimensionless, with $(\mu/\rho_s\kappa) \sim 0.3(1 - T/T_c)^{\frac{1}{2}}$ near T_c at melting pressure.[19] If there are L/λ twists in ℓ with L the length of the flow channel, this estimate leads to

$$\Delta P \propto v_s^3(1 - T/T_c)^{-\frac{1}{2}} \tag{13}$$

for $v_s \gg v_c$. I would expect a similar result from the equations of Bagley et al.[17] This result should be compared with the observation by Paalanen and Osheroff[18] of a linear dependence on $v_s > v_c$, with $d\Delta P/dv_s = 40(1 - T/T_c)^{\frac{1}{2}}$ dynes sec/cm^3, which has the opposite temperature dependence near T_c. It seems hard to reproduce this dependence from (12), since other lengths one might substitute for λ (e.g., ξ_D or d) are also temperature independent. This discrepancy seems to be a serious argument against the idea that dissipation for $v_s > v_c$ in this experiment is due to simple coherent textural motion.

An alternative approach, based on the notion that the motion of boojums and their associated non-singular vortices is important, gives a result more in accord with experiments. Noting that the energy dissipated by the motion of unit length of such a vortex is

$$\dot{E} = \mu \int (\dot{\hat{\ell}})^2 d^2r = \mu V^2 \int \left[\frac{\partial \hat{\ell}}{\partial x} \right]^2 d^2r \tag{14}$$

with V the velocity of the vortex relative to the normal fluid, we see, since the integral in (14) is scale independent, that the orbital viscosity μ plays the role of the friction coefficient for vortex motion. Equating the frictional force to the Magnus force gives

$$\rho_s(2\kappa)\hat{\kappa}x(\vec{V}-\vec{v}_s) + \gamma\mu\vec{V} = 0 \tag{15}$$

with γ a numerical constant. For small $(\mu/\rho_s\kappa)$ the component of $\vec{V}$ perpendicular to $\vec{v}_s$ that leads to phase slip is given by

$$\delta V \sim (\mu/\rho_s\kappa) v_s \tag{16}$$

leading to a rate of phase slip $\delta V/d$ per vortex line. This expression leads to a temperature dependence in agreement with the results of Paalanen and Osheroff, although a velocity independent number of vortices of order 10^4 is needed to agree with the measured pressure. Such a large number suggests a better approach may be to consider a vortex tangle following Schwartz,[4] noting that $(\mu/\rho_s\kappa)$ takes the role of his dimensionless mutual friction coefficient α, although the lack of a well defined core in our problem may make his approach inapplicable.

It seems important to do more experiments on flow in the A phase to test some of these theoretical ideas. In particular some sort of local probe seems necessary so that we may know whether we are dealing with a turbulent tangle or the coherent motion of a simple texture. Nuclear magnetic resonance provides some information (indeed the well characterized results of Paalanen and Osheroff[18] remain unexplained). Unfortunately, in the magnetic field usually present to make such measurements, one expects a dipole locked critical velocity $v_c \gg (\kappa/d)$ so that a tangle is to be expected (because the length scale at $2v_c$ for example, κ/v_c is $\ll d$). Ultrasound measurements across the flow path in zero magnetic field suggested coherent textural motion at these lower flow velocities, and have already provided a great stimulus to theory.[20]

B PHASE FLOW DISSIPATION

The B phase is a more conventional superfluid, with a well defined phase variable, separate from other broken symmetries. It is not strictly isotropic, but instead shows a broken symmetry of relative rotations between spin and orbit spaces, specified by a directrix $\hat{n}$ (the axis of the rotation). However, the anisotropy is usually manifested only through very small energies, such as the dipole energy, and seems unlikely to be important in flow dissipation involving much larger energies.

There is rather good experimental evidence[21–23] for a critical mass current in d.c. flow experiments with the temperature dependence

$$j_c = j_{co}(1 - T/T_c)^{3/2} \qquad (17)$$

The corresponding critical velocity, typically $5(1 - T/T_c)^{1/2}$cm/sec, is generally smaller than the value (7) for singular vortex formation transferred from ^{4}He. Above the critical current (17), a dramatic increase in the dissipation is seen. There is less agreement on whether the superflow is truly dissipation free at smaller currents - this question will be discussed further below.

The temperature dependence (17) strongly suggests an ideal critical velocity, in which the flow energy $\frac{1}{2}\rho_s v_s^2$ competes with the bulk condensation energy $N(0)T_c^2(1 - T/T_c)^2$. The behavior expected as the superfluid current is increased is sketched in fig. 1.

The details of this picture were worked out in the Ginzburg-Landau regime by Fetter.[24] Here, all strong coupling effects may be incorporated in terms of a few parameters - the standard "β" parameters of the condensation energy,[5] and (to be completely general, not included in ref. 24) two independent stiffness constants giving the tensor $\vec{\rho}_s$. Experimental observation of these effects then provides information on these parameters (in new combinations due to the distorted gap at j_c). The situation at lower temperatures was studied by Kleinart[25] and by Vollhardt et al.[26] The result (17) is found to be a rather good approximation for nearly all temperatures, and is independent of Fermi-liquid corrections. The velocity v_c however has more complicated dependences.

The experimental situation for the parameter j_{co} is less clear. Hutchins et al. observed little pressure dependence, inconsistent with weak coupling predictions and recent estimates of strong coupling corrections.[21] Manninen and Pekola, using tubes of smaller cross-section, found a pressure dependence in good agreement with the expected trend.[22] The overall agreement of these two experiments with those of Eisenstein and Packard at s.v.p. is rather good. The torsional pendulum experiments of ref. 27 do not fit in so well, showing a rather lower critical velocity, and a less strong temperature dependence near T_c.

The question of whether flow dissipation exists below these critical velocities seems to be rather complicated. The interesting theoretical question, with the prejudice of this presentation, is whether there is dissipation for flow of the superfluid component alone, at a velocity $v < v_c$. This question can be answered experimentally using a "four terminal" approach in which the pressure across the superleak is measured directly by a device involving no flow.[28] In the "two terminal" approach more often used, end effects may lead to dissipation that is not intrinsic to the superflow. Thermal damping effects are well known,[29] but are less important in ^{3}He since there is less entropy around. Brand and I[30] suggested a new mechanism, involving normal flow and dissipative interconversion of normal and superflow near the liquid free surfaces, to explain the large linear dissipation encountered in the U-tube experiments of Eisenstein and Packard. The general point is, just as for the Kapitza resistance, that there may be a chemical potential drop (and additional temperature drop) at the liquid-solid or liquid-vapour interface proportional to the counterflow velocity $v_s - v_n$ approaching the surface. For ^{3}He we were able to calculate this effect from the macroscopic two fluid equations (I refer you to the paper for details), since only changes over a macroscopic healing length were involved. A similar calculation in ^{4}He would lead, inconsistently, to a microscopic length, so that a fully microscopic calculation would be needed there. The pressure

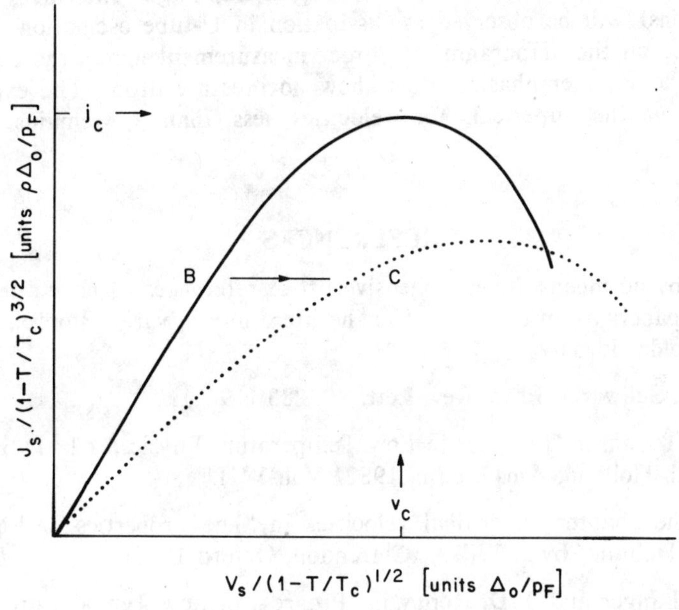

Fig. 1. Current velocity characteristic for B phase (solid) and another phase (dashed).

At first the curve is linear. As the velocity becomes comparable with Δ/p_F, with Δ the energy gap, the gap and superfluid density are depressed - and become anisotropic in the previously isotropic B phase. Nuclear magnetic resonance should be sensitive to this. In the simplest case the current eventually saturates at (17). Beyond this point the state becomes unstable, presumably to phase slipping fluctuations to the normal state. Since normal flow is strongly pinned, and does not provide a good parallel flow channel in ^{3}He, a very strong increase in dissipation is expected for current driving at $j > j_c$, or a rapid decay for velocities $v > v_c$. Below these critical values, the flow should be completely dissipation free in this picture. An alternative behavior is that a first order transition could occur to a different phase, such as the line BC in fig. 1. The flow properties would then depend on the new state - and would be highly dissipative for the A phase at these velocities.

drop near the free surface (or near liquid-diaphragm interfaces in other configurations) will be observed as dissipation in U-tube oscillations or in the driving force on the diaphragm. A direct measurement across the ends of the superleak should, I emphasize, then show no pressure drop. The existence of dissipation *in* the superleak for velocities less than v_c remains, I think, unproven.

REFERENCES

This is by no means a comprehensive list of references. I have attempted to cite recent papers as an entry point to the literature on various topics, or review articles on older ideas.

1. K. W. Schwartz, Phys. Rev. Lett. *49*, 283 (1982).

2. J. T. Tough, in "Progress in Low Temperature Physics," ed. D. F. Brewer (North-Holland, Amsterdam, 1982) Vol. VIII.

3. See the chapter on critical velocities in "The Properties of Liquid and Solid Helium," by J. Wilks, (Clarendon, Oxford 1967).

4. J. S. Langer and J. D. Reppy in "Progress in Low Temperature Physics," ed. C. J. Gorter (North Holland Amsterdam 1970) Vol. VI.

5. For a review of the phases see A. J. Leggett, Rev. Mod. Phys. *47*, 331 (1975).

6. N. D. Mermin and T-L Ho, Phys. Rev. Lett. *36*, 594 (1976).

7. M. Bagley, P. C. Main, J. R. Hook, D. J. Sandiford and H. E. Hall, J. Phys. C *11*, L729 (1978).

8. T-L Ho, Phys. Rev. Lett. *41*, 1473 (1978).

9. G. E. Volovik, JETP Lett. *27*, 605 (1978).

10. H. E. Hall and J. R. Hook, J. Phys. C *10*, L91 (1977).

11. P. W. Anderson and G. Toulouse, Phys. Rev. Lett. *38*, 508 (1977).

12. P. Battacharyya, T-L Ho and N. D. Mermin, Phys. Rev. Lett. *39*, 1290 (1977).

13. M. C. Cross and M. Liu, J. Phys. C *11*, 1795 (1978).

14. A. L. Fetter, Phys. Rev. B *20*, 303 (1979); A. L. Fetter and M. Williams, Phys. Rev. B *23*, 2186 (1981).

15. N. D. Mermin, Physica B&C *90*, 1 (1977).

16. T-L Ho, Phys. Rev. B *18*, 1144 (1978).

17. R. Gay, H. E. Hall, J. R. Hook and D. J. Sandiford, Physica B&C *108*, 797 (1981).

18. M. Paalanen and D. D. Osheroff, Phys. Rev. Lett. *45*, 362 (1980).

19. J. C. Wheatley, in "Progress in Low Temperature Physics," ed. D. F. Brewer (North-Holland, Amsterdam, 1982) Vol. VIIa.

20. D. N. Paulson, M. Krusuis and J. C. Wheatley, Phys. Rev. Lett. *37*, 599 (1976).

21. A. J. Dahm, D. S. Betts, D. F. Brewer, J. Hutchins, J. Saunders and W. S. Truscott, Phys. Rev. Lett. *45*, 1141 (1980); J. D. Hutchins, D. S. Betts, D. F. Brewer, A. J. Dahm and W. S. Truscott, Physica B&C *108*, 1159 (1981).

22. J. P. Eisenstein, G. W. Swift and R. E. Packard, Phys. Rev. Lett. *43*, 1676 (1979); J. P. Eisenstein and R. E. Packard, Phys. Rev. Lett. *49*, 564 (1982).

23. M. T. Manninen and J. P. Pekola, Phys. Rev. Lett. *48*, 812, 1369E (1982).

24. A. L. Fetter, in "Quantum Statistics and the Many-Body Problem," eds. S. B. Trickey, W. P. Kirk and J. W. Duffy, (Plenum, New York 1975).

25. H. Kleinert, Jour. Low Temp. Phys. *39*, 451 (1980).

26. D. Vollhardt, K. Maki and N. Schopohl, Jour. Low Temp. Phys. *39*, 79 (1980).

27. J. M. Parpia and J. D. Reppy, Phys. Rev. Lett. *43*, 1332 (1979).

28. D. D. Osheroff, private communication. Such a configuration was used in ref. 18.

29. J. E. Robinson, Phys. Rev. *82*, 440 (1951).

30. H. Brand and M. C. Cross, Phys. Rev. Lett. *49*, 1959 (1982).

CRITICAL CURRENTS AND DISSIPATION IN SUPERFLUID ^{3}HeA AND ^{3}HeB

D.F. Brewer*
Physics Laboratory, University of Sussex,
Brighton, Sussex, England.

ABSTRACT

This paper summarises results of previous low frequency oscillatory flow experiments of ^{3}HeA and B through a rectangular cross-
section superleak, and outlines new results on D.C. flow. Both sets
of measurements show that at least two distinct flow regimes exist,
separated by sharply defined critical currents. Saturation currents
are observed, in general agreement with theory, but it seems possible
to drive the flow beyond them in the D.C. experiments. Like other
flow experiments, these cannot demonstrate the existence of "ideal"
superflow such as exists in helium II, but there is evidence for
long-lived dissipative structures that decay in times of the order of
several hours. Significant differences are observed in the temperature dependence and magnitude of critical currents in ^{3}HeA and B.

The complex structure of superfluid ^{3}He makes it probable that
its hydrodynamic properties will be even more interesting but difficult to clarify than in superfluid ^{4}He. Experimentally, the obvious
approach is to establish the systematics of flow in various geometries and experimental conditions, and in Sussex we have now carried
out a number of experiments on low frequency oscillatory and D.C.
flow.

The purpose of this short paper is to summarize results previously obtained at Sussex on the flow and dissipation of superfluid
^{3}He and ^{3}HeB in low frequency ($\sim$ 40 Hz) oscillatory motion through a
superleak of well-defined geometry, and to outline some of our more
recent work on D.C. flow. Since time is limited I shall not compare
these in detail with flow results in other laboratories nor with
experiments such as fourth sound, torsional pendulums, vibrating
wires, or heat flow which may take place, as in helium II, by counterflow of normal and superfluid components. Some of our results have

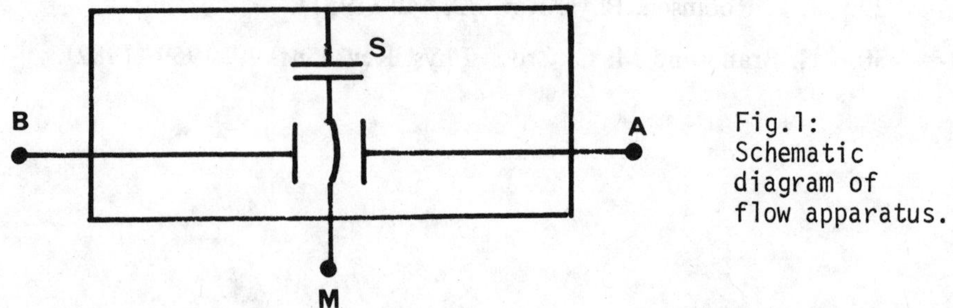

Fig.1:
Schematic
diagram of
flow apparatus.

*S.E.R.C. Senior Fellow.

been described briefly in refs. 1-3, and in more detail in theses[4], where they are also compared with other work.

Our experimental arrangement is shown schematically in Fig. 1. Two vessels containing superfluid ^{3}He are joined by a superleak S of rectangular cross-section 48 μm $\times$ 2.86 mm and 9.16 mm long. Fluid can be driven through the superleak by means of a flexible mylar diaphragm which is coated with gold on each side and connected to terminal M, so that a voltage applied between it and an electrode A can be used to draw the diaphragm to one side, and the displacement or velocity of the diaphragm (and hence the fluid flow current) can be measured sensitively by measuring the capacitance change between it and electrode B. This design has the advantage over the U-tube type driven by gravitational potential in that measurements can be made at pressures above zero, including the zero-field A phase.

Observations with the apparatus can be made in two ways, by application either of a step voltage between AM or of a ramp. In the former, the system will respond (for small displacements) as a damped harmonic oscillator with damping dependent on losses occurring through dissipation of the superfluid flow through the superleak, and also possibly elsewhere. In these experiments the superfluid mass current density through the superleak (assuming no normal fluid flow) is given by

$$J_s = \frac{8dA\rho}{14a} \frac{\dot{C}}{C_0} \tag{1}$$

where A is the diaphragm area, d the equilibrium distance between the diaphragm and the electrode, C_0 the equilibrium capacitance and a the superleak cross-sectional area. The change in capacitance is proportional to the diaphragm displacement. The time constant for decay of superfluid oscillation is

$$\tau = \frac{2Q}{\omega_0} \tag{2}$$

where Q is the quality factor and ω_0 is the oscillation frequency, given by

$$\omega_0^2 = \frac{8\pi\sigma}{\rho A^2} \cdot \frac{a}{\ell} \cdot \left(\frac{\rho_s}{\rho}\right)_{ch} \tag{3}$$

in which σ is the diaphragm tension per unit length, ℓ the superleak length and $(\rho_s/\rho)_{ch}$ the superfluid fraction in the flow channel. These three equations allow determination of critical currents, dissipation and the superfluid fraction within the accuracy of determination of the apparatus constants, which in our case is not high: the current is probably uncertain to within about 20%, for example, although the sensitivity to <u>changes</u> in current which occur on making transitions between different flow regimes is much higher.

The second way of making observations is to apply not a step voltage, but a ramp. It is most convenient to use a ramp voltage such that $V^2 \propto t$, for the following reason. With a constant voltage V applied and the diaphragm in static equilibrium, the electrostatic pressure on it is just balanced by that due to its tension which is proportional to its displacement x:

$$\alpha V^2 - \beta x = 0 \qquad (4)$$

where α and β are constants determined by the apparatus design. The same equation holds in ideal superflow with zero dissipation, so that if we apply a voltage $V^2 = \gamma t$ where γ is constant, the diaphragm displacement will be linear in time and the superfluid current will be constant (apart from an acceleration term which is negligible in most of our experiments):

$$\frac{dx}{dt} = \frac{\alpha \gamma}{\beta} \qquad (5)$$

In this equation the ratio α/β is, essentially, known accurately from the calibration equation (4) without knowing α and β separately, and also γ is accurately known, and hence the existence of ideal superflow can be detected. The sensitivity is not, however, particularly high, and it turns out that pressure drops less than about 1 μbar cannot be accurately observed by this method. It is much less sensitive than the oscillatory method using equation (2), which can detect pressure differences less than 10^{-3} μbar.

If ideal D.C. superflow does not occur a pressure difference Δp develops across the channel, and the corresponding equation is

$$\frac{dx}{dt} = \frac{\alpha \gamma}{\beta} - \frac{1}{\beta} \frac{d(\Delta p)}{dt} \qquad (6)$$

This equation allows one to measure the build-up of the dissipative pressure Δp, but now the apparatus constants need to be determined and the accuracy is much less, although again sensitivity to changes in displacement, current and Δp are much higher. In the following paragraphs we outline the results of both types of experiment, restricting the discussion mainly to ^{3}HeB unless otherwise stated.

Oscillatory Measurements

If the diaphragm is initially in a stationary equilibrium position, application of a step voltage $V_i \to V_f$ will change the equilibrium position and the diaphragm will begin to move at $t = 0$ towards the new position in different ways depending on the size of the step. For small steps, as already noted, it will start off with $dx/dt = 0$ at $t = 0$ and oscillate with damped harmonic motion about the new equilibrium position, with amplitude initially proportional to $(V_f^2 - V_i^2)$ (see equation (4)) but decaying according to equation (2). It is predicted[5-7], however, that there should be a maximal current J_s^{max} in superfluid ^{3}He above which the fluid cannot be driven in superflow: this arises because $J_s = \rho_s v_s$ and $\rho_s \propto (1 - a v_s^2)$ (where v_s is the superfluid velocity and a is a constant) so that there is a maximum in the $J_s \sim v_s$ curve. The maximum corresponds closely, though not exactly, with the pair-breaking velocity at which the kinetic energy becomes equal to the gap energy and Cooper pairing of the 3He atoms can be disrupted. Thus as the step size increases the behaviour should cross over from the damped harmonic response with initial amplitude proportional to $(V_f^2 - V_i^2)$ to one where the diaphragm moves off with a velocity which is independent of step size. We observe this general behaviour, but with some unanticipated complications. The principal results using this technique can be summarised as follows (see ref. 1 for further details):

(i) For large voltage steps, the saturation current just discussed is observed in 3HeB but not in 3HeA[8].

(ii) The temperature dependence of the saturation current agrees with theoretical prediction at not too high pressures,

$J_s^{max} = A(1 - T/T_c)^{3/2}$ where A is constant. At higher pressures, the power index is nearer 2 (see the discussion of DC results below).

(iii) The prefactor A is in reasonable agreement with theory at p = 0, but it is almost pressure-independent[2], in contradiction to theory.

(iv) At lower flow currents oscillatory behaviour is observed in both A and B phases, the Q in 3HeA being typically around 5. A phase results are independent of the direction of a 98G NMR field, consistent with textures in which the ℓ vector is directed along the flow channel (except, of course, at the walls).

(v) The oscillatory results in 3HeB are current-dependent, with a sharp and strongly temperature dependent transition between two flow regimes. A higher current regime has a low Q ($\lesssim 10$) and a period which decreases with decreasing current, and a lower current regime has a higher Q (see (vi)) and a period which is independent of current.

(vi) The Q of the low current, lower dissipation regime increases very rapidly with decreasing temperature, the largest observed Q being about 150. This is probably the most sensitive test of dissipation in any experiment so far done: in a DC flow experiment it would correspond to a pressure drop due to viscous dissipation of $\lesssim 10^{-9}$ bar, which is about three orders of magnitude more sensitive than the DC flow type of experiment discussed below.

(vii) Hysteresis effects are observed in 3HeA but not in 3HeB (ref.3)

Some of these observations are illustrated in Fig. 2, taken from ref. 1, which shows typical diaphragm displacements as a function of time for 3HeA and B for various voltage steps. Fig.2(b) shows that at high enough voltage steps the slope of the graphs in 3HeB is the same, establishing clearly the existence of a saturation current. It may be important to note, in relation to the saturation of the current, that the voltage is changed only at t = 0, and is thereafter constant. The DC measurements described below seem to suggest that if the voltage is changing rapidly during a ramp

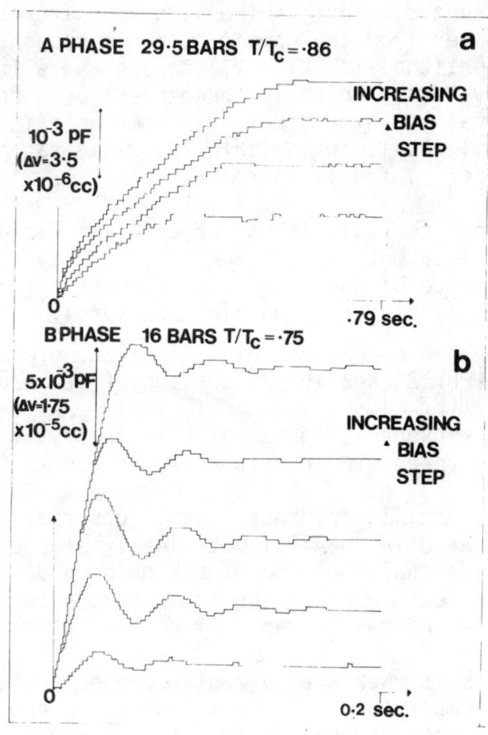

Fig. 2: Diaphragm displacement ($\propto$ capacitance change) as a function of time for various voltage steps in (a) ^{3}HeA and (b) ^{3}HeB.

excitation it is possible to exceed the "saturation" current.

As the diaphragm approaches its new equilibrium position the current decreases and it overshoots the equilibrium position into the oscillatory regime. On the scale of this figure, the transition from the low Q to the high Q regimes cannot be discerned, but extensive data[4], taken at several different pressures and temperatures, establishes this second most important observation, namely the existence of two distinct dissipative flow regimes for superfluid ^{3}HeB in this geometry, with a sharp transition between them ((v) above). This clearly marks the existence of two different types of dissipation in ^{3}HeB flow, but we shall not speculate here on that they may be.

The third important conclusion is that this experiment, in common with all other flow experiments on ^{3}HeA and B that we are aware of, does not demonstrate "ideal" superflow to the degree that it exists in superfluid ^{4}He. However, the very rapid increase in Q as the temperature decreases gives hope that this may be observed at lower temperatures in persistent current experiments which are being prepared in this laboratory.

Figure 2(a) shows the failure to reach a saturation current in ^{3}He, although this does not preclude the possibility of such an effect if larger voltage steps could be used. Again, the scale is too large to show the oscillations that take place about the final equilibrium position of the diaphragm.

D.C. measurements using a V^2 ramp excitation (equations (4)-(6)) with the same apparatus confirm the above conclusions where they overlap and provide additional information[9]. We shall discuss first the B phase. A qualitative observation which should first be

mentioned concerns the equilibrium time needed in order to achieve
reproducible results. This is found to be in the region of five
hours, about an order of magnitude longer than the calculated relaxation time and during which period the thermometers indicate that
temperature equilibrium exists in the liquid ^{3}He. The necessary
standard practice has now become to demagnetise the nuclear stage
over $\sim$ 10 hours and then leave the apparatus at apparently constant
low temperature for a further five hours or so before starting
measurements. This effect could be due to the time necessary for
dissipative structures such as vortex lines to decay; similar effects
have been found before in superfluid ^{4}He. It is interesting to note
that in the oscillation experiments described above the resonance
frequency ω_0 (equation (3)) is very reproducible in the initial periods but the Q is not so much so: if the irreproducibility is due
to remanent vorticity it does not much effect $(\rho_s/\rho)_{ch}$ although it
does affect the dissipation.

The most interesting initial quantitative observation concerns
the response of the diaphragm to the V^2 ramp. In the low velocity
"subcritical" region the response is just as expected from equations
(4) and (5), and is shown schematically in Fig. 3. This is a norm-
alised plot of V^2 and displacement x as a function of time. (Note
that the experimental procedure is to change the voltage from 15 to

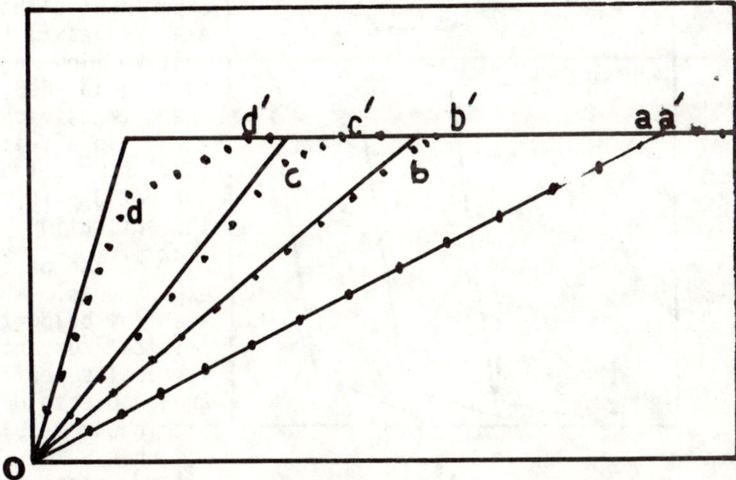

Fig. 3: The displacement of the diaphragm (dotted lines) in response
to a $V^2 = \gamma t$ ramp (full lines) as a function of time. The
V^2 ramp is switched off when V becomes 500 volts at a,b,c
and d (schematic diagram). Dots show displacement.

500 volts and vary the ramp γ by varying the time for the voltage
change.) On this diagram, a subcritical flow rate is represented
by the coincidence along (for example) Oa of the V^2 ramp line
(solid line) with the diaphragm displacement (dotted line). As the
ramp rate γ increases the diaphragm velocity increases proportionally

(equation (5)) and the lines still coincide. In the supercritical state - for example, Ob, Oc and Od - the diaphragm velocity lags behind (equation (6)) and a pressure difference builds up proportional to the vertical displacement between the dotted and solid lines. On switching off the ramp when the new voltage reaches a constant value of 500V, the diaphragm relaxes towards its new equilibrium position along bb', cc', dd' ... etc.

An interesting point to note is that in all regions of this graph (at least at the lower pressures) the diaphragm response is linear in time. We expect this naturally in the subcritical region of equations (4) and (5) for ideal superflow or when the pressure drop across the superleak is less than about 1 μbar (the resolution of our measurements). The observation that Ob, Oc, Od... are also linear gives us the new experimental information that the dissipative pressure difference (equations (6)) is also building up linearly with the flow current, and in the region of our experiments it shows no sign of saturating to a state where a constant current is driven by a constant pressure difference. In this region (where, it must be remembered, the drive voltage is constantly increasing) the rate of increase of pressure difference is shown schematically in Fig.4.

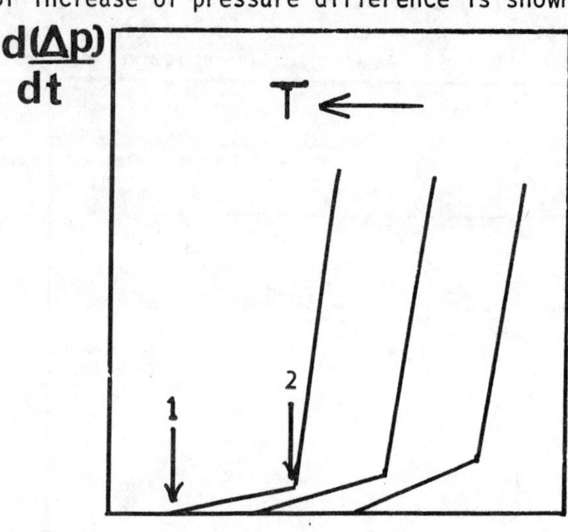

It confirms the conclusion of the oscillation experiments, that there are (at least) two dissipative flow regimes in superfluid ^{3}HeB, with a sharp transition between them. To a fair approximation, the transition current J_{c2} is close to the saturation current J_s^{max} of the oscillation experiments. Whether the flow below the lower critical current, J_{c1}, is ideal superflow we are unable to determine within our resolution of $\sim$ 1 μbar which, as already mentioned, is three orders of magnitude less sensitive than can be inferred from the oscillation experiment. J_{c1} is found to vary with temperature as $(1 - T/T_c)^{3/2}$ at low pressures, but as $(1 - T/T_c)^2$ above the

Fig. 4: Schematic diagram of the rate of creation of dissipative pressure in ^{3}HeB as a function of flow current J_s at different temperatures. Points such as 1 and 2 represent lower and upper critical currents J_{c1} and J_{c2}.

polycritical pressure. The change in the power dependence takes place very sharply - over a pressure range of 2 bar at most - even well into the B phase below the polycritical temperature.

The final observation to be made about these remarkable B phase flow phenomena is the behaviour at the end of the ramping periods, when the voltage becomes constant at 500V. In a supercritical experiment, a pressure difference has built up at this point and it relaxes to zero as the diaphragm moves to its equilibrium position. The motion is indicated by bb', cc', dd'..., all of which are linear, indicating again that the flow current is independent of the pressure difference. Moreover, all these relaxation currents are the same (at a given temperature and pressure) independent of the initial ramp rate, i.e. of the pressure which has built up, Δp_f, at the end of the ramping period, so that the lines bb', cc', dd' are parallel. Thus a graph of Δp_f versus current taken from plots such as Fig. 3 shows an initial subcritical region with $\Delta p_f \lesssim 1$ μbar, then a critical current at which Δp_f rises vertically, like a true maximal saturation current as in the oscillation experiments. On the other hand, it is clear that as long as the ramp is switched on and the voltage is changing it is possible to drive the current beyond the saturation value. The reason for this apparent inconsistency is not known.

A number of similar measurements have been made in the A phase. Present indications are that the existence of distinct flow regimes such as those in Fig. 3 is not so clear, that there is a sharp drop in the value of J_{cl} on passing from the B to the A phase, and that the temperature dependence in ^{3}HeA is much slower.

References

1. A.J. Dahm, D.S. Betts, D.F. Brewer, J. Hutchins, J. Saunders and W.S. Truscott, Phys. Rev. Letters 45, 1411 (1980).
2. J.D. Hutchins, D.S. Betts, D.F. Brewer, A.J. Dahm and W.S. Truscott, Physica 108B, 1159 (1981).
3. J. Hutchins, D.S. Betts, D.F. Brewer and W.S. Truscott, Physica 108B, 1157 (1981).
4. J.D. Hutchins, D. Phil Thesis (Sussex University 1980), unpublished (oscillatory measurements); Ling Renzhi, D. Phil Thesis (Sussex University, to be completed (D.C. measurements).
5. A.L. Fetter, Quantum Statistics and the Many-Body Problem, Ed. S.B. Trickey, W.P. Kirk and J.W. Dufty (Plenum, N.Y. 1975), p.127.
6. D. Vollhardt, K. Maki and N. Schopol, J. Low Temp. Phys. 39, 79 (1980).
7. H. Kleinert, J. Low Temp Phys. 39, 451 (1980).
8. J.R. Hook (private communication) has put forward an explanation of the A-phase damping.
9. Similar measurements are described elsewhere in these Proceedings by J. Pekola.

FLOW DISSIPATION IN B-PHASE U-TUBE EXPERIMENTS

J. P. Eisenstein[a]
Department of Physics
University of California
Berkeley, CA   94720

ABSTRACT

We present a brief summary of B-phase flow experiments performed
at saturated vapor pressure in a U-tube device. Dissipation is found
over the entire range of accessible flow velocities. At low
velocities the losses seem to be accounted for by the mechanism
proposed by Brand and Cross[1] whereas higher velocity non-linear
dissipation may be related to the depairing critical current.

A number of flow experiments have been performed in the
superfluid B-phase of liquid ^{3}He to date. A variety of effects have
been observed including critical currents,[2,3] non-linear
dissipation,[3,4,5] hydromagnetic distortions of the order parameter,[6]
and perhaps even dissipationless superflow.[7] The latter is an
essential ingredient of our present understanding of the B-phase
state. As yet, however, there has not been a direct confirmation
that persistent currents can be set up. We summarize here results of
U-tube experiments in the B-phase in which dissipation is found over
the entire range of accessible flow velocities. These results
include non-linear damping at large velocity which gives way, at a
well-defined critical velocity, to linear losses persisting to the
lowest velocities investigated.

The U-tube device used in these experiments has been described in
detail elsewhere.[5,8] The device, shown in Fig. 1, consists of a
central reservoir connected to four outer reservoirs by cylindrical
flow channels of radii 102, 126, 177, and 227 μm, with lengths of
0.5, 0.5, 1.0, and 1.0 cm, respectively. It is possible to produce
flow in any given channel independent of the others by a technique
described elsewhere.[8] The fluid reservoirs are actually the annular
gaps of concentric cylinder capacitors. These capacitors are used to
electrostatically drive and detect the motion of the liquid surface
and thus the mass current in the flow channel. The superfluid
velocity is at least 20 to 100 times (depending on flow channel
radius) higher in the flow channel than it is in the reservoir. The
data consists of digitized records of the liquid surface position as
a function of time resulting from a step in the DC potential applied
to one of the capacitors.

We have examined superfluid velocities (in the flow channel) over
the range from about 7×10^{-4} to 1 cm/sec resulting from pressure
differences across the channels varying from about 10^{-9}Bar to 16 μ
Bar. These measurements are made at saturated vapor pressure at
temperatures down to 0.5 T_c ($T_c \sim 1$ mK). As mentioned above, two

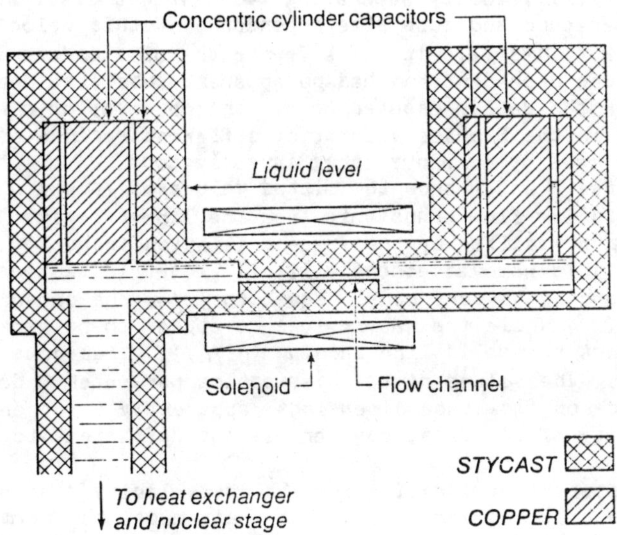

Fig. 1. Schematic diagram of one arm of the U-tube device. The solenoid was not used in these experiments.

distinct regimes of dissipation have been found separated by a critical velocity. We begin by discussing the low velocity linear regime.

To systematically analyze the low velocity flow transients we have modeled the system as a damped simple harmonic oscillator where the liquid displacement, x, in the reservoirs is assumed to follow the differential equation

$$\ddot{x} + 2L\dot{x} + \omega^2 x = 0 \tag{1}$$

The liquid level transient resulting from a step force is fitted by a solution of Eq. (1) using a standard least square method. The results of the fitting give values for the damping and frequency parameters, L and ω, respectively, and a reduced χ^2 goodness-of-fit parameter. At a given temperature a sequence of flow transients resulting from various size step forces is recorded. The fitting procedure reveals that for forces below a certain critical value the reduced χ^2 is near unity indicating a good fit. Over essentially this same range the extracted L and ω parameters are constant to within error. We interpret this as indicating the fit function is statistically good and that losses at low velocity are consistent with a linear dissipation.

From the value of the step force where the fit begins to fail, and the fitted values of L and ω in the linear regime, we can calculate the maximum flow-channel velocity in a transient at the upper limit of the linear range. We have interpreted this velocity as a critical velocity separating two distinct dissipation regimes. The temperature and tube size dependence of this velocity are shown in Fig. 2. This velocity of a few tenths of a cm/sec is weakly temperature dependent and has no apparent tube size dependence, at least for the data presented here. The observation of a critical velocity in the B-phase separating different regimes of dissipation has been made by numerous other investigators.[3,4,9]

Figures 3 and 4 show the fitted values of ω and L for representative flow transients from the linear regime. The ω values have been normalized by the free-oscillation frequency, ω_0, of the U-tube. This removes essentially all geometric factors and, in fact, $(\omega/\omega_0)^2$ should be very nearly equal to ρ_s/ρ, the superfluid density fraction.[10] While the ω/ω_0 values do appear to be geometry-independent the solid line showing $(\rho_s/\rho)^{1/2}$ lies considerably above the data. The values of the dissipation parameter L do show some dependence on flow tube dimensions, apparently L increases with ω_0. The quality of the data, however, is not sufficient to quantify this trend.

An apparent explanation for the low velocity dissipation has been presented by Brand and Cross.[1] In their model the normal fluid component, while immobilized almost everywhere by its large viscosity must, in fact, be in motion near the liquid surfaces in the capacitor reservoirs. This must be, Brand and Cross argue, so that the elementary excitations comprising the normal fluid remain in equilibrium with the moving fluid meniscus. This normal fluid velocity decays away to zero below the surface in a characteristic

length about equal to the capacitor gap, .02cm. Dissipation arises from the finite value of ζ_3, a second viscosity coefficient, and the consequent divergences in v_s and v_n in the surface layer. Brand and Cross assume steady flow and offer no estimate of relaxation times for changes in the quasi-particle distribution accompanying accelerations of the surface. Improved numerical estimates[11] of the parameters involved, accounting especially for mean free path effects, show Brand and Cross' model agrees well with the data presented here. The model predicts L to be proportional to ω_0^2 and although L does apparently

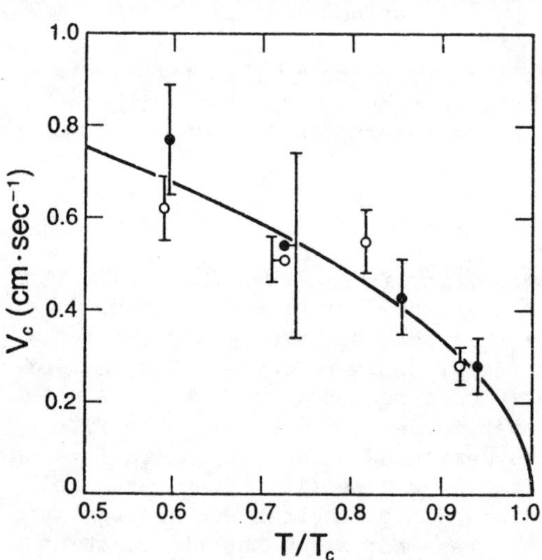

Fig. 2. Dissipation critical velocity. Open dots, 126 µm radius tube; closed dots, 227 µm radius tube. Solid line is a guide to the eye.

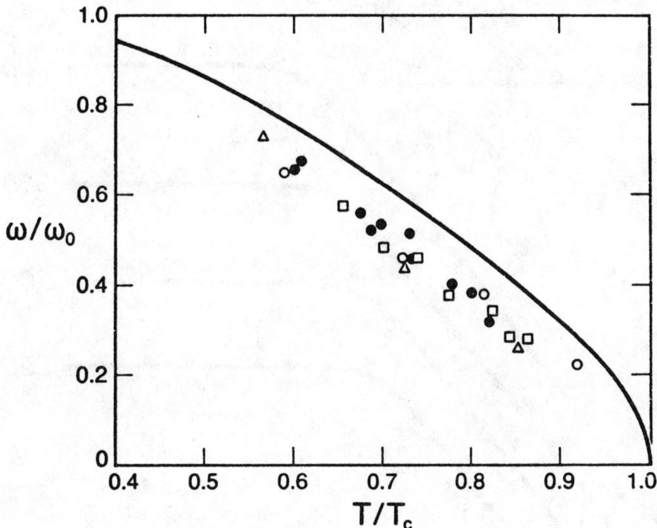

Fig. 3. Fitted values of ω/ω_0 vs. T/T_c. Solid circles, 102 μm radius tube, ω_0=6.1 sec^{-1}; open circles, 126 μm radius tube, ω_0=7.6 sec^{-1}; boxes, 177 μm radius tube, ω_0=7.5 sec^{-1}; triangles, 227 μm radius tube, ω_0=9.6 sec^{-1}.

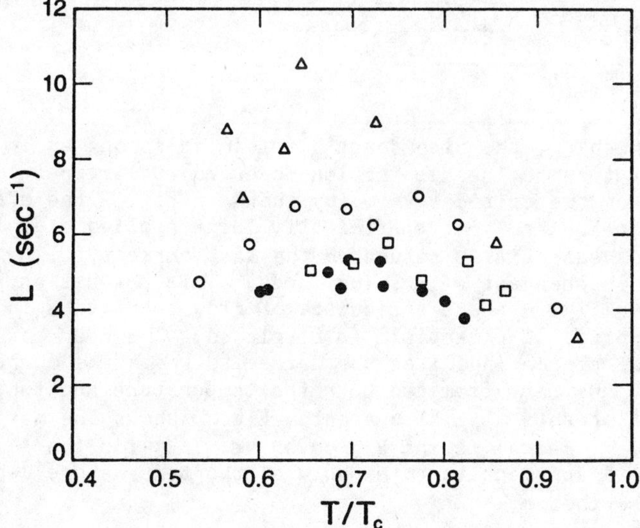

Fig. 4. Fitted values of L vs. T/T_c. Symbols as in Fig. 3.

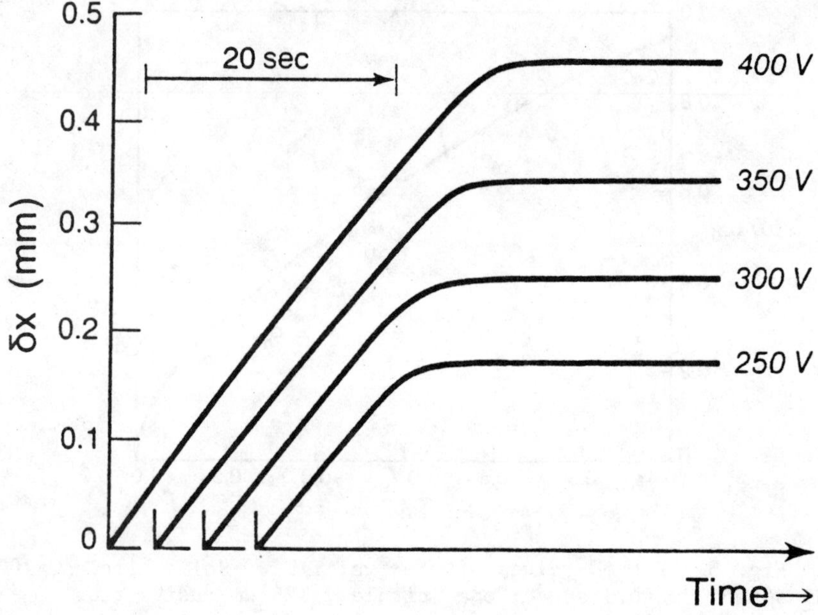

Fig. 5. Four transients taken at fixed temperature. Note near parallelism of initial slopes.

increase with ω_0, the rise doesn't appear as strong as ω_0^2.

We now turn to the dissipation observed at larger velocities. Beginning at the critical velocity shown in Fig. 2 the effective drag becomes non-linear. For sufficiently large applied step forces, the dissipation essentially saturates the mass current, $J_S = \rho_S v_S$, making it nearly independent of driving force. This saturation phenomena is revealed in Fig. 5 where a sequence of flow transients induced by various applied DC potentials is displayed. The near parallelism of the initial slopes indicates the degree to which the current has saturated. We have examined both the temperature and tube size dependence of this maximal current. Fig. 6 shows the maximal current, J_{sc}, raised to the 2/3 power vs. T/T_c for the 177 μm radius tube. It is apparent that the bulk of the temperature dependence of J_{sc} is described by

$$J_{sc} = A(1-T/T_c)^{3/2} \qquad (2)$$

Allowing for a small ($\sim$1%) uncertainty in our determination of T_c we have determined the coefficient A in Eq. 2 by making a least squares fit of $J_{sc}^{2/3}$ vs. T/T_c. The results are summarized in Table I.

Table I

Tube Radius (micron)	A (g-cm^{-2}-sec^{-1})
126	.22
177	.18
227	.18

The above picture of large driving forces leading to saturated supercurrents with very simple temperature dependence ignores a number of small scale effects which we now briefly mention. First, the saturation we observe is not complete. Close examination of Fig. 5 shows the initial sections of "saturated" transients to be slightly curved. This simply means that larger initial pressure heads give slightly greater mass current. By examining this effect quantitatively[12] we find that doubling the maximum pressure head applied (16μ Bar) would produce less than a 10% increase of the mass current. Interestingly, this effect becomes smaller nearer T_c. One possible source of this lack of complete saturation would be a small amount of normal component motion in the flow channel. In addition to this there are also small deviations from the simple $(1 - T/T_c)^{3/2}$ temperature dependence. Various dips and rises (of a few percent) about a straight line were frequently seen and were often reproducible from run to run and in the various U-tubes. Generally, the larger the pressure head used to initiate the flow, the smaller these deviations were.

Lastly, the maximal currents showed some history dependence. For example, on occasion the size of the currents would be different on cooling as opposed to warming, even though the rates of temperature change were kept small. These effects rarely exceeded 5-10%.

The simplest theoretical model that can be applied to these observations is the so-called depairing critical current originally calculated by Fetter[13] in the Ginzburg-Landau regime. Extension of this theory to lower T has been performed.[14] The central result is that due to a depression of the superfluid density with increasing superflow the mass current, $\rho_s v_s$, exhibits a maximum value. Over a wide range of temperature (to below $T/T_c = 0.5$) the critical current J_{sc} follows $(1 - T/T_c)^{3/2}$. The prefactor (to be compared with A in Table I) is 0.30 assuming weak coupling. Thus our observed current is about 60% of this value. Other investigators[3,7] have also found a value for A well below the weak coupling value at saturated vapor pressure. The theory assumes that there are no transverse variations of the order parameter, an assumption which may not be valid in our rather large geometry.

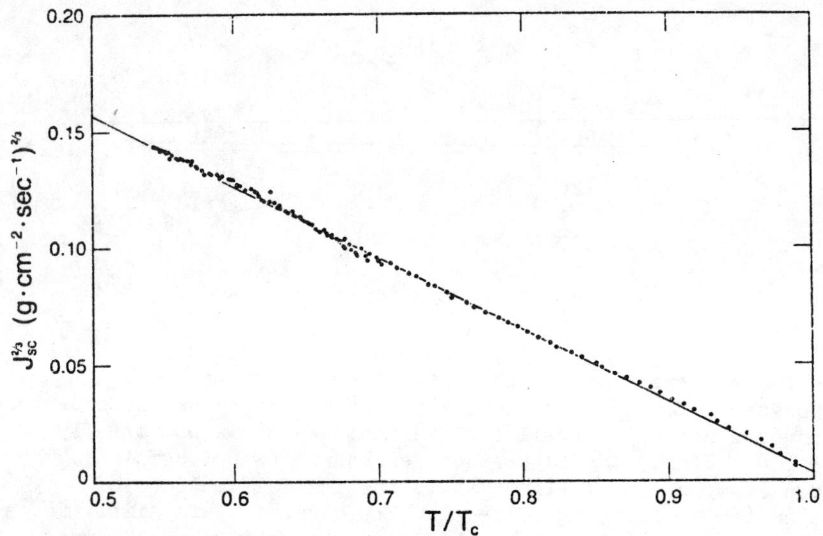

Fig. 6.   The maximal current J_{sc} (raised to the 2/3 power) vs. T/T_c for the 177 µm radius flow tube. All points result from flow induced by 550v step.

In summary, our experiments on B-phase flow at saturated vapor pressure have revealed an interesting variety of dissipative effects. At low velocity the linear dissipation observed may be due, if Brand and Cross are right, to a purely two-fluid effect involving the second viscosity. This dissipation, vanishingly small in ^{4}He, may also be present in a wide variety of other flow experiments and it would be interesting to see if it is consistent with the levels of dissipation seen in these other works.[3,4,7]   It remains to be seen then, whether one can indeed detect persistent flow in the B-phase without this effect or some other causing undue dissipation.

At larger velocities we have found the linear losses give way at a critical velocity to non-linear effects which eventually become sufficiently large to nearly saturate the current. The origin of the critical velocity and the subsequent saturation is not clearly understood. It is not known, for example, if they are distinct effects or are together part of the depairing mechanism outlined above. The approximate geometry independence of both may be a clue that they are, in fact, related.

It is a pleasure to acknowledge the assistance of Professor Richard Packard, without whom these experiments could not have been performed. Greg Swift was also instrumental. This work was supported by NSF Grant No. DMR-79-25098.

REFERENCES

a) Present address: Department of Physics and Astronomy
   Williams College
   Williamstown, MA   01267

1. H. Brand and M. C. Cross, Phys. Rev. Lett. $\underline{49}$, 1959 (1982)
2. J. P. Eisenstein, G. W. Swift, and R. E. Packard, Phys. Rev. Lett. $\underline{43}$, 1676 (1979)
3. A. J. Dahm, D. S. Betts, D. F. Brewer, J. Hutchins, J. Saunders, and W. S. Truscott, Phys. Rev. Lett. $\underline{45}$, 1411 (1980)
4. J. M. Parpia and J. D. Reppy, Phys. Rev. Lett. $\underline{43}$, 1332 (1979)
5. J. P. Eisenstein and R. E. Packard, Phys. Rev. Lett. $\underline{49}$, 564 (1982)
6. J. P. Eisenstein, G. W. Swift, and R. E. Packard, Phys. Rev. Lett. $\underline{45}$, 1569 (1980)
7. M. T. Manninen and J. P. Pekola, Phys. Rev. Lett. $\underline{48}$, 812 (1982)
8. J. P. Eisenstein, G. W. Swift, and R. E. Packard, Phys. Rev. Lett. $\underline{45}$, 1199 (1980)
9. R. T. Johnson, R. L. Kleinberg, R. A. Webb, and J. C. Wheatley, J. Low Temp. Phys. $\underline{18}$, 501 (1975)
10. C. N. Archie, T. A. Alvesalo, J. D. Reppy, and R. C. Richardson, Phys. Rev. Lett. $\underline{43}$, 139 (1979)
11. John Hook, unpublished remarks at this conference.
12. J. P. Eisenstein, thesis, Berkeley, 1980, unpublished.
13. A. L. Fetter, in Quantum Statistics and the Many Body Problem, eds. S. B. Trickey, W. Kirk, and J. Dufty (Plenum, New York, 1975)
14. D. Vollhardt, K. Maki, and N. Schopohl, J. Low Temp. Phys. $\underline{39}$, 79 (1980)

A SEARCH FOR THE AC JOSEPHSON EFFECT IN SUPERFLUID ^{3}He

O.V. Lounasmaa, M.T. Manninen, S.A. Nenonen, J.P. Pekola, R.G. Sharma, and M.S. Tagirov

Low Temperature Laboratory, Helsinki University of Technology, 02150 Espoo 15, Finland

Experiments searching for the AC Josephson effect in superfluid ^{3}He, analogous to phenomena observed in superconducting microbridges, have been performed. Measurements of this type test the fundamental properties of the superfluid described by a macroscopic wave function. Small orifices were employed as the weak link between two reservoirs filled with ^{3}He.

^{3}He is a better candidate for observing the Josephson effect than He-II because of its much longer coherence length ξ_o (0.076 μm at P = 0), which makes the preparation of suitable weak links easier. Two basically different weak link geometries were used. The first type consisted of polycarbonate membranes (5 μm or 10 μm thick) containing a large number of cylindrical channels. The channel diameter varied from 0.03 μm to 0.08 μm. The second type of weak link was a 8 μm-diam. single hole, prepared in a 6 μm thick Mylar foil.

An aluminized Mylar diaphragm divided our ^{3}He chamber into two compartments, and formed, with electrodes on both sides, a double capacitor. One side of this was employed for producing a pressure difference, ΔP, by applying a DC bias voltage, and the other side was employed for measuring the displacement of the diaphragm. A simple capacitive transducer was used to produce an AC pressure modulation across the weak link.

Simple model calculations suggest that steps in the flow characteristics should be observable with our resolution when, $\Delta P = (h\rho/2m) \cdot nf$, where m is the mass of the ^{3}He atom, ρ is the density of ^{3}He, and f is the frequency of the modulation. Several types of experiments were performed. No sign of the behavior expected for the Josephson effect was observed in any of the measurements.

Our data does not prove that the AC Josephson effect does not exist in ^{3}He but shows that its observation is not possible with our resolution and straightforward technique. There may be reasons why the membranes with large number of channels are not suitable; phase slippage in each channel separately may smear out the possible step structure. A single small hole is a more ideal weak link, but the resolution becomes the limiting factor because the total current is proportional to the area of the hole.

PRECISE ZERO-SOUND VELOCITY
MEASUREMENTS IN THE A AND A_1
PHASES OF ^{3}HE NEAR T_c *

R.F. Berg and G.G. Ihas

Department of Physics
University of Florida
Gainesville, Florida 32611

We have made phase-velocity change measurements for 5 and 15 MHz zero sound within a few microkelvin of the ^{3}He superfluid transition, T_c, at 31.1 bar. Our sample cell contains a quartz transducer pair separated by the former for an RF coil thus enabling simultaneous sound and NMR measurements to be performed on the same volume of ^{3}He. The known shifts of the transverse NMR resonant frequency in the A^1 and Al^2 phases enable us to measure temperatures relative to T_c to within 50 nK in the A-phase. Our sound spectrometer, which uses phase-sensitive detection of 2 microsecond pulses, gives us a resolution of about 5 parts per million. Data were taken every two minutes while drifting at the rate of 2 nK per second toward T_c.

Our results show no marked feature at $h\omega = 2\Delta(T)$. However, there is a marked reduction in the slope of dc/dT upon passing from the A-phase into the Al-phase. Also, for at least $T_c - T < 3$ micro-kelvin, dc/dT is greater for the 5 MHz data than for the 15 MHz data. This difference may be due to either pair-breaking effects or the less-developed zero-sound character of the 5 MHz sound.

1. A.I. Ahonen, M. Krusius and M.A. Paalanen, Journ. Low Temp. Phys. __25__, 421 (1976)

2. D.D. Osheroff and P.W. Anderson, Phys. Rev. Lett. __33__, 686 (1974)

*Partially supported by NSF Grant DMR-8006929.

ZERO-SOUND MEASUREMENTS IN SUPERFLUID ^{3}He

M.W. Meisel, B.S. Shivaram, B.K. Sarma, J.B. Ketterson
and W.P. Halperin
Department of Physics and Astronomy and Materials
Research Center, Northwestern University,
Evanston, Illinois 60201

Pulsed transmission and acoustic impedance techniques have been
used to perform zero-sound measurements in superfluid ^{3}He. In the A-
phase, a new collective mode of the order parameter has been observed
and is identified as the super-flapping mode. Anomalies have also
been observed at the A-phase to B-phase transition. These results
are believed to be associated with the changing order para-meter
boundary conditions at the transducer-liquid interface. In the B-
phase, the squashing mode has been studied in magnetic fields up to
160 mT. The J_z = 0 state is observed to shift to lower temperatures
in a field. Our observations are discussed with respect to recent
theoretical expressions. In our pulsed experiments, the pair-
breaking attenuation has been completely resolved for P<2bar where
anamolous group velocity shifts near 2Δ are also observed. Finally
low pressure, low frequency attenuation oscillations have been
resolved and do not have any straightforward relationship with the
phase velocity.

COLLECTIVE MODES IN ^{3}He-B

B.S. Shivaram, M.W. Meisel, B.K. Sarma, W.P. Halperin
and J.B. Ketterson

Department of Physics and Astronomy and Materials
Research Center, Northwestern University,
Evanston, Illinois 60201

A high resolution acoustic impedance technique has been used to
investigate the order parameter collective modes in superfluid ^{3}He-B
in magnetic fields up to 160 mT. The Zeeman splittings of the J = 2
real squashing mode have been investigated. In the T→0 limit, the
magnetic splittings are approximately linear; however, a new splitting of the J_z = 0 mode is observed. The doublet structure of the
J_z = 0 state is not understood. In the T→T_c limit, the magnetic
splittings are extremely non-linear. These results are consistent
with recent theoretical work and offer a possible method to evaluate
Fermi liquid parameters. In addition, a new mode has been observed
between the pair-breaking edge and the squashing mode. The position
of the mode at $T/T_c \sim 0.52$ is, a = 1.54 for $\omega = a\Delta_{BCS}(T)$. The temperature of this new mode appears to have a weak magnetic field depen-
dence but does not split in fields up to 160 mT.

^{3}HE IN VYCOR GLASS: A NEW SUPERFLUID PHASE?*

Gary Ihas and Greg Spencer

Department of Physics
University of Florida
Gainesville, Florida 32611

We report indirect measurements on the behavior of ^{3}He inside Vycor glass which indicates the possible existence of a superfluid phase there. The observations were made in an epoxy NMR/flow cell with the ^{3}He maintained at 2.89 bar and 28.4 mT. Afterwards, the λ transition in the cell was measured showing that the Vycor (which was epoxied into the flow channel) had a characteristic dimension of 56Å. There was thus no flow path parallel to the Vycor. An isochoric bellows pump, to be described elsewhere, was used to establish flow. An NMR coil mounted on the flow tube connecting to the Vycor allowed measurement of the B-phase flare-out texture.

While the cell, filled with bulk B-phase liquid, was warming towards T_c the pump was first stroked to produce a constant mass flow for 3 mins., and then allowed to freely relax to its initial condition. However, rather than return to the original equilibrium position C_0, the pump relaxed to a new equilibrium C_1 ($\Delta C=.090pf$)in about 2 min.. After 10 min., the pump was stroked a second time (starting from C_1). At the beginning of this stroke the entire sample immediately warmed through bulk T_c. After being held in position for over 1 hour with $T > T_c$, the pump was allowed to relax, returning again to C_1. We interpret this result as the passage of ^{3}He through the Vycor during the first stroke. During the pump relaxation, the cell warmed through the depressed T_c in the Vycor, preventing 3 millimoles of transferred mass from returning to its original side of the Vycor. The pump was held away from C_0 by this trapped mass. Had the coincidence of passing through the depressed T_c (ΔT_c = -0.18 mK) not occurred when it did, the pump would have simply returned to its original equilibrium. If we assume zero mass transfer and use accepted values of the thermodynamic parameters of ^{3}He, the liquid in the NMR region should have cooled causing textural broadening of the spectrum. This was not observed. Alternately, if one assumes normal liquid can pass through the Vycor during the first stroke, the amount which could be transferred through this impedance in the given time is 100 times too small to hold the pump at C_1. Also, there is no mechanism to trap the normal liquid on one side if it can be transferred. Thus, one would expect the pump to eventually return to C_0, which it did not. We believe this result indicates the existence of superfluid ^{3}He in the Vycor. As the BW state is 3 dimensional, it is presumed to be suppressed due to the restricted geometry. One candidate is the polar phase, which is one dimensional in character.

*Partially supported by NSF Grant DMR-8006929.

DISSIPATIVE FLOW IN ^{3}He-B BELOW THE CRITICAL CURRENT*

V.K. Samalam, Y.L. Liu, and P. Kumar
Department of Physics
University of Florida
Gainesville, Florida 32611

We present a theory of the spatially inhomogeneous fluctuations that can cause supercurrents to decay. The fluctuations are similar to the well known phase slip centers in a one dimensional superconductor in that they are responsible for dissipative flow below the critical current. Their complexity appears from the flow caused anisotropic distortion of an otherwise isotropic order parameter of superfluid ^{3}He-B. We analyze the topography of the order parameter space to find the energy minimum phase slip center. The dissipative effects are however confined to a very narrow region of the current values, $0.95\ j_c < j < j_c$.

*Work supported by a grant from NSF DMR-8006311.

DYNAMICS OF THE HELIUM SOLID-LIQUID INTERFACE

D. Thoulouze, B. Castaing and L. Puech
Centre de Recherches sur les Très Basses Températures, C.N.R.S.,
Cédex 166, 38042 GRENOBLE, FRANCE

ABSTRACT

The very high mobility of the ^{4}He liquid-solid interface for rough parts leads to a Kapitza thermal resistance $R_K \propto T^{-5}$, depending on its crystallographic orientations, in opposition to the classical Khalatnikov behaviour $R_K \propto T^{-3}$ observed for facets. This mobility is limited by an interface mass equivalent to about 1/20 of an atomic layer. A critical slowing down has been observed when approaching a roughening transition.

For ^{3}He, the growth kinetic coefficient is found to be more than two orders of magnitude larger than predicted.

As Metallurgists and Chemists tell us, in ordinary materials, melting and crystal growth kinetics are very often controlled by the dissipation of the latent heat transformation. The growth kinetics also depends on the nature of the interface :
- fast if the interface is rough, by spontaneous formation of kinks and steps ;
- slow if the interface is smooth since then nucleation occurs on crystal defects.

In Helium, the idea (Andreev[1]) was that the interface was always rough, due to the very large zero point motion. Growth and melting occur in a coherent manner, without dissipation, except friction of the interface with elementary excitations.

HELIUM 4

Elements of growth kinetics

The melting pressure is almost constant below 1 K, so that no latent heat is developped. Furthermore, any heat flows very easily through the superfluid liquid. Finally, no consequent excitation, except phonons, remains below about 0.6 K, in either the liquid or the solid.

The first experimentalist (Shalnikov[2], Greywall[3], Simmons[4]) noted that "it was difficult not to obtain single crystals", when growing the solid, with a low density of defects (10^6, 10^7/cm^2). These crystals reach very quickly their equilibrium shape, within some seconds, and melting is very fast, some mm/second for a difference from the melting pressure much less than 1 mbar[5].

This general behaviour makes Helium 4 the ideal case for studying as well the growth kinetics than the properties of an interface or the general trends of roughening transition.

358

Recent experiments[6,7,8] have shown the appearance of three roughening transitions, for three different facet orientations, at 1.1 K, 0.9 K and 0.36 K. At the roughening transition, a facet appears at the free interface for simple cristallographic orientations. When the part in contact with the walls are rough, the meniscus has been observed not to wet[9] the walls.

On growing, the surface of facet is enhanced due to their low mobility, while the curvature of the rounded parts is related to the difference :

$$\delta P = P_{solid} - P_{liquid}$$

through the surface tension $\tilde{\alpha}$:

$$\delta P = \frac{\tilde{\alpha}}{R}$$

When the facet is large enough to occupy all the interface area, further growing can only occur by jumps which restore the rounded parts. On melting, the meniscus only consists in a rounded rough part as a consequence of its much faster kinetics. Thus, in the absence of optical measurement, one has to be careful about the conditions for working either on a facet or on a rough interface.

Dynamics of the interface

In absence of any dissipative phenomena, it becomes possible to observe on the solid-liquid interface all phenomena characteristic of the surface of ordinary liquids. In particular, capillary waves are here crystallization waves. Predicted by Andreev and Parshin[1], they have been evidenced by Keshishev[8]. Exciting the interface by an a.c. electrical field, they observed by laser beam reflexion both the expected dispersion relation in the kHz region :

$$\omega^2 \simeq \frac{\alpha\rho}{(\delta\rho)^2} k^3$$

and a damping due to the excitations.

Another possibility proposed by Castaing and Nozieres[10] consisted in driving the interface under the influence of an acoustic pressure wave. Because of its high mobility, the interface maintains constant pressure (the melting pressure Pm) and nearly no stress is transmitted to the solid : an anomalous transmission of the sound was observed at 1 MHz, almost all the sound intensity being reflected[11].

The question was then : up to what frequencies is the interface mobile ? In the thermal frequencies range, the answer may be given by the knowledge of the Kapitza resistance. The Kapitza conductance is known to be proportional to the phonon transmission rate, τ :

$$R_K^{-1} \propto C_v \cdot c \cdot <\tau>$$

where C_v is their specific heat and c the sound velocity. If the interface is bound or tight to mass transfer, the transmission rate is constant, and the Kapitza resistance is expected to be given by the well-known Khalatnikov-Little formula :

$$R_K T^3 = Cte$$

However, if the interface is mobile, Marchenko and Parshin[12] emphasized that, for perfect melting ($\Delta\mu = 0$), for an oblique incidence of the acoustic wave, the modulation of the interface which follows the velocity modulation of the liquid restores a force on the solid through the surface tension :

$$\tau \propto \tilde{\alpha}^2 \omega^2$$

In that case :

$$R_K T^3 \alpha T^{-2}$$

Experiments performed by Huber and Maris[13] show a clear increase of $R_K T^3$ when the temperature is lowered, by an amount much lower than predicted by Marchenko and Parshin.

Simultaneously, by a different technique, we give further evidence for a $R_K \propto T^{-5}$ variation depending on the crystallographic orientation, in opposition with the classical $R_K \alpha T^{-3}$ behaviours observed for a facet[14,15].

Experimental Arrangement

The cell consists in two cylindrical CuNi tubes between which the crystal is grown at constant pressure (fig. 1). Four Matshushita carbon resistors were disposed in order to measure the temperature gradients produced by the heat flow. Slow temperature drifts and the very low value of the Kapitza resistance made it necessary to use a dynamical method. It consists in slowly moving the interface in the cell and measuring the change in the ratio of two subsequent resistors when the interface crosses their level. In melting the crystal, the temperature difference goes from its "all solid" value to its "all liquid" value. When the interface lies between R_2 and R_3, for instance, the Kapitza resistance gives rise to a temperature jump δT_K.

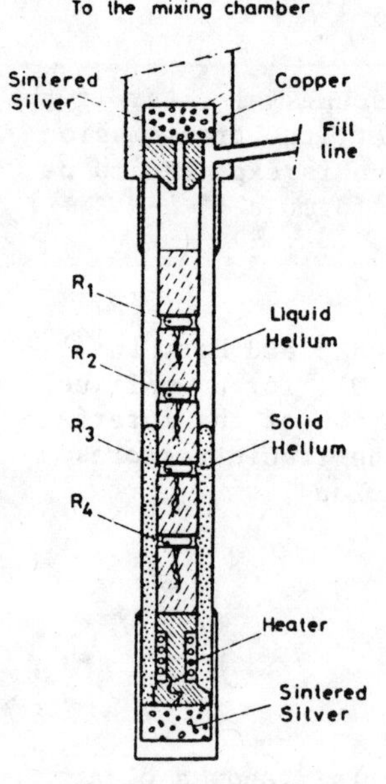

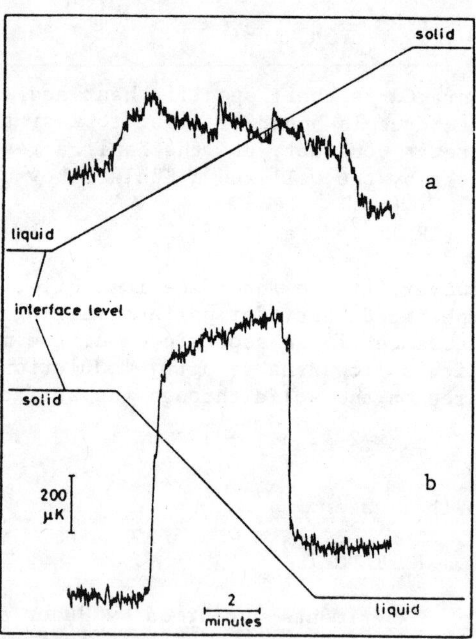

Fig. 1. The experimental cell. The interface level is given by the capitance between the two concentric CuNi tubes.

Fig. 2. An experimental recording of the ratio R_3/R_2 at T = 200 mK corresponding to the temperature difference $\delta T = T(R_3) - T(R_2)$ as a function of the interface positions. The Kapitza temperature jump δT_K is clearly seen.

Fig. 2 shows a typical signal obtained :

a) When growing the crystal. The temperature difference is rather small. Also evident are the jumps of the facet described earlier.

b) On melting, the rough interface resistance is much larger by a factor five at this temperature.

<u>Results and interpretations</u>

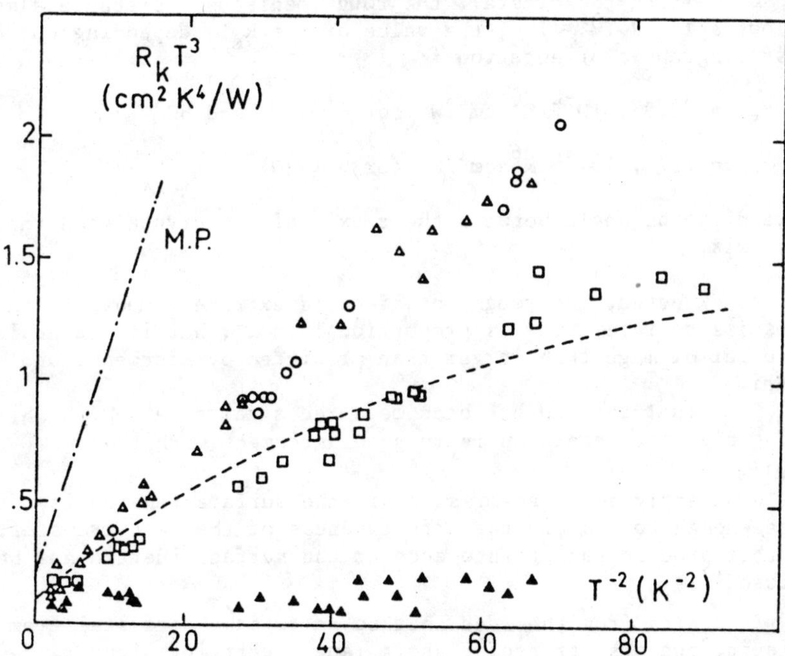

Fig. 3. $R_K T^3$ as a function of T^{-2}
on growing the crystal ▲
on melting the crystal △ $\theta \sim 45°$
o $\theta \sim 80°$
□ $\theta \sim 30°$.

The results are plotted on fig. 3. The minimum values are obtained on growing the crystal :
as expected, $R_K T^3$ is a constant for facets. In that case :

$$R_K T^3 \simeq 0.1 \ cm^2 K^4/W$$

while calculation[13] from Khalatnikov acoustic mismatch theory leads to :

$$R_K T^3 = 0.18 \ K^4 \ cm^2/W$$

Thermalization processes of the phonons in both solid and liquid can explain the difference between those two values[15]. For comparison, the value of the Kapitza resistance between copper and superfluid helium is known to be a thousand times larger :

$$R_K T^3 = 100 \ cm^2 K^4/W$$

362

On melting the crystal, the rough meniscus resistance almost follows a law $R_K T^3 \propto T^{-2}$, the value of $\gamma_5 = R_K T^5$ depending on the cristallographic orientation :

$$\gamma_5 = 2.8 \cdot 10^{-2} \ K^6 \ cm^2/W \quad \text{for } \theta \sim 45° \text{ and } 80°$$

$$\gamma_5 = 1.7 \cdot 10^{-2} \ K^6 \ cm^2/W \quad \text{for } \theta \sim 30°$$

where θ is the angle between the c axis of the crystal and the cell axis.

As expected, the rough interface is extremely mobile ; the transmission rate $< \tau >$ is proportional to ω^2, but its value is one order of magnitude larger than predicted by Marchenko and Parshin.

Note that this R_K has been measured a third way which while giving a poorer precision is in good agreement with these results[16].

Apparently the transmission via the surface tension is not large enough to explain the effectiveness of the heat transfer. An other process taking into account the surface inertia has been proposed[17].

It starts from the idea that so as to make a crystal from a liquid, one must transofmr short range order into long range order ; this occurs through small displacements of atoms, which give rise to a surface kinetic energy proportional to J^2, where J is the mass flow across the interface :

$$\Delta \mu \propto \frac{\sigma}{\rho_L \rho_s} \frac{\partial J}{\partial t} \Rightarrow \tau \propto \omega^2$$

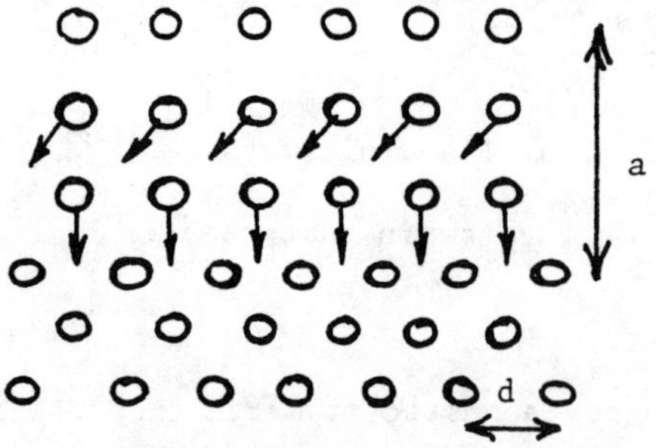

A simple idea may be pictured in the following way :
If crystallization occurs on a distance a, the time of crystallization for an interface moving at a velocity v_{int} is :

$$\tau = \frac{a}{v_{int}}$$

The velocity of an atom moving over an interatomic distance d is :

$$v_{at} = \frac{d}{\tau_{at}} = \frac{d}{a} v_{int}$$

It gives rise to a supplementary surface kinetic energy

$$E_c = \rho_s a \ v_{at}^2 = \rho_s d \left(\frac{d}{a}\right) v_{int}^2$$

and an interface mass per unit area :

$$\sigma \;=\; \rho_s \, d \left(\frac{d}{a}\right)$$

A fraction of an atomic layer $\dfrac{d}{a} \sim \dfrac{1}{20}$ is large enough to explain the supplementary transmission rate which is compared on fig. 4 with the surface tension process, for which $\sigma = 0$.

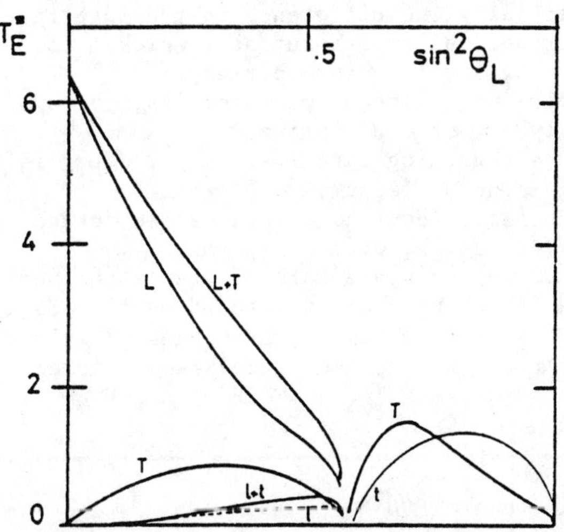

Fig. 4. Energy transmission rate as a function of the incident angle θ. L and T refer to longitudinal and transverse phonons with $\sigma = 1/20$, ℓ and t with $\sigma = 0$.

This interpretation leads to some particular points :

- with $\sigma \neq 0$, the transmission rate, even at normal incidence is finite ;

- the dispersion relation for crystallization waves becomes linear at high frequency, $\omega = C_o K$.

Roughening transition : critical slowing down

Although the kinetics of the facets may be qualified of slow in comparison to that one of the rough parts, we have seen that, as a general feature, the helium 4 growth dynamic is faster by orders of magnitude than for any ordinary material. This makes possible the study of the roughening transition : the crystal

364

shapes may be investigated by optical measurements[18], and its dynamics by thermal methods[19].

We have used the method previously described to measure the relaxation of the interface when approaching a roughening transition.

The idea is to measure the time of return to equilibrium of a facet. A perturbation (growing, then melting) eliminates the facet which, then, builds again to its initial size with a time constant which may diverge at the roughening temperature.

It is essentially the difference in the interface thermal resistances of rough and smooth surfaces which allows to determine their respective area within some percent.

In fact, with this type of measurements, the system we measure is doubly out of equilibrium :
- first, a constant growing rate ($\sim$ 1 mm per hour) is necessary to impose a large size of facet ;
- second, a heat flow, about 25 μW, is needed for the experiment, to get enough precision, with $\Delta T \sim$ 2 mK.

It may be useful to state more precisely the perturbation supplied to the facet which are stretched on fig. 5. It consists in closing a valve at room temperature, thus growing the crystal by about 50 mm^3 ; the facet size increases, by forced nucleation,

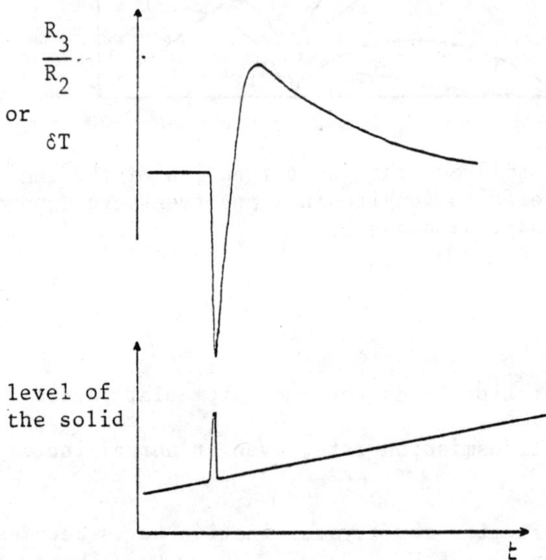

Fig. 5. The level of the solid represents the perturbation applied to the meniscus, as a function of time.
The higher curve is the thermal answer ; the last relaxation part is studied.

while moving forward : the temperature difference δT between solid
and liquid decreases. Then the valve is open ; the dead volume
which is decompressed corresponds to a volume of solid larger than
that necessary to recover the initial size of the facet, since the
facet advanced during that time. Thus the crystal melts and the
facet size becomes smaller than it was at the beginning : δT
increases, getting larger than originally.

Then, the interface relaxes to its stationnary size, as well
as δT. The measurement which is shown in fig. 5 corresponds to
this relaxation. The inverse relaxation times measured in that way
are reported on fig. 6 as a function of the temperature. In fact,

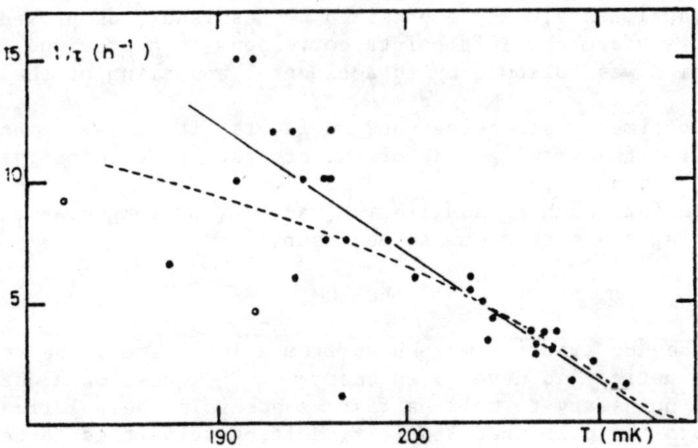

Fig. 6. The inverse relaxation time as a function of temperature
     o   before annealing
     •   after   annealing.

τ varies between 1 minute at 150 mK to 38 minutes at 211 mK. We
interpret this as a critical slowing down when approaching the
roughening transition. Also plotted on the figure are two theoretical curves :
- the continuous straight line corresponds to a Landau type model
  for a second order phase transition ;
- in the discrete Gaussian SOS model[20], ξ, and then τ, diverge as

$$\tau \; \alpha \; \exp \frac{b}{\sqrt{(T_R - T)/T_R}}$$

The dotted line correspond to $b = 0.4$ and $T_R = 215$ mK.
    This value of T_R strongly differs from the value recently
obtained by Wolf and Balibar[18]. In the present status of our knowledge two explanations are possible :

366

- either we are in presence of two different transitions. In this
respect, note that the orientations of these facets respective to
the c axis are equal within the experimental uncertainties ;
- or the linear extrapolation in our diagram is far below the
actual roughening transition as suggested by plotting the exact
result presented by Saam[21]. In that case relaxation times would
become so long nearer the transition that it would be hopeless
to perform real equilibrium measurements.

In conclusion, we have proved for the first time the existence
of a critical slowing down at the approach of the roughening temperature. Some remarks may be important :
- the roughening temperature appears to depend on the quality of
the crystal, as already emphasized by Keshishev, as proved also
by fig. 6 where the left points correspond to a previous experiment which was followed by an accidental annealing of the sample
up to 0.7 K ;
- when experiments are being done on facets, it is very important
to control the growing rate of the crystal, at a velocity lower
than 0.5 mm/hour ;
- the time for reaching equilibrium may be very long near the
roughening transition, up to one hour !

HELIUM 3

If the ^{4}He kinetics was known from a long time to be very fast,
the ^{3}He kinetics had never been observed. The question arose when
it became necessary to melt as fast as possible the polarized solid
^{3}He to form the polarized liquid[22]. Melting velocities faster than
1 cm/sec were obtained for $\Delta p \lesssim 20$ bars, without precise measurement
of the experiment.

The only prediction was due to Andreev[1] ; according whom at
finite temperature, the mobility of the interface was limited only
by dissipative effects due to its interactions with the Fermi excitations : the kinetic growth coefficient K is defined as

$$V_{int} = K \Delta \mu$$

where $\Delta \mu$ is the chemical potentiel difference between liquid and
solid and V_{int} the velocity of the boundary.

The dissipated energy is :

$$\dot{E} = n\, V_{int}\, \Delta \mu = n\, \frac{V_{int}^2}{K}$$

where nV is the flow of atoms.

The "Doppler effect" type energy dissipated by the interface
when moving is

$$nV_{int} \cdot E_F\, \frac{V_{int}}{v_F} = np_F\, V_{int}^2$$

$$K \propto \frac{1}{p_F}$$

Other reasoning[23] may lead to take into account multiplicative terms in T/T_F or $(T/T_F)^2$.

Experimentally, $\Delta\mu$ is given by a pressure drop Δp such that :

$$\Delta\mu = m(\frac{1}{\rho_L} - \frac{1}{\rho_s})\ \Delta p$$

$$= K^{-1}\ V_{int}$$

$$\delta P = \frac{\rho^2}{\delta\rho}\ \frac{P_F}{m}\ V_{int}$$

Numerically

$$\delta P(Pascal) = 4\ .\ 10^5\ V\ (m/s)$$

Then a pressure drop δp = 0.1 bar was expected to lead to interface velocity of the order of 1 cm/sec.

The experiment[24] was done in the cell described previously at the minimum of the melting curve for commodity reasons :
- the latent heat L is null ;
- the problems of blocking the capillary are avoided.

In fact, the melting was carried out by opening a valve at room temperature and the velocity of the interface was limited by the impedance of the filling capillary to a value of 0.6 cm/sec.

The pressure drop was measured through a capacitive gauge located in a secondary cell connected on the top of the main cell. It may be emphasized that such experiments on ^{3}He are much more difficult than those performed on ^{4}He : the solid ^{3}He has a tendancy to grow everywhere except where we should like it ; it is only by a very slow transfer, typically 24 hours at temperatures lower than the minimum of the melting curve and slightly heating the bottom of the cell that it becomes possible to get what we think to be a rather good crystal ; its thermal conductivity is comparable to an Helium 4 one at the same temperature.

In those conditions, it was possible to stabilize the temperature at the minimum of the melting curve, as measured with the pressure gauge, slightly heating the bottom of the cell to favour stability of the system.

The results of a decompression are shown on fig. 7. As may be seen, no pressure drop is measured when the level of the interface goes down, except the hydrostatic pressure decrease due to lowering of the interface level. The sensitivity of the measurement is about 0.1 mbar.

Thus, the velocity of the interface appears to be faster than expected by more than two orders of magnitude.

The first question that may arise concerns the nature of the crystal, and of the interface itself. The experimental conditions to grow the crystal were optimized as exposed and the results were reproducible. It is difficult to characterize the interface as well as the ^{4}He one.

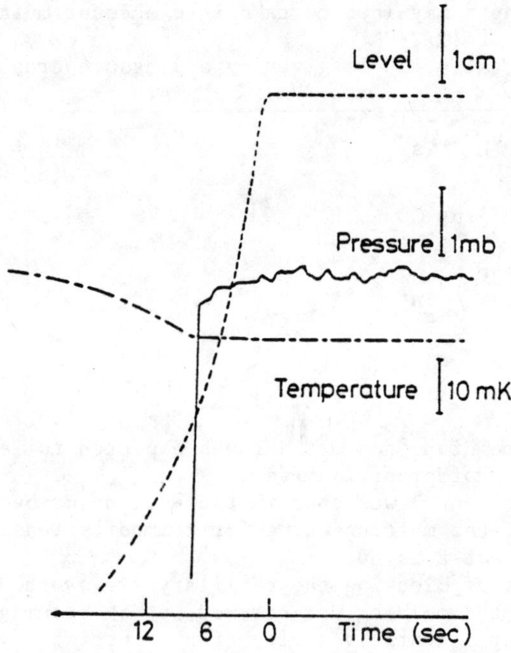

Fig. 7. Characteristic diagram for a decompression at the minimum of the melting curve:level of the solid, pressure, temperature. The final pressure decrease corresponds to the hydrostatic pressure.

Note that in our case no process seems to favor the formation of dendrites. The temperature gradient stabilises the interface position and the ^{4}He impurity concentration is very low (less than 30 ppm). Just a slightly higher temperature was triggering instabilities which were easy to detect. Furthermore, the bottom heating was suppressed some seconds before decompression.

Another point which could help in the sense of these results is the fact that the liquid ^{3}He is not degenerate at 0.3 K. This may increase the mobility, but seems not to justify such a difference.

We may however notice that Andreev estimation refers to the ballistic regime and then could be an over estimation as remarked on general grounds by Bowley and Edwards[25]. We thus need a more precise calculation taking into account the strong interactions between quasi-particles.

CONCLUSION

The helium 4 interface is characterized by a high mobility up to thermal frequencies, of the order of $\theta_D/10$, which leads to an increased Kapitza resistance $R_K T^5 \sim$ Cte depending upon the cristallographic orientation. This mobility is limited by an interface mass of the order of 1/20 atomic mass. ^{4}He is a very good opportunity for studying the roughening transition.

The growth kinetics coefficient of Helium 3 is more than two orders of magnitude faster than expected. In this respect, it remembers the ^{4}He case and Kapitza resistance experiments are being performed for comparison with the Helium 4 kinetics.

REFERENCES

1. A.F. Andreev and A. Ya. Parshin, Sov. Phys. JETP $\underline{48}$, 763 (1978).
2. A.I. Shal'Nikov, Sov. Phys. JETP $\underline{14}$ (1961), n° 4.
3. D.S. Greywall, Phys. Rev. A3 ($\overline{1970}$), vol. 3, n° 6.
4. B.A. Fraas, S.M. Heald and R.O. Simmons, J. Cryst. Growth $\underline{42}$, 370 (1977).
5. B. Castaing and P. Nozières, J. Physique $\underline{41}$, 701 (1980).
6. J.E. Avron, L.S. Balfour, C.G. Kuper, J. Landau, S.G. Lipson and L.S. Schulman, Phys. Rev. Lett. $\underline{45}$. 814 (1980).
7. S. Balibar and B. Castaing, J. Physique Lettres $\underline{41}$, 329 (1980).
8. K.O. Keshishev, A.Ya. Parshin and A.B. Babkin, Sov. Phys. JETP $\underline{53}$, 362 (1981).
9. S. Balibar, D.O. Edwards and C. Laroche, Phys. Rev. Lett. $\underline{42}$, 782 (1977).
10. B. Castaing and P. Nozières, J. Physique $\underline{41}$, 701 (1980).
11. B. Castaing, S. Balibar et C. Laroche, J. Physique $\underline{41}$, 897 (1980).
12. V.I. Marchenko and A.Ya. Parshin, JETP Lett. $\underline{31}$, 767 (1980).
13. H.J. Maris and T.E. Huber, J. Low Temp. Phys. $\overline{48}$ (1982).
14. L. Puech, B. Hebral, D. Thoulouze and B. Castaing, J. Physique Lettres $\underline{43}$, 809 (1982).
15. L. Puech, Thesis, University of Grenoble (1983).
16. P.E. Wolf, S. Balibar and D.O. Edwards, J. Low Temp. Phys., to be published.
17. L. Puech and B. Castaing, J. Physique Lettres $\underline{43}$, 601 (1982).
18. S. Balibar, P.E. Wolf, D.O. Edwards, this issue.
19. L. Puech, B. Hebral, D. Thoulouze and B. Castaing, J. Physique Lettres $\underline{44}$, 159 (1983).
20. J.D. Weeks, Ordering in strongly fluctuating condensed matter systems, Ed. T. Riste, Plenum Press 1980, p. 293.
21. C. Jayaprahash, W.F. Saam and S. Teitel, preprint.
22. B. Castaing and P. Nozières, J. Physique $\underline{40}$, 257 (1979).
    G. Schumacher, D. Thoulouze, B. Castaing, Y. Chabre, P. Segransan and J. Joffrin, J. Physique Lettres $\underline{40}$, 143 (1979).
23. P. Nozières and M. Cross, Private Communications.
24. L. Puech, B. Castaing, D. Thoulouze , to be published.
25. R.H. Bowley, D.O. Edwards, J. Physique, to be published.

UNIVERSAL FEATURES OF ROUGHENING AND FACET FORMATION IN CRYSTALS; APPLICATION TO ^{4}He

W. F. Saam, C. Jayaprakash and S. Teitel
Department of Physics, Ohio State University, Columbus, Ohio 43210

ABSTRACT

An explicit connection between roughening and facet formation is made, showing that there is a universal jump in the crystal surface curvature at the roughening transition. A hyperscaling argument provides facet sizes exhibiting universal temperature dependence near the roughening transition. Within a model for the a-facet of hcp ^{4}He calculations yield additional information, including detailed facet shapes and a universal exponent governing crystal shapes near a facet. Earlier work on facet formation in ^{4}He is briefly reviewed.

INTRODUCTION

In this paper we discuss the problem of facet formation in crystals. An explicit connection between roughening and facet formation is made, resulting in a prediction of a universal jump in the curvature of a crystal surface at the temperature (roughening temperature) T_R at which the facet begins to form. Below but near T_R the facet size has a universal temperature dependence. In addition, the shape of the crystal near the edge of a facet is found to be governed by a universal exponent, reflecting an underlying Pokrovsky-Talapov[1] transition in the surface free energy. These predictions emerge from our recent work[2] on solid-on-solid models. The focus here will be on a generalization of these earlier results applied to the a-facets of hcp solid ^{4}He. The generalization is expected to be appropriate for the wider class of facets possessing a rectangular symmetry. It is based on the six-vertex model[3,4] for antiferroelectrics.

To put the present work in perspective, we will begin with a short review of earlier work on solid ^{4}He, followed by a brief explication of the theory of equilibrium crystal shapes. We then present a simple argument leading to a universal discontinuity in crystal surface curvature at a roughening transition. Subsequently, we introduce the model for rectangular facets and discuss in detail its application to the a-facets of hcp ^{4}He. We conclude with further remarks on the limitations of the model, other applications of our ideas, and the possibility of first-order transitions associated with crystal shapes.

REVIEW OF HELIUM WORK

The current high interest in the surface properties of helium crystals began with the work of Andreev and Parshin,[5] who suggested that the surface of solid ^{4}He would be atomically rough even at $T = 0$ and would thus support melting-freezing waves roughly similar to capillary waves on the surface of a liquid. Shortly thereafter such

waves were observed by Keshishev _et al._,[6] and it appeared that
Andreev and Parshin's suggestion might be correct. However, stable
facets of hcp ^{4}He were soon observed by Landau _et al._[7] That roughening was involved here was soon suggested by Balibar and Castaing.[8]
At about the same time the c-facet and what may be the a-facet were
observed by Avron _et al._[9] to appear (with the lowering of T) at well-
defined temperatures which they identified as roughening temperatures.
The disappearance of the c-facet was later studied by Keshishev _et
al._[10] Roughening of a third facet has been observed by two groups.[11]
At this point it is consistent with experimental observations to suggest that low index facets of hcp ^{4}He possess roughening temperatures
$T_R > 0$, but that high index facets may have $T_R = 0$. Quite recently,
however, Fisher and Weeks[12] have presented rather general arguments
leading to the conclusion that all stable faces of any crystal, classical or quantum, will have non-zero roughening temperatures. Further
experimental work, at temperatures lower than those studied to date,
is needed in order to clarify this point for ^{4}He.

THEORY OF EQUILIBRIUM CRYSTAL SHAPES

This theory begins with the observation that the shape of a
crystal is determined from minimization of the surface energy

$$\Omega = \int d\vec{x}\,[f(\vec{s}) + 2\lambda z(\vec{x})] \quad , \tag{1}$$

in which $z(\vec{x})$ gives the height of the surface with respect to the
plane $\vec{x} = (x,y)$, and $f(\vec{s}) = \alpha(\vec{s})\sqrt{1 + |\vec{s}|^2}$, where $\alpha(\vec{s})$ is the sur-
face tension depending, in general, on the slope $\vec{s} = \nabla z(\vec{x})$. In
Eq.(1), 2λ is a Lagrange multiplier included to account for the
constraint of constant volume. The minimization leads to[13] the
expression

$$f(\vec{s}) = \lambda[z(\vec{x}) - \vec{x} \cdot \vec{s}] \quad . \tag{2}$$

Rewritten in the form

$$\alpha(\vec{s}) = \frac{\lambda[\vec{e}_z - \vec{s}] \cdot \vec{r}}{\sqrt{1 + |\vec{s}|^2}} \quad , \tag{3}$$

this result provides the Wulff construction. Eq.(3) describes a
family of planes found by constructing a polar plot ($\vec{s}$ is expressed
in terms of angle variables) of $\alpha(\vec{s})$, drawing a radius vector to
each point of the plot, and constructing a plane perpendicular to the
tip of the vector. The inner envelope of all such planes is the
equilibrium crystal structure.[14]
If $\alpha(\vec{s})$ possesses cusps as a function of angle, these will
lead to facets in the equilibrium shape. Landau[15] was the first to
understand the physical origin of cusps. The amplitude of the cusp
is essentially the energy per unit length of a step on the surface.
A sequence of widely separated steps forms a facet at a slight angle
θ from the original facet. The energy of the stepped configuration
is independent of whether or not the steps go up or down and thus is

proportional to $|\theta|$; hence the cusp. It is useful to realize that the size of a facet is proportional to the associated step energy.[15] It is quite helpful to regard the step energy as the surface tension of the one-dimensional surface separating two "phases" of the surface consisting of flat planes displaced normally to one another by the step height. When this surface tension disappears the phases become indistinguishable and the surface wavy, i.e., rough.

The Wulff construction in the forms (2) or (3) will be inconvenient for us here. A Legendre transform of $f(\vec{s})$, first used by Andreev,[16] will be most convenient. Let $\vec{\eta} = \partial f(\vec{s})/\partial \vec{s}$. The Legendre-transformed free energy is the

$$\tilde{f}(\vec{\eta}) = f(\vec{s}) - \vec{\eta} \cdot \vec{s} \quad . \tag{4}$$

From Eq.(2) it follows that $\vec{\eta} = -\lambda\vec{x}$ for the equilibrium shape. Combination of this fact with Eqs.(2) and (4) yields

$$\lambda z(\vec{x}) = \tilde{f}(-\lambda\vec{x}) \quad . \tag{5}$$

One may thus compute the crystal shape directly from knowledge of $\tilde{f}(\vec{\eta})$. For later reference we pause to define a curvature tensor

$$\varkappa_{ij} = \left| \frac{\partial^2 \lambda z(\vec{x})}{\partial(\lambda x_i)\partial(\lambda x_j)} \right|_{\vec{x}=0} = \left| \frac{\partial^2 \tilde{f}(\vec{\eta})}{\partial\eta_i\partial\eta_j} \right|_{\vec{\eta}=\lambda\vec{x}=0} \tag{6}$$

for the center of the face of interest, chosen to be at $z(\vec{x}=0)$.

ROUGHENING AND THE UNIVERSAL DISCONTINUITY IN CURVATURE

The crucial results from the standard theory of roughening[17] of an infinite crystalline surface are couched in terms of the behavior of the correlation function

$$g(\vec{x} - \vec{x}') = \langle [z(\vec{x}) - z(\vec{x}')]^2 \rangle \tag{7}$$

for large values of $\rho = |\vec{x} - \vec{x}'|$. In particular, for a (100) surface of a simple cubic lattice,

$$g(\vec{\rho}) \underset{\rho \to \infty}{\longrightarrow} \frac{K(T)}{\pi} \ell n(\rho/a_o); \qquad T \geq T_R \tag{8}$$

where $K(T)$ is a temperature-dependent constant and a_o is the lattice constant. Equation (8) indicates a surface with very large waves and an infinite mean square thickness, i.e., a rough surface. For $T < T_R$, K vanishes, and (8) is replaced by

$$g(\vec{\rho}) \underset{\rho \gg \xi}{\longrightarrow} (\text{constant}) \ell n \xi/a_o \qquad T < T_R \quad , \tag{9}$$

where $\xi \sim \exp(\text{const.}/|T_R - T|^{\frac{1}{2}})$ is a correlation length measuring the size of valleys or plateaus on a facet. The constant K takes on the value

$$K(T_R) = 2/\pi \tag{10}$$

at T_R and thus has a universal discontinuity at T_R. This discontinuity is reflected in the surface curvature at a crystalline face. We now present a simple argument leading to this conclusion; a more rigorous argument is presented elsewhere.[2]

Consider for simplicity a theory where $z(\vec{x})$ and $\vec{x}$ are continuous variables. To compute the curvature in the $\hat{e}_1$ direction add a term $-\eta_1 \int d\vec{x}\, s_1(x)$ to the hamiltonian. The free energy in this ensemble is just $\tilde{f}(\eta_1 \hat{e}_1)$, and standard techniques[13] yield

$$\begin{aligned}
\partial_1^2 f(\eta_1 \hat{e}_1) \Big|_{\eta_1 = 0} &= -\beta \int d\vec{x}\,\langle s_1(\vec{x}) s_1(0) \rangle \\
&= -\beta \int d\vec{x}\,\langle \partial_1 z(\vec{x}) \partial_1 z(0) \rangle
\end{aligned} \tag{11}$$

using $\vec{s} = \nabla z$. Using the fact that $\langle s_1(\vec{x}) s_1(\vec{x}') \rangle$ depends only on $|\vec{x} - \vec{x}'|$ one can put Eq.(11) in the form

$$\partial_1^2 \tilde{f}(\eta_1 \hat{e}_1) \Big|_{\eta_1 = 0} = \beta \int d\vec{x}\, \partial_1^2 \langle z(\vec{x}) z(0) \rangle \quad . \tag{12}$$

The integral may be evaluated using Eq.(8) and exercising care[18] in the way in which the system size is taken to infinity. Using Eq.(6), we find

$$\varkappa_{11} = \frac{K(T)a_o^2}{k_B T} \equiv \varkappa \quad . \tag{13}$$

Note that the principal curvatures $\varkappa_{11} = \varkappa_{22} = \varkappa$ are the same in this case. From Eqs.(10) and (13) it follows that $\varkappa$ undergoes the universal jump

$$\Delta\varkappa = 2a_o^2/\pi k_B T_R \tag{14}$$

at the roughening transition. Put another way, the curvature takes on a universal value $2a_o^2/\pi k_B T_R$ at the center of the face at T_R, whereas below T_R one has a flat facet centered at $\vec{x} = 0$.

Using the above results, we may write Eq.(5) in the form

$$z(\vec{x}) = \begin{cases} z_o - \dfrac{1}{2}\dfrac{\varkappa \tilde{f}(0)\vec{x}^2}{z_o} + \ldots; & T \geq T_R \\[3ex] z_o; & T < T_R \end{cases} \tag{15}$$

for small $\vec{x}$. Here z_o is the distance from the center of the crystal to the center of the face in question and $\tilde{f}(0)$ is the surface free energy per unit area of that face. In terms of λ, $z_0 = \tilde{f}(0)/\lambda$. The radius of curvature $1/R \equiv \partial^2 z/\partial|\vec{x}|^2$ thus undergoes the universal jump

$$(z_0/R) = \frac{2}{\pi}\left(\tilde{f}(0)a_0^2/k_B T_R\right) \tag{16}$$

at T_R. In mean-field theory[16] the curvature vanishes at T_R.

As remarked earlier, the linear size L of a facet is proportional to the step energy on the facet. As the step energy η_0 may be regarded as a surface tension between flat surface phases displaced normal to one another by the step, a standard hyperscaling argument[19] gives $\eta_0 \xi \sim k_B T$; we thus conclude that

$$L \sim \xi^{-1} \sim \exp\left[-(\text{const.})/\sqrt{T_R - T}\right] . \tag{17}$$

The corresponding mean-field result[16] is $L \sim [T_R - T]^{\frac{1}{2}}$.

THE SIX-VERTEX MODEL FOR FACET FORMATION

Here we consider a surface with rectangular symmetry, such as the a-facet of hcp ^{4}He. At $T = 0$ the facet will be perfectly flat, reflecting an absence of excitations. As we raise the temperature excitations in the forms of bumps and depressions will occur. We assume, as a consequence of the symmetry, that these can be constructed from small steps that raise or lower the local surface of the crystal by a characteristic atomic distance. Voids and overhangs will not be allowed. The orientations of the elementary steps will be determined by the rectangular symmetry and thus will be parallel to the two principal directions on the crystalline face. The energy of a given elementary step will be determined by the step energy per unit cell for a step in the given direction. In Fig. 1(a) we show the a-facet of an hcp crystal with the c-axis oriented upwards. In Fig. 1(b) we look down along the c-axis at the atomic arrangement in the crystal. The open circles represent atoms displaced down along

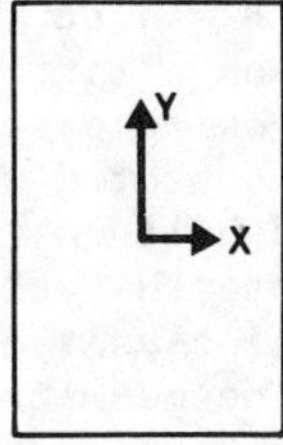
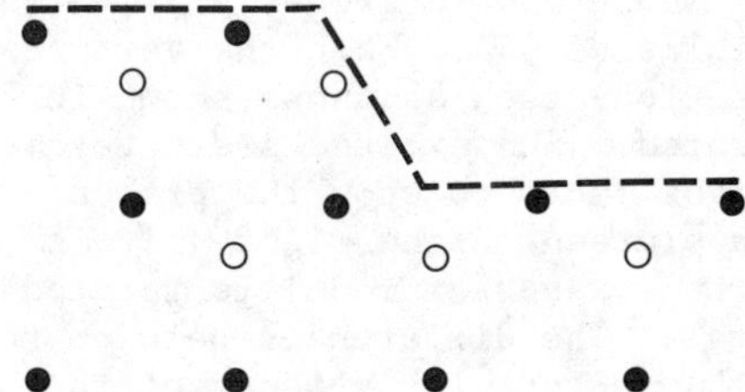

Fig. 1(a). a-facet.   Fig. 1(b). View along the c-axis.
The c-axis is vertical.   The step is parallel to c-axis.

the c-axis by a distance $c/2$ from the closed circles. The top of the figure is the a-facet in cross-section, and a single elementary step parallel to the c-axis is shown. If $-K/2$ is the nearest neighbor attractive coupling in the crystal (further neighbor interactions are ignored), the elementary step energy here is $2K$ from missing bonds in the unit cell at the step. For steps along the

orthogonal direction, the analogous elementary step energy is K.
Note that the surface states separated by the step in Fig. 1(b) are
displaced with respect to one another. The surface thus has two
ground states, a fact which is important for the correspondence we
will make with the six-vertex model.

The six-vertex model[3,4] is a model constructed by placing one of
six vertices at each point of a square lattice. There are four arrows
going into or out of each vertex, and the number of incoming arrows is
equal to the number of outgoing arrows at each vertex [a condition
which guarantees a unique $z(\vec{x})$ everywhere]. The vertices are depicted
in Fig. 2, along with the energies associated with them. Vertices one

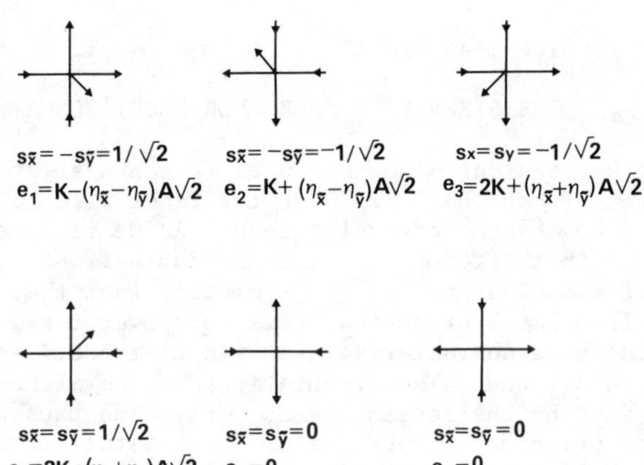

Fig. 2. Vertices and energies for the six-vertex model.[3]
The small arrows indicate upward slopes of elementary steps.

through four are excited states. These vertices represent elementary
steps whose upward slopes are indicated by the extra short arrows at
multiples of 45°. When the vertices are arranged on a lattice, configurations such as those shown in Fig. 3 (on following page) are
generated. The excited state vertices have been connected with
straight lines to show the presence of an isolated hill on the left
and a plateau on the right. It is important to note that the lattice
for the six-vertex model is rotated by 45° from the c-axis of the
crystal. The differences between the square lattice geometry of Fig. 3
and the rectangular a-facet of the hcp crystal can be accounted for by
inserting appropriate geometric factors which make the slopes of the
two kinds of steps correspond to those of the hcp crystal. Specifi-
cally, the replacements of the fields $(\eta_{\tilde{x}}+\eta_{\tilde{y}})$ and $(\eta_{\tilde{x}}-\eta_{\tilde{y}})$ for
the model of Fig. 3 according to

$$\left(\eta_{\tilde{x}}+\eta_{\tilde{y}}\right) \to \sqrt{3}\left(\eta_{\tilde{x}}+\eta_{\tilde{y}}\right)$$
$$\left(\eta_{\tilde{x}}-\eta_{\tilde{y}}\right) \to \left(3/2\sqrt{8}\right)\left(\eta_{\tilde{x}}-\eta_{\tilde{y}}\right), \tag{18}$$

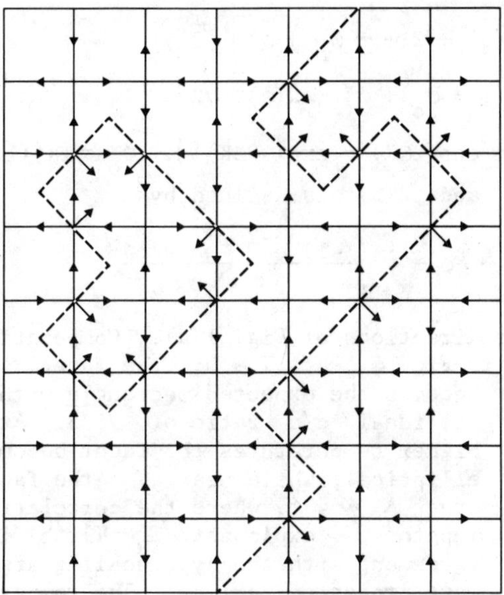

Fig. 3. A hill (left) and a plateau (right) in the six-vertex model.

the $\tilde{x}$- and $\tilde{y}$-directions here referring to directions on the square lattice of the six-vector model, accomplish this task. Now, note that if we can compute $\tilde{f}(\vec{\eta})$ for the six-vertex model, we can compute the crystal shape via Eq.(5). Fortunately, Sutherland, Yang and Yang (SYY)[4] have obtained most of the necessary results.

Consider first the case $T > T_R = K/\ell n\left(\frac{1 + \sqrt{5}}{2}\right)$, the transition temperature of the model. From the results of SYY and the correspondences of Fig. 2 and Eq.(18) one readily finds

$$\tilde{f}(\vec{\eta}) = \tilde{f}_o(T) - \frac{s_-^2 A}{2a(1 - b)}\,\eta_x^2$$

$$- \frac{s_+^2 A}{2a(1 + b)}\,\eta_y^2 \qquad (19)$$

for small $\vec{\eta}$. The subscripts x and y refer to the directions of Fig. 1(a), $s_+ = \sqrt{3}$ and $s_- = (3/2\sqrt{8})$ [as in Eq.(18)], $\tilde{f}_o(T)A$ is the zero-field free energy of SYY, and A is the area of a unit cell for the a-facet. The quantities a and b are given by

$$a = \frac{\pi - \mu}{4\cos\left(\frac{\theta_o}{2\mu}\right)}\,k_B T; \qquad b = \sin\left(\frac{\theta_o \pi}{2\mu}\right), \qquad (20)$$

where θ_o and μ are temperature-dependent quantities defined by SYY. For Eq.(19) the curvature tensor of Eq.(6) is diagonal, and the geometric mean of the principal curvatures is easily found to be[19a]

$$\sqrt{\varkappa_x \varkappa_y} = \frac{s_+ s_- A}{(\pi - \mu)k_B T} = \frac{2d^2}{(\pi - \mu)k_B T}, \qquad (21)$$

where d is the height of the step in Fig. 1(b). As $\mu(T) \to 0$ smoothly as $T \to T_R$, the geometric mean has the universal value $2d^2/k_B T_R \pi$ found earlier.

Below T_R, $\tilde{f}(\vec{\eta})$ is a constant independent of $\vec{\eta}$ over a specified region in the η-plane. From Eq.(5) this means that the surface is flat over the corresponding region in the $\lambda\vec{x}$-plane, indicating a facet. Using the results of SYY, the boundary is specified by the parametric equations

$$A\lambda s_+ x = -k_B T\left[Z(\phi) + Z(\omega - \phi_0 + \phi)\right]\sqrt{2}$$

$$A\lambda s_- y = -k_B T\left[Z(\omega - \phi_0 + \phi) - Z(\phi)\right]\sqrt{2} \tag{22}$$

where $\omega = \cosh^{-1}\left(\tfrac{1}{2}[\exp(6\beta K) - \exp(2\beta K) - \exp(-2\beta K)]\right)$, $\phi_0 = \ell n\,[\,(1 + \eta e^{\omega})/(\eta + e^{\omega})\,]$, $\eta = \exp(\beta K)$, and $Z(\phi)$ is defined by[3]

$$Z(\phi) = \ell n\frac{\cosh\tfrac{1}{2}\,(\omega + \phi)}{\cosh\tfrac{1}{2}\,(\omega - \phi)} - \tfrac{1}{2}\phi - \sum_{n=1}^{\infty}\frac{(-1)^n e^{-2n\omega}\sinh n\phi}{n\cosh n\omega} \quad . \tag{23}$$

In Eq. (22) x and y are the directions of Fig. 1(a). Computation yields the facet shapes shown in Fig. 4. At $T \to 0$, the facet becomes the expected rectangle with the ideal c/a ratio of $\sqrt{8/3}$. At higher temperatures the facet becomes elliptical, while near T_c the facet area $A_F \sim \xi^{-2}$, where the correlation length $\xi \sim \exp(\text{const}/\sqrt{T_R - T})$,[20] in agreement with the hyperscaling argument presented earlier. The temperature dependence of A_F should be universal.

The shape of the crystal near the facet is described by a universal exponent. This can be demonstrated for special directions[2] using results of Lieb and Wu.[3] It is true for general directions on the facet, as will be shown in a forthcoming publication.[21] For example, in the x-direction on the facet, the shape is described in the form

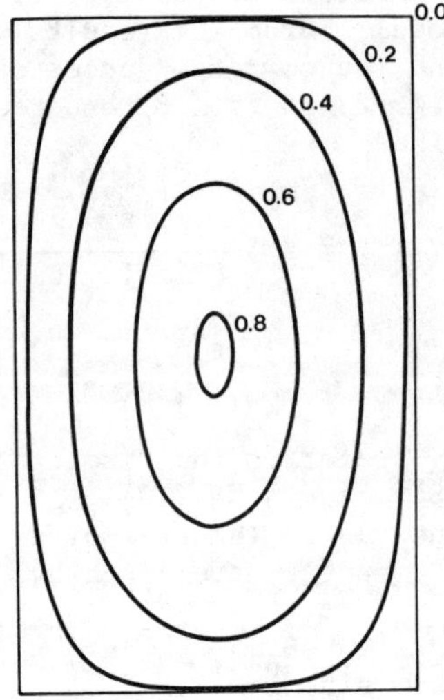

Fig. 4. Some a-facet shapes. Temperatures are indicated in units of T_R.

$$\lambda z(x,0) = \begin{cases} \tilde{f}(0) - \alpha_x(T)\left[|\lambda x| - \eta_x(T)\right]^{\frac{3}{2}}; & |\lambda x| > \eta_x(T) \\[2ex] \tilde{f}(0); & |\lambda x| \le \eta_x(T) \quad . \end{cases} \tag{24}$$

Here $\eta_x(T)$ is the boundary of the facet in the x-direction as found from Eq. (22). The exponent $\tfrac{3}{2}$ is universal, reflecting a Pokrovsky-Talapov[1] transition in the underlying free energy. The transition occurs at a given temperature T when the field $\vec{\eta}$ is sufficiently large that it becomes free-energetically favorable for an infinitely long step normal to $\vec{\eta}$ to appear on the crystal surface. In mean-field theory,[16] the $\tfrac{3}{2}$ is replaced by 2.

FURTHER REMARKS

We point out an important limitation of solid-on-solid models of the type considered here. The conclusions apply only for the facets for which a given model has been constructed and for surfaces at small angles with respect to this facet. For large angles they can and do fail. For example, a sequence of the steps shown in Fig. 1(b) can produce at most a 60° angle. This is a consequence of the neglect of overhangs in solid-on-solid models. Further, the six-vertex model used here excludes certain excitations that are most certainly present on crystal surfaces. In fact, the model as illustrated in Fig. 3 only allows troughs having widths which are odd multiples of the unit distance and terraces which are even multiples. This neglect will certainly affect any non-universal predictions of the model. However, the universal predictions result from phenomena occuring over large length scales and are probably not affected by such microscopic details. The problem with trough and terrace lengths does not occur for surfaces with a body-centered symmetry.

The six-vertex model used here as well as the body-centered solid-on-solid (BCSOS) model[2,22] employed by us in earlier work apply only to cases with particular symmetries. The BCSOS version applies directly to the (100) and (110) facets of both fcc and bcc crystals. The six-vertex model is designed for all facets with a rectangular symmetry. The c-facet of an hcp crystal has a different symmetry. The relevant solid-on-solid model here corresponds to a twenty-vertex model,[23] which has not, unfortunately been solved. However, the model does possess transitions of the type found in the six-vertex model.

It is quite possible that our principal results are independent of facet symmetry. Detailed experimental work would be of great value in this regard.

We conclude by remarking that in certain situations first-order transitions may be associated with crystal shapes. This is a possibility first brought to our attention by M. Wortis and C. Rottman[24] via arguments based on mean-field calculations. For the (110) facets of fcc crystals we note that the step energies in one direction are governed by the nearest-neighbor coupling while step energies in the orthogonal direction are governed by the next-nearest neighbor coupling. If the former coupling is attractive and the latter repulsive (as might occur in metals), the appropriate solid-on-solid model corresponds to a ferromagnetic six-vertex model.[3] Such models have a first-order transition associated with an edge between two facets intersecting at an angle. This topic will be explored in a future publication.

ACKNOWLEDGMENTS

It is a pleasure to thank D. O. Edwards, M. Wortis, C. Rottman, S. Balibar and D. Thoulouze for very useful conversations. One of us (CJ) is grateful to the A. P. Sloan Foundation for support. One of us (ST) was supported in part by NSF Grant No. DMR 81-14842.

REFERENCES

1. V. L. Pokrovsky and A. L. Talapov, Phys. Rev. Lett. 42, 65 (1979). See also H. W. J. Blöte and H. J. Hillhorst, J. Phys. A15, L631 (1982). The connection between the Pokrovsky-Talapov transition and crystal shapes has been made independently by C. Rottman and M. Wortis (private communication).

2. C. Jayaprakash, W. F. Saam and S. Teitel, submitted for publication.

3. E. H. Lieb and F. Y. Wu, in Phase Transitions and Critical Phenomena, edited by C. Domb and M. S. Green (Academic, London, 1972), Vol. 1.

4. B. Sutherland, C. N. Yang and C. P. Yang, Phys. Rev. Lett. 19, 588 (1967). See also C. P. Yang, Phys. Rev. Lett. 19, 586 (1967) and C. Jayaprakash and A. Sinha, Nucl. Phys. B210 [FS6], 93 (1982).

5. A. F. Andreev and A. Ya. Parshin, Zh. Eksp. Teor. Fiz. 75, 1511 (1978) [Sov. Phy. JETP 48, 763 (1976)].

6. K. O. Keshishev, A. Ya. Parshin and A. V. Babkin, Pis'ma Zh. Eksp. Fiz. 30, 63 (1979) [JETP Lett. 30, 516 (1979)].

7. J. Landau, S. G. Lipson, L. M. Määtänen, L. S. Balfour and D. O. Edwards, Phys. Rev. Lett. 45, 31 (1980).

8. S. Balibar and B. Castaing, J. de Phys. Lett. 41, L-329 (1980).

9. J. Avron, L. S. Balfour, C. G. Kuper, J. Landau, S. G. Lipson and L. S. Schulman, Phys. Rev. Lett. 45, 814 (1980).

10. K. O. Keshishev, A. Ya. Parshin and A. B. Babkin, Zh. Eksp. Teor. Fiz. 80, 716 (1981) [Sov. Phys. JETP 53, 362 (1981)].

11. L. Puech, B. Hebral, D. Thoulouze and B. Castaing, J. Physique Lett. 44, L-159 (1983), and S. Balibar, P. E. Wolf and D. O. Edwards, private communication. See also the articles by S. Balibar and D. Thoulouze in these Proceedings.

12. D. S. Fisher and J. D. Weeks, Phys. Rev. Lett. 50, 1077 (1983).

13. See, e.g., L. D. Landau and E. M. Lifshitz, Statistical Physics (Addison-Wesley, Reading, Mass., 1969), Sec. 143.

14. A classic paper on the subject is C. Herring, Phys. Rev. 82, 87 (1951).

15. Collected Papers of L. D. Landau, (Pergamon Press, Oxford, 1965), p. 540.

16. A. F. Andreev, Zh. Eksp. Teor. Fiz. 80, 2042 (1981) [Sov. Phys. JETP 53, 1063 (1982)]. This paper presents a mean-field theory of facet formation. The results disagree with ours in major respects.

17. H. J. F. Knops, Phys. Rev. Lett. 39, 766 (1977), S. T. Chui and J. D. Weeks, Phys. Rev. B14, 4978 (1976), T. Ohta and K. Kawasaki, Prog. Theor. Phys. 60, 365 (1978).

18. See, e.g., T. Ohta and D. Jasnow, Phys. Rev. B20, 139 (1979). The point is that one takes the limit $W \to \infty$ before $L \to \infty$, where W is the width of the system perpendicular to $\vec{\eta}$ and L is the length parallel to $\vec{\eta}$.

19. See B. Widom in Phase Transitions and Critical Phenomena, edited by C. Domb and M. S. Green (Academic, London, 1972), Vol. 2.

19a Transforming (20) back to the potential $f(\vec{s})$ yields

$f(\vec{s}) = f_o + \frac{1}{2}[\Gamma_x s_x^2 + \Gamma_y s_y^2]$ where the geometric mean of the "macroscopic surface stiffness" $\sqrt{\Gamma_x \Gamma_y} = \dfrac{(\pi - \mu)k_B T}{2D^2}$ has a universal jump to infinity at T_R. This equivalent manifestation of the Kosterlitz-Thouless nature of roughening was realized and used to argue the stability of the smooth phase in the presence of quantum fluctuations by D. S. Fisher and J. D. Weeks in Ref. 12.

20. See R. J. Baxter, <u>Exactly Solved Models in Statistical Mechanics</u>, (Academic, London, 1982), Ch. 8. The validity of the particular hyperscaling argument used here is demonstrated explicitly by Baxter for the six-vertex model.
21. C. Jayaprakash and W. F. Saam (unpublished).
22. H. van Beijern, Phys. Rev. Lett. <u>38</u>, 993 (1977).
23. S. B. Kelland, Austral. J. Phys. <u>27</u>, 813 (1974).
24. M. Wortis and C. Rottman, private communication.

RECENT EXPERIMENTS ON THE SOLID-LIQUID ^{4}He INTERFACE:
ONSAGER COEFFICIENT AND ROUGHENING TRANSITIONS

S. Balibar, D.O. Edwards[*], F. Gallet and P.E. Wolf
Groupe de Physique des Solides de l'Ecole Normale Supérieure
24 rue Lhomond, 75231 Paris Cedex 05, France

ABSTRACT

After a general introduction which describes the history of
the study of rough and smooth parts on the interface between superfluid and crystalline Helium, we focus on two experiments performed
recently at the E.N.S., Paris, namely the measurement of the Onsager
coefficient for rough surfaces and the discovery of a new roughening
transition at .365 K which is related to the present controversy on
the shape of He crystals at T = 0.

HISTORICAL INTRODUCTION

Since the measurement of the interfacial tension between
superfluid and solid ^{4}He[1], the understanding of the surface
properties of these crystals has greatly improved, due to both
theoretical and experimental work in various laboratories. The
interest in the field was successively stimulated by a prediction
by Andreev and Parshin[2] that this interface has to be rough down
to T = 0 as a consequence of the zero point motion of the atoms, the
observation of very high mobilities of these interfaces leading to
the existence of melting freezing waves[6] and anomalous reflection
of sound waves[3], and then the discovery of roughening transitions
(4) to (7) at which smooth facets appear on the equilibrium shape
of the crystals. Both dynamic and static properties of the solid
surface depend drastically on whether it is rough (T > T_R) or smooth
(T < T_R), consequently on its orientation since T_R is an anisotropic
quantity. The dynamic properties of rough surfaces have been
analysed by introducing a matrix which relates ΔT and $\Delta\mu$, the
temperature and chemical potential differences across the interface,
to both mass and energy currents through it[8]. The three
coefficients of this symmetric matrix have been calculated(9) to (12);
two of them are closely related to simple quantities (the Kapitza
resistance and the growth coefficient of the interface); the third
one, being the cross-term, is called the Onsager coefficient for
this problem. These quantities have been successively measured[6]
(13) to (16) and we focus in the first part of this article on the
most recent progress, i.e. the measurement of the Onsager
coefficient we made last year in Paris[16]. In the second part we
describe our recent discovery[17] of a third roughening transition.
This latest development extends previous experimental studies of
the static properties of the crystals and should help answering a

* Permanent address : Ohio State University, Columbus, Ohio, USA.
Partially supported by NSF Grant #DMR 7901073.

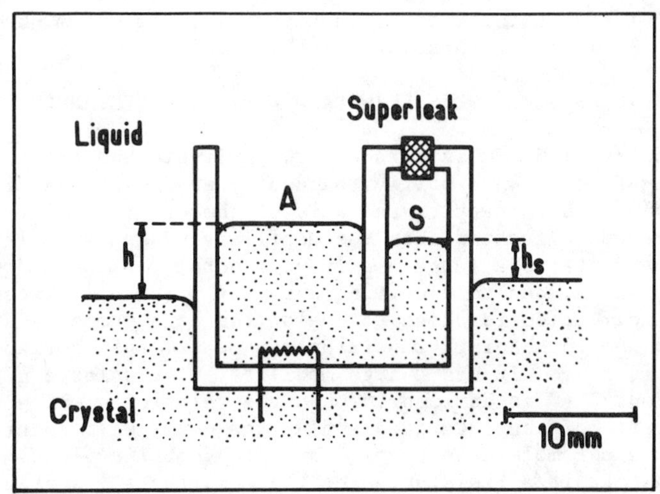

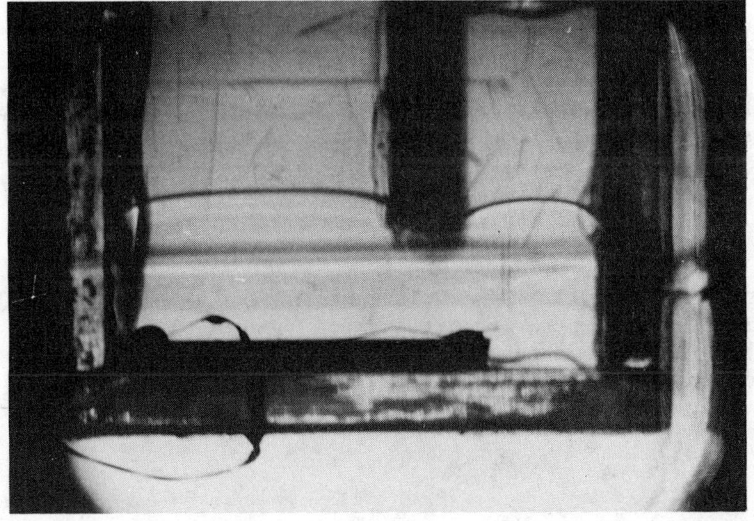

Fig.1 : Schematic drawing and photograph of the box in which the measurement of b was performed. The transparent box contains a single crystal with a mainly rough surface and it is partly immersed in another crystal of the same orientation. The level differences h and h_s are produced by turning on the electric heater. The photograph shows the real box with J_E = .503 mW cm^{-2} and T = .27 K. The depth of field is too small to show the level of the crystal outside the box very clearly, but it is about 2 mm below those inside.

384

question which remains open : does quantum roughening exist according to A.P.[2], or not, according to Weeks and Fisher (W.F.) [18] ? Related to this point, what is the equilibrium shape of He crystals as T goes to absolute zero ?

ONSAGER COEFFICIENT AND KAPITZA RESISTANCE

The equilibrium interface between liquid and crystalline ^{4}He is partly facetted and partly rounded depending on the direction of its normal with respect to the axis of the crystals. For crystals with dimensions greater than the capillary length ($\ell \sim 1.5$ mm), the interface contains a large nearly horizontal portion and is bend near the walls, due to a wetting angle of $\sim 140°$[1]. Except if the vertical direction is close to one of the high symmetry directions in which the roughening transition may occur, this horizontal surface is rough. It has a high mobility, i.e. a large growth coefficient K (K is defined through the relation $v = K\Delta\mu$ where v is the velocity of the interface and $\Delta\mu$ the chemical potential difference per mole across it). This high mobility leads to an anomalous Kapitza resistance for these surfaces : instead of being small and proportional to T^{-3} as expected for two media with very similar acoustic impedances, it varies as T^{-5} below $\sim .4$ K, due to the fact that the solid and the liquid side can exchange mass very rapidly[2] [8] to [15]. We will be concerned here with these rough surfaces.

As Castaing and Nozières[8] pointed out there must be a 2x2 matrix of coefficients describing the dissipation on this interface for small departure from equilibrium :

$$\frac{\Delta\mu}{T} = aJ + b_1 \, J_E$$

$$\frac{\Delta T}{T^2} = b_2 J + c \, J_E$$

$\Delta\mu = \mu_L - \mu_C$ and $\Delta T = T_L - T_C$ are respectively the chemical potential and temperature differences between the liquid and the crystal, J and J_E are the mass and heat current densities from liquid to crystal. The coefficients a and c are respectively related to the growth coefficient K and the Kapitza resistance R_K :

$$a = (Km)^{-1}/\rho_c T$$

$$c = R_K/T^2$$

An Onsager relation gives $b_1 = b_2 = b$ and thermodynamic stability requires $b^2 < ac$.

We report here on the measurement of b_1 which was performed with the aid of a little plexiglas box shown on Fig.1. Growth or melting is controlled by adding and removing ^{4}He through the capillary to the pressure cell attached to a dilution refrigerator with optical windows. The electric heater produces a heat current J_E through the interface at A, causing a level difference h between

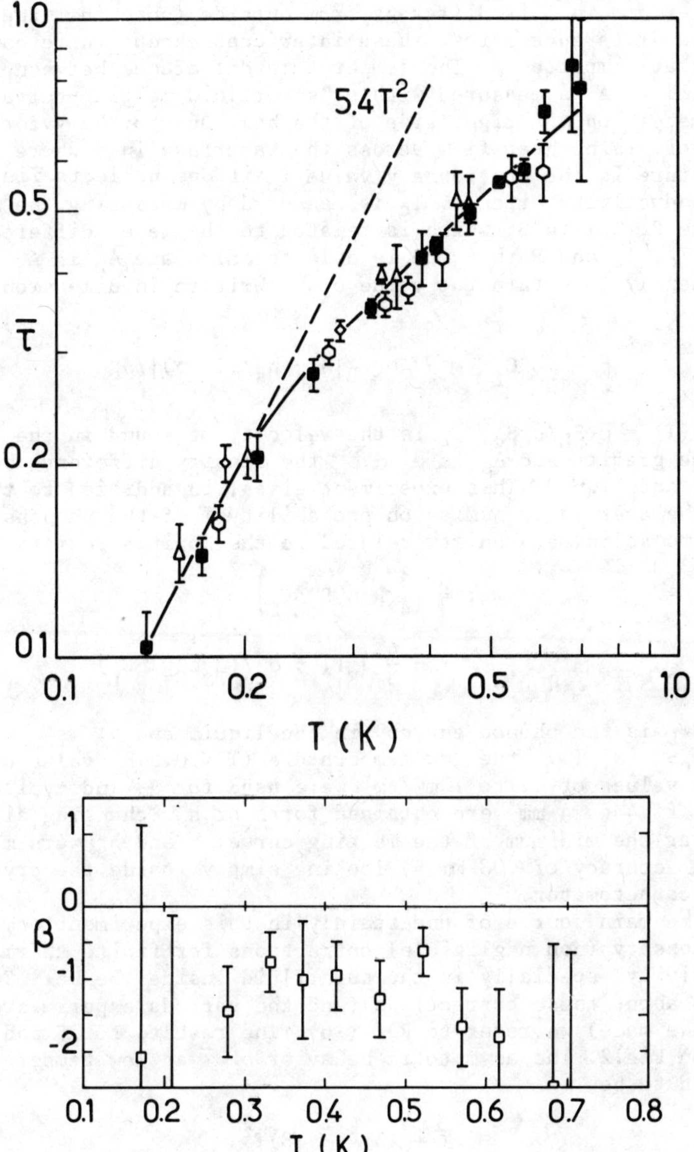

Fig.2 : The phonon transmission probability τ̄ and the values of the
dimensionless Onsager coefficient β vs. temperature. Various symbols
correspond to various crystals.

386

the crystal inside and outside the box meaning that the pressure at
the interface in A is different from outside (note that there is no
pressure difference across these interfaces except where capillary
effects are important). The temperature difference between liquid
and solid in A is measured with a "superfluid melting-curve
thermometer" on the right side of the box. Due to the vycor superleak there is no heat flow across the interface in S where the
temperature is the stationnary value T_c if one neglects finite
heat conductivity effects. T_c is measured by measuring the melting
pressure $P_m(T_c)$ in S, which is related to the level difference h_S.
Having T_c, T_L and P in A one is able to calculate Δ_μ in A,
consequently to obtain the value of b. Written in dimensionless
form

$$\beta = b\rho_c C_L T = - \Delta\rho g C_L |h - h_S/(1-\sigma)|/J_E$$

where $\sigma(T) = \rho_L S_L/\rho_c S_c$, C_L is the velocity of sound in the liquid,
g is the gravity and Δ_ρ is $\rho_c - \rho_L$ the density difference between
crystal and liquid. This experiment gives, in addition to the value
of b, the average transmission probability $\overline{\tau}$ of the phonons from
liquid to solid, a quantity related to the Kapitza resistance, R_K :

$$\overline{\tau} = \left[\frac{3}{4} R_K \rho_L S_{\phi L} C_L\right]^{-1}$$

$$= - \frac{4}{3} J_E (1 - \sigma)/(\Delta_\rho h_S g \sigma_o C_L)$$

where $S_{\phi L}$ is the phonon entropy in the liquid and σ_o is
$\rho_L S_{\phi L}/\rho_c S_c$ is .157, the low temperature (T < 0.5 K) value of σ.
Typical values of .1 to 1 mW/cm^2 were used for J_E and typical
values of .4 to 4 mm were obtained for h or h_S, changing sign at
.8 K near the minimum of the melting curve. h and h_S were measured
with an accuracy of .08 mm by looking simply inside the cryostat
with a cathetometer.

   The main source of uncertainly in this experiment came from
the necessary (non negligible) corrections for finite thermal
conductivity, specially in the superfluid inside the box. For
details about these corrections (and the various experimental
checks we made) we refer to Ref.(16). The results for $\overline{\tau}$ and β are
shown on Fig.2. The asymptotic behavior of $\overline{\tau}$ at low temperature
is found to be

$$\overline{\tau} = (5.4 \pm .8)T^2$$

a value in agreement with previous results obtained with different
methods[13][14][15]. This value means that phonons are anomalously
reflected by the interface as expected since it is highly mobile
but seems to differ substantially from the value predicted by a
theory due to Marchenko and Parshin[10]. An interpretation was
proposed by Puech and Castaing[11] who introduced a surface inertia
to describe the response of the interface to an incident phonon.

The result for the coefficient $\beta = b\rho_c C_L T$ can be written as

$$\beta = -1.1 \pm 0.4$$

This value has to be compared with the theory of Bowley and Edwards [9] who predicted a value of -1.5 or $-.6$ depending on whether the propagation of the phonons is ballistic both in the liquid and solid or only in the liquid, the propagation in the solid being then hydrodynamic. In our experimental conditions (T > .2K ; box size = 10 mm), we should probably have found $\beta = -.6$ according to this theory. The experimental value, intermediate between $-.6$ and -1.5, has however both right sign and right order of magnitude. It confirms that the dissipation of energy in this problem is controlled, as suggested first by Andreev and Parshin [2], by the transmission and reflection of elementary excitations at the interface, an idea on which the theory by Bowley and Edwards is based.

ROUGHENING TRANSITIONS AND CRYSTAL SHAPES NEAR T = 0

We describe here some preliminary results obtained for crystal shapes down to 70 mK. Part of the interest in this problem came from A.P.'s suggestion [2] that, due to the quantum motion of atoms, the kinks on the steps at the interface should be delocalized, leading eventually to a negative energy for isolated steps and a finite number of these steps at T - 0. These zero point steps were supposed to create a "quantum roughness" of the interface, a property which was proposed as an explanation of the fact that, at this time (1978), no facets were observed yet on the surface of the crystals. Two or three years later, after the discovery of the appearance of facets in two of the highest symmetry directions of the crystals around 1 K [4][5][6], several authors [6][20][21] were still thinking that A.P.'s arguments should still apply to all the other directions and that these two kinds of facets were the only two one could ever observe. However, having transformed our dilution refrigerator in order to make optical measurements below .4 K, we detected the existence of a third kind of facets on growth shapes below .35 K. At the same time, Weeks and Fisher (W.F.) [18] suggested that A.P.'s arguments are wrong. We report here the details of the observation of this third kind of facets and discuss its implication in the frame of A.P.'s and W.F.'s theories.

The two photographs on Fig.3 show two typical growth shapes obtained at 0.35 K and 0.4 K. The one at 0.4 K contains (0001) or c planes perpendicular to the six fold symmetry axis and a planes which are parallel to this axis and could be either $(10\bar{1}0)$ or $(11\bar{2}0)$ type planes. We checked that these planes appear both on growth and equilibrium shapes respectively at $T_{R1} = 1.2 \pm .1$ K and $T_{R2} = .9 \pm .1$ K in agreement with previous measurements [4] to [7]. The new facet, which we call "s" and which is present on the photograph at .35 K makes an angle of $60 \pm 3°$ with the c plane and

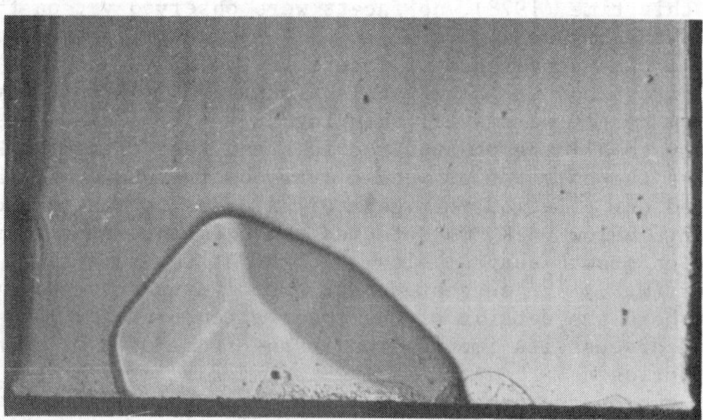

Fig.3 : Two typical growth shapes obtained (A) at .35 K and (B) at .4 K. The growth velocity of facets without contact with walls is ∿ .5 mm/mn. These photographs were taken with a Sodium lamp and an optical technique close to strioscopy.

can be either of $(10\bar{1}1)$ type if the a facets are $(10\bar{1}0)$ or $(11\bar{2}2)$ type if the a facets are $(11\bar{2}0)$ (see table below). No other facet were observed down to 70 mK. In order to be sure that we observed a third roughening transition, and since it is always difficult to distinguish on a growth shape between zero and small curvature in a given direction, we wanted to see these facets on equilibrium shapes. This appeared impossible in our experiment probably because their equilibrium size is very small and that c facets (a fortiori a facets) are already difficult to observe at the equilibrium, even using interferometric techniques, especially if the light beam is not at grazing incidence with respect to them. However, we succeeded to perform a static measurement, showing that there is no observable difference between the temperature at which "s" facets appear on equilibrium shapes and growth shapes respectively. This temperature varies slightly however from one crystal to another, a property already observed for other faces and attributed to an eventual dependence of T_R on the number of dislocations[6]. Our measurements for this third roughening transition are well represented by $T_{R3} = 0.365 \pm 0.03$ K.

The method we used to observe equilibrium properties of crystal surfaces is similar to the one used in the "bubble" experiment by Balibar et al[1] in 1978. A transparent box closed at the bottom by a thin plate in which a small hole (radius r = .8 mm) was drilled is used to form a nearly free standing crystal whose equilibrium shape depends on the pressure difference across the interface and the interfacial tension. The effect of gravity is small since the crystal size is less than the capillary length and the absence of contact with walls makes easier to look a many different crystalline orientations on the interface. The hole has sharp enough edges to make the angle of contact of the bend meniscus with the bottom plate arbitrary, in first approximation. It is also thin enough (.1 mm) to avoid the complete blocking of its area by a facet as was observed in the 1978 experiment[1] and interpreted later[4] by suggesting the existence of a roughening transition around 1 K. The pressure difference across this single crystalline bubble is obtained by measuring the height h_1 of the interface outside the box with the usual relation

$$\Delta P = \Delta\rho \; g \; h_1$$

It is monitored by adding or removing Helium in the cell as in the b-measurement described above. The height h_g of the bubble is a function of ΔP, and, at constant ΔP and for some orientation, it was observed to change at the various roughening transitions, a property which provides a way to measure the various T_R at the thermodynamic equilibrium. This is illustrated by Figs 4 and 5 in the case of T_{R3}. For a physical explanation of this change we refer to Ref. 17. Finally, if ΔP exceeds a critical value, the bubble becomes unstable and acts as a nucleus for the further growth of a crystal inside the box (Fig.3).

In order to discuss the implications of the existence of this

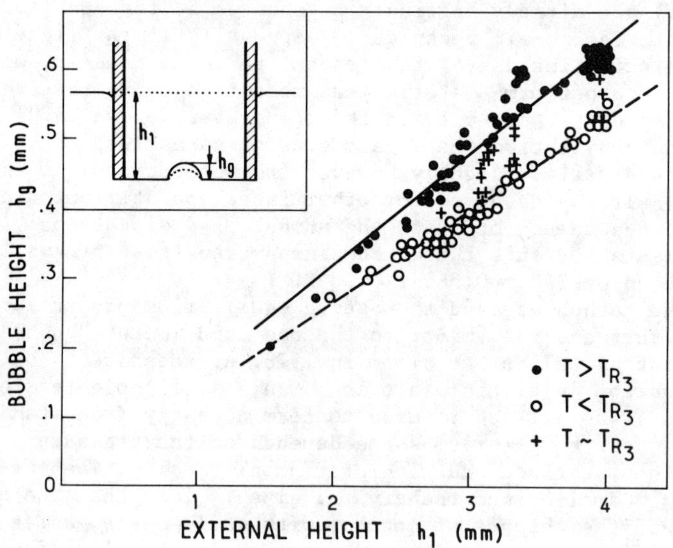

Fig.4 : The variation of the height h_g of the crystalline bubble with the height h_1 of the level outside the transparent box. Open circles correspond to temperatures $T < .34$ K and black circles to $T > .38$ K; crosses correspond to $T \simeq T_{R_3} = .365$ K. The solid and broken lines are guides to the eye.

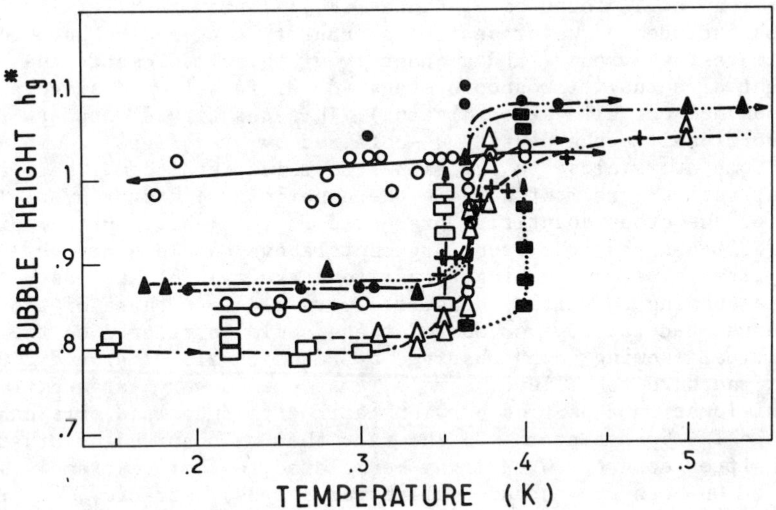

Fig.5 : The variation of h_g^* with temperature T. h_g^* is the value of
the actual bubble height h_g normalized by using the upper solid
line of Fig.4. A change in h_g^* is observed when the temperature is
raised through the temperature T_{R3} which varies slightly from one
crystal to another. This third roughening transition temperature is
$T_{R3} = .365 \pm .03$ K.

hierarchy of roughening transitions, let us make a crude estimate of the various T_R. One can suppose that T_R is proportional to the low temperature value of the step energy E_s and that E_s is proportional to $\alpha_{1s}n$ or $\alpha_{1s}d$ where α_{1s} is the interfacial energy, n the surface density in the planes parallel to the facet under consideration and d their mean spacing. Remarking then that α_{Ls} is not very anisotropic since $\tilde{\alpha}_{1s} = \alpha_{1s} + \alpha''_{1s}$ varies only by a factor two with orientation[6][7], we find that T_R is proportional to d. The table below shows a comparison between the hierarchy of distances between planes in the hcp structure and the actual hierarchy of T_R found here. Since W.F.[18] propose a formula for T_R which includes d^2 we present this quantity too, and we have added estimates by Avron[5] and a quantity d^* which represents the maximum height of actual microscopic steps (d^* differs from d since parallel planes are not always equidistant). This quantity d^* appears to be proportional to the density n^* obtained by Drechsler[22] who allows for some atoms close to the planes to relax into them. There is no quantitative agreement between the experimental hierarchy of T_R and any of the other quantities presented in the table. Furthermore if one believes that the crude arguments above should apply qualitatively, two other roughening transitions are missing at least, corresponding to surface densities very close to those of the observed facets. However one of these could be related to the "critical showing down" observed by Puech et al[23] around 200 mK (we find hard to believe, namely, that even a very large difference in dislocation densities between their experiment and ours can change T_{R3} by a factor ~ 2. The fact that we observed no other transition down to 70 mK might be related to their extremely small effect on both growth and equilibrium shapes). Moreover, the ratios presented in the table are in qualitative agreement with the hierarchy of T_R and one could conclude that Weeks and Fisher are right and that a complete analysis down to even lower temperatures should show the appearance of more and more facets as T goes to absolute zero. But their arguments apply only to the ideal case of an infinite surface; on the other hand quantum motion seems necessary to consider in order to interpret the very high mobilities observed for rough surfaces, and Kapitza resistances were observed to vary as T^{-5} down to 100 mK even on curved interfaces[14]. One aspect of the problem being to know if all T_R are substantially displaced towards $T = 0$ (and eventually into the unphysical $T < 0$ region) by the effect of zero point motion, as is sometimes suggested[24], we think that a calculation of the various T_R is now needed. Three different experimental values T_{R1}, T_{R2}, T_{R3} are known, now, with some accuracy. Such a calculation should tell us the exact role of zero point motion in the problem of roughening at the surface of Helium crystals.

T_R (K) (this experiment)	$T_{R1} = 1.2 \pm .1$	$T_{R2} = .9 \pm .1$		$T_{R3} = .365 \pm .03$	
corresponding face	(0001)	$(10\bar{1}0)$ or $(11\bar{2}0)$		$10\bar{1}1$ or $11\bar{2}2$	
angle with (0001)	0	90°	90°	58.5°	62.1°
na^2 or $d\sqrt{2}/a$	$2/\sqrt{3} = 1.155$	$\sqrt{3}/2\sqrt{2} = .612$	$1/\sqrt{2} = .707$	$\sqrt{12/41} = .541$	$2/\sqrt{11} = .603$
max step height d^*/a	$\sqrt{2}/3 = .816$	$1/\sqrt{3} = .577$	$1/2 = .5$	$10/\sqrt{246} = .637$	$\sqrt{2}/11 = .426$
$n/n_1 = d/d_1$	1	.53	.612	.468	.522
$(d/d_1)^2$	1	.28	.374	.219	.272
d^*/d^*_1	1	.707	.612	.78	.522
Avron et al[5]	1	.35 or .2	.5 to 1	.34 or .16	
Experimental hierarchy T_R/T_{R1}	1	.75		.3	

TABLE

TABLE :

A comparison between the hierarchy of roughening transitions observed and various quantities, namely the surface density n of the face, the mean spacing d between parallel planes, the maximum height d^* of the microscopic steps (d^* differs from d since parallel planes are not always equidistant. It appears proportional to the density obtained by allowing for some surface relaxation in the work of Drechsler[22]). a is the distance between nearest neighbours. The agreement between these various hierarchies is qualitative. Other planes, such as $(10\bar{1}2)$ $(30\bar{3}2)$ $(11\bar{2}1)$ and $(10\bar{1}3)$ which are tilted respectively at 43°, 70.5°, 73° and 32° have slightly smaller densities (na^2 = .42; .38, .34 and .33 respectively).

REFERENCES

1. S. Balibar, D.O. Edwards and C. Laroche, Phys. Rev. Lett. 42, 782 (1979).
2. A.F. Andreev and A.Y. Parshin, Sov. Phys. JETP 48, 763 (1978).
3. B. Castaing, S. Balibar and C. Laroche, J. Phys. 41, 897 (1980).
4. S. Balibar and B. Castaing, J. Phys. Lett. 41, L 329 (1980).
5. J.E. Avron, L.S. Balfour, C.G. Kuper, J. Landau, S.G. Lipson and L.S. Shulman, Phys. Rev. Lett. 45, 814 (1980).
6. K.O. Keshishev, A.Y. Parshin and A.B. Babkin, Sov. Phys. JETP 53, 362 (1981) and JETP Lett. 30, 56 (1979).
7. S. Ramesh and J.D. Maynard, Phys. Rev. Lett. 49, 47 (1982).
8. B. Castaing and P. Nozières, J. Phys. 41, 701 (1980).
9. R.M. Bowley and D.O. Edwards, J. Phys. (Paris) to be published.
10. V.I. Marchenko and A.Y. Parshin, JETP Lett. 31, 724 (1980).
11. L. Puech and B. Castaing, J. Phys. Lett. 43, L 601 (1982).
12. A.M. Kosevitch and Y.A. Kosevitch, Physica 108b, 1195 (1981); Sov. J. Low Temp. Phys. 7, 203 (1981).
13. T.E. Huber and H.J. Maris, J. Low Temp. Phys. 48, 99, 463 (1982).
14. L. Puech, B. Hebral, D. Thoulouze and B. Castaing, J. Phys. Lett. 43, L 809 (1982).
15. T.E. Huber and M.J. Maris, Phys. Rev. Lett. 47, 1907 (1981).
16. P.E. Wolf, D.O. Edwards and S. Balibar to appear in J. Low Temp. Phys. may/june (1983).
17. P.E. Wolf, S. Balibar and F. Gallet to be published.
18. J. Weeks and D.S. Fisher, Phys. Rev. Lett. 50, 1077 (1983).
19. J. Landau, S.G. Lipson, L.M. Maattanen, L.S. Balfour and D.O. Edwards, Phys. Rev. Lett. 45, 31 (1980).
20. S.G. Lipson, Proc. of LT 16 Physica 109-110 B, 1805 (1982).
21. A.Y. Parshin, Proc. of LT 16 Physica 109-110 B, 1819 (1982).
22. M. Drechsler, Proc. Intern. Field Emission Symp. Göteborg 1982 Almquist and Wiksell Intern., Stockholm.
23. L. Puech, B. Hebral, D. Thoulouze and B. Castaind, J. Physique Lett. 44, L 159 (1983).
24. K.O. Keshishev, communication at the "Workshop on 2D problems" Les Houches (France) feb. 1983 for example.

WETTING OF fcc ^{4}He GRAIN BOUNDARIES BY FLUID ^{4}He

J. P. Franck
University of Alberta, Edmonton, Alta. Canada T6G 2J1

ABSTRACT

Thin films of solid ^{4}He were visually observed and photographically recorded during the initial stages of solidification. The films are formed on sapphire single crystal surfaces, their thickness is of the order of 50μ or less. It was found that the morphology of these films depends on whether deposition takes place in the existence range of the fcc ^{4}He phase (P≥1.13 kbar), or the hcp ^{4}He phase (P≤1.13 kbar). Grain boundaries in the fcc phase show a network of polygons, approaching a hexagonal network. Melting of this structure occurs preferentially at the grain boundaries, with the liquid wetting the grain boundaries. Deposition in the hcp ^{4}He phase shows films that are traversed by a system or several systems of parallel grain boundaries. Melting of these structures never resulted in the wetting of the grain boundaries. When deposition takes place near the triple point hcp-fcc-fluid ^{4}He (P=1.13 kbar, T=15.0 K), both morphologies are observed. During isothermal holding of such two-phase structures one can observe the slow transformation from fcc into hcp ^{4}He.

EXPERIMENTAL DETAIL

The experiments were conducted in the optical high pressure cell described by Franck and Daniels.[1] Each closure employs a sapphire window of 4.76 mm diameter and 2.5 mm thickness. The clear optical path is restricted to a cross-section of 1.58 mm diameter. The pressure cell was mounted on the temperature-controlled stage of a cryostat. White light from a standard source was used to illuminate the cell. Apertures restricted the opening angle of the ingoing light to 0.44°. Outgoing apertures, provided by the pressure cell, stopped any ray with a deviation of more than 3.3° (back window) and 5.7° (front window). Such ray deviations were produced at the steep parts of surface grooves where grain boundaries meet the fluid. The contrast which allowed the grain boundaries to be observed in spite of the very small difference in refractive index between solid and fluid helium is therefore produced by a Schlieren method. The observed structures were imaged with a lens of focal length 15 cm, and photographed on Kodak Ektachrome ASA 400 colour film. The optical arrangement had a depth of focus of about 50μ. It was found to be necessary to photograph the structures slightly out of focus to increase contrast. The resulting blurring is produced by the ingoing opening angle, and is not an indication of the actual thickness of the observed grain boundary structures.

EXPERIMENTS

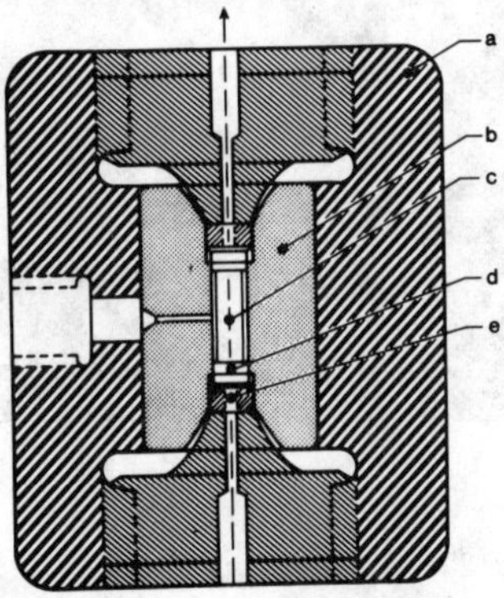

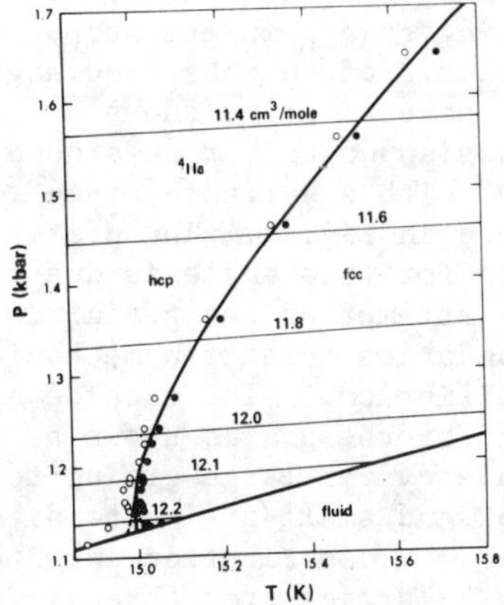

Fig. 1. The pressure cell.
The helium space is at c,
and the sapphire windows
are at d.

Fig. 2. The phase diagram
of ^{4}He in the vicinity of
the hcp-fcc-fluid triple
point.

All experiments described here
were conducted with high purity ^{4}He.
The total impurity content was 1.1 ppm
with major impurities of H_2O (0.6 ppm)
and N_2 (0.38 ppm). These impurities
were further reduced through absorption
and liquid nitrogen traps. The iso-
topic ^{3}He concentration was less than
1.7 ppb.

The crystal growth was initiated
by slightly exceeding the melting
pressure (about 15 bar) or slightly
lowering the temperature below the
melting point (by about 50 to 100 mK).
Holding the external conditions constant, one frequently observed after
several minutes the appearance of
polycrystalline films on both sapphire
windows of the cell. The contrast of
the grain boundaries tended to
increase for awhile, then remained
constant. Under constant external
conditions these films could be
observed for times of one to two hours.
Slight heating, at rates around 1 mK/
min resulted in melting, and eventual
disappearance of the layers. Continued
cooling resulted in the loss of optical contrast, and eventually in the
production of solid throughout the
cell. The thickness of the films was
less than the depth of field of 50μ.
The films were in general very sensitive to any change in external conditions and were easily lost.

In Fig. 2 we show the phase
diagram of ^{4}He in the vicinity of the
the hcp-fcc-fluid triple point. The
temperature hysteresis of the transition is indicated by the full circles
(transition in heating) and the open
circles (transition in cooling). The morphology of the films
depended on the range of the phase diagram in which we operated.

Fig. 3 shows three stages in the development of a solid film,
produced in the existence range of fcc ^{4}He. Fig. 3(a) shows the
typical polygonal structure always observed for this phase. One
usually observes only three point vertices, although four point
vertices have on rare occasions also been seen. The structure, when
it first comes into view after several minutes at the correct
external conditions, is already fully polygonal, with sharp angles.

398

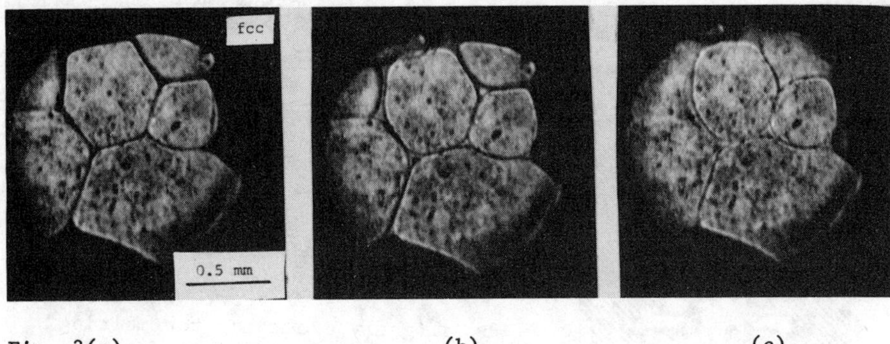

Fig. 3(a)   (b)   (c)

Polygonal structure observed in a film at 2.8 kbar. a) Structure as first deposited. b) and c), progressive melting with the development of fluid at the vertices and grain boundaries.

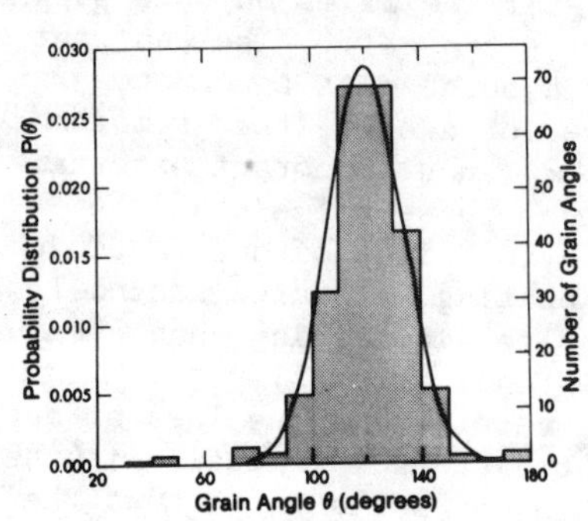

Fig. 4. Distribution of observed angles in several polygonal structures.

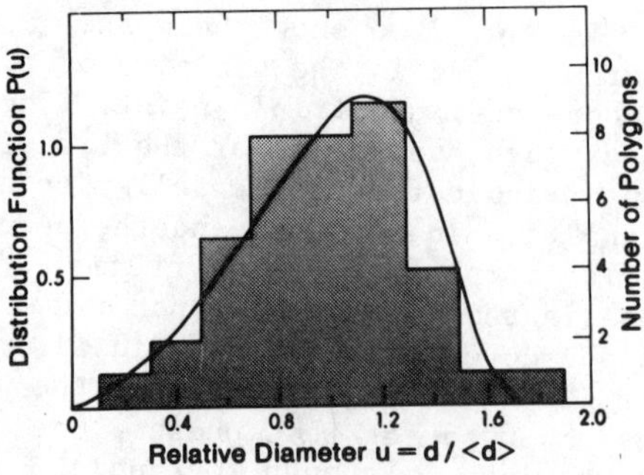

Fig. 5. Size distribution of grains for three different polygonal films.

It continues to change with time, at first rather fast (say for the first 20 minutes), then slowing down, without ever coming to a complete stop. An analysis of the observed angles for several films shows a Gaussian distribution around 120° with a variance of 14°, as shown in Fig. 4. The preference for this angle is due to the attempt of the grain boundaries to attain mechanical equilibrium.[2] The slow development of the grains shows a tendency for large grains to grow and small grains to disappear. A distribution in size for 39 grains from three films is shown in Fig. 5. It shows the expected bias toward larger grains. Comparison is made with the grain size distribution in two dimensions suggested by Lifshitz and Slezov[3] and by Hillert[4]:

$$P(u) = \frac{8u}{(2-u)^4} \cdot e^{\frac{2u}{2-u}}, \qquad (1)$$

where $u = r/<r>$, r is the effective diameter of a grain, and $<r>$ the average effective

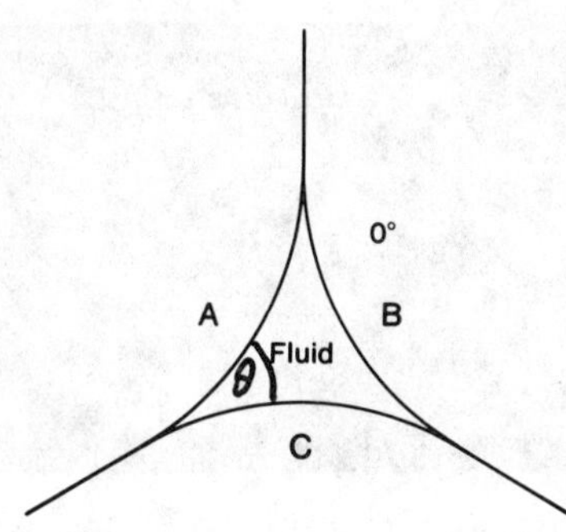

Fig. 6. Appearance of the vertex between three grains A, B, C, with fluid interspersed. Dihedral angle 0°.

diameter. The agreement with the theoretical distribution is apparently quite good.

From the analysis described we conclude that it is the surface energy between grains, γ_{gg}, which is the determining factor in the mechanical equilibrium of the structure. Grain boundary energies are in general quite isotropic, with the exception of certain special directions. The preference for hexagonal grains is an obvious consequence of this. Because of the different sizes of the individual grains, no perfect hexagonal structure can be achieved, the resulting structure retains therefore a certain amount of stress. This remaining stress drives the slow evolution in time of the structure.

In Figs. 3(b) and 3(c) we show the result of the slow heating of the polygonal structure of Fig. 3(a). One observes the appearance of fluid, first at the grain boundary vertices, and finally between individual grains, separating them. The fluid gives the appearance of small triangular shapes replacing the three point vertices. The angle of these triangles is the dihedral angle. Mechanical equilibrium is obtained at each point of the triangle if

$$\gamma_{gg} = 2\gamma_{gf} \cdot \cos \frac{\theta}{2}, \tag{2}$$

where γ_{gg} is the grain-grain surface energy and γ_{gf} the grain-fluid surface energy. A determination of the dihedral angle θ therefore gives γ_{gg}/γ_{gf}. Fig. 6 shows the situation for a dihedral angle of 0°, the experimental observations are quite close to this. We can estimate that the dihedral angle lies between 0 and 30°, which puts the ratio γ_{gg}/γ_{gf} in the range between 1.9 and 2.0. A grain boundary of this type can be expected to be incoherent, with only a very thin layer of amorphous helium interspersed.[5] The observations indicate that fluid [4]He is very close to the point of complete wetting of the fcc [4]He grain boundaries.

When the solid helium film is observed in the existence range of the hcp [4]He phase, i.e. below 1.3 kbar and 15 K, one never finds the polygonal structure. The films in this case display a system of parallel grain boundaries, which tend to grow and multiply with time. The time development is again on the same scale as for the polygonal structure, i.e. the development of the morphology takes place over a period of about one hour. Fig. 7 shows the development of such a structure. In Fig. 8 we show a film in which several parallel grain boundary systems are observed. The thin helium film is apparently yielding along easy glide planes. It is also probable that at least some of the structures represent twinning.

400

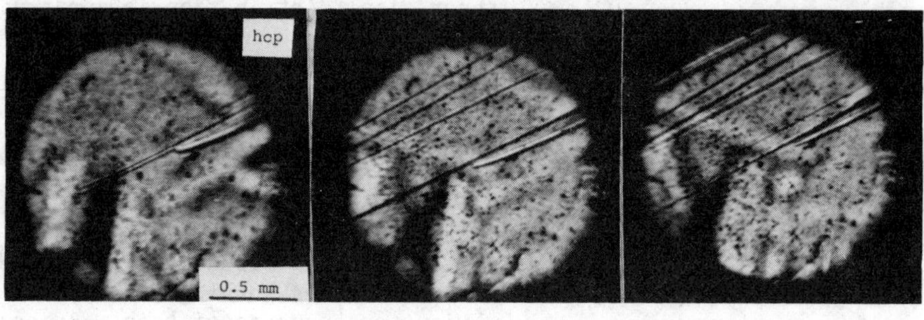

Fig. 7(a)   (b)   (c)

Grain boundary structure observed in a film at 960 bar. The
structure develops in time from left to right for about 30 minutes.

Heating of these structures at a low rate (ca. 1 mK/min) leads
again to melting of the structure. This is shown in Fig. 9. In
contrast to the polygonal structure, one does not observe any
intrusion of liquid among the grains along the parallel grain
boundaries. The grain structure becomes thinner and shrinks away
from the grain-fluid interface, without, however losing its
connection. This observation shows that the grain boundaries in
this case have low surface energy, so that it is not energetically
favorable for the fluid to replace these boundaries. This low
surface energy is typical for coherent twin boundaries.

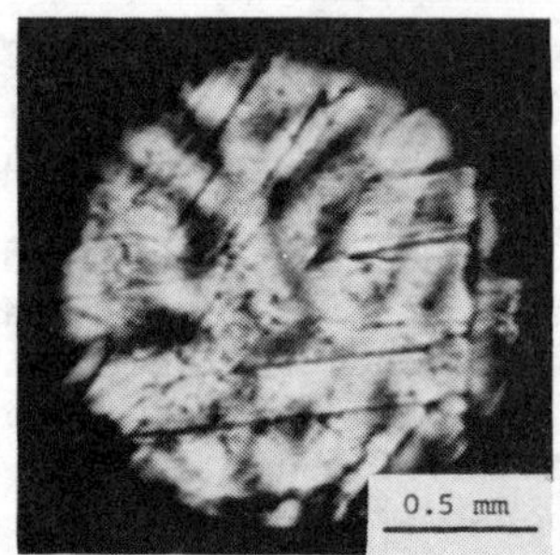

Fig. 8. Grain boundary
structure in a film at
960 bar.

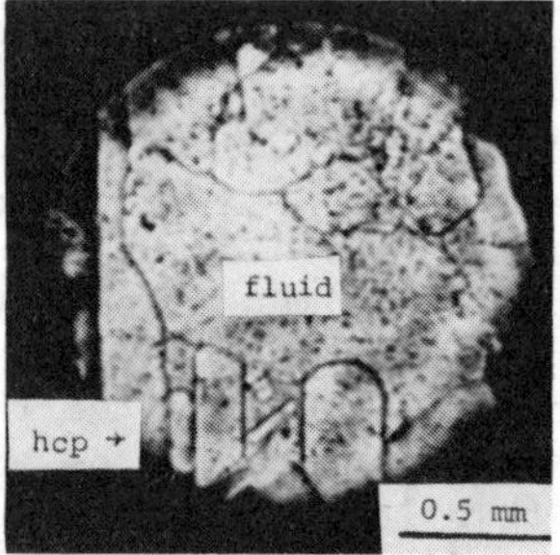

Fig. 9. Partial melting of
an hcp film at 990 bar.

The fact that different macroscopic morphologies are observed
for the hcp and the fcc ^{4}He phase makes it possible to identify
these phases simply by their morphology. It was therefore decided
to study films deposited at the hcp-fcc-fluid ^{4}He triple point. In
this case one always observes that both the hcp and the fcc ^{4}He
phases are laid down side by side on the sapphire windows.
Examples of this are shown in Figs. 10 and 11. It appears therefore
that the choice of the phase that is nucleated and grows to

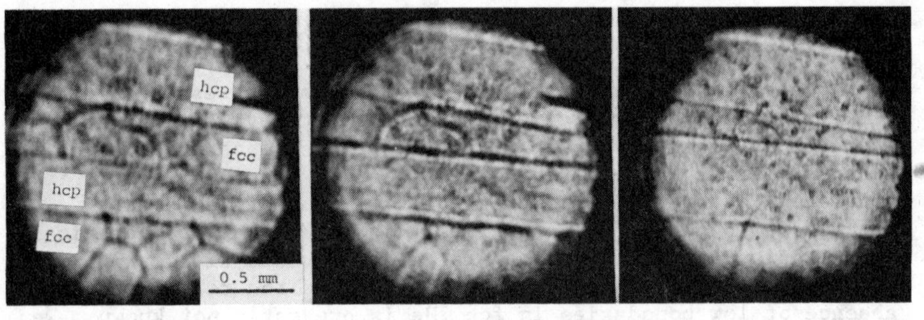

Fig. 10   (a)   (b)   (c)

a) Helium film deposited at 1.1 kbar, showing both hcp and fcc
phases. b) and c), transformation from fcc to hcp ^{4}He during
isothermal holding, time scale about 20 minutes.

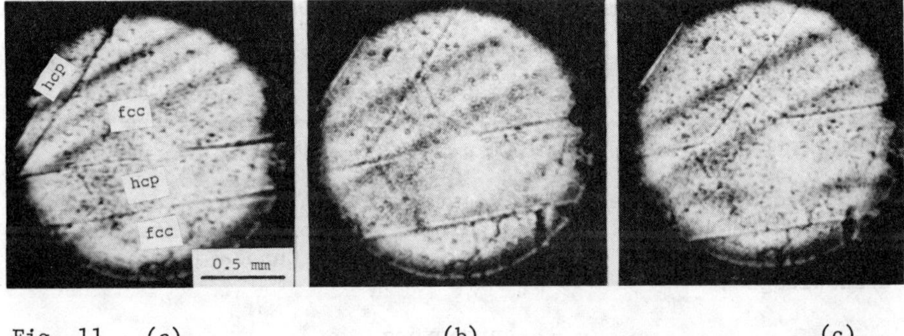

Fig. 11   (a)   (b)   (c)

Transformation from fcc to hcp ^{4}He, conditions similar to those of
Fig. 10.

macroscopic size is determined by the thermodynamical stability
conditions. Prolonged holding of this structure at constant
temperature shows, however, in all cases that the hcp ^{4}He phase
grows   at the expense of the fcc ^{4}He phase. One is therefore able
under these special circumstances to observe the fcc $\rightarrow$ hcp ^{4}He
transition in situ.   In Figs. 10 and 11 we show several stages of
this process. One notices that the transformation takes place by
the parallel displacement of the low energy hcp ^{4}He grain boundaries. The process always takes place in this direction for
isothermal holding. For the fcc $\rightarrow$ hcp ^{4}He transition, high energy
grain boundaries are being gradually eliminated. The process is
therefore energetically favorable. The driving force is apparently
provided by the surface tension of the fcc grain boundaries.
Attempts to observe the reverse transformation (hcp $\rightarrow$ fcc ^{4}He) have
so far failed. Slight heating will be required for this, which
always resulted in the loss of the structure.

DISCUSSION

The observations described here show that stresses in thin films of solid ^{4}He are relieved in different ways for the fcc and the hcp ^{4}He phase. For hcp ^{4}He relief takes place by slipping along low energy grain boundaries. In fcc ^{4}He this process is apparently not possible so that the film is relieved by the creation of high energy boundaries. It is known that mechanical twinning is rare in fcc materials, where certain dislocation climb mechanisms do not operate.[6] Whether this is the reason for the absence of low boundaries in fcc ^{4}He is presently not known. We also do not know whether epitaxial growth on the sapphire windows plays any role.

Melting of these structures shows that the high energy boundaries are very close to the condition of perfect wetting by the fluid. This observation shows that these boundaries are essentially diffuse. The low energy boundaries are not wetted by the fluid. No estimate of the surface energy of these boundaries can be given, apart from the statement that it is much lower than that of the high energy boundaries.

The transformation from fcc to hcp ^{4}He has been observed to take place via the parallel displacement of low energy boundaries. The driving force is apparently the surface energy of the fcc ^{4}He grain boundaries.

Acknowledgments:- These experiments were conducted in cooperation with John Manuel and Kevin Kornelsen. Support by grants from the Natural Sciences and Engineering Council of Canada is gratefully acknowledged.

REFERENCES

1. J. P. Franck and W. B. Daniels, Phys. Rev. B 24, 2456 (1981).
2. C. S. Smith, Trans AIMME 125, 15 (1948); Met. Rev. 9, 1 (1964).
3. I. M. Lifshitz and V. V. Slezov, Zh. Eksp. Teor. Fiz. 35, 479 (1958).
4. M. Hillert, Acta Met. 13, 227 (1965).
5. J. E. Hilliard and J. W. Cahn, Acta Met., 6, 772 (1958).
6. A. H. Cotrell and B. A. Bilby, Phil. Mag. 42, 573 (1951).

USING THE SURFACE TENSION TO ESTIMATE
THE CONDENSATE DENSITY OF SUPERFLUID ^{4}He*

L. J. Campbell
Theoretical Division
Los Alamos National Laboratory
Los Alamos, NM 87545, USA

ABSTRACT

Distortion of the condensate wavefunction at the free surface of superfluid ^{4}He contributes to the surface tension in proportion to the condensate fraction $n_o(T)$. Using this to resolve the present discrepancy between the measured and predicted temperature dependencies of the surface tension gives $n_o(T)$ in good agreement with results from neutron and x-ray scattering measurements. This picture is also consistent with the measured ^{3}He-^{4}He interfacial tension.

INTRODUCTION

The existence of a Bose condensate in superfluid ^{4}He is widely accepted although its direct experimental measurement has proven to be elusive. Currently, the best estimates of the condensate density from neutron and x-ray scattering measurements are forced to rely on assumptions of strong intuitive appeal. Even though the results over the last four years from different workers using different methods (including different assumptions) have been in mutual agreement, independent corroboration is still worthwhile.

On an entirely different subject, the surface tension of superfluid ^{4}He, a puzzle has existed for the last six years since accurate measurements revealed significantly more temperature dependence to the surface tension than can be understood theoretically. It is proposed here that this additional temperature dependence arises from the distortion of the condensate density at the free surface. This leads to an absolute prediction of the condensate density as a function of temperature, which agrees well with the results from scattering measurements mentioned above.

This proposed relationship, although it successfully correlates rather disparate data, also depends on assumptions which, I hope, are not without intuitive appeal.

THE CONDENSATE SURFACE ENERGY

The starting point is the picture of superfluidity represented in the papers of Penrose and Onsager,[1] Yang,[2] and Anderson[3] which emphasize the importance of macroscopic occupation, of off-diagonal long-range order (ODLRO) in the one-particle density matrix, and of broken gauge symmetry. Macroscopic occupation permits the identification of a macroscopic wavefunction in the one-particle density

*Supported by US DOE.

matrix, and the extensive nature of this wavefunction, in particular, of its phase, represents a broken gauge symmetry. The unambiguous normalization of this macroscopic wavefunction is the square root of the condensate density[4] $\sqrt{\rho_0}$.

To avoid confusion it should be pointed out that this macroscopic wavefunction is logically distinct from the more phenomenological order parameter introduced to describe the second-order phase transition at the Lambda point. This order parameter has <u>no absolute</u> normalization; it is an expansion parameter of the thermodynamic potential at T_λ. The form of the phenomenological thermodynamic potential is strongly influenced by the existence of large fluctuations near T_λ and is not expected to be valid beyond 0.1K of T_λ. Usually, the thermodynamic potential is taken to have a form whose variation gives an equation for the order parameter (Ginzburg-Landau) that is formally identical to the Gross-Pitaevskii equation for the condensate. This expansion parameter is sometimes called a macroscopic wavefunction and its structure (a complex scalar) is the same as the condensate wavefunction, but it is not of interest here because the fluctuation-dominated region at T_λ will not be discussed. The present state of the theory near T_λ was recently reviewed by Ginzburg and Sobyanin.[5]

If the macroscopic wavefunction represents only the condensate how can the entire superfluid density ρ_s participate in superflow? This can be understood by the following argument. Include in the single-particle density matrix a term which does not have ODLRO but is phase coherent with the condensate

$$\rho(r,r') = \Psi(r)\Psi^*(r') + g(r-r')e^{i\phi(r)-i\phi(r')} + \rho_n(r,r') \qquad (1)$$

Here g is a real function which goes to zero as the particle separation increases, and $\phi(r)$ is the phase of Ψ: $\Psi(r) \equiv \rho_0^{\frac{1}{2}}(r)e^{i\phi(r)}$. The non-superfluid part of the system is represented by $\rho_n(r,r')$. The current density is

$$j(r) = \frac{1}{2m} \lim_{r'\to r} \frac{\hbar}{i}(\nabla_r - \nabla_{r'})\rho(r,r')$$

$$= \frac{\hbar}{m} (\rho_0(r) + g(0))\nabla\phi(r) + j_n \qquad (2)$$

which shows the total superfluid density associated with the irrotational velocity $\hbar/m\,\nabla\phi$ is

$$\rho_s(r) = \rho_0(r) + g(0) \qquad (3)$$

Note that spatial variation occurs <u>only</u> through the condensate.
(The $g'(0)$ terms are zero because $g(r)$ is symmetric about $r = 0$.)
The normal component $\rho_n(r,r')$ was studied by Fröhlich,[6] who recovered the basic equations of ordinary hydrodynamics. In the
limit that $r-r'$ is large only the condensate part of $\rho(r,r')$
survives and the derivation of a Gross-Pitaevskii type equation for
the condensate goes through as before. (See, for instance, the
text by Fetter and Walecka.[7])

The energy functional for the condensate is taken to be

$$E = \frac{\hbar^2 \rho_0 \xi}{2m^2} \int d^3x \left[|\vec{\nabla}\Phi|^2 + \frac{1}{2}(|\Phi|^2 - 1)^2 \right] \tag{4}$$

where ξ is a coherence length, $x = r/\xi$, and $\Phi = \Psi/\sqrt{\rho_0}$. Minimizing
this energy with respect to Φ gives the Gross-Pitaevskii equation
which, as mentioned before, can be independently derived from the
single-particle density matrix (at least by using some short-cuts,
such as delta function interparticle potentials). Also, and of
some profundity, is the fact that this form of the energy is the
simplest which contains the essential physics of broken gauge
symmetry of the complex scalar variety[8-11] including quantized
vortices.

To calculate the condensate energy associated with a surface
one solves the variational equation (Gross-Pitaevskii) for Eq. (4),
then substitutes it into Eq. (4) to see how it differs from the
uniform case, $\Phi(x) = 1$. Note, there is neither time dependence nor
phase dependence in this exercise, and the outcome is determined
simply from the boundary condition of $\Phi(x)$ at the surface. It is
assumed here that $\Phi = 0$ at the surface (and, of course, $\Phi \rightarrow 1$ in
the fluid) because the existence of a surface implies localization
of atoms in the direction perpendicular to the surface, which is
incompatible with their participation in the condensate. The
depletion of the condensate at the surface implies nothing about
the behavior of the total density there.

It turns out that just this problem was solved (approximately)[12,13] years ago and even appears (exactly) in Chapter 14 of
Fetter and Walecka's text.[7] However, it was solved for a macroscopic wavefunction normalized to the total fluid density (at
$T = 0$). As explained above, normalization of Ψ to ρ_s is not necessary to achieve the full superfluid current, nor is it justified
for ^{4}He.

The answer for the surface energy arising solely from the
condensate is[7]

$$\sigma_c = \frac{\sqrt{2}}{3}\left(\frac{\hbar}{m}\right)^2 \frac{\rho_0}{\xi} \tag{5}$$

where both ρ_0 and ξ are temperature dependent. A more careful calculation of σ_c could take account of the surface profile of the total density, but the present method seems consistent with other uncertainties, as discussed below.

NORMAL SURFACE ENERGY

Absolute values of the surface tension are difficult to calculate. Calculations from first principles use wavefunctions consistent with the existence of a condensate and so automatically and inextricably contain the condensate contribution, but there is no current prospect for such calculations to be done except for T = 0, and so the question of temperature dependence must be addressed by identifying the responsible physical processes and calculating them piecemeal. That the various contributions are simply additive is an assumption.

Atkins[14] identified the most important normal contribution, surface vibrational modes, and from the classical dispersion relation for an incompressible nonviscous fluid he obtained

$$\sigma_n(T_\lambda) - \sigma_n(T) = (-39.7 + 6.50\ T^{7/3}) \times 10^{-3} \text{erg cm}^{-2} \qquad (6)$$

for the temperature dependence of the surface tension. This theory was extended by Edwards, Eckardt, and Gasparini[15] to include effects of compressibility, phonon dispersion, and Gibb's "surface mass," still within a hydrodynamic framework. Two parameters that relate the dependence of surface tension to surface curvature are needed in their theory. These parameters were subsequently calculated by Ebner and Saam,[16] using a density functional theory, and then employed by Eckardt, Edwards, Shen, and Gasparini[17] to obtain the surface entropy $S^s(T)$ for T < 1.7K. In Fig. 1 this is shown, relative to the Atkins entropy, as the straight solid line between 0. and 1.7K. (The slight s-shaped curvature of S^s_{EESG}/S_{Atkins} is not numerically significant here and was eliminated.)

The hydrodynamic theory of Edwards et al. is not expected to remain accurate for temperatures much above 1K, where the effects of surface modes with wave numbers larger than 1A^{-1} begin to affect the entropy. The energy of such modes should be calculated microscopically, and a promising effort in this direction was made by Chang and Cohen,[18] who encountered some difficulties, however, with calculating the surface excitation spectrum at low wave numbers. Their approach, if it could be implemented for a wider range of wave numbers, would have the advantage of eliminating "the question of how classical hydrodynamics should be modified to take account of quantum effects."

Physically, the energy of high wave number surface modes, like all vibrational modes, is expected to decrease compared to the ideal incompressible values. This causes more modes to be excited at a given temperature, thereby increasing the surface entropy. The procedure used here to find the surface entropy will take, at low temperatures, the hydrodynamic theory of Edwards et al. with the Ebner and

Saam parameters, and interpolate across the terra incognito of higher temperature to the terra firma above T_λ, where superfluid effects are gone and the slope of the experimental surface tension can obviously be equated to the slope of the normal surface tension. This interpolation is shown in Fig. 1 as the solid curve for $T > 1.7K$. The experimental value $(d\sigma_e/dT)_{T_\lambda^+} = -0.074$ erg/cm^2°K was taken from

Atkins and Narahara.[19] The more precise data of Magerlein and Sanders[20] do not extend far enough above T_λ to give an unambiguous slope of the normal surface tension. The value they chose, -0.0875 erg/cm^2°K, only represents the slope relative to which $\sigma_e(T)$ appears symmetric around T_λ.

As a sort of sensitivity check, another extrapolation, having no physical signficance, is shown by the long-dash line in Fig. 1.

The normal surface energy is found from the surface entropy by integrating:

$$\sigma_n(T_\lambda) - \sigma_n(T) = -\int_T^{T_\lambda} S^s(T)\, dT \tag{7}$$

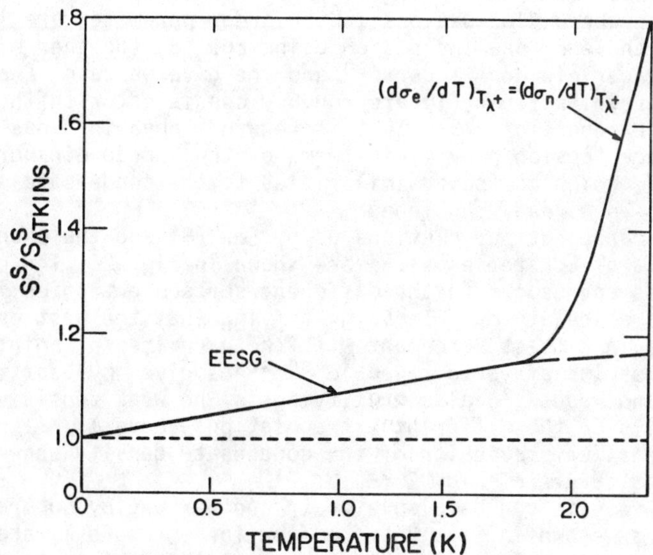

Fig. 1. Surface entropy normalized by the Atkins value, Ref. 14. Solid line below 1.7K: theory of Edwards et al., Ref. 15, using parameters from Ref. 16 as shown in Ref. 17. Solid line above 1.7K: quadratic interpolation to match slope of surface tension above T_λ. Long-dash line: alternative interpolation to T_λ.

RESULTS

Setting the experimental surface tension equal to a sum of the condensate and normal components, subtracting the T_λ values, substituting for σ_c, and introducing the condensate fraction $n_o = \rho_o/\rho$, gives

$$n_o(T) = \frac{3}{\sqrt{2}} \left(\frac{m}{\hbar}\right)^2 \frac{\xi(T)}{\rho} \left[\sigma_e(T) - \sigma_e(T_\lambda) + \sigma_n(T_\lambda) - \sigma_n(T)\right] \qquad (8)$$

For the coherence length ξ the value 1.10^{-8} cm, inferred from the measured dynamics of vortex rings,[21-23] will be used. There is some uncertainty in this value arising from differences in experimental results and from the classical vortex models used to analyze the data (contrasted with Gross-Pitaevskii or Bose vortex rings[12,24,25]). However, any second thoughts about ξ would merely rescale the results for n_o. Supporting this value of ξ are variational calculations of vortex rings using a model many-body wavefunction.[26] These calculations also agree with the dynamical experiments.[27]

Very near T_λ $\xi(T)$ diverges[28] as $(1-T/T_\lambda)^{-2/3}$, but this region is excluded from consideration here. Even though curves for $n_o(T)$ and $\sigma(T)$ will be drawn through T_λ they cannot be taken seriously in the region where fluctuations in the order parameter are large. If, nevertheless, one insists on using Eq. (8) together with the data of Magerlein and Sanders,[20] and the divergence of ξ at T_λ, then the results for $n_o(T)$ are roughly consistent with the expected critical exponent of $2/3$. In this regard, Sobyanin[29] has discussed the surface tension near T_λ in terms of the Landau-Ginzburg order parameter, which has some similarities to the condensate wavefunction, as already mentioned.

The condensate predictions using Eq. (8) and the experimental $\sigma_e(T)$ data of Eckardt et al.[17] are shown in Fig. 2. The different line types correspond to the different surface entropies in Figure 1 used to calculate σ_n. It is gratifying that the best estimate for S^s is in closest agreement with the experimental points, which are the results of various people,[30-32] as given by Sears, Svensson, Martel, and Woods.[32] Also gratifying is the weak sensitivity of the results to the different extrapolations above 1.7K.

An earlier prediction of the condensate density showed only the dashed curves of Fig. 2.[33]

The results can be displayed in another way by subtracting σ_c from σ_e, as shown in Fig. 3. The data for σ_e above T_λ are taken from Atkins and Narahara[19] and consist of their highest temperature result (the point shown) and the associated slope. Although fluctuations may affect the local slope of $\sigma_e(T)$ near T_λ, the apparent cause of the change of average slope across T_λ is the existence of the condensate surface energy.

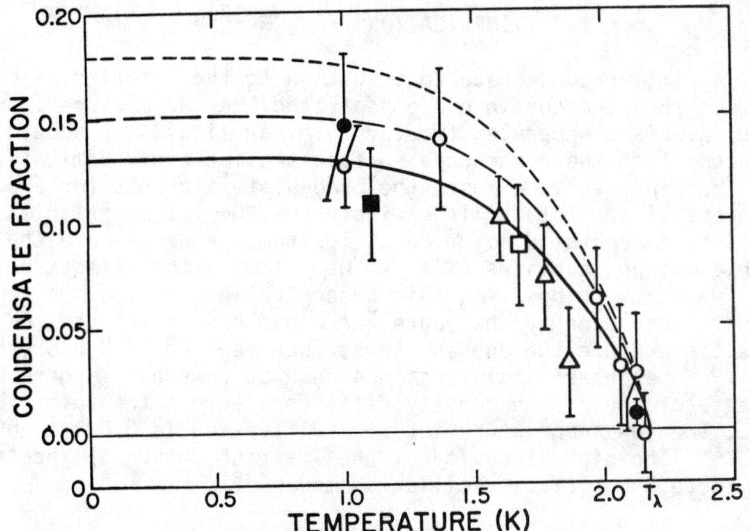

Fig. 2. Predictions of the condensate fraction using the entropies in Fig. 1 with matching line type. The solid line corresponds to the entropy which reproduces the slope of the surface tension above T_λ. The experimental points, taken from Ref. 32 are the results of x-ray (open square), Ref. 30, and neutron scattering (other symbols) measurements, Refs. 31 and 32.

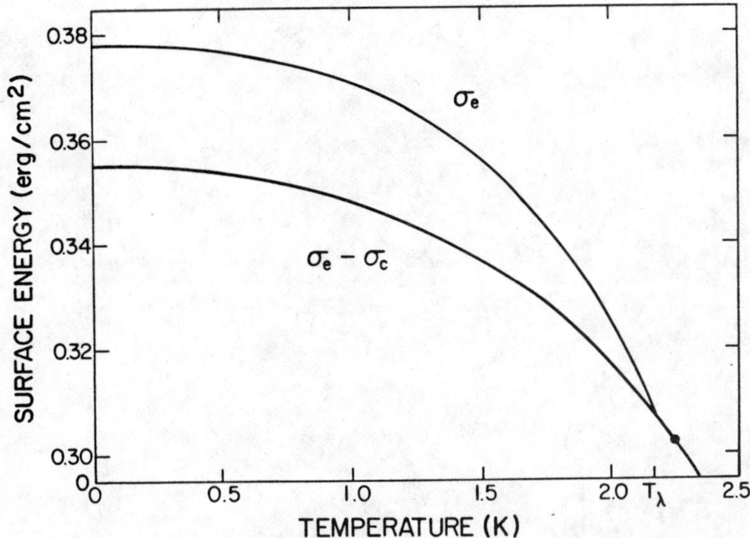

Fig. 3. Experimental surface tension σ_e of Ref. 17 (below T_λ) and Ref. 19 (above T_λ). Also, the difference $\sigma_e - \sigma_c$ using the solid line condensate prediction in Fig. 2.

IMPLICATIONS FOR ^{3}He-^{4}He

A similar condensate contribution to the interfacial surface tension should occur in phase separated ^{3}He-^{4}He mixtures. Because that interface separates two very similar liquids it should be expected that the condensate surface tension would dominate, unlike the ^{4}He-vapor surface where the condensate accounts for about 6%, at most, of the total. To estimate the ^{3}He-^{4}He interfacial tension, consider a separated solution at low temperature, where the ^{4}He-rich component contains 6.4% of ^{3}He. Taking the effective mass of the ^{3}He atoms to be $2.4m_3$ this concentration corresponds to the normal fluid density that pure ^{4}He would have at about 1.6K. At this temperature the condensate surface tension is 0.018 erg cm^{-2}. The ^{3}He-^{4}He interfacial tension should be somewhat larger than this because of the finite density difference across the interface. In fact, the low temperature interfacial tension is 0.023 ± .003 erg cm^{-2}.[34] The high viscosity of the ^{3}He-rich phase is expected to eliminate the surface oscillation modes.[35]

ACKNOWLEDGEMENTS

The stimulation and encouragement Eric Svensson and John Wheatley have given me during this work are gratefully acknowledged.

REFERENCES

1. O. Penrose, and L. Onsager, Phys. Rev. $\underline{104}$, 576 (1956).
2. C. N. Yang, Rev. Mod. Phys. $\underline{34}$, 694 (1962).
3. P. W. Anderson, Rev. Mod. Phys. $\underline{38}$, 298 (1966).
4. E. M. Lifshitz and L. P. Pitaevskii, Statistical Physics, Part 2 (Pergamon, New York, 1980), Sec. 26.
5. V. L. Ginzburg and A. A. Sobyanin, J. Low Temp. Phys. $\underline{49}$, 507 (1982).
6. H. Fröhlich, Physica $\underline{37}$, 215 (1967).
7. A. L. Fetter and J. D. Walecka, Quantum Theory of Many-Particle Systems (McGraw-Hill, New York, 1971), Chaps. 13 and 14.
8. A. Jaffe and C. Taubes, Vortices and Monopoles (Birkhäuser, Boston, 1980), Chap. 1.
9. H. B. Nielsen and P. Olesen, Nuc. Phys. $\underline{B61}$, 45 (1973).
10. J. Chela-Flores, J. Low Temp. Phys. $\underline{21}$, 307 (1975).
11. M. Ichiyanagi, J. Phys. Soc. Jpn. $\underline{43}$, 1125 (1977).
12. D. Amit and E. P. Gross, Phys. Rev. $\underline{145}$, 130 (1966).
13. W. Brouwer and R. K. Pathria, Phys. Rev. $\underline{163}$, 200 (1967).
14. K. R. Atkins, Can. J. Phys. $\underline{31}$, 1165 (1953).
15. D. O. Edwards, J. R. Eckardt, and F. M. Gasparini, Phys. Rev. A $\underline{9}$, 2070 (1974).
16. C. Ebner and W. F. Saam, Phys. Rev. B $\underline{12}$, 923 (1975).
17. J. R. Eckardt, D. O. Edwards, S. Y. Shen, and F. M. Gasparini, Phys. Rev. B $\underline{16}$, 1944 (1977).
18. C. C. Chang and M. Cohen, Phys. Rev. B $\underline{11}$, 1059 (1975).
19. K. R. Atkins and Y. Narahara, Phys. Rev. $\underline{138}$, A437 (1965).
20. J. H. Magerlein and T. M. Sanders, Phys. Rev. Lett. $\underline{36}$, 258 (1976).
21. G. W. Rayfield and F. Reif, Phys. Rev. $\underline{136}$, A1194 (1964).
22. G. Gamota and T. M. Sanders, Phys. Rev. A $\underline{4}$, 1092 (1971).
23. M. Steingart and W. I. Glaberson, J. Low Temp. Phys. $\underline{8}$, 61 (1972).
24. P. H. Roberts and J. Grant, J. Phys. A $\underline{4}$, 55 (1971).
25. C. A. Jones and P. H. Roberts, J. Phys. A $\underline{15}$, 2599 (1982).
26. G. V. Chester, R. Metz, and L. Reatto, Phys. Rev. $\underline{175}$, 275 (1968).
27. V. Chiu and G. V. Chester, Phys. Rev. A $\underline{1}$, 1549 (1970).
28. G. G. Ihas and F. Pobell, Phys. Rev. A $\underline{9}$, 1278 (1974).
29. A. A. Sobyanin, Zh. Eksp. Teor. Fiz. $\underline{61}$, 433 (1971) [Sov. Phys. JETP $\underline{34}$, 229 (1972)].
30. H. N. Robkoff, D. A. Ewan, and R. B. Hallock, Phys. Rev. Lett. $\underline{43}$, 2006 (1979).
31. V. F. Sears and E. C. Svensson, Phys. Rev. Lett. $\underline{43}$, 2009 (1979).
32. V. F. Sears, E. C. Svensson, P. Martel, and A. D. B. Woods, Phys. Rev. Lett. $\underline{49}$, 279 (1982).
33. L. J. Campbell, Phys. Rev. B $\underline{27}$, 1913 (1983).
34. H. M. Guo, D. O. Edwards, R. E. Sarwinski and J. T. Tough, Phys. Rev. Lett. $\underline{27}$, 1259 (1971).
35. D. O. Edwards and W. F. Saam, in Progress in Low Temperature Physics, edited by D. F. Brewer (North-Holland, Amsterdam, 1978), Vol. VIIA, Chap. 4.

UNDERSTANDING EXPERIMENTALLY THE TRANSITION TO CHAOS
IN A CONVECTING DILUTE SOLUTION OF ^{3}He IN SUPERFLUID ^{4}He

Yoshiteru Maeno,* Hans Haucke,* and John Wheatley
Los Alamos National Laboratory, Los Alamos, NM 87545

ABSTRACT

After a brief review of Bénard convection experiments at low
temperatures, experiments with dilute solutions of ^{3}He in
superfluid ^{4}He in small-aspect-ratio cells are described. In
both rectangular and cylindrical geometries, two states with
distinctively different thermal conductances are observed. The
square of the frequency of the oscillation found in the state with
higher conductance is observed to vary linearly with ε =
$(\Delta T/\Delta T_c)$-1. The transition to chaos has been studied using a
solution of 1.6 mole % ^{3}He in a rectangular cell with an aspect
ratio Γ = 1.00 at 0.7 K. Techniques used to characterize the
transition include the time series, power spectral density, phase
space reconstruction of the trajectory, and Poincaré sections. The
analyses reveal clearly that the chaotic state is described by a
strange attractor.

I. BÉNARD CONVECTION AT LOW TEMPERATURES

Consider a horizontal layer of fluid in a gravitational field.
The upper boundary is maintained at temperature T_0 and the lower
boundary is heated to temperature $T_0 + \Delta T$. When the temperature difference ΔT exceeds a certain value ΔT_c, the conductive state of the fluid becomes unstable and convection starts,
forming rolls such as the ones shown in Fig. 1. This phenomenon,
called Bénard convection, is a typical example of dissipative
structures seen in non-equilibrium thermodynamical systems. The
convective onset can be treated as a phase transition of the second
kind, with the ordered phase appearing at larger ΔT and with the
amplitude of the convective velocity as an order parameter. For
some ΔT greater than ΔT_c, steady convective rolls may become
unstable and start to oscillate. The oscillatory convection may be

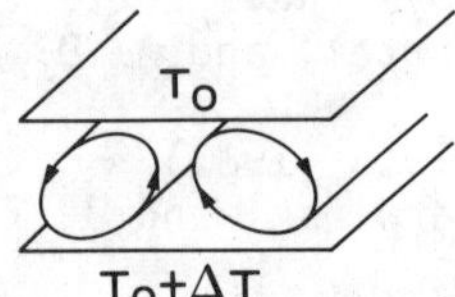

Fig. 1 Bénard convection in
the form of rolls.

*Also at: University of California at San Diego, La Jolla,
California 92093

viewed as an example of an intrinsically irreversible heat engine.[1] A transition to turbulence of the oscillatory convection as ΔT is further increased provides an excellent opportunity to study the behavior of a non-linear dynamical system.

There are three important parameters in this problem. The Rayleigh number R, which reflects the ratio of buoyant to viscous forces and is proportional to ΔT, measures the driving strength. The Prandtl number $Pr = \nu/\kappa$, where ν is the kinematic viscosity and κ is the thermal diffusivity, characterizes the fluid. For a given geometry, it is the ratio of characteristic times for diffusion of heat to that of vorticity. The amplitude of convection and the time-dependent behavior depend strongly on the Prandtl number. The third parameter is the aspect ratio $\Gamma = L/2d$, where L is the diameter of the cell for cylindrical geometry and longer of the horizontal side lengths for rectangular geometry, and d is the vertical height of the cell. The aspect ratio characterizes the geometry, especially how strongly the lateral walls affect the convection.

Study of Bénard convection at low temperatures was pioneered by the work of Ahlers using normal liquid ^{4}He.[2] Behringer and Ahlers have emphasized certain advantages of using a low-temperature environment, in which a precise experiment with a high signal-to-noise ratio can be performed.[3] Libchaber and Maurer have done a series of experiments with normal liquid ^{4}He confined in small-aspect-ratio rectangular cells.[4] By using temperature-sensing local probes, they observed various routes to chaos predicted by dynamical system theory,[5] including the period-doubling cascade. Lucas, Pfotenhauer, and Donnelly have been investigating the convective instability in a rotating cylindrical cell using normal liquid ^{4}He.[6]

Figure 2 shows the Prandtl number of various cryogenic fluids for which Bénard convection can be studied. No one seems to have studied the convection in normal liquid ^{3}He to date. It may be an interesting system to study to complement the experiments with normal liquid ^{4}He.

The convective instability in dilute solutions of ^{3}He in superfluid ^{4}He has been recently investigated both experimentally[8] and theoretically.[9,10] This fluid has various peculiar properties associated with superfluidity. Because of the ^{3}He impurities, the heat flows primarily diffusively, so that the fluid can sustain a steady temperature gradient under which $\partial \vec{v}_s/\partial t$ is zero and both $\vec{v}_s$ itself and the chemical potential μ_4 of ^{4}He per unit mass of the solution are spatially homogeneous.

This situation imposes a well-defined relationship between temperature T and molar concentration x_3, such that at zero pressure

414

gradient and in the ideal dilute solution limit

$$s_{40}dT + \frac{k_B}{m_4} d(Tx_3) = 0 , \qquad\qquad (1)$$

where s_{40} is the entropy per unit mass of pure [4]He, k_B is the Boltzmann constant, and m_4 is the [4]He atomic mass. Because of this relationship, [3]He quasiparticles tend to accumulate where the temperature is lower; this behavior is known as the heat-flush effect. Thus heating from above can cause the fluid to become gravitationally unstable. As far as the convective instability is concerned, what is implied by Eq. (1) is that x_3 does not vary independently of T, so that the system is well approximated by a

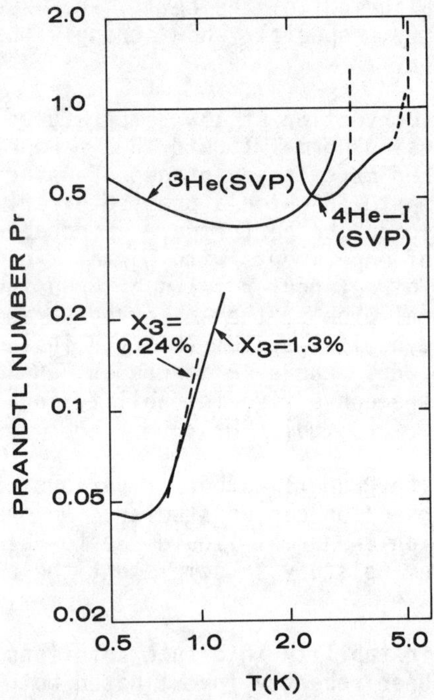

Fig. 2. The Prandtl number of various cryogenic fluids. Data used are taken from Refs. 2 and 6 for [4]He-I, and from Ref. 7 for [3]He. Vertical dashed lines indicate the liquid-vapor critical temperatures at 5.20 K for [4]He and at 3.32 K for [3]He.

one-component fluid with the Rayleigh number[9]

$$R = g \alpha_{p \mu_4} d^3 \Delta T / \nu \kappa_{eff} , \qquad (2)$$

where g is the acceleration due to gravity, $\alpha_{p \mu_4} = -\rho^{-1} (\partial \rho / \partial T)_{p \mu_4}$ is the effective thermal expansion coefficient, $\kappa_{eff} = k_{eff} / \rho C_{p \mu_4}$ is the effective thermal diffusivity with k_{eff} being the effective thermal conductivity, and $\nu = \eta / \rho$ is the kinematic viscosity.

Since curl $\vec{v}_s = 0$, the superfluid cannot participate in convective motion unless quantized vortices are produced. When only the normal component fluid (phonons, rotons, and ^{3}He quasiparticles) is convecting, the appropriate Prandtl number is

$$Pr = \rho \nu / \rho_n \kappa_{eff} ,$$

which is shown in Fig. 2. This is an order of magnitude smaller than that of normal liquid ^{4}He or ^{3}He. If it is possible for the superfluid to participate in convective motion by generating quantized vortices, the Prandtl number can be as small as

$$Pr = \nu / \kappa_{eff} ,$$

which is 0.0011 at 0.60 K for $x_3 = 1.3\%$. No other known terrestrial convecting fluid has such a small Prandtl number. (Liquid metals have low Pr because heat can be transported very effectively by electrons. For mercury Pr = 0.025, and for liquid Potassium at 700°C Pr = 0.0035.)

II. EXPERIMENTS WITH A DILUTE SOLUTION OF ^{3}He IN SUPERFLUID ^{4}He

We will now describe experiments with a solution of 1.6 mole % ^{3}He in superfluid ^{4}He, and discuss various aspects of this problem. A cross-sectional drawing of the convection cell is schematically shown in Fig. 3. The cell was constructed with top and bottom plates of copper, and a spacer fitting between the plates made of a graphite-resin material, Vespel, of high thermal resistivity.[11] This spacer fitted snugly into a 0.20-mm wall thickness stainless steel tube, 4.34 cm i.d., which served to confine the fluid. The shape and size of the hole in this removable spacer, together with the top and bottom plates, defined the cell geometry. The cell height was d = 0.80 cm; for most of the data presented here the cell geometry was rectangular with aspect ratios $\Gamma = L/2d = 1.00$ and $\Gamma' = W/2d = 0.70$, where L and W are the longer and shorter horizontal side lengths, respectively. A small copper piece was inserted into a hole in the center of the top plate and thermally isolated from the rest of the top plate by a

0.05-mm-thick Mylar sheet. This piece, which we call the "probe", had a diameter of 0.30 cm. Two thermocouple differential thermometers connected to SQUID ammeters[12] were used to infer the fluid motion. One was attached between the top and bottom plates (top-bottom TC) and the other between the probe and the top plate (probe TC). Near the transition to convection the response time of the top plate temperature to a perturbation of the bottom plate temperature is typically 4 s. The top-bottom TC is therefore sensitive to fluctuations in bottom-plate temperature with shorter periods. In contrast, the thermal response time of the probe TC system is only about 15 ms at 1 K because of the very small mass of the probe. Hence, any low frequency noise common to the top plate and the probe is not seen in the probe TC. This situation makes the probe TC an ideal sensor to study time-dependent convection with very large signal-to-noise ratio.

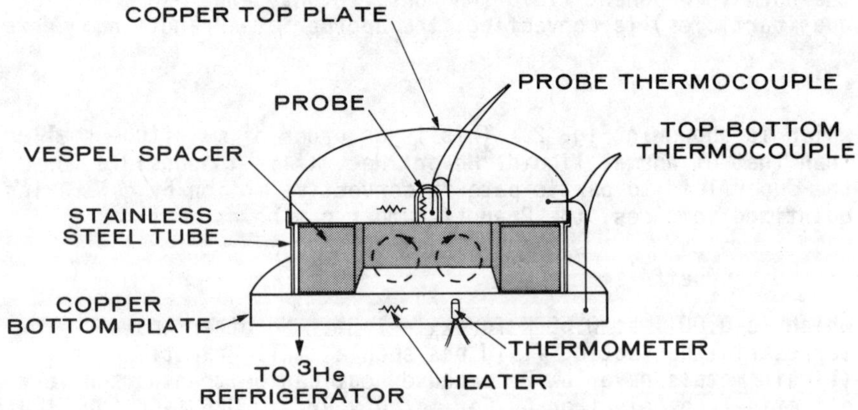

Fig. 3 Cross-sectional schematic of the convection cell.

The bottom plate was connected to a ^{3}He evaporation refrigerator and was maintained at a fixed temperature by means of a germanium resistance thermometer and a resistance heater mounted in the bottom plate. The bottom plate temperature could be chosen to have any value between 0.5 K and 1.2 K. Over this temperature range, the Prandtl number of the fluid increases by about a factor of five. A constant amount of heat $\dot{Q}_{TOP}$ was applied to the top plate, while the outputs of the top-bottom and probe TC's were recorded. $\dot{Q}_{TOP}$ was divided into two parts: $\dot{Q}_P$, applied to a heater mounted in the probe, and $\dot{Q}_{TOP} - \dot{Q}_P$, applied to a heater attached to the rest of the top plate. By considering the fluid conductance, the wall conductance, and the Kapitza resistance between the fluid and the top and bottom plates, we chose $\dot{Q}_P/\dot{Q}_{TOP}$ to be a fixed value of 3.7% at 0.70 K for the rectangular geometry.

The onset of convection may be detected by measuring the effective thermal conductance $K = \dot{Q}_{TOP}/\Delta T$ as a function the time-averaged temperature difference ΔT between the top and bottom plates. In Fig. 4 the normalized conductance K/K_C for a

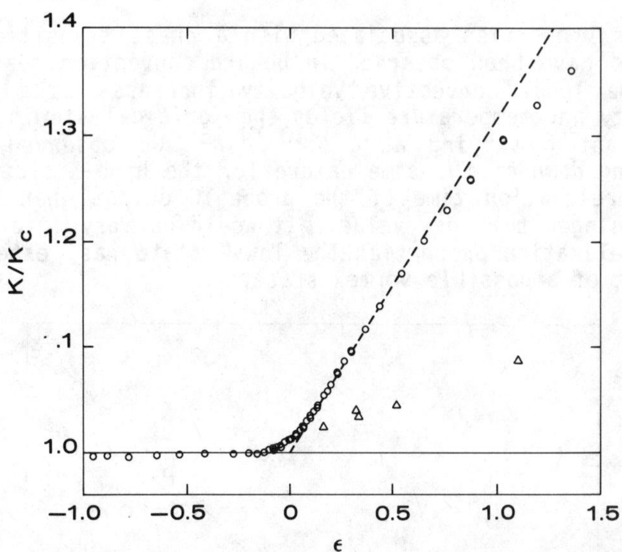

Fig. 4. The normalized thermal conductance K/K_C for a
rectangular cell with an aspect ratio $\Gamma = 1.00$ at 0.70 K,
plotted against $\varepsilon \equiv (\Delta T/\Delta T_C) - 1$. $K_C = 3.660$ mW/K
and $\Delta T_C = 7.706$ mK. o:high-K state, $\triangle$:low-K state. Dashed
line shows a fit to the high-K state conductance between ε =
0.134 and 0.438.

rectangular cell with an aspect ratio $\Gamma = 1.00$ at 0.70 k is
plotted against a quantity $\varepsilon \equiv (\Delta T/\Delta T_C) - 1$, which we will
use as a driving parameter. A critical temperature difference for
convection, ΔT_C, was defined by fitting K for the diffusively
conductive state and for a region of the convective state where K
increases linearly with ΔT to straight lines and finding the intersection. The effective conductance at the onset, K_C, was
taken as the value of the fitted lines at ΔT_C. The rounding of
K seen near the onset is likely due to the imperfection in cell
geometry. Heat conduction through the walls contributes about 12%
of K and is mostly due to the stainless steel tube, rather than the
Vespel spacer. A correction due to the Kapitza boundary resistance
between the copper plates and the fluid is about 10%. In both
cylindrical geometry ($\Gamma = 1.00$ and 1.20) and rectangular geometry
($\Gamma = 1.00$) and in all solutions studied with ^{3}He molar concentration ranging from 0.24 to 1.6%, there exist two distinctive
states of convection with significantly different conductance values, as indicated in Fig. 4 by circles for a high-K state and by
triangles for a low-K state. In contrast to behaviors of the
high-K state, which will be discussed below, the low-K state does

not have a clear onset of oscillation and is much noisier.[8] They
may be understood by the concept of the two possible convective
states discussed in the previous section.

Various properties associated with a phase transition of the
second kind have been observed in Bénard convection near its onset.[13] The local convective velocity increases as $\varepsilon^{1/2}$ and
the velocity and temperature fields show critical slowing down with
time constant τ varying as ε^{-1}.[13,14,15] We observed critical slowing down of the same nature for the high-K state by measuring the relaxation time of the probe-TC output when $\dot{Q}_{TOP}$ was
suddenly changed to a new value. It would be very interesting to
see what relaxation properties the low-K state has, especially in
the context of a possible vortex state.

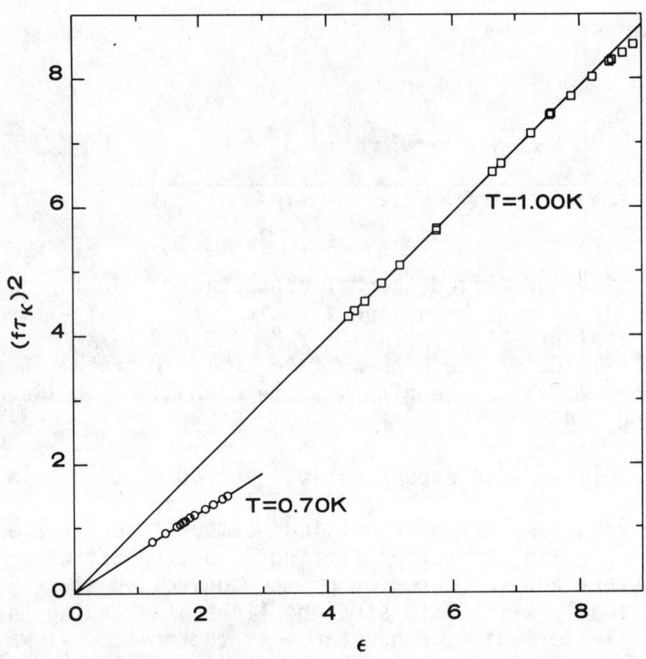

Fig. 5. The square of the frequency of the oscillations
observed in a cylindrical cell with an aspect ratio Γ = 1.20,
plotted against ε. The frequency is normalized by the
vertical diffusion time τ_K. Solid lines show fits to the
data below ε = 8.

The high-K state exhibits a clear onset of oscillations. This
instability is interpreted as the "oscillatory instability" expected to occur in low Prandtl number fluids.[16] The square of
the frequency of oscillations observed in a cylindrical cell with
an aspect ratio Γ = 1.20 is plotted against ε in Fig. 5. The
frequency is normalized by τ_K^{-1}, where τ_K =
$4d^2/\pi^2\kappa_{eff}$ is the vertical thermal-diffusion time, and

ε has been corrected for the wall conductance and the Kapitza boundary resistance. Solid lines show least-square fits to the data below $\varepsilon = 8$. The period of the convective-roll oscillation is expected to be closely related to the turnover time of the rolls; the square of the frequency is expected to vary as ε, as long as the convective velocity varies as $\varepsilon^{1/2}$. But the fact that the observed square of the frequency varies very closely as ε all the way up to $\varepsilon = 8$ is rather surprising.

III. TRANSITION TO CHAOS

In this section we concentrate on time-dependent convection in a 1.6 mole % solution near 0.70 k for a rectangular cell with aspect ratios $\Gamma = L/2d = 1.00$ and $\Gamma' = W/2d = 0.70$. Our intention is to explain several techniques available to analyze a complicated dynamical behavior of the system. The onset of an oscillatory state is at $\varepsilon = 3.77$. As the temperature difference, or the strength of the drive, is increased, the fluid eventually starts to show a more and more complicated periodic motion, then displays a chaotic motion. We will see that the "trajectory" of the time-dependent temperatures in a "phase space" can be used to represent the time evolution of the state of the system, and that even in the chaotic state the trajectory is confined to a definite geometrical structure, called an "attractor". In the chaotic state, motion on the attractor is so sensitively dependent on initial conditions that one cannot predict the evolution of the state far into the future. Such an attractor is known as a "strange attractor", or a "chaotic attractor".[17]

In Figs. 6 and 7, the output of the probe TC, $X(t)$, and top-bottom TC, $Y(t)$, are plotted as functions of time for three closely spaced values of the driving parameter ε. The amplitude of ca. 0.3 mK p-p seen in both thermocouple sensors may be compared to $\Delta T \approx 40$ mK. The power spectral density (PSD) of the "time series" $X(t)$ is shown in Fig. 8. At $\varepsilon = 4.358$, only the fundamental frequency at $f_0 = 0.1085$ Hz and its harmonics are present. As seen in Fig. 6, the probe TC output has a large frequency component at $4f_0$, caused by the complicated motions of the rolls both along and transverse to the roll axes. At $\varepsilon = 4.369$ a second frequency f_1, incommensurate with f_0, appears with the ratio $f_1/f_0 = 1.893/3$. As ε is increased, f_1/f_0 approaches the value 2/3. At $\varepsilon = 4.381$ these frequencies are locked to the ratio of two to three, and the $f_0/3$ difference-frequency component exhibits a "period-doubling", as $f_0/6$ is seen in Fig. 8(b). Subsequently, there is a period-doubling cascade; $f_0/12$ can be seen in Fig. 8(c). So far we have been able to obtain a PSD with a $f_0/24$ frequency component, which is 1/8 of $f_0/3$. The existence of the $f_0/6$ component is easily seen in the time series in Fig. 6(b), in which maxima appear, as indicated by arrows, at intervals which are six times the fundamental period. Both a significant rise in the noise floor and a

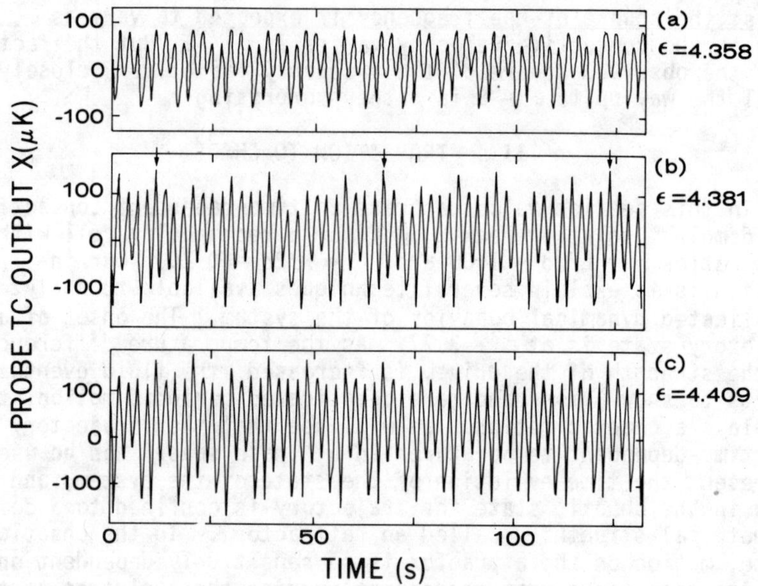

Fig. 6. The probe-TC output, $X(t)$, for three closely-spaced values of $\varepsilon \equiv (\Delta T/\Delta T_c)-1$. The output was filtered by a two-pole low-pass filter with a cutoff frequency of 10 Hz.

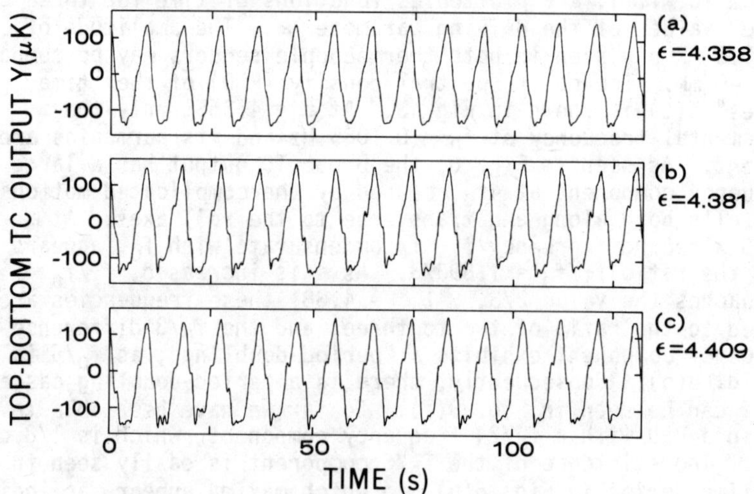

Fig. 7. The top-bottom-TC output, $Y(t)$, corresponding to $X(t)$ in Fig. 6. A 10 Hz low-pass filter was used.

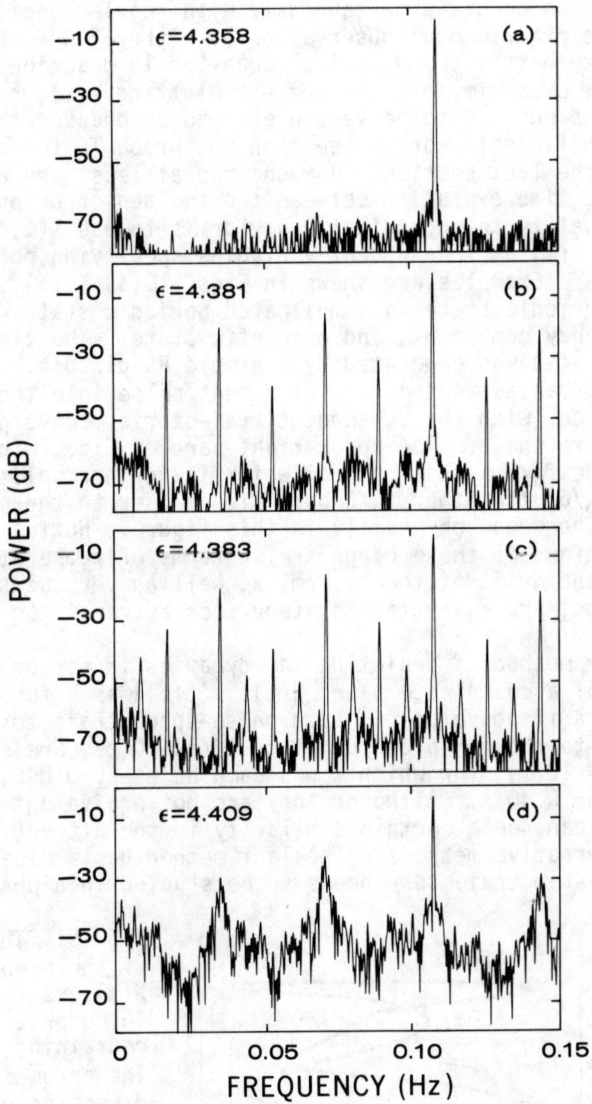

Fig. 8. The power spectral density of the probe-TC output. The sampling time was 0.308s, and the output was filtered by a two-pole low-pass filter with a cutoff frequency of 0.6 Hz.

broadening of the $f_0/3$ peak and its harmonics can be seen in Fig.
8(d) for ϵ = 4.409 and indicate that the motion is chaotic.
Notice the PSD peak at f_0 is still very sharp; the motion is only
weakly chaotic in this sense.

Both the time series (Fig. 6) and the PSD's (Fig. 8) show that
chaotic motion occurs simultaneously with regular oscillation, im-
pairing the experimental observation of either. One approach to
characterize better the dynamical behavior is described below in
the context of a simple phase space. Plotting X(t) vs Y(t) as in
Fig. 9 turns out not to be very useful to us because the top-bottom
TC has significantly more noise than the probe TC for reasons discussed in the last section. But one can at least see a coherence
in periodic time evolution between the two sensor outputs. Another
way to visualize the dynamical behavior is to use X(t) and its time
derivative $\dot{X}(t)$ as independent variables specifying phase-space
coordinates. Examples are shown in Figs. 10(a)-10(c), for a
"simple" periodic state, a complicated periodic state with the
$f_0/6$ frequency component, and a chaotic state. The time
derivative $\dot{X}(t)$ was generated by a simple RC circuit. In Fig.
10(a) disturbances in the form of a heat pulse into the probe were
applied twice, with the subsequent trajectories converging onto the
attractor are shown. The upper-right part of Fig. 10(b) shows two
of the three $f_0/3$ bands, each of which is further split into two
bands of $f_0/6$; the other $f_0/3$ band lies closer to the center
and cannot be seen very easily in this figure. Notice how small
the separations of these bands are in terms of temperature. Very
low noise and drift of the system, as well as very sensitive means
of detection, are essential to study such a transition to chaos.

Another method of depicting the dynamics is to see the development of a coordinate pair $\{ X(t), X(t+\tau) \}$ as a function of
time t for a fixed value τ. Such phase-space trajectories, corresponding to the X,$\dot{X}$ plots of Figs. 10(a)-10(c), are shown in
Figs. 10(d)-10(f), for which X was sampled every 0.040 s and τ is
chosen to be 0.40 s. Although they are not equivalent to the X,$\dot{X}$
plots, one can see a certain similarity in the attractors generated
by the alternative methods. The last method has a clear advantage
when the system trajectory needs to be studied in a phase space

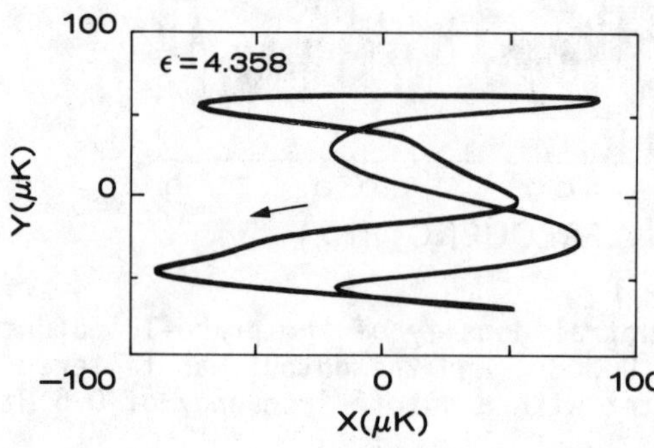

Fig. 9. The trajectory
of the probe-TC output,
X(t), vs the top-bottom-
TC output, Y(t),
containing 15 orbits.
The arrow indicates the
direction of flow of
the trajectory. The
cutoff frequencies of
the low-pass filters
used are 10 Hz for X(t)
and 0.1 Hz for Y(t).

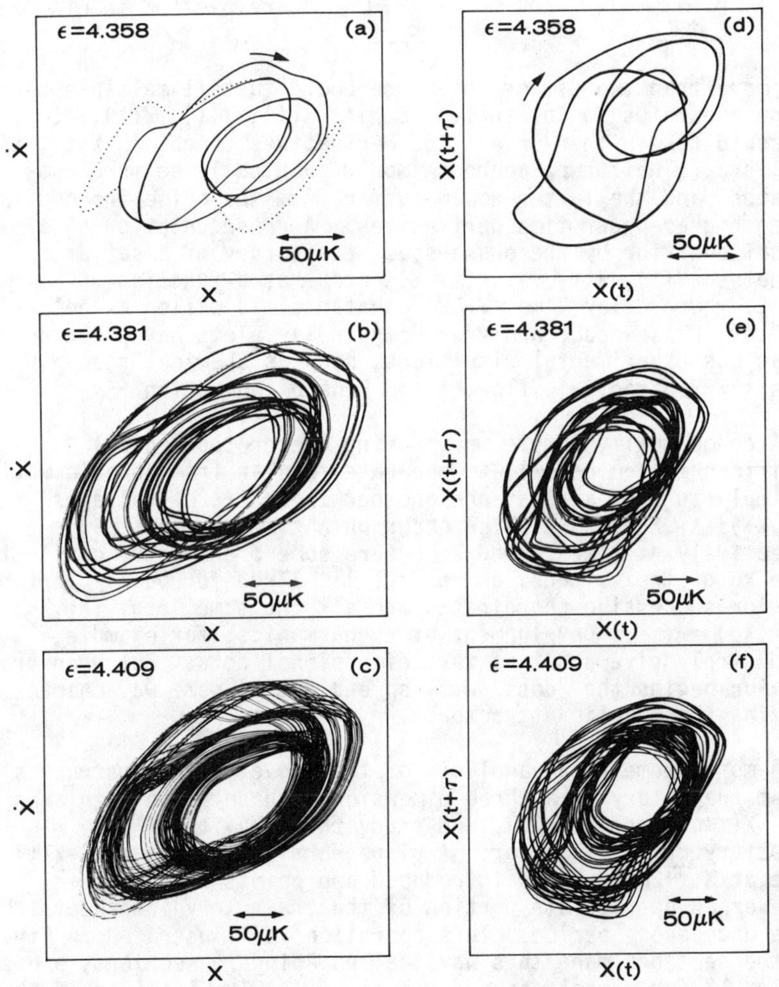

Fig. 10. Alternative ways of representing the phase-space trajectories. (a)-(c): the probe-TC output X(t) vs its time derivative $\dot{X}(t)$ (X,$\dot{X}$ plots). (d)-(f): X(t) vs X(t+τ) with τ = 0.40s (delay plots). The two dotted lines show the trajectory after disturbances were applied (see text).

424

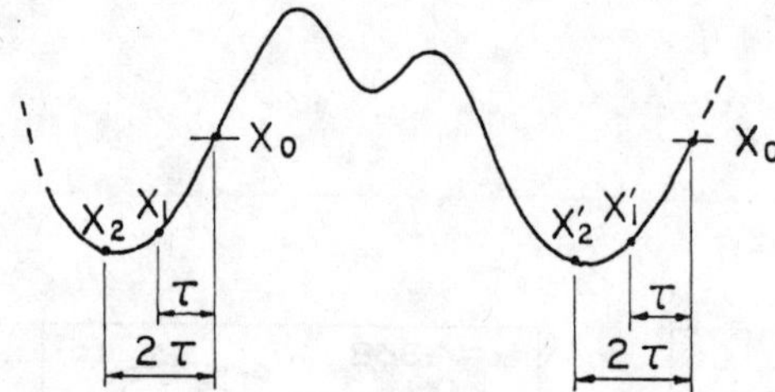

Fig. 11. A method to separate regular from chaotic behaviors in a time series X(t). Co-ordinate pairs (X_2,X_1) are plotted in Fig. 13.

with more than two dimensions. One could install multiple sensors in the apparatus to obtain time series X(t), Y(t), Z(t), etc., or one could take higher-order time derivatives to obtain X(t), Ẋ(t), Ẍ(t), etc. The former approach would eventually be more complicated, and the latter would suffer from the noise induced in taking higher-order time derivatives. A representation of a dynamical behavior by the phase-space trajectory of a set of co-ordinates $X(t)$, $X(t-\tau_1)$, $X(t-\tau_2)$, etc. as a function of time t, where delay time τ_i are constants, is called a "delay plot".[18] Phase-space analyses using delay plots have been done in various experimental situations, such as chemical reactions,[19,20] baroclinic flow,[21] and Bénard convection.[22]

A conceptually simple way, having deeper topological significance (see below), to separate regular from chaotic motion is simply to plot against one another the pairs of voltages $[(X_1,X_2);(X'_1,X'_2); \cdots]$ which occur on a time series respectively at times τ and 2τ before some preselected constant value x_0 of the voltage, as in Fig. 11. Then for purely periodic behavior successive coordinates are all the same, e.g. $(X_1,X_2) = (X'_1,X'_2) = \cdots$. Development of subharmonics, for example, could simply give a set of zero dimensional dots. But when chaotic behavior begins the "dots" will spread out in some way characterizing the chaotic attractor.

A more geometrical analysis of the above is to construct a system trajectory in a three-dimensional phase space with axes $X(t)$, $X(t+\tau)$, and $X(t+2\tau)$, and study the behavior of the trajectory on an attractor. A plane parallel to the $X(t)$-$X(t+\tau)$ plane at $X(\overline{t+2\tau}) = X_0$ is introduced and points are recorded whenever an appropriate portion of the trajectory intersects this plane once every period. This operation is illustrated in Fig. 12, and the sections made this way, called Poincaré sections, are shown in Fig. 13 for closely spaced values of ϵ. The location of the section along $X(t+2\tau)$ axis, X_0, is chosen carefully so that each Poincaré section is made consistently and nearly transverse to the flow of the trajectory. The cluster in Fig. 13(a) contains 43 points and is indicative of the noise level in our experiment. It shows how this method of describing the data removes sensitivity to the fundamental oscillation frequency. Figure 13(b) is for a complicated periodic state with the $f_0/6$ frequency component. The numbers in the figure indicate the order in which points appear on

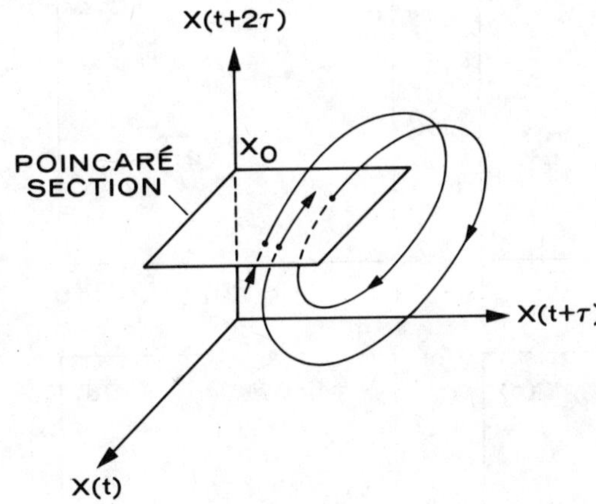

Fig. 12. Illustration of a Poincaré section.

the section. After six periods, a point appears in the same cluster. In Figs. 13(b)-13(d), each of which contains 131 points, the 3n+1st, 3n+2nd, and 3n+3rd points appearing on the section are represented by crosses, closed circles, and closed triangles, respectively. At ε = 4.399 the state is only slightly chaotic and there is no "mixing" between the three clusters of points, each of which represents the $f_0/3$ frequency component; however, within each cluster the points appear in an erratic manner. At ε = 4.409 there is not only overlapping but also mixing between the clusters. The fact that the corresponding PSD in Fig. 8(d) shows broad, but distinctive $f_0/3$ frequency peaks suggests there exists at least a short-time correlation as to where the successive points appear.

Consider an observer finding points on a Poincaré section. For a periodic state as in Fig. 13(b), he could predict accurately by using his knowledge of the initial-point location in which region of the section a point would appear, say, 100 periods later. For a slightly chaotic state as in Fig. 13(c), he would not be able to make an accurate statement, for a given cluster, as to the locations within the cluster where points will appear in the future. For a more chaotic state as in Fig. 13(d), knowledge about an initial-point location very soon loses its value and it is impossible to predict correctly the location of the points appearing many periods later. However, the points in the Poincaré section are not distributed at random, as we might suppose would characterize "noise". Rather, they are confined to a definite part of the space, reflecting the qualities of a chaotic attractor.

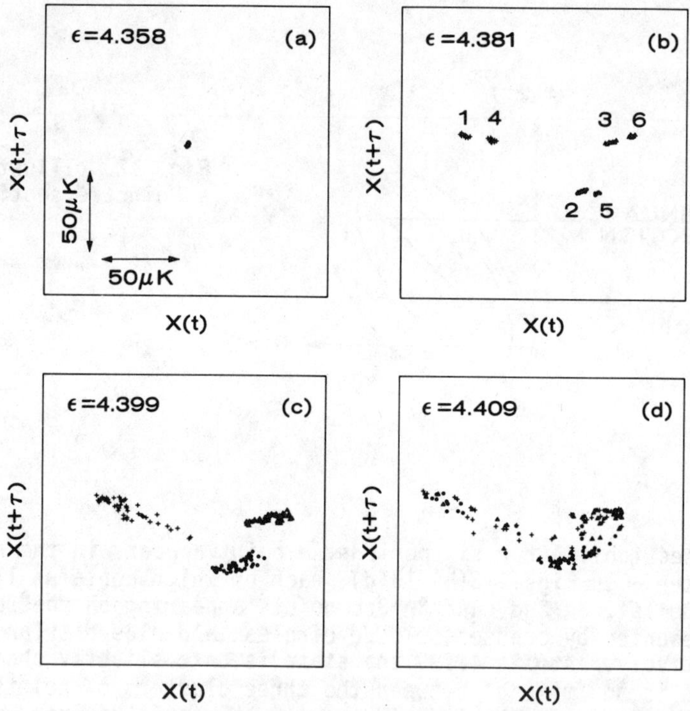

Fig. 13. Poincaré sections made by the method illustrated in Fig. 12. (a): a periodic state with a single frequency f_0. (b): a complicated periodic state with an $f_0/6$ frequency component. (c) and (d): chaotic states. The 3n+1st, 3n+2nd, and 3n+3rd points appearing on the section are represented by +, ●, and ▲, respectively. The temperature scales are the same for all the figures.

The qualities of our thermal oscillator as it approaches a chaotic state have been presented here primarily in a graphical and intuitive way. There exist in principle a variety of ways quantitatively to characterize a chaotic attractor. Quantities such as fractal (or Hausdorff) dimension, information dimension, a correlation exponent,[23] as well as a Liapunov characteristic exponent, topological entropy, and metric entropy[24] are examples. The evaluation of these quantities in practice require a very high signal-to-noise ratio and very low drift of the observations, a large number of data points, and a relatively simple structure of the chaotic attractor. We are aware of only one instance in which any of these quantities has been successfully evaluated for a real physical system to date.[25] In addition to the quantities mentioned above, we are currently attempting to develop other quantitative means which are experimentally more accessible to characterize the evolution of the chaotic attractor observed in Bénard convection.

ACKNOWLEDGEMENTS

We have greatly benefited from discussions with L. Campbell, A. Fetter, G. Baym, and R. Donnelly on superfluid hydrodynamics, and with J. D. Farmer and N. Packard on dynamical system theory. We also wish to thank A. Migliori and G. Swift for their many helpful suggestions and W. E. Keller for his encouragement.

REFERENCES

1. J. Wheatley, T. Hofler, G. Swift, and A. Migliori, Phys. Rev. Lett., 50, 499 (1983).
2. G. Ahlers, Phys. Rev. Lett. 33, 1185 (1974).
3. R. Behringer and G. Ahlers, J. Fluid Mech. 125, 219 (1982).
4. A. Libchaber, Physica 109 & 110B, 1583 (1982), and references to their work therein.
5. J. Eckmann, Rev. Mod. Phys. 53, 643 (1981).
6. P. Lucas, J. Pfotenhauer, and R. Donnelly, J. Fluid Mech. 129, 251 (1983).
7. J. Wilks, The Properties of Liquid and Solid Helium (Oxford University Press, London, 1967).
8. P. Warkentin, H. Haucke, and J. Wheatley, Phys. Rev. Lett. 45, 918 (1980), and H. Haucke, Y. Maeno, P. Warkentin, and J. Wheatley, J. Low Temp. Phys. 44, 505 (1981).
9. A. Fetter, Phys. Rev. B26, 1164 and 1174 (1982).
10. V. Steinberg, Phys. Rev. A24, 975 (1981).
11. Vespel SP-22 polymide resin, DuPont Company.
12. Y. Maeno, H. Haucke, and J. Wheatley, to appear in Rev. Sci. Instrum. (1983).
13. J. Weisfreid, Y. Pomeau, M. Dubois, C. Normand, and P. Berge, J. de Phys. Lett. 39, 725 (1978).
14. Y. Sawada, Phys. Lett. 65A, 5 (1978).
15. R. Behringer and G. Ahlers, Phys. Lett. 62A, 329 (1977).
16. F. Busse, Rep. Prog. Phys. 41, 1929 (1978).
17. For a good introductory review of strange attractors, see R. Shaw, Z. Naturforsch. 36a, 80 (1981).
18. N. Packard, J. Crutchfield, J. D. Farmer, and R. Shaw, Phys. Rev. Lett. 45, 712 (1980).
19. J. Roux, A. Rossi, S. Bachelart, and C. Vidal, Physica 2D, 395 (1981).
20. R. Simoyi, A. Wolf, and H. Swinny, Phys. Rev. Lett. 49, 245 (1982).
21. J. D. Farmer, J. Hart, and P. Weidman, Phys. Lett. 91A, 22 (1982).
22. H. Haucke and Y. Maeno, to appear in Physica D (1983).
23. P. Grassberger and I. Procaccia, Phys. Rev. Lett. 50, 346 (1983).
24. J. D. Farmer, Z. Naturforsch. 37a, 1304 (1982).
25. R. Shaw has successfully evaluated the metric entropy in his dripping faucet experiment. To be published.

CONVECTION IN A SUPERFLUID SOLUTION*

V. Steinberg
Department of Physics
University of California
Santa Barbara, CA 93106

ABSTRACT

The convective instability of superfluid ^{3}He-^{4}He mixtures
confined between horizontal planes is discussed. It is emphasized
that this system is one of the most convenient objects for the
study of nonlinear effects which are Prandtl number dependent
because of the wide variation of the properties of the mixture
with temperature and concentration and because of the advantages
of the low temperature environment and technique. Linear stability
analysis for stationary and oscillatory instabilities is reviewed
for bulk solutions as well as for solutions in porous medium.
The influence of non-classical two-fluid effects on the onset of
convection is also discussed. The relationship of these theoretical results to recent experiments and possible new experiments
is considered.

INTRODUCTION

When a horizontal layer of fluid is heated from below, convection will start if the temperature gradient exceeds a critical
value.[1] Recent interest in this phenomenon is provoked by the
fact that such a system is one of the most convenient objects for
the study of the nonlinear phenomena in dynamical systems, including pattern and wave number selection in nonlinear processes
and the transition to turbulence.

The onset of convection in a laterally infinite horizontal
layer of a one-component fluid depends only on one non-dimensional
parameter, namely the Rayleigh number which is proportional to
the temperature difference between the top and bottom plate.
Insofar as the non-linear convection. is concerned, another parameter, namely the Prandtl number, should be considered. The
Prandtl number is characteristic of the fluid and is the ratio
between the energy diffusion time and the momentum diffusion time.
The possibility of varying the Prandtl number has very essential
and interesting implications in non-linear problems. The competition between two non-linear terms, convection of fluid vorticity
in the Navier-Stokes equation and heat convection in the Fourier
equation, are responsible for all non-linear behavior of the
systems and the path to turbulence.[2] The ratio between these
two terms is the Prandtl number, P. Until now, this dependence
has been studied on several systems with very different Prandtl
numbers, and qualitatively different behavior has been observed.
An experiment with continuous change of P between 0.4 and 0.8 has

*This work was supported by NSF Grant #MEA81-17241

been performed on normal liquid ^{4}He by changing the temperature
and the pressure.[3]

A low temperature system that shows a much wider variation
of the Prandtl number is a ^{3}He-^{4}He superfluid mixture. The Prandtl
number changes for this system from about 0.02 to 2.3 mostly in
a dilute region.[4] Therefore, by changing temperature and concentration, it is possible to vary P from water-like values to values
typical of liquid-metals. Since in a superfluid solution different mechansims are responsible for momentum and energy transfer, their interplay leads to a strong temperature and concentration dependence of the Prandtl number.

On the other hand, from an experimental point of view, cryogenic studies of convection have certain experimental advantages
especially for investigation of non-linear behavior where the
signal-to-noise ratio becomes crucially important.[5] This makes
the superfluid mixtures probably one of the most suitable objects
for studying non-linear convection.[4,6]

Although ^{3}He-^{4}He mixtures become superfluid below the λ-line,
they have finite effective thermal conductivities. Then, in
analogy with classical fluids, a uniform temperature gradient can
exist in a steady state,[7] and this conducting state should be
stable up to a critical temperature difference.[4,8] Recent experiments have confirmed the existence of a convective threshold
in ^{3}He-^{4}He superfluid mixture.[6,9] This temperature gradient is
balanced by the concentration gradient that leads to a steady
counter flow. Since the ^{3}He atoms are part of the normal fluid,
they tend to accumulate in the colder region which becomes less
dense than the rest of the fluid. Similar concentration distributions occur in classical binary mixtures with abnormal thermodiffusion effects ($k_T > 0$).[10-12] Such systems are unstable
with respect to stationary convection when heated from above and
with respect to oscillatory convection when heated from below,[10-12]
in contrast to the Rayleigh-Bénard case.

The superfluid hydrodynamics differ greatly from classical
hydrodynamics and manifests itself in the persistence of a counter
flow with a non-zero normal fluid velocity in the initial conducting state.[13] On the other hand, two-fluid effects also lead
to additional dissipation which mechanically stabilizes the system.[4]

Another possibility for the study of high Prandtl number
fluids is to consider ^{3}He-^{4}He superfluid mixtures in a porous
medium. Since a porous medium suppresses the vorticity diffusion, it leads to effectively high Prandtl number fluid dynamics.
On the other hand, in this system, the inclusion of two-fluid effects becomes important for the convective instability consideration.

Thus, I would like to review the results of the linear stability analysis for a superfluid mixture in bulk as well as in
porous media and to show that this system presents a wide variety
of convective instabilities.

HEAT CONDUCTING STATE.
LINEAR STABILITY ANALYSIS AND APPROXIMATIONS

We consider a laterally infinite horizontal layer of a super-fluid mixture with ideally heat conducting boundaries separated vertically by a distance ℓ. This system is subjected to a vertical temperature gradient. We will treat the two-fluid hydrodynamics in the Boussinesq approximation. This implies that the perturbations of the total mass density ρ are small and can be neglected in the hydrodynamic equations except in the buoyancy term describing the influence of the gravitational field in the Navier-Stokes equation. This set of equations in a general form was first derived by A. Parshin.[8]

In contrast to the heat conducting state in a classical fluid, we have in a superfluid mixture non-zero values for both the normal and superfluid velocity, V_{no} and V_{so}, but with zero total mass flux. Then, from the heat conduction equation and in a steady state

$$\chi_{eff}\Delta T_0 = 0. \tag{1}$$

From the mass conservation equation

$$\vec{i}+\rho C\vec{V}_{no} = 0, \quad \vec{i}=-\rho D(\nabla C_0 + \frac{k_T}{T}\nabla T_0) \tag{2}$$

one may obtain

$$\nabla T_0 = A\vec{z} \text{ and } \vec{V}_{no} \sim \nabla T_0 \sim \nabla C_0, \tag{3}$$

where A denotes the applied temperature gradient $(A = \Delta T/\ell)$, $\vec{z}$ is the unit vector along the vertical axis and the remaining notation is the same as in Ref. 4. As a result of Eq. (3) and zero mass current $\vec{j}_0 = 0$, one obtains

$$\text{div}\vec{V}_{no} = 0, \quad (V_{no}\nabla)V_{n0} = 0 \text{ and } (V_{so}\nabla)V_{so} = 0, \tag{4}$$

and, finally

$$\nabla P_0 = \rho_0\vec{g}, \quad \nabla\mu_{40} = \vec{g}. \tag{5}$$

This set of equations has the solution

$$\nabla C_0 = \gamma\nabla T_0 \tag{6}$$

and

$$V_{no} = \frac{D}{C}(\gamma + \frac{k_T}{T})\nabla T_0, \tag{7}$$

where

$$\gamma = (\frac{\partial C}{\partial T})_{p,\mu_4} \quad , \quad \vec{g} = -g\vec{z}.$$

Since $\vec{V}_{no}$ should be directed opposite to the temperature gradient, the coefficient in Eq. (7) is negative, $\gamma + k_T/T < 0$. On the other hand, γ is always negative as a result of the thermodynamic stability condition. As first mentioned by Fetter,[13,14] non-zero hydrodynamic flow in the conducting state reflects the superfluid nature of the fluid.

The small amplitude convection equations are obtained by linearizing about this conducting state with finite V_{no}. It is convenient to use μ_4' and $\sigma' = S'/\rho C$ as appropriate variables to simplify the corresponding convection equations.[4,15] Since we keep in general both variables σ' and μ_4', the convection equations are rather complicated for a stability analysis. However, it is possible to proceed when the two dimensionless parameters

$$m = \rho_n \frac{\zeta_4 - \rho\zeta_3}{\eta} \quad . \tag{8}$$

and

$$(\frac{\ell_0}{\ell})^3 = \frac{\eta\kappa}{g\rho_n\ell^3} \tag{9}$$

have extreme values. The parameter m gives the ratio between dissipation of superfluid motion due to second viscosity and dissipation of normal motion due to shear viscosity. The second parameter $(\ell_0/\ell)^3$ reflects the relative values of the dissipation length of normal motion and the layer thickness ℓ. This is the analogy of the boundary layer in classical hydrodynamics.

Using data for the shear viscosity and the first- and second-sound absorption coefficients, it is possible to estimate the values of these parameters as functions of temperature and concentration in the superfluid region of the ^{3}He-^{4}He phase diagram (Fig. 1). Two lines $m \simeq 1$ and $(\ell_0/\ell)^3 \simeq 1$ divide this phase diagram onto several regions.

Small dissipation due to superfluid motion (m < 1) occurs at temperatures below 1 K at all concentrations. In this region, the superfluid motion is important and weakly damped; hence, perturbations of the chemical potential can be neglected, and the temperature and concentration fluctuations are coupled via the thermodynamic parameter γ. Only the thermodynamic variable σ has to be considered in the convection equations as only the temperature has to be considered in a pure classical fluid.

In a major part of the phase diagram, the condition m > 1 is fulfilled. It implies large dissipation due to superfluid motion, the chemical potential perturbations become essential and relax diffusively. Since in this case two thermodynamic variables σ and μ_4 have to be considered, the situation is similar to a normal binary mixture.[10-12]

In a major part of the phase diagram, the dissipation length ℓ_0 is rather small so that the condition $(\ell_0/\ell)^3 < 1$ is fulfilled

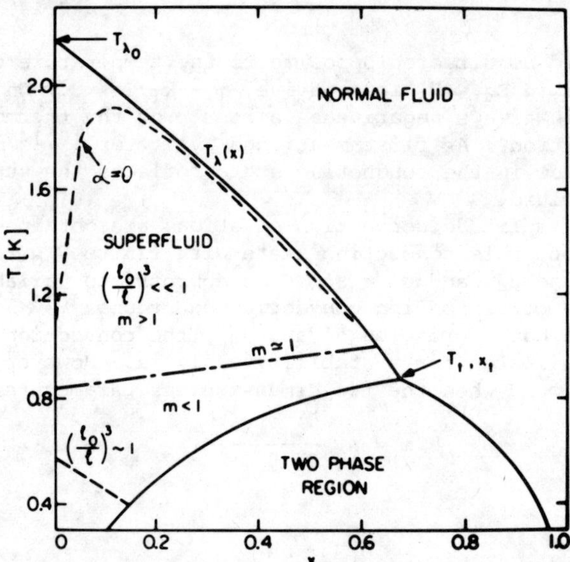

Fig. 1. The phase diagram of He^3–He^4 mixtures in the T–x plane and different regions of parameter values [m and $(\ell_0/\ell)^3$] affecting the onset of stationary convection.

for a realistic layer thickness ℓ. However, for very low temperatures (below 0.5 K) and small concentrations, the dissipation length increases drastically due to a strong increasing of the kinetic coefficients D and κ and decreasing of the normal component density ρ_n. Then, the condition $(\ell_0/\ell)^3 > 1$ can be fulfilled for realistic layer thicknesses. In this region, two-fluid effects play a crucial role in the hydrodynamical stability of the superfluid mixture.

STATIONARY INSTABILITY CRITERIA

In the different regions discussed above, different criteria for the occurrence of stationary and oscillatory instabilities exist and depend on the magnitude of the parameters m and $(\ell_0/\ell)^3$. Let us discuss first the onset of steady convection.

In the case when superflow is relevant and weakly damped (m << 1), the perturbations of the chemical potential can be neglected, i.e. $\mu_4' = 0$. Then, one can show from the equation for superfluid motion that

$$C' = \gamma T'. \qquad (10)$$

Eq. (10) is identical to the approximation made by A. Fetter[13,14] and earlier by A. Parshin.[8] I would like to point out here that this approximation is valid only in a certain concentration and temperature range of the phase diagram (Fig. 1). Without

inclusion of two-fluid effects, this approximation leads to the
same set of linearized convection equations which appear for a
classical fluid.[1] The corresponding instability criterion
is[8,4,13]

$$R = \begin{cases} \dfrac{27\pi^4}{4} & \text{free-free boundaries} \\[2mm] 1708 & \text{rigid-rigid boundaries} \end{cases} \qquad (11)$$

where

$$R = - \frac{\rho g \alpha_{p,\mu_4} \ell^3 \Delta T}{\eta \kappa} \;,\quad \kappa = \frac{\chi_{eff}}{\rho C_{p,\mu_4}} \;,\quad C_{p,\mu_4} = CT\left(\frac{\partial \sigma}{\partial T}\right)_{p,\mu_4} ,$$

and

$$\alpha_{p,\mu_4} = - \frac{1}{\rho}\left(\frac{\partial \rho}{\partial T}\right)_{p,\mu_4}$$

is the thermal expansion coefficient at constant μ_4.

Thus, the only difference from a classical pure liquid is
that instead of

$$\alpha = - \frac{1}{\rho}\left(\frac{\partial \rho}{\partial T}\right)_{p,C}$$

we have

$$\alpha_{p,\mu_4}.$$

It is possible to show that the last quantity is negative, reflecting the physical fact that the light ^{3}He impurities tend to move
toward the cooler regions through heat flush. Indeed, one has

$$\alpha_{p,\mu_4} = \alpha(1 + \phi_0) \qquad (12)$$

where $\phi_0 = \beta/\alpha\,\gamma$, $\beta = -1/\rho\,(\partial\rho/\partial C)_{p,T}$, and, in the superfluid part
of the phase diagram, one has $\gamma < 0$ and $|\phi_0| > 1$. Therefore,
the density gradient is mainly determined by the concentration
gradient just like in a normal binary mixture with large abnormal
thermal diffusion.[10-12] Therefore, steady convection appears
only when heated from above.

In the case of large dissipation due to superflow ($m \gg 1$),
the fluctuations of the chemical potential have to be taken into
account. Since the chemical potential perturbations destabilize
the system when it is heated from above, the mechanical stability
should be reduced, as in a normal binary mixture with abnormal
thermal diffusion.[10-12] The corresponding criterion is similar
to (11); however, R should be replaced by

$$\tilde{R} = R\psi_1, \qquad (13)$$

where

$$\psi_1 = \frac{1+\phi+\phi_0 \frac{\kappa}{D}}{1+\phi_0}$$

and

$$\phi = -\frac{\beta}{\alpha}\frac{k_T}{T}.$$

For the same reasoning that led to a negative α_{p,μ_4} can be used to show that $\psi_1 > 1$ in the region of the phase diagram under consideration.

For a dilute solution, the expression (13) is considerably simplified and becomes

$$R = -\frac{g}{\eta}\frac{\rho_n}{D}\frac{M_3}{M_4}\gamma \ell^3 \Delta T, \tag{14}$$

$$\psi_1 = \frac{4\alpha_{40}-k_T/T+\gamma\kappa/D}{4\alpha_{40}-k_T/T} \quad \text{and} \quad \gamma = -\frac{S_{40}M_3}{R_B T},$$

where R_B is the gas constant, M_3 and M_4 are the molecular weights of ^{3}He and ^{4}He, respectively, S_{40} is the ^{4}He entropy per gram and α_{40} is the ^{4}He thermal expansion coefficient. Here we used

$$\chi_{eff} = \frac{\rho S_{40}^2 M_4 D}{R_B C}, \tag{15}$$

for the effective thermal conductivity.[7] This expression is valid only in temperature range between 1 K and T_λ. The approximation used in Eq. (13) is also valid in the same temperature range.

TWO-FLUID EFFECTS IN THE
STATIONARY CONVECTIVE INSTABILITY

As was mentioned above, we did not take into account two-fluid effects when we derived the criterion (11). There are two different types of terms and mechanisms responsible for the influence of two-fluid hydrodynamics on the stability. One type is associated with the non-zero values of the normal and superfluid velocities in the heat conducting state. These terms are proportional to V_{no} and have been taken into consideration first by Fetter.[13,14] Another type of terms is proportional to $\mathrm{div}\vec{V}_n$ and has been considered in Ref. 4. It turns out that they are of the same order of magnitude as the first terms.[15] Both terms stabilize the system, but the physics is different. It is known for classical pure fluid that a non-zero value of the velocity in the initial state stabilizes the system,[16] e.g. in the case of convection with penetrative flow through the horizontal boundaries. The second terms stabilize the system due to additional dissipation that has two-fluid origin. The perturbations of thermodynamic variables cause the superfluid flow which, in turn, dissipates through interaction with the normal motion. Both mechanisms become significant in the region of small dissipation of superfluid motion and large dissipation of normal motion. This is possible at low temperatures (below 0.5 K) and small concentration where the dissipation length increases drastically and the condition $(\ell_0/\ell)^3 > 1$ can be fulfilled for a realistic layer height (Fig. 1). Inserting numbers appropriate for the experiments in Ref. 6 (T = 0.8 K, C = .0024, P = .66), we obtain an enhancement of the Rayleigh number by about .01%[15]

ESTIMATES OF STATIONARY STABILITY CRITERIA
AND COMPARISON WITH EXPERIMENTS

The result of calculations of the critical temperature gradient are represented in Fig. 2. There are two experimental investigations devoted to the study of the convection onset in a superfluid solution heated from above.[6,19,17]

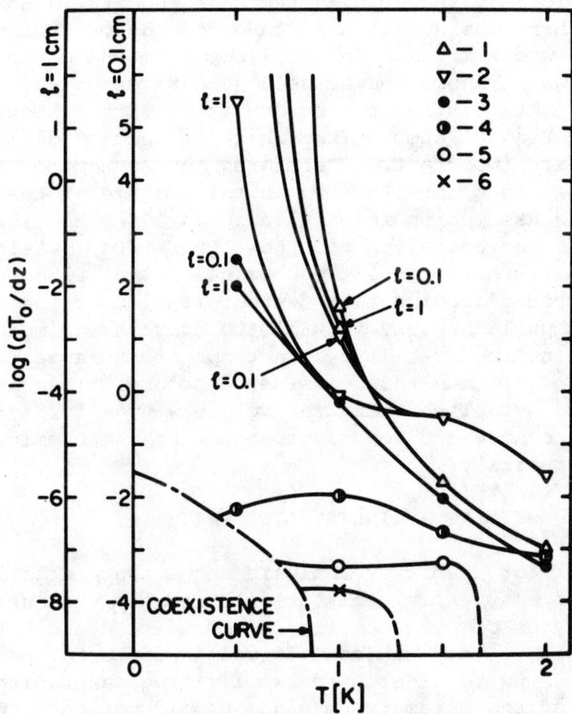

Fig. 2. Temperature dependence of the critical temperature gradient defining the onset of stationary convection for different He^3 concentrations: 1 - .01% mol He^3, 2 - 0.1% mol He^3, 3 - 1% mol He^3, 4 - 10% mol He^3, 5 - 30% mol He^3, 6 - 50% mol He^3.

436

In Ref. 9, the onset of stationary convection was observed
in a 15% ^{3}He mixture with a temperature difference between the
plates of up to about 0.1 K. The experiment was performed in a
cylindrical cell with aspect ratio $\Gamma \simeq 6$ at the averaged temperature close to T_λ. The critical temperature difference was found
to be of the order of 10 μK.[9] The temperature and concentration
range at which the experiments were run corresponds to values of
the parameters which yield m >> 1 and $(\ell_0/\ell)^3$ << 1 as seen from
Fig. 1. Using the corresponding criterion (13) and the eigenvalue
for rigid-rigid boundary conditions 1708, we obtain for the critical temperature difference 8 μK, consistent with the experiment.
It is worthwhile to mention that taking into account the perturbations of chemical potential reduces the stability about five
times. In other words, using the criterion (11) instead of (13)
as suggested by several authors[6,8,13,14,18,25] leads to a critical temperature difference about five times higher than observed
in the experiment. The experiments in Refs. 6 and 25 also show
a decrease in the critical temperature difference from what one
would expect for a pure fluid in the same geometry. However,
there is another consideration in favor of the criterion (13)
in the region where m > 1. This follows from the diagram in Fig.
4 of Ref. 25 which shows the Nusselt number versus the effective
temperature. This diagram can be converted into a plot of Nusselt
number versus Prandtl number from which it becomes clear that mixture effects are important. Therefore, the assumption that T'
is proportional to C' breaks down in this regime of the phase
diagram. This assumption is only valid in a certain range of
values for the concentration and the temperature; outside this
regime, the perturbations of the chemical potential need to be
taken into account leading to a destabilization of the system.
Therefore, it would be very valuable to check experimentally for
dilute solutions the stability in a temperature range much wider
(e.g. 0.1 K to T_λ) than that considered before. Then it will be
much easier to establish quantitatively how two-fluid effects
stabilize the system and how fluctuations of the chemical potential reduce stability.

OSCILLATORY INSTABILITY CRITERIA[20,21,22]

The result of stability analysis shows that an oscillatory
instability in a superfluid mixture occurs only when heated from
below. Two types of overstability are predicted and have different
physical origins. They can be observed in different parts of the
phase diagram. We consider here two limiting cases with respect
to the value of the parameter m since we did not take into account
two-fluid effects at all. In the case of large dissipation due
to super flow, the oscillatory instability criterion and the neutral
frequency are similar to the corresponding ones in a normal binary
mixture with a large abnormal thermodiffusion effect.[10,12] The
stability onset is determined by the temperature gradient as well
as by the ratio of the temperature and concentration relaxation
rates. The calculations show that this branch of convective

instability can be observed in a rather narrow temperature range
between curves $m \simeq 1$ and $\alpha = 0$ (in the region where $\alpha > 0$) (Fig. 1).
The result of calculations of the critical temperature difference
and the neutral frequency at 1 K and 1.5 K versus concentration
are presented in Fig. 3.

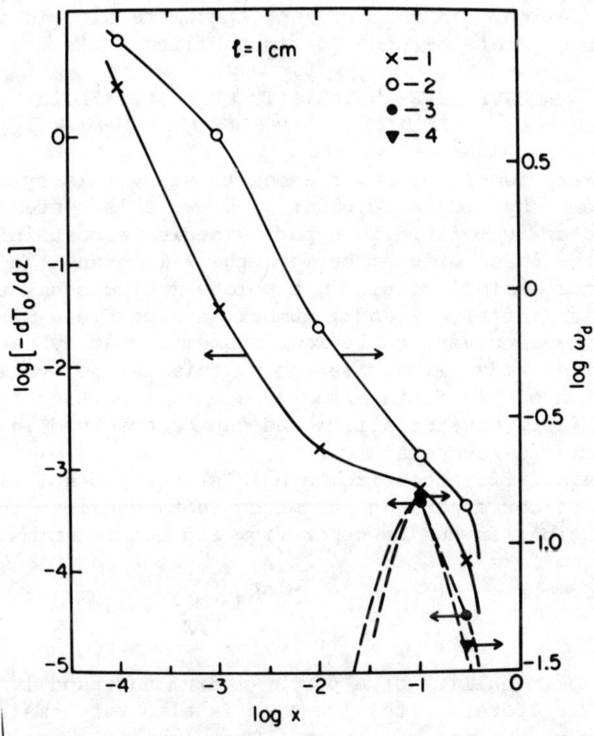

Fig. 3. Concentration dependence of the critical temperature
gradient and the neutral frequency at $m > 1$ for two
different temperatures. 1 - 1K and 3 - 1.5K (the critical
temperature gradient (K/cm)); 2 - 1K and 4 - 1.5K (the
neutral frequency (sec^{-1})).

A fundamentally new types of oscillatory instability (which,
in fact, is an undamped standing second-sound wave) takes place
in the regime of small dissipation due to super flow. When the
second-sound wave velocity becomes rather small and comparable
to the velocity of internal gravity waves, one expects the rate
of energy supply by the gravitational field to balance the rate
of wave dissipation. Then the excitation of undamped second-
sound waves, in fact, becomes possible. Estimates show that the
only possibility to observe the predicted instability is in the
vicinity of the tricritical point.

I would like to emphasize here that in superfluid solutions
the stationary convection occurs only when heated from above

while the oscillatory convection occurs only when heated from below. Therefore, there is no intersection between these two branches of instability. In a classical binary mixture, this intersection occurs at a small value of the separation parameter (small abnormal thermodiffusion coefficient) while a superfluid solution behaves like a normal binary mixture with large abnormal thermal diffusion. In that region, there are also in a classical mixture two separate branches of instability.[10-12]

CONVECTIVE INSTABILITY OF A SUPERFLUID
SOLUTION IN A POROUS MEDIUM[23]

There are two important reasons to study this system. First of all, from a hydrodynamic point of view, this system behaves as a pure classical fluid in a porous medium almost in the whole region of the phase diagram between the λ-line and the coexistence curve. Since a simple fluid in a porous medium behaves as a bulk fluid with an infinite Prandtl number, a superfluid mixture in a porous medium is a very convenient system for studying high Prandtl number nonlinear dynamics. Secondly, this is the system where the inclusion of two-fluid effects becomes important for convective instability considerations and can be observed in experiments at moderately low temperatures.

The main differences from a bulk superfluid solution are as follows. The corresponding parameter that describes the ratio between dissipation due to super flow and due to normal flow has the following form

$$m = \rho_n \frac{\zeta_4 - \rho\zeta_3}{\eta} \cdot \frac{K}{\ell 2} , \qquad (16)$$

where K is permeability of a porous medium and usually is very small.[24] Therefore, in this case, m is also very small in the whole range of the phase diagram. That implies a small dissipation of a superfluid motion. As was already mentioned above for bulk solutions, this case corresponds to pure fluid like behavior. The convection onset now determines for realistic boundary conditions as follows

$$R = 4\pi^2$$

and $\qquad\qquad\qquad\qquad\qquad\qquad\qquad\qquad\qquad\qquad$ (17)

$$R = - \frac{\alpha_{p,\mu_4} \rho g K \ell \Delta T}{\eta \kappa}$$

Secondly, the stationary instability when heated from above is predicted.

Third, the two-fluid effects for reasonable values of the parameters enhance the stability criterion by about 1% at moderate low temperatures.

CONCLUSION

I have reviewed here recent developments in the theoretical
study of the convective onset in superfluid ^{3}He-^{4}He mixtures.
Due to its superfluid nature, this system has unique properties
that make it extremely suitable for the study of nonlinear effects
which are Prandtl number dependent. Further, there are additional
variables in the system, namely the concentration of ^{3}He atoms
and the two fluid components which lead to additional types of
phenomena such as the oscillatory instability and the undamped
second-sound waves. These phenomena can be observed in different
parts of the phase diagram. The influence of nonclassical two-
fluid effects on the onset was also discussed. This wide variety
of the phenomena and also the experimental advantage of the low
temperature environment make the superfluid solutions very attractive and, probably, fruitful systems for convection experiments.
The theoretical calculations reviewed here thus should be considered as a useful guide for future experiments.

REFERENCES

1.   C. Normand, Y. Pomeau and M. G. Velarde, Phys. Mod. Phys.
      49, 581 (1977).
2.   F. H. Busse, Rep. Prog. Phys. **41**, 1929 (1978).
3.   J. Maurer and A. Libchaber, J. Phys. (Paris) Lett. **41**, L-515
      (1980).
4.   V. Steinberg, Phys. Rev. A **24**, 975 (1981).
5a.   G. Ahlers, Fluctuations, Instabilities and Phase Transitions,
      T. Riste, ed (Plenum, NY, 1975) p. 181.
5b.   R. Behringer and G. Ahlers, J. Fluid Mech. **125**, 219 (1982).
6.   H. Haucke, Y. Maeno, P. Warkentin and J. C. Wheatley, J.
      Low Temp. Phys. **44**, 505 (1981).
7.   I. M. Khalatnikov, Introduction to the Theory of Superfluidity
      (Benjamin, NY, 1965).
8.   A. Ya. Parshin, Zh. Eksp. Teor. Fiz. Pis'ma **10**, 567 (1969).
      [JETP Lett. **10**, 362 (1969)].
9.   G. Lee, P. Lucas, A. Tyler and E. Vavasour, J. Phys. (Paris)
      Collog. **39**, C6-178 (1978).
10.   V. Steinberg, J. Appl. Math. Mech. **35**, 335 (1971).
11.   D. T. J Hurle and E. Jakeman, J. Fluid Mech. **47**, 667 (1971).
12.   M. G. Velarde and R. S. Schechter, Phys. Fluids, **15**, 1707
      (1972).
13.   A. Fetter, Phys. Rev. B **26**, 1164, 1174 (1982).
14.   A. Fetter, Physica, **107B**, 149 (1981).
15.   V. Steinberg and H. Brand, Phys. Rev. A, submitted for publi-
      cation.
16.   G. Z. Gershuni and E. M. Zhukhovitskii, Convective Stability
      of Incompressible Fluids (Israel Program for Scientific
      Translations, Jerusalem, 1976).
17.   P. Warkentin, H. Haucke and J. C. Wheatley, Phys. Rev. Lett.
      45, 918 (1980).

440

18. G. Lee, P. Lucas and A. Tyler, Physica, <u>107B</u>, 153 (1981).
19. P. Warkentin, Ph.D. Thesis, 1980, University of California,
     San Diego, unpublished.
20. V. Steinberg, Phys. Rev. Lett. <u>45</u>, 2050 (1980).
21. V. Steinberg, Phys. Rev. A <u>24</u>, 2584 (1981).
22. V. Steinberg, Physica, <u>107B</u>, 151 (1981).
23. V. Steinberg, J. Low Temp. Phys. submitted for publication.
24. F. A. L. Dullien, Porous Media, Fluid Transport and Pore
     Structure (Academic Press, NY, 1979).
25. P. Warkentin, H. Haucke, P. Lucas and J. C. Wheatley, Proc.
     Nat. Acad. Sci. USA, <u>77</u>, 6983 (1980).

NUCLEAR COOPERATIVE PHENOMENA IN COPPER
AT NANOKELVIN TEMPERATURES

M.T. Huiku, T.A. Jyrkkiö, M.T. Loponen, and O.V. Lounasmaa

Low Temperature Laboratory, Helsinki University of Technology
Espoo 15, Finland

ABSTRACT

We have studied nuclear cooperative phenomena in metallic copper
by demagnetizing highly polarized spins to low fields where spin-spin
interactions dominate. In zero external field, the critical temperature T_c of the antiferromagnetic phase was 60 ± 10 nK, and during the
first order transition the entropy changed from 0.49Rln4 to 0.65Rln4
The critical field $B_c = 0.25 \pm 0.01$ mT. Evidence for another transition at 0.09 mT was also seen. The entropy of nuclear spins in zero
external field is presented as a function of temperature; our S vs. T
diagram, however, is provisional especially in the ordered phase.

INTRODUCTION

Electronic magnetism shows a wide spectrum of different ordering
phenomena, extending from room temperature and above in iron to a few
millikelvins in CMN. Because the nuclear magnetic moments μ are
about 1000 times smaller than their electronic counterparts and
because the dipolar interaction is proportional to μ^2, analogous
phenomena can be expected in a nuclear spin system only at microkelvin temperatures and below. Solid ^{3}He is an exception owing to
the strong quantum mechanical exchange force, enhanced by the large
zero point motion, and so are the van Vleck paramagnets, such as
$PrNi_5$,[1] in which a considerable hyperfine enhancement of the magnetic
field occurs.

The pioneering experiments of Kurti and his coworkers at Oxford
in 1956 established the feasibility of the nuclear demagnetization
method; a spin temperature of 1.2 μK was reached for a short time in
a copper specimen. The first studies of nuclear cooperative phenomena were made by Abragam and Goldman at Saclay about 15 years later.
Our work on copper, which was started in Helsinki in 1974, has now
produced the first results on spontaneous nuclear ordering in a
simple metal.[2] Owing to many experimental difficulties our progress
was rather slow and only since the spring of 1982 have we definitely
been reaching the nuclear antiferromagnetic phase in copper. Our
early work, which will not be discussed here, has been described in a
number of publications,[3] both experimental and theoretical. For a
recent review of the nuclear demagnetization method we refer to
Andres and Lounasmaa.[4]

The Hamiltonian of the nuclear spin system in copper has three
terms: the Zeeman interaction H_Z with the external magnetic field,

the magnetic dipolar interaction H_D between spins, and the indirect Ruderman-Kittel exchange interaction H_{RK} between spins which is mediated by conduction electrons. The internal field felt by an individual spin in copper is 0.34 mT; in external fields above this value the Zeeman interaction dominates. Therefore, in order to study spontaneous nuclear ordering in copper, demagnetization must be carried to zero field or, at least, well below 1 mT. This causes practical problems with the external heat leak because the amount of energy that can be absorbed by the nuclear spin system is very small during warm-up in a low or zero field.

At very low temperatures it is meaningful to speak about two distinct temperatures, the nuclear spin temperature T_n and the conduction electron temperature T_e. The nuclei reach local thermal equilibrium among themselves in a time characterized by τ_2, the spin-spin relaxation time, whereas the approach to equilibrium between nuclear spins and conduction electrons is governed by the spin-lattice relaxation time τ_1. At low temperatures $\tau_2 \ll \tau_1$; this makes a separate nuclear spin temperature meaningful. In our experiments. T_e and T_n can differ by several orders of magnitude.

In a metal, the Korringa relation is given by $\tau_1 = \kappa/T_e$, where κ is Korringa's constant which depends on the external magnetic field and on impurities. For pure copper at zero field $\kappa = 0.5$ sK; at 10 mT and above, $\kappa = 1.2$ sK. This means that at $T_e = 0.5$ mK, which was a typical conduction electron temperature during our early experiments, the warm-up of the nuclear spin system after demagnetization to a low field is characterized $\tau_1 = 1000$ s as heat flows from the relatively hot conduction electrons to the very cold nuclei. Electronic magnetic impurities can assist in the relaxation process at low fields and further shorten τ_1. A longer relaxation time would obviously be highly desirable. To achieve this it is necessary to lower the conduction electron temperature and to eliminate the possible effects of magnetic impurities.

EXPERIMENTAL

For our experiments we constructed a cryostat which consisted of a dilution refrigerator and two copper nuclear stages, all operating in series. In the latest version of our apparatus, illustrated in Fig. 1, the first nuclear stage was made of a piece of bulk copper into which vertical grooves were machined to reduce eddy current heating during demagnetization. The first stage contained 10 mol of copper in the high field region. Between the mixing chamber of the dilution refrigerator and the upper nuclear stage there was a superconducting heat switch made of tin. The second nuclear stage, which was also our sample, was much smaller, typically 2 - 3 g of copper. There was no heat switch between the nuclear stages which were simply welded together; the conduction electron temperature was thus approximately the same throughout. The superconducting solenoids

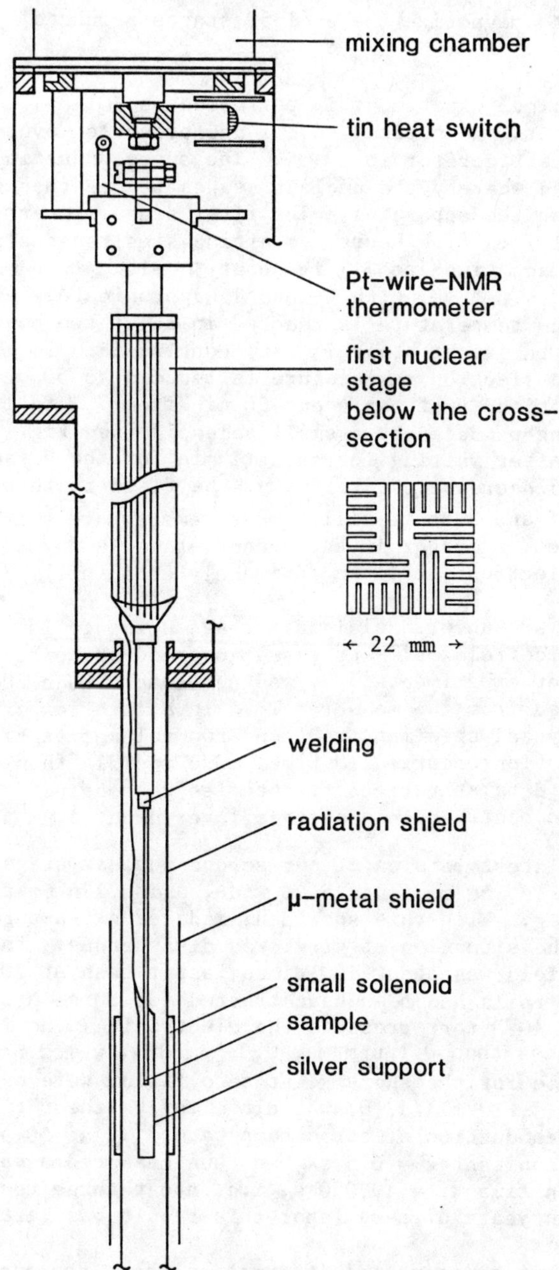

Fig. 1. The latest version of our cascade nuclear demagnetization refrigerator. Superconducting solenoids surrounding the nuclear stages and their field compensating coils are not shown.

employed to magnetize the nuclear stages produced 8 T and 7 T fields, respectively.

A typical experiment is performed as follows. The upper nuclear stage is first magnetized to 8 T and precooled overnight by the dilution refrigerator to 12 mK. The superconducting heat switch is then opened whereby the nuclear stages become thermally isolated from the rest of the apparatus. The first stage is next demagnetized from 8 T to 0.1 T in five hours. Starting simultaneously, the second stage is magnetized to 7 T in about 50 min. Both nuclear stages then cool to 0.2 mK, with T_e and T_n approximately equal; this equilibrium temperature is reached in about two hours. Demagnetization of the first stage is then continued to 20 mT, whereby the conduction electron temperature is reduced to 50 µK. The second stage is demagnetized next, in about 20 min from 7 T to 10 mT, the latter field being produced by a small solenoid (see Fig. 1) around the sample. After waiting for about 1 min for the noise in the SQUID circuit to decrease sufficiently, the field is further reduced well below 1 mT and susceptibility measurements are started. The nuclear spin system of copper in the second stage is now below 50 nK, with conduction electrons still approximately at 50 µK; T_e/T_n is thus 10^3!

The warm-up of nuclei in the second stage is regulated by the spin-lattice relaxation process, governed by Korringa's law. During our earlier experiments,[3] τ_1 was never more than 20 min, faster than anticipated from the measured T_e. In a later experiment, with a single crystal specimen, we found two relaxation times: about 90% of the relaxation occurred 10 times more rapidly than expected. By comparing data at various frequencies we concluded that the slow relaxation occurred in a surface layer about 5 µm thick.

The latest version of our second nuclear stage consisted of eight copper foils, 50 mm long, 5 mm wide, and 0.125 mm thick, weighing about 2.5 g. With this specimen, made of extra-pure Marz grade copper, the situation at first was disasterous: the relaxation time in zero field was about 100 times faster than at 20 mT! However, after the foils had been heat treated for three hours at 950°C in dry air under 10^{-4} torr pressure the electronic magnetic impurities (0.8 ppm Fe, less than 0.1 ppm Mn, 0.3 ppm Cr), which presumably were responsible for the short τ_1 at zero field, were oxidized, whereafter the ratio $\tau_1(20 \text{ mT})/\tau_1(0)$ was close to its theoretical value 2.4. With the conduction electron temperature T_e at 50 µK and with the Korringa constant $\kappa = 0.5$ sK, we then measured a spin-lattice relaxation time $\tau_1 = 10,000$ s, i.e. about three hours. After more than seven years of hard labor this result was certainly gratifying!

All our experimental information about the nuclear spin system of copper has been obtained by means of SQUID NMR or susceptibility measurements. For excitation and pick-up of the signal two saddle-shaped coaxial coils were used; the pick-up coil has a figure-eight configuration.

Both the absorptive and dispersive parts of the complex magnetic susceptibility $\chi = \chi' - i\chi''$ can be used for calculating the nuclear spin polarization and hence the entropy, provided that the external field is 1 mT or above where the equations of the paramagnetic state apply. In our earlier experiments χ'' was employed. With 0.125 mm thick high purity copper foils, measurements could be made only below 50 Hz because higher frequencies would not penetrate to the interior of the specimen owing to eddy current shielding. The frequency actually used was 10 Hz and the quantity measured was $\chi'(10\ \text{Hz})$ because χ'' becomes very small at low frequencies. In fact, $\chi'(10\ \text{Hz})$ was practically equal to the static susceptibility $\chi'(0)$.

SUSCEPTIBILITY VS. TIME AT VARIOUS EXTERNAL FIELDS

Before presenting and discussing the new data, it should be mentioned that our earlier experimental results and theoretical calculations[3] had shown that the nuclear spin system of copper will order antiferromagnetically at the lowest temperatures. On an inverse susceptibility vs. temperature plot experimental points above 300 nK fell on a straight line, which intercepted the temperature axis at $\theta = -150$ nK. Below 200 nK, some curvature was found in the $1/\chi'(0)$ vs. T plot, indicating that we were probably approaching the transition region. In addition, measurements of line shifts and of the relative magnitude of the NMR signals from the two copper isotopes both gave the same negative value for the Ruderman-Kittel exchange interaction constant, thus clearly pointing to antiferro-magnetic order.

Some of our recent experimental data, are shown in Fig. 2, plotted as $\chi'(0)$ versus time. The sample was demagnetized to the final field, and the susceptibility was measured as the spin system slowly warmed under the influence of the spin-lattice relaxation process. In zero field, $\chi'(0)$ first increased, reached a broad maximum and finally began to decrease exponentially with time as expected under a constant heat leak in the paramagnetic region. The initial increase was about 7%. By linear extrapolation, the reduction of $\chi'(0)$ at $t = 0$ was found to be about 15%. In non-zero fields, this reduction first decreased and then increased with increasing fields, as shown by the curves at 0.065 mT and 0.14 mT. Above 0.2 mT the $\chi'(0)$ curve started to approach paramagnetic behavior; at 0.3 mT there was no longer any deviation from exponential decay.

The initial increase of the $\chi'(0)$ with time implying the same qualitative behavior with temperature, clearly shows, in analogy with data on electronic systems, that a magnetic phase transition is taking place near the maximum, with the copper nuclear spin system in the ordered state on the left and in the para-magnetic state on the right. On the basis of our earlier

data, confirmed by the new measurements and by mean field calculations, we conclude that the order phase is antiferromagnetic.

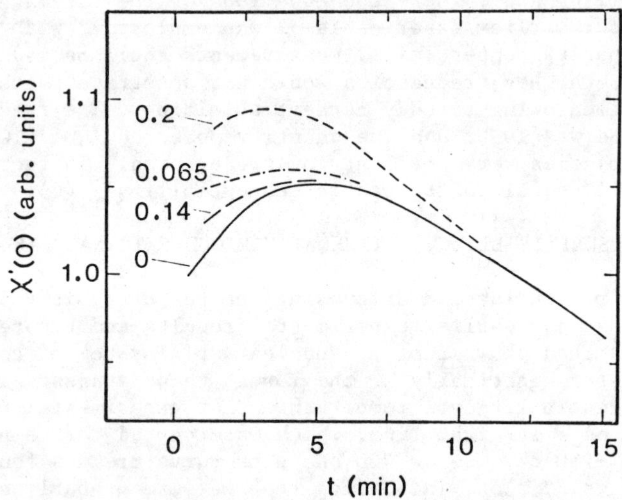

Fig. 2. The measured static susceptibility $\chi'(0)$ of copper nuclear
spins as a function of time during warm-up after demagnetization to
the final field indicated for each curve in mT.

In Fig. 3, the results of Fig. 2 have been plotted as a function
of the external magnetic field. Two distinct regions can be identified, in addition to the paramagnetic regime denoted by P. In the A-
region, near zero field, the initial susceptibility χ'_{in} increases
with increasing field as noted before. Going from the A- to the B-
region there is a sudden decrease in χ'_{in} at about 0.09 mT. In the
B-region χ'_{in} is first roughly constant, up to about 0.15 mT, whereafter it starts to increase before the paramagnetic behavior appears
around 0.25 mT. The maximum value of the susceptibility, χ'_{max}, remains
approximately constant up to 0.15 mT, whereafter it starts to increase.

The anomaly observed at 0.09 mT might indicate a spin-flop
transition which is possible in antiferromagnetic systems if one
assumes an axis of asymmetry along which ordering is preferred. In
metallic copper, which has an fcc structure, favored directions may
be caused by the 10% nonsphericity of the Fermi surface. In low
external fields the preferred crystallographic direction determines
the orientation of antiferromagnetic sublattices. With increasing
field, the direction perpendicular to $\vec{B}$ becomes more favorable
because the spins then can lower their energy by tilting towards the
field direction. The transition occurs at a field which is a
function of the angle between $\vec{B}$ and the preferred axis. In the
vicinity of the transition the susceptibility is enhanced and,

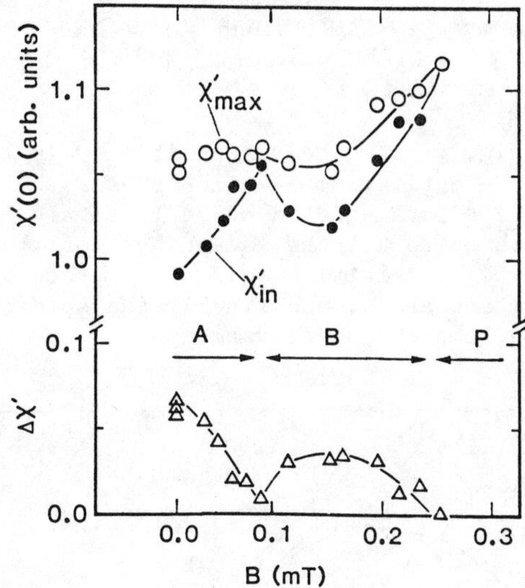

Fig. 3. The initial susceptibility χ'_{in}, obtained by extrapolating the $\chi'(0)$ vs. time curves of Fig. 2 to t = 0, as a function of the external magnetic field. The maximum susceptibility χ'_{max} and the difference $\Delta\chi' = \chi'_{max} - \chi'_{in}$ are also shown. The different phases are marked by A, B, and P.

therefore, the spin-flop proposal agrees with our experimental observations.

The picture presented so far is, however, oversimplified because of some metastabilities and supercooling effects that were observed. In fact, the ordered state was often difficult to reach. Demagnetization frequently resulted in a susceptibility above χ'_{max}; this effect could be seen at low entropies when the field was swept for a short time above 0.3 mT, which brought the sample into the paramagnetic region, and then back to zero again. The observed susceptibility was then above χ'_{max} and the ensuing relaxation was exponential in time, indicating that the sample had remained in the paramagnetic region. These metastabilities show that the transition between the ordered and paramagnetic states is of first order, with an associated latent heat.

ENTROPY VS. SUSCEPTIBILITY

Fig. 4 shows some additional features of the susceptibility vs. time curve. S_i is the initial entropy before demagnetization from 1 mT to zero field. S gives the entropy corresponding to the lower-most curve at the minimum χ' (25.5% of Rln4), at the maximum χ' (49%), and when metastability could no longer be observed (65%). There is an irreversible entropy increase of 12%, from 13.5% to 25.5%, which occurs during demagnetization from 1 mT to zero

448

field. The highest initial entropy from which the ordered phase could
still be reached is $S_i(max) = 0.37R\ln4$ and all values of S_i below
$S_i(max)$ must be increased by 12% in order to obtain the actual nuclear
spin entropy after demagnetization.

Dashed lines in Fig. 4 show the susceptibility in the metastable
paramagnetic state. The antiferromagnetic A-phase, the transition
region at $T = T_c$, and the paramagnetic phase P are also shown. The
spin system first warms up to T_c, the latent heat of the transition
is then supplied via the spin-lattice relaxation process while the
temperature stays constant at T_c, and finally the spin system, now in
the paramagnetic state, continues its warm-up.

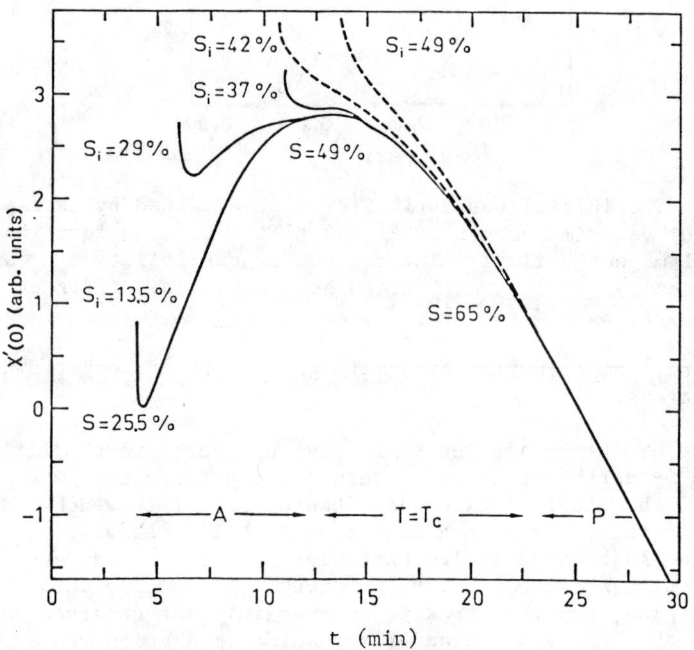

Fig. 4. The nuclear susceptibility as a function of time with the
final demagnetization from 1 mT to zero performed at different degrees of polarization. S_i is the initial entropy at 1 mT, varying
from 13.5% to 49% of $R\ln4$ in the different experiments. For further
explanations, see text.

The slowness of the ordering process is clearly seen from Fig.
4; the minimum susceptibility is reached approximately 15 s after
the external field is reduced to zero. This shows that during
demagnetization the nuclear spin system apparently first remains in
the supercooled paramagnetic region, with the transition to the
ordered state taking place a bit later. This phenomenon may be the

reason for the 12% increase in entropy during the final stage of
demagnetization. This also explains why we did not reach the
antiferromagnetic state in our earlier experiments: at that time
the spin lattice relaxation was too rapid to obtain initial entropies
below $S_i(max) = 0.37 Rln4$.

In Fig. 5 the nuclear spin entropy of copper has been plotted as
a function of $\Delta\chi' = \chi'(0) - \chi'_{min}$, where χ'_{min} is the minimum value of
susceptibility near $t = 0$ as shown in Fig. 4. The relation between
entropy and susceptibility was found in a series of demagnetizations
to zero field: After recording the increase of $\chi'(0)$ in the ordered
phase for some time, the field was swept back to 1 mT to determine
the entropy which could be calculated from the measured susceptibility in the paramagnetic region. Since the field sweep from zero
to 1 mT was found to be adiabatic, this value also gave the entropy
in the ordered phase immediately before the field sweep. By varying
the time spent in $B = 0$, and thereby the value of $\chi'(0)$, the relation
S vs. $\chi'(0)$ was established.

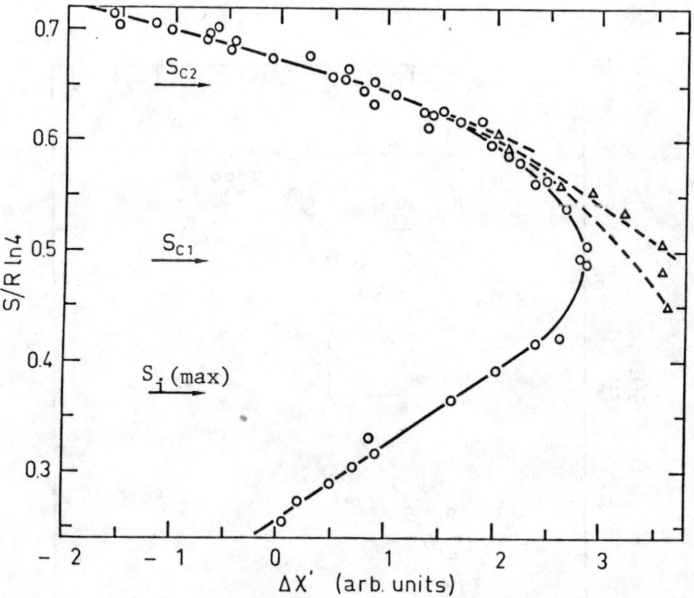

Fig. 5. The nuclear spin entropy of copper, in units of Rln4, as a
function of $\Delta\chi' = \chi'(0) - \chi'_{min}$, where χ'_{min} refers to the minima of
the susceptibility curves in Fig. 4. Dashed lines illustrate
experiments in the overcooled paramagnetic state. S_{c2} and S_{c1} are
the upper and lower critical entropies, respectively, and $S_i(max)$ is
the maximum initial entropy at 1 mT from which the ordered phase can
still be reached. For further explanations, see text and compare
with Fig. 4.

450

The critical entropies of the first order transition were
found from the curve shown in Fig. 5. The upper critical entropy S_{c2} was set to the value at which metastability disappeared;
this happened at $(0.65 \pm 0.02)R\ln 4$. The lower critical entropy S_{c1}
was found as the sum of the measured 12% nonadiabaticity of the field
sweep from 1 mT to zero and the maximum initial entropy $S_i(\max) =$
$0.37R\ln 4$ which still led to the ordered state after demagnetization;
we thus obtained $S_{c1} = 0.49R\ln 4$, which was also the entropy at the
maximum of $\chi'(0)$ in Fig. 5.

ENTROPY VS. TEMPERATURE

So far we have not discussed the experimentally determined
temperature of the nuclear spin system. This is because measurements
of T in the nanokelvin region have proven to be the most difficult
task in these experiments. In principle, the temperature of the
nuclear spin system can be found by employing directly the second law
of thermodynamics, $T = dQ/dS$: the entropy increase ΔS caused by a
small heat input ΔQ must be measured. In the ordered state, because

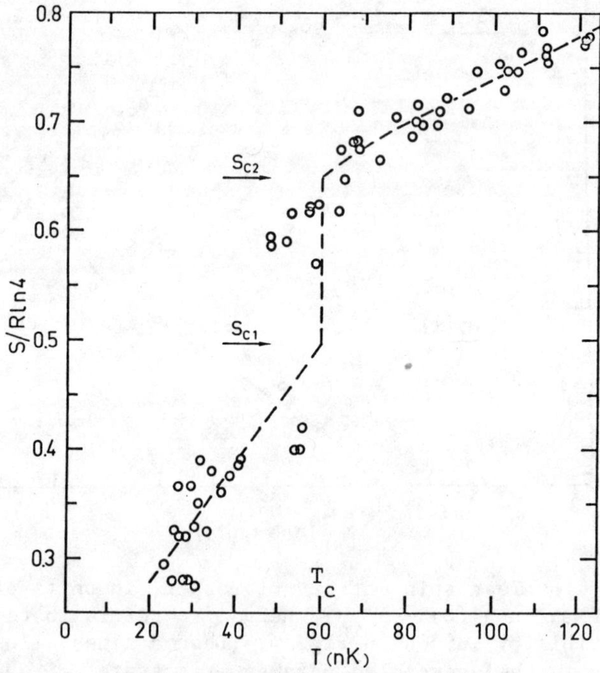

Fig. 6. Provisional entropy diagram of the nuclear spin system of
metallic copper below 120 nK. The antiferromagnetic transition
occurs at $T_c = 60$ nK. Dashed lines are drawn only to emphasize the
first order phase transition.

of the nonadiabaticity during the final demagnetization from 1 mT to zero, these two quantities could not be obtained during the same experimental sequence and the temperature had to be found by employing the susceptibility as the thermometric parameter, i.e. $T = (dQ/d\chi')/(dS/d\chi')$. $(dS/d\chi')$ was obtained from Fig. 5; a similar graph was made for $(dQ/d\chi')$. However, owing to several experimental complications a satisfactory temperature scale has not been established; changes are to be expected especially in the ordered phase.

In any case, Fig. 6 shows the entropy, in zero external field, as a function of temperature for the region below 120 nK. The discontinuous drop in entropy at the first order transition is clearly seen, even though the data show considerable scatter. The antiferromagnetic transition in the nuclear spin system of copper occurs at $T_c = 60 \pm 10$ nK; the latent heat $L = T_c(S_{c2} - S_{c1}) = 0.11$ µJ/mole.

CONCLUSIONS

We find that the nuclear spin system of copper orders antiferromagnetically, with the transition temperature $T_c = 60$ nK in zero external field. Transition to the paramagnetic state begins at $S_{c1} = 0.49R\ln4$ and it is completed at $S_{c2} = 0.65R\ln4$. The latent heat of the first order transition, calculated from these experimental values, is 0.11 µJ/mole. The critical field above which the system is paramagnetic is 0.25 mT. Two different, antiferromagnetically ordered phases seem to exist: the A-phase between 0 and 0.09 mT, and the B-phase between 0.09 and 0.25 mT. It is possible that the latter is the so-called spin-flop phase. Metastability is observed: the spin system may remain in the paramagnetic state below T_c. This shows that the transition is of first order.

More experiments of the type described here are still needed before the entropy and susceptibility diagrams, as functions of temperature, can be presented with confidence. This work is in progress and the final results will be submitted to the Journal of Low Temperature Physics, hopefully during the fall of 1983. However, only by means of neutron diffraction experiments will it be possible to obtain detailed information about the nuclear spin arrangements in the ordered phases of metallic copper. Our preliminary feasibility studies indicate that such measurements are possible.

ACKNOWLEDGEMENTS

Work reported in this paper was done by the YKI-group. Over the years many persons have participated; Gösta Ehnholm and Jouko Soini, in particular, should be mentioned. We thank A.S. Oja for help with the new measurements. This research has been financially supported by the Academy of Finland.

REFERENCES

1. M. Kubota, H.R. Folle, Ch. Buchal, R.M. Mueller, and F. Pobell,
    Phys. Rev. Lett. $\underline{45}$, 1812 (1980).

2. M.T. Huiku and M.T. Loponen, Phys. Rev. Lett. $\underline{49}$, 1288 (1982);
    M.T. Huiku, T.A. Jyrkkiö, and M.T. Loponen, Phys. Rev. Lett. $\underline{50}$,
    1516 (1983).

3. For a list of papers prior to 1982 see: O.V. Lounasmaa, Physica
    $\underline{109}$ & $\underline{110B}$, 1880 (1982); K.J. Niskanen, L.H. Kjäldman, and
    J. Kurkijärvi, J. Low Temp. Phys. $\underline{49}$, 241 (1982); M.T. Huiku and
    J.K. Soini, J. Low Temp. Phys. $\underline{50}$, 523 (1983); K.J. Niskanen and
    J. Kurkijärvi, J. Phys. C (to be published).

4. K. Andres and O.V. Lounasmaa, Progress in Low Temperature
    Physics, Vol. 8, 221 (Editor D.F. Brewer, North Holland
    Publishing Company, 1982).

CURRENT SENSING NOISE THERMOMETRY:
SOME RECENT IMPROVEMENTS

Richard A. Webb and Sean Washburn
IBM T. J. Watson Research Lab., Yorktown Heights , NY 10598

ABSTRACT

We report preliminary measurements on several current sensing noise thermometers presently under study in our laboratory. The emphasis of our work is to produce a fast practical absolute thermometer that can be used over the entire temperature range from 4.2 K to .05 mK. The first part of the paper is devoted to demonstrating a practical thermometer based on commercially available rf SQUIDs useful in the temperature range 4.2 K to 2 mK that allows a 1% determination of the absolute temperature with 200 sec of averaging the Johnson noise currents. A comparison of the Johnson noise temperature with the NBS low temperature fixed points is presented. A discussion of the improvements in noise thermometry that are currently possible and under development in our laboratory using low noise dc SQUIDs is given. Using existing dc SQUIDS, it is possible to have a device noise temperature of .01 mK and to determine the absolute temperature to within 1% in 15 sec.

INTRODUCTION

The recent controversy over the heat capacity of normal liquid ^{3}He[1-7] has sparked renewed concern about absolute thermometry below 0.3 K. Although the reason for the discrepancy is still not completely clear, differences in the temperature scales used in the various laboratories can not account for the entire problem[7,8]. The assumption that the thermometry was the source of the discrepancy underlines the need for an internationally accepted temperature scale below 0.5 K. Using two absolute thermometers, the National Bureau of Standards (NBS) has developed a preliminary temperature scale that covers the region from 10.2 mK to 519 mK[9] and is based on two absolute thermometers. The first is a noise thermometer that uses a resistive SQUID[10] and the second is a nuclear orientation thermometer based on a ^{60}CoCo single crystal[11]. This scale is commercially available as a self-contained assembly of coils and five superconducting samples which have been assigned transition temperature values by using the two primary standards[12]. It is claimed that these standards when used properly allow the user to achieve a temperature reproducibility of ±0.2 mK. However, recently the NBS has found an error in the method of calibration for their lowest temperature fix point (the Tungsten transition) and have now assigned a temperature of 15.6 mK $\pm.1$ mK to this device. This changes the previous temperature assignment by as much as 5% for some units.

Even if we assume the current fixed point devices give an accurate value for the thermodynamic temperature down to 15.6 mK the problem of absolute temperature measurement below this temperature remains. Many secondary thermometers do not have a simple relationship to the absolute temperature and become insensitive below about 1 mK. Most absolute thermometers such as vapor pressure, ideal gas and nuclear orientation have only a limited range of usefulness. One of the most practical thermometers currently available is the ^{3}He-melting curve thermometer[13]. It offers many advantages such as high precision, small size, field independent up to 5 kG and useful over the temperature range 1mK to 320 mK. However, the Johnson noise thermometer in principle is the only absolute thermometer that we are aware of with the capability of spanning the entire range from 0.05 mK to 5 K, and the practical realization of this performance is the subject of this paper.

BASIC PRINCIPLES OF OPERATION

The use of Johnson noise for thermometry relies upon the statistical relationship (first derived by Nyquist[14]) between the mean square value of the random voltages appearing across the ends of a metallic resistor and the absolute temperature of the conduction electrons in the resistor. The Nyquist theorem states that the mean square open circuit voltage across a passive two-terminal network at a temperature T is

$$<V^2> \; = \; 4 \, kT \, R(f) \, \delta f \tag{1}$$

where k is Boltzmann's constant, f is the frequency and R(f) is the real part of the impedance. Lawson and Long[15] suggested over 35 years ago that Johnson noise might be an absolute thermometer in the milli-Kelvin temperature range. The problem has been that at low temperatures these noise voltages have to be amplified, and the amplifiers introduce additional noise into the measurement. However, superconducting devices can be used to reduce the effective noise temperature of the amplifiers to the microdegree region.

There are two approaches to noise thermometry currently in use below 4.2 K. The first, originally introduced by Kamper and Zimmerman[10], uses a superconducting ring containing both a Josephson junction and a resistor. The circuit is biased by a dc current from a low noise source. The resulting voltage drop across the resistor sets up an oscillating signal at frequency f given by the ac Josephson relation $f = V/\Phi_o$, where Φ_o is the flux quantum (h/2e). This ac signal is coupled to an rf tank circuit operating around 24 MHz, and using parametric up conversion provided by the junction, the output voltage, after suitable demodulation, is an exact replica of the input signal[16]. In the absence of thermal noise from the resistor, the output signal is at a single frequency usually adjusted to be between 1 and 100 kHz. However, due to the random Johnson noise in the resistor, the the output frequency fluctuates about the center frequency f. From many measurements of the frequency, the variance can be computed and in the absence of any other noise sources, can be related to the mean square value of the Johnson noise voltage. Of course there are other contributions to the voltage across the the Josephson junction, but careful analysis of many of the additive noise contributions have been made[16], and it is claimed that these introduce no more than 0.1 % error at 10 mK. The attractive features of this type of noise thermometry are that the device noise is low and the properties of the amplifiers do not effect the magnitude of the measured variance. However, in order to obtain a 1% precision for the absolute temperature long averaging times (on the order of 20,000 sec) have been required (with the precision increasing with the number of averages as $N^{-1/2}$). Recently[17], it has been pointed out that that 1% precision may be obtained in 200 sec by operating the frequency counter used to make the measurements at a faster rate but with an additional 5% correction to the measured variances. In our opinion, another disadvantage of this method of thermometry is that the entire SQUID and rf tank circuit must be well thermally connected to the temperature reservoir whose temperature is to be determined.

The second type of noise thermometer which was first introduced by Giffard et al.[18] and later treated more completely by Webb et al.[19] is shown schematically in figure 1. A resistor R is placed in thermal contact with the temperature reservoir whose temperature is to be determined. Superconducting leads are connected to the ends of the resistor, thermally grounded to the reservoir, and connected to a coil that is tightly coupled to an rf SQUID. The SQUID can be located in a different temperature reservoir such as 4.2 or 1K. The random voltage across the ends of the resistor generates a random current in the input coil to the SQUID. The mean square current at a frequency f is given by

456

$$\langle I^2(f)\rangle \; = \; \frac{\langle V^2(f)\rangle}{R^2 + \omega^2 L^2}\,\delta f \; = \; \frac{4kT\,\delta f}{R\,(1 + \omega^2\tau^2)} \tag{2}$$

where $\omega = 2\pi f$, $\tau = L/R$, and L is the total inductance of the low temperature input circuit. The output of the flux-locked loop is a voltage that is linearly proportional to the input current and thus is a random function of time. This output voltage is amplified and passed through an active band-pass filter to an integrating mean square voltmeter. The relationship between the measured mean square voltage and the absolute temperature is given by

$$\langle V^2(t)\rangle \; = \; \int_0^\infty \frac{4kT}{R}\,Z^2(f)\,\frac{F(\omega,\omega_L,\omega_H)}{1 + \omega^2\tau^2}\,df\;, \tag{3}$$

where $F(\omega,\omega_L,\omega_H)$ is the power transfer function for the active filter, and $Z(f)$ is the current to voltage transfer function of the SQUID system and any room temperature amplifiers. In order to use this type of system as an absolute thermometer, one must know accurately the value of the resistance, the L/R time constant and amplifier gain as well as the transfer function of the filters. Alternatively, by measuring

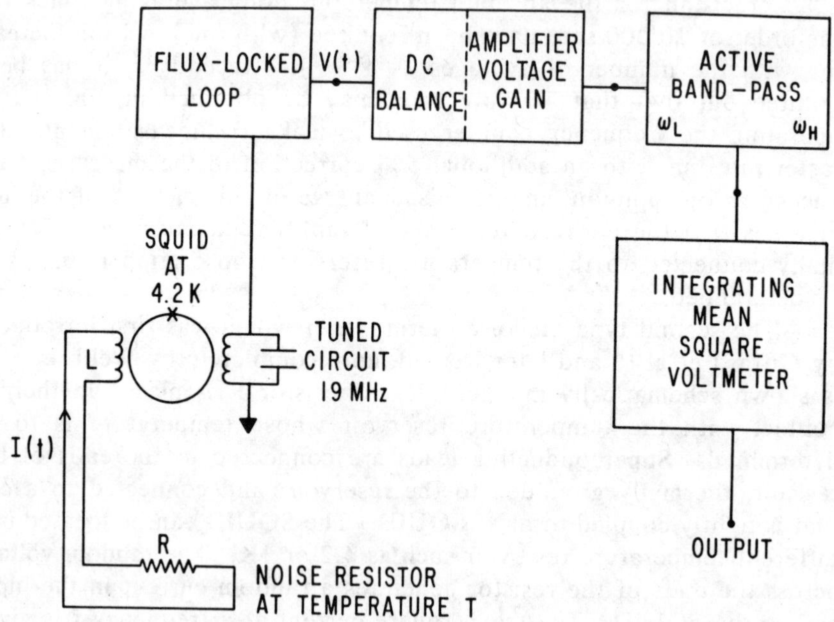

Fig. 1. Schematic of the basic circuitry used to measure the Johnson noise in a low temperature resistor.

the mean square voltage at a single known temperature (such as 4.2 K
or at the superconducting transition of one of the fixed point devices
supplied by NBS) a complete calibration constant can be determined,
and one can use the noise thermometer as an accurate secondary thermometer. In addition, the extra noise introduced into the measurement
by either the SQUID or the amplifiers must be known. In general, this
device noise can be measured accurately as shown below and should
not pose a serious problem.

We believe that for the noise thermometer to become a widely
accepted form of thermometry the length of time that one must average
the noise in order to achieve a 1% precision in temperature will have to
be 30 seconds or less. Webb et al.[19] have shown that the precision in
general is a complicated function of the L/R time constant and active
filter settings. A rough estimate can be obtained in the limit where the
bandpass is opened up to allow all the noise to pass through. In this
case the precision is given by

$$\frac{\Delta T}{T} = [(2\tau/t) - (\tau^2/t^2)]^{1/2}. \tag{4}$$

If we let $\tau = 0.5$ msec, which corresponds to a half power roll off
frequency in the noise spectrum at 300 Hz, then a only a 10 sec averaging interval is required to obtain a 1% precision in the measurement
of temperature. Due to the additional noise from the SQUID, the
general practice has been to reduce the bandwidth of the measurement
so that it coincides with the L/R time constant of the input circuit. In
the above example, this would increase the time for a 1% measurement
of the temperature to approximately 15 seconds. Clearly, the L/R time
constant of the input circuit should be as large as possible. However,
the maximum value of the mean square noise current in the input coil is
given by the equipartition theorem and is equal to

$$<I^2> = \int_0^\infty \frac{4kT/R}{1 + \omega^2\tau^2} = \frac{kT}{L}. \tag{5}$$

Thus without changing the input coil inductance of the SQUID (which
is fixed in commercially available devices), the only way to increase the
bandwidth of the measurement is to increase the resistance. This, of
course, couples more of the SQUID device noise into the measurement
and raises the device noise temperature. In the frequency range of
most noise thermometry measurements, the flux noise in the SQUID
can be characterized by a white noise spectrum and is given by
$\Phi_N = \varepsilon\Phi_o$, where $\varepsilon = 7x10^{-5}$ is typical of commercial SQUID
systems. The device noise temperature, T_D, can be expressed in terms
of this flux noise and the circuit parameters shown in figure 1 and is

458

given by[20]

$$T_D = \frac{\varepsilon^2 \Phi_o^2 R}{4\,kM^2} \frac{B_D}{B_R}, \qquad (6)$$

where B_D and B_R are effective passbands of the device noise and
Johnson noise from the resistor R. Thus, after a measurement of both
the flux noise in the SQUID and the mutual inductance, M, of the input
coil to the SQUID, the device noise temperature can be selected by
simply choosing R.

MEASUREMENTS WITH A COMMERCIAL RF SQUID

Our first experiments on noise thermometry were designed to
produce a fast responding thermometer suitable for use in our dilution
refrigerator with a large but hopefully accurately measurable device
noise temperature using a commercially available SQUID system[21]. We
made two noise thermometers from brass foil 0.038 mm thick. One
thermometer (4cm long and 1cm wide) was mounted on the outside of
the mixing chamber of our dilution refrigerator. The foil was in contact
with a copper post that was screwed to the mixing chamber. Superconducting leads were soft soldered to the foil[19], and the entire assembly
was sheathed in Nb. The leads were brought out to the SQUID (in the
^{4}He bath) in Nb tubes. The second noise thermometer (the inside
thermometer) was fabricated in essentially the same manner except
after being folded, with mylar to separate the two halves, it was rolled
into the shape of a cylinder and placed inside a small, magnetically
shielded epoxy appendage to the mixing chamber of the dilution refrigerator. Thermal contact to the thermometer was provided by dilute
solution, and this thermometer was intentionally placed approximately
13 cm away from the phase boundary of the mixing chamber. The
leads for this thermometer were also brought out in Nb tubes and
thermally grounded to the outside of the mixing chamber and connected
to the input terminals of a second SQUID system held at 4.2 K. The
low temperature NBS fixed point device (SRM 768 - 59) was screwed
to the outside of our mixing chamber in good thermal contact with the
outside noise thermometer. An S.H.E. SQUID based CMN thermometer was located on the outside of the mixing chamber but with the
CMN in thermal contact with dilute solution. A SQUID based diluted
LaCMN thermometer was placed 13 cm below the mixing chamber in a
column of dilute solution and located next to the inside noise thermometer. Unfortunately, a wire broke in this thermometer, and it could not
be used for the measurements reported below. The entire refrigerator

was surrounded by mu-metal shields that provided an ambient field of less than 1 mG.

After cooling the refrigerator to 1.6 K, noise spectra were obtained on both noise thermometers and are shown in figures 2 and 3. From these spectra, it was decided to operate the active filters at low pass settings of 49 Hz and 65 Hz for the outside and inside noise thermometers respectively. The high pass setting was chosen to be .5 Hz for both thermometers since this is just above the frequency where $1/f$ noise begins to be important in the SQUID. The power transfer function $F(\omega,\omega_L,\omega_H)$ for the filters was then measured at the settings used to take all the noise data. The cryostat was cooled to the temperatures of the first three superconducting fixed points (208.94, 162.68 and 101.5 mK), and the value of the mean square noise voltages from both thermometers at each temperature were measured several times using 800 second averaging intervals. The NBS fixed point standards were operated in the standard fashion[12] with the primary excitation current always below 10 μA rms. The CMN thermometer was also calibrated using only the first three fixed points. The cryostat was then cooled to a low temperature and the complete noise spectrum was again

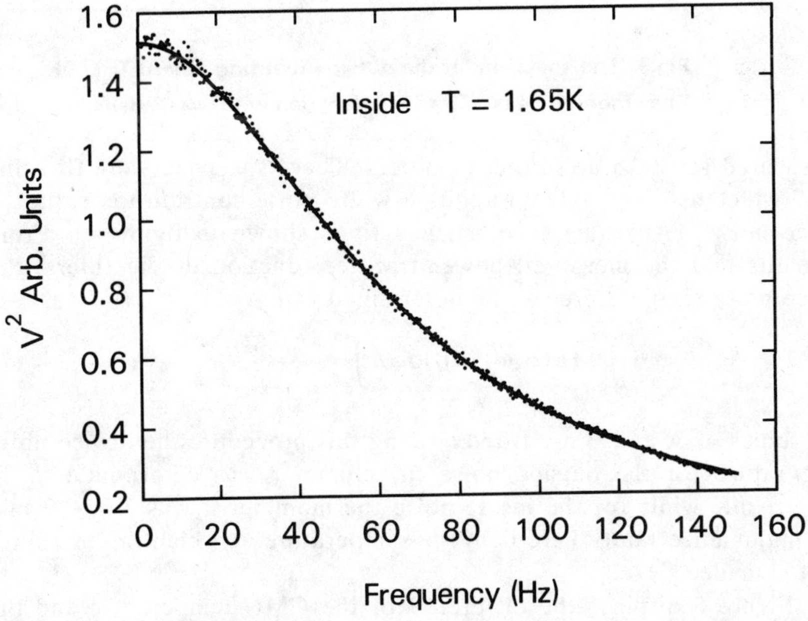

Fig.2. The spectrum of the inside thermometer at T=1.65K. The smooth line is a fit to a Lorentzian with τ=2.52msec.

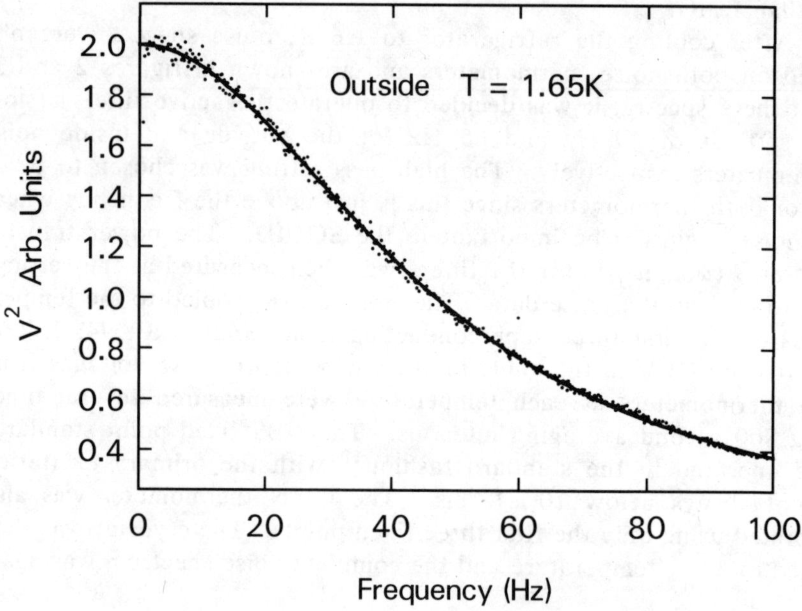

**Fig.3. The spectrum of the outside thermometer at T=1.65K.
The smooth line is a fit to a Lorentzian with τ=3.49msec.**

remeasured for both noise thermometers. These spectrum were fit using
Eq. 7 which assumes only an additive white noise contribution from the
device noise. An example of such a fit is shown in figure 4. From
these fits and the measured power transfer function of the filters, the
device noise temperature can be determined using

$$T_D \; = \; T\int_0^{\infty} AF(\omega,\omega_L,\omega_H)d\omega \Big/ \int_0^{\infty} \frac{BF(\omega,\omega_L,\omega_H)}{1+\omega^2\tau^2}\, .d\omega \qquad (7)$$

The values of A and B are fitted. Using this procedure the device noise
temperature for the outside noise thermometer was determined to be
1.6 $\pm$.1 mK while for the inside noise thermometer it was 3.0 $\pm$.2 mk.
The major uncertainty here is in the temperature at which the measurement is made.

Figure 5 displays the difference of the CMN temperature and the
Johnson noise temperature of the outside thermometer (determined
using the above procedure) as a function of the temperature of the
outside noise thermometer from 1.0 K to 4.8 mK. Above 1.6 K, our
CMN thermometer is too insensitive to make a meaningful comparison.

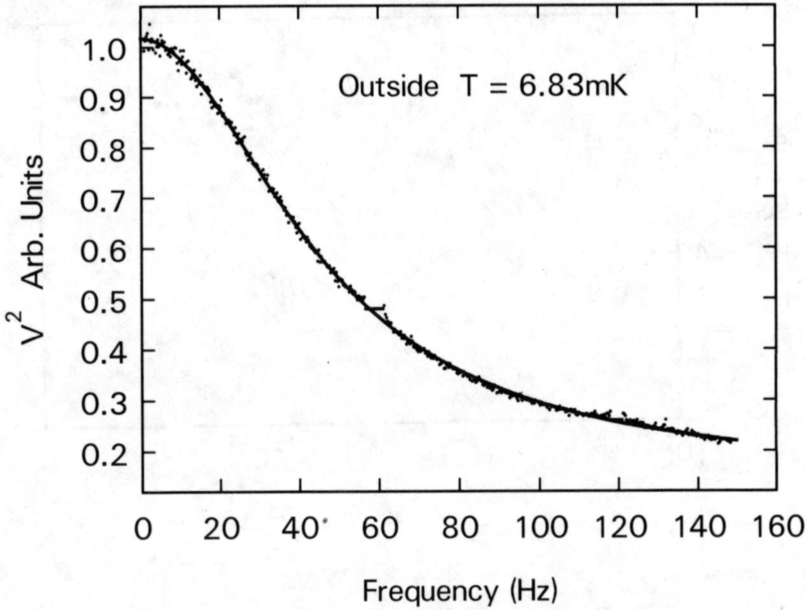

Fig.4. The spectrum of the outside thermometer at 6.8mK.
The smooth line is a fit by Eq.7 with A=0.147 and B=0.870.

Above 10 mK, this CMN thermometer seems to be described by the relation $T_J = T^* + \Delta$ with $\Delta = +0.25 \pm 0.1$ mK. Below 10 mK, the CMN thermometer is systematically indicating a lower temperature; probably this is because it is in better thermal contact with the inside of the mixing chamber than with the outside.

Fixed pt.	T_{NBS}	T_{out}	T_{in}	T^*	$T^* + \Delta$
$AuIn_2$	208.9	209.4	211.0	208.41	208.66
$AuAl_2$	162.7	162.4	160.5	162.51	162.76
Ir	101.5	101.4	101.4	101.5	101.75
Be	23.16	22.93	23.66	22.82	23.07
W	15.03	15.60	16.2	15.43	15.68

Table I. Comparison of the NBS fixed points with the CMN and and noise thermometers.

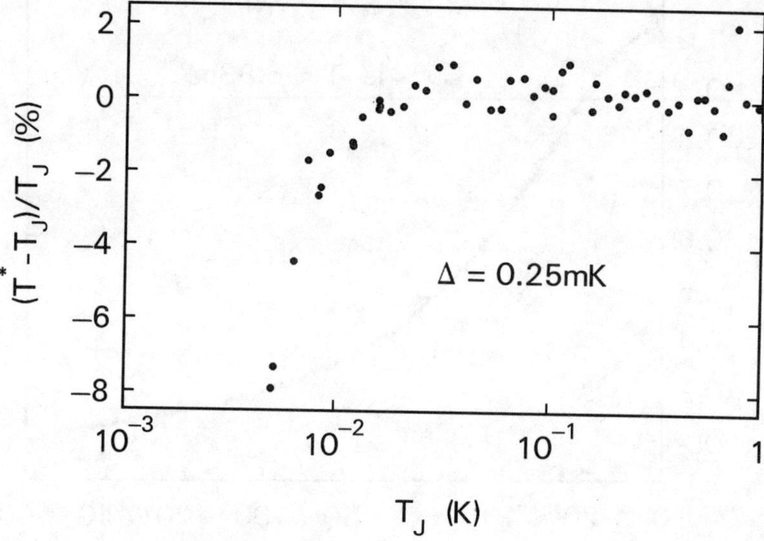

Fig.5. **The reduced difference between the CMN thermometer and the outside noise thermometer.**

After completing the above calibrations, the Johnson noise temperature at the lowest two fixed points (23.16 and 15.03 mK) was measured. The results are summarized in Table I. We include the results for the inside noise thermometer although we expect that there could be sizable temperature differences at low temperatures between the outside thermometer and the inside thermometer; the results tend to support this. Basically, we find that the first four fixed points and the outside noise thermometer are in good agreement with one another, but the NBS temperature assigned to the W transition appears to be low by 0.6 mK. However, the measured value of 15.60 mK is currently in excellent agreement with the revised transition temperature[22] (15.6 ±.1 mK) currently quoted for all Tungsten fixed points.

The length of time necessary to average the mean square voltage in order to obtain any desired precision can be experimentally determined. Figure 6 demonstrates a typical time dependence of the integrated mean square voltage for the outside thermometer and figure 7 demonstrates the same for the inside thermometer. We have also calculated the theoretical averaging time using the measured values for the filter transfer function and L/R time constants using a procedure

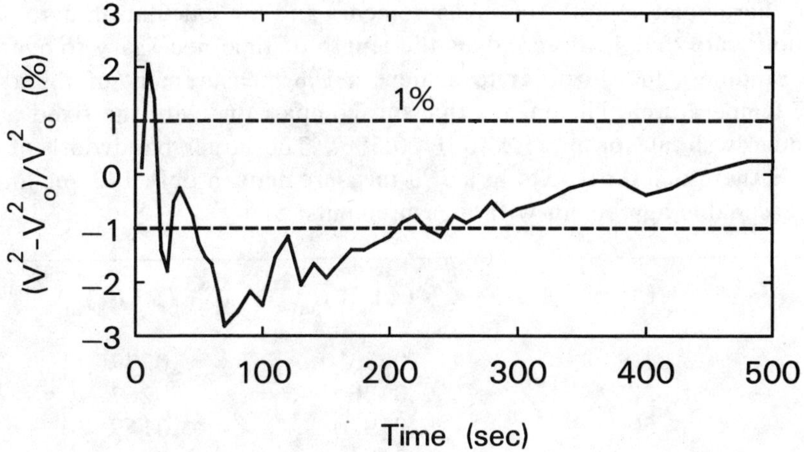

Fig.6. Deviation of the inside noise thermometer signal vs time. T=1.65K

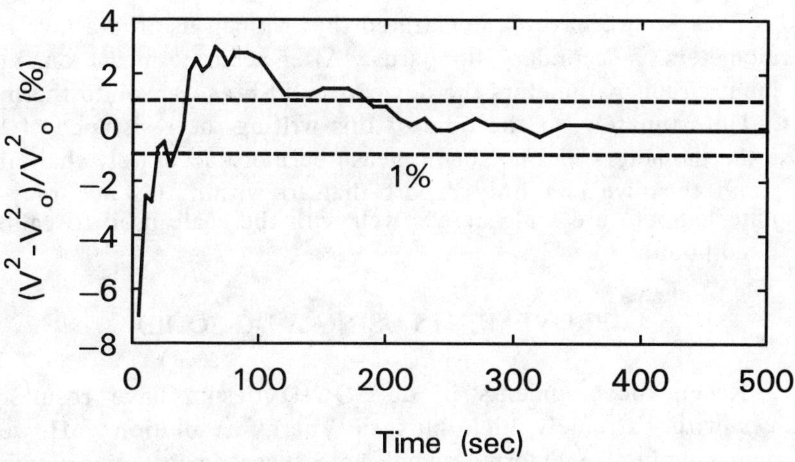

Fig.7. Deviation of the outside noise thermometer signal vs time. T=7.75mK.

outlined in Ref. 19 and the results are given in Table II. For the outside thermometer, both the measurements and the calculation agree and demonstrate that 190 seconds is the length of time necessary to average the random signal in order to achieve a 1% measurement of the absolute temperature. This means that our comparisons with the fixed point standards should be precise to $\pm$ 0.5%. The larger bandwidth of the inside thermometer results in a 1% measurement in only 142 sec and is in reasonable agreement with measurements.

t, sec	$(\Delta T/T)_{out}$	$(\Delta T/T)_{in}$
10	.0433	.0377
20	.0306	.0267
50	.01920	.0169
100	.0136	.0119
200	.00967	.00843
500	.00619	.00533
1000	.00433	.00377

Table II. Averaging time versus precision for both the inside and outside noise thermometers.

Thus far we have demonstrated that we can use the above noise thermometers as secondary standards. After accurate measurements of the input circuit parameters the device should be an absolute thermometer. Unfortunately, at the time of this writing the resistance of both noise thermometers had not been measured more accurately than about 5%, and thus we can only report that to within this accuracy our absolute temperature scale agrees well with the scale used to calibrate the fixed points.

IMPROVEMENTS USING A DC-SQUID

Recent developments in dc SQUID design have resulted in devices with extremely high intrinsic energy resolution. However, efficient coupling of external signals into these devices has only been achieved recently[23]. As was experimentally shown in the last section, the main impediment to developing a fast responding low device noise temperature noise thermometer is finding a SQUID system with very low device noise temperature. We will not discuss the principals of operation of dc-SQUIDS here and refer the reader to an article by

Ketchen[24]. We have been working with the dc-SQUID reported in Ref 23. The coupled energy resolution of this SQUID was measured to be 5×10^{-33} J/Hz at 1.5 K in the white portion of the spectrum above 1KHz. Below this frequency the noise is $1/f$ in character and, at 10Hz, has a value of 4.4×10^{-32} J/Hz. This is to be compared with 7×10^{-29} J/Hz for the rf SQUIDS used in the last section. The input coil inductance of this SQUID is 0.19 μH, considerably smaller than that of commercial rf SQUIDS, and the mutual inductance of the input coil to the SQUID is 3.8 nH. We have chosen to use a resistance for our first noise thermometry attempt of 1 $m\Omega$ and we estimate that the leads will add an extra 0.2 μH of inductance resulting in a L/R time constant of 0.39 msec. Using a conservative estimate for the energy resolution of this SQUID of 4.5×10^{-32} J/Hz over the frequency range 10-410 Hz and assuming the active filter is set at the L/R time constant of the input circuit, we compute that the device noise temperature of this noise thermometer should be 12 μK while the averaging time for a 1 % precision should be 15 seconds. This indeed would be a practical thermometer. However we wish to emphasize that these are calculated values and not measured in an actual experiment. We hope to be testing this thermometer in the near future. Also, recent improvements in the noise thermometer used at NBS have been made, and it is hoped that 1% precision in approximately 20 seconds of averaging will soon be realized[22].

REFERENCES

1. A.C. Anderson, W. Reese, and J.C. Wheatley, Phys. Rev. **130**, 495 (1963).

2. W.R. Abel, A.C. Anderson, W.C. Black, and J.C. Wheatley, Phys. Rev. **147**, 111 (1966).

3. A.C. Mota, R.P. Platzeck, R. Rapp, and J.C. Wheatley, Phys. Rev. **177**, 266 (1969).

4. T.A. Alvesalo, T. Haavasoja, M.T. Manninen, and A.T. Soinne, Phys. Rev. Lett. **44**, 1076 (1980).

5. E.K. Zeise, J. Saunders, A.I. Ahonen, C.N. Archie, and R.C. Richardson, Physica (Utrecht) **108B**, 1213 (1981).

6. B.Hebral, G. Frossati, H. Godfrin, and D. Thoulouse, Phys. Lett. **85A**, 590 (1981).

7. Dennis S. Greywall and Paul A. Busch, Phys. Rev. Lett. **49**, 146 (1982).

8. E. Lhota, M.T. Manninen, J.P. Pekola, A.T. Soinne, and R.J. Soulen Jr., Phys. Rev. Lett. **47**, 590 (1981).

9.   R.J. Soulen, Jr. and H. Marshak, Cryogenics **7**, 408 (1980).

10. R.A. Kamper and J.E. Zimmerman, J. Appl. Phys. **42**, 132 (1971).

11. See for example H. Marshak, Temperautre its Measurement and Control in Science and Industry **5** Part 1, 147 (1982).

12. R.J. Soulen, Jr. and R.B. Dove, NBS Special Publication 260-62.

13. Dennis S. Greywall and Paul A. Busch, J. of Low Temp. Phys. **46**, 451 (1982).

14. H. Nyquist, Phys. Rev. **32**, 110 (1928).

15. A.W. Lawson and E.A. Long, Phys. Rev. **70**, 220 (1946).

16. R.J. Soulen, Jr. and Deborah Van Vechten, Temperature its Measurement and Control in Science and Industry **5** Part 1, 115 (1982).

17. R.J. Soulen,Jr., Deborah Van Vechten and H. Seppa, to be published in Review of Scientific Instruments.

18. R.P. Giffard, R.A. Webb and J.C. Wheatley, J. Low Temp. Phys. **6**, 533 (1972).

19. R.A. Webb, R.P. Giffard and J.C. Wheatley, J. Low Temp. Phys. **13**, 383 (1973).

20. Our measurements of the device noise spectrum indicate that the value of M_3 discussed in reference 19 is very small and its effects on the frequency dependence of the device noise have not been included here.

21. A model 330/hybrid SQUID system purchased from S.H.E. Corporation.

22. Private communication with R.J. Soulen, Jr.

23. M.B. Ketchen and J.M. Jaycox, Appl. Phys. Lett. **40**, 736 (1982).

24. M.B. Ketchen, IEEE Trans. Magn. **Mag-17**, 387 (1981).

Simple Differential Thermometer Using a Thermocouple with a SQUID Detector

Y. Maeno,* H. Haucke,* and John Wheatley
Los Alamos National Laboratory
Los Alamos, New Mexico 87545

Details of design and construction of a simple and rugged thermocouple differential thermometer are presented. Gold + 0.03 at. % iron wire is used, with niobium leads connected to a SQUID. It is demonstrated that at 1 K the response time is about 15 ms and that the temperature sensitivity of 10^{-7} K with a 10 Hz filter is Johnson-noise limited. Additional qualities such as small size, little self-heating, and high reproducibility make this device ideal for differential thermometry at low temperatures.

REFERENCES

1. H. Armbrüster and W. Kirk, Physica 107B, 335 and 385 (1981): in the proceedings of LT-16. An application of the device down to 20 mk.
2. H. Armbrüster, W. Kirk, and D. Chesire, "Temperature, Its Measurement and Control in Science and Industry, Volume Five," (American Institute of Physics, New York, 1982), p. 1025: a review of their work.
3. Y. Maeno, H. Haucke, and John Wheatley, to appear in Rev. Sci. Instrum. (1983): a detailed description of the design considerations.

*Also at: Department of Physics, University of California, San Diego, La Jolla, California 92093.

DIRECT COMPARISON OF COPPER, SILVER, AND PALLADIUM HEAT EXCHANGERS *

R.F. Berg and G.G. Ihas

Department of Physics
University of Florida
Gainesville, Florida 32611

In order to consistently compare various metal powders used in heat exchangers for cooling liquid helium samples, five different miniature exchangers were constructed and tested between 7 and 70 millikelvin. Each exchanger consisted of a cylindrical copper cell containing the exchanger material in the lower half and a heater and thermometer suspended in the bulk ^{3}He in the upper half. Carbon resistors ground flat to increase surface area and reduce bulk proved to be economical and effective thermometers down to 5 mK. Boundary resistances were calculated using the measured thermal time constants of the exchangers and the heat capacity data of Greywall and Busch.[1]

The copper powder we used[2] was about .07 micron in average particle diameter and included a 60-80% volume fraction of oxide. Simply pressing this powder into a cell with no heat treatment produced a very poor exchanger as did pressing at room temperature after reduction of the powder. An effective exchanger was produced by first pressing the untreated powder into the cell and then gently heating it at 200°C in hydrogen until the oxide fraction was removed.

A palladium exchanger was constructed by sintering pressed 1 micron palladium powder according to a recipe first used at the Ohio State University.[3] This typically had a boundary resistance two to three times higher than the successful copper or silver exchangers.

The most effective exchanger, in terms of boundary resistance per unit sinter volume and ease of preparation, was constructed simply by pressing .07 micron silver powder[2] into a silver-plated copper cell with no heat treatment whatsoever. A large version of this "cold-pressed" silver exchanger is now operating successfully in our nuclear demagnetization cryostat.

1. D.S. Greywall and P.A. Busch, Phys. Rev. Lett. __49__, 146 (1982)

2. Vacuum Metallurgical Company, Ltd., Tokyo, Japan

3. Thesis by Kevin A. Muething, Ohio State University, Columbus, Ohio 43210 USA (1979)

*Partially supported by NSF Grant DMR-8006929.

PERFORMANCE OF DILUTION REFRIGERATORS

R.L. Fagaly, D.N. Paulson, and W.L. Zoeckler

S.H.E. Corporation
4174 Sorrento Valley Blvd.
San Diego, California 92121

In order to increase the available surface area, one can utilize sintered metallic powers. Frossati[1] showed for a T^{-2} Kapitza law

$$T_{mc} = AR_{km} \, \dot{n}_3/\sigma + \dot{Q}_{mc}/(B\dot{n}_3) \tag{1}$$

for submicron silver powders and low temperatures where $A = 27$ and $B = 82$. Examining the numeric factors A and B from an enthalpy standpoint and taking into acconut the proper temperature dependances, the actual values can be otained from Radebaugh[2]. Using eq. 1 and knowing T_{mc}, n_3, and $\dot{Q}$, one can determine σ. These σ were found to be consistent for several SHE machines with Frossati's effective area of 2 m^2/gm. As can be seen, the first term (finite surface area) in eq. 1 has an increasing effect below 40 mK. If we increase surface area, we can improve the low temperature performance. A determination of the effective surface area (σ) can serve as a useful diagnostic tool when operating a dilution refrigerator. For example, improper placement of the phase boundary will cause the effective surface to be significantly less.

Using the data from a 1982 SHE Model DRI-550 Dilution Refrigerator to calculate $\dot{Q}_{int}$ vs $\dot{n}_3$ (at temperature of 3.5 mK) we have plotted $\dot{Q}_{int}$ vs $\dot{n}_3$. As could be seen, $\dot{Q}_{int}$ can be described as $\dot{Q}_{int} = \dot{Q}_0 + D\dot{n}_3^2$. Using $\sigma = 500\ m^2$, we found $\dot{Q}_0 = 1.24$ erg/sec, and $D = 7.87 \times 10^{-6}$ erg sec/mol^2. The experimentally observed value for D was used to calculate the effective impedance and was subsequently identified with the dilute line exiting the mixing chamber.

In summary, we have shown that when analyzing the performance of a dilution refrigerator proper attention must be paid to enthalpy considerations. The use of the proper relationship in determining cooling powers and effective surface area can serve not only as a diagnostic tool but as an aid in design of dilution refrigerators.

REFERENCES

1. G. Frossati, J. Physics (Paris) 39, Suppl 8, C6-1578 (1978).
2. R. Radebaugh, NBS Tech Note 362 (1967).

THE QUANTIZED HALL EFFECT

P. Berglund
Francis Bitter National Magnet Laboratory, MIT,
170 Albany Street, Cambridge, Massachusetts   02139

At low temperatures and high magnetic fields the ratio of hall voltage to electric current is quantized[1] to $\rho_{xy} = h/ie^2$ to better than 2 parts in 10^7 (h is Planck's constant, e is the electron charge and i is an integer)[2]. Simultaneously the electrical resistance of the sample drops to a very low value. A state reminiscent of super-conductivity is obtained. For high mobility samples[3] such as modulation doped GaAs-(AlGa)As heterojunctions quantization effects have been observed for fractional quantum numbers (i = 1/3, 2/3, etc.).[4,5] The mechanism for this effect does not appear to be the same as for the integral quantum numbers.[6] Studies of the fractional quantized Hall effect will require higher fields (up to 30T) and lower temperatures (down to 50 mK).

REFERENCES

1. K.V. Klitzing, G. Dorda and M. Pepper, Phys. Rev. Lett. 45, 494 (1980).
2. D.C. Tsui, A.C. Gossard, B.F. Field, M.E. Cage and R.F. Dziuba, Phys. Rev. Lett. 48, 3 (1982).
3. H.L. Stormer, R. Dingle, A.C. Gossard, W. Wiegmann and M.D. Sturge, Solid State Comm. 29, 705 (1979).
4. D.C. Tsui, H.L. Stormer and A.C. Gossard, Phys. Rev. Lett. 48, 1559 (1982).
5. H.L. Stormer, A. Chang, D.C. Tsui, J.C.M. Hwang, A.C. Gossard and W. Wiegmann, submitted to Phys. Rev. Lett.
6. R.B. Laughlin, to be published in Phys. Rev. Lett.

EPILOGUE

SUMMARY OF 1983 SANIBEL SYMPOSIUM ON QUANTUM FLUIDS AND SOLIDS

R.C. Richardson

Laboratory of Atomic and Solid State Physics
Cornell University, Ithaca, NY 14853

ACKNOWLEDGEMENT

I would like to begin the conference summary in the usual
way. That is, I want to make certain that we thank our hosts
before I run out of time and before everyone else in the crowd
evaporates to catch transportation to the airport. Let's give the
great gang from Florida State....I mean the University of Florida a
big hand. [The response of the audience to this attempted jest was
nervous laughter. Earlier in the week a session chairman, not from
the United States, had introduced Neil Sullivan as the newest
faculty member of Florida State University. He has really joined
the faculty of the University of Florida.]
     Unless you have experienced the joy of planning and running a
meeting such as this you are probably unaware of how much we owe
our hosts. Their work did not end with the selection of this
rather spectacular location for the meeting. Before the meeting
there was all of the work of raising the funds, planning the
program, mailing out the announcements, and much more. During the
meeting there were myriads of details which were handled behind the
scene of our sessions. It has all gone very smoothly for the
participants but this is only because of the conscientious work of
our hosts.
     Our hosts sacrificed something else beyond the time which they
invested in planning this meeting. On the basis of some mistaken
notion about conflicting interests they disqualified themselves
from making presentations in the formal part of the program. This
is a pity because they have contributed so much to the field:
Dwight Adams with his work on solid ^{3}He; Gary Ihas with his work on
the flow of superfluid ^{3}He; and Pradeep Kumar with his theoretical
work on superfluid textures and on magnetic order. I hope that at
least some of their work will be included in the conference
proceedings.
     On the first two transparencies I have listed the people who
have been doing all of the work, both that which has been obvious
and that behind the scenes. The research associates, graduate
students, and visitors at the University of Florida who have worked
to make certain that everything in the session rooms worked were
called by the demeaning title "gofer". Unless you speak American
English, the source of the word is not obvious until you see it in
the written form. It comes from "go for this...go for that."

472

Table I. List of University of Florida Hosts who planned and
          conducted the Meeting

Gofers:

        Kurt Uhlig
        Brad Engel
        Greg Spencer
        Simon Philpot
        Deepak Srivastava
        Ralph Rosenbaum
        Bobby Berg
        Greg Haas
        Vijay Samalam
        Ying Lin

Program and Planning:

        Pradeep Kumar
        Dwight Adams
        Neil Sullivan
        Gary Ihas
        Sam Trickey

Conference Secretary:

        Sheri Hill

A "PITCH"

One of the reasons I agreed to give the conference summary was
that it gives me the opportunity to preach a sermon. Let me give
you my "pitch". It is about communicating our results to the
larger scientific community and the general public. The origin of
the word pitch, as I am using it here, probably lies with the
advertising talks given by the men standing outside of circus or
sideshow tents to entice the crowd to pay the admission fee to see
what was going on inside. The "pitchman" would say things like,
"Step right inside and see the two headed orangutang..." The pitch
I am making is intended to persuade all of you to help in the
business of explaining the new things in low temperature physics to
your colleagues in other fields.

In the hundred years since air was first liquefied in Poland,
we have been repeatedly called upon to answer the question: "What
are low temperatures good for?" My roommate during this week has
been Professor Masuda who told me an interesting story about the
argument used against funding low temperature research in Japan
(until very recently). It is likely that all of us have faced the
same argument although it might not have been stated quite as

explicitly. Low temperature physics was being condemned through Equation 1, the Third Law of thermodynamics:

$$\lim_{T \to 0} S = 0 \qquad\qquad (1)$$

The Japanese administrators of the science budget were quite convinced of the correctness of the Third Law. Therefore, since the condition of zero entropy must be utterly boring and quite useless they chose not to invest in the research. Why pay more and more to see the condition of almost nothing? Most attempts at answering such a question rely upon pointing out the history of past achievements in the field. It is almost imposssible to predict the discovery of something which has never before occurred to anyone. We can not be certain that the equivalent of the discovery of superconductivity will happen again. Even without finding a new superconductivity the science in our field is very exciting. However, it does not do us much good from the standpoint of research support unless we tell the rest of the world about it.

The entire collection of low temperature physicists in the world is quite small. The number in any one country is even smaller. We cannot form an effective lobby to support the research anywhere, even if it were appropriate. Nevertheless, at the same time that we are competing with each other it is important to identify the significant progress which has been made by others. It is the only way to demonstrate the continued fertility of the field. We should all go back to our own countries and rave about the things that are happening elsewhere. A good example of something to talk about is found in the lecture which Olli [Lounasmaa] has just given us. A very short time ago I would have never imagined that by 1983 someone would be seriously arguing whether or not a nuclear magnetic phase transition measured at 60 ± 10 nK occurred at a temperature in disagreement with theory. If even we are surprised by such technical progress, think how it must appear to people outside of our field.

In a sense, the low temperature research community is like the plants in a hothouse. We are protected from the rest of the environment by funding agencies. What we do is not of any immediate practical consequence. We do it because we find it intrinsically interesting. If we wish to have continued financial support it is not sufficient that we only discuss our results with each other. We must respond to the broader community. We are perceived by many people as having a very narrow specialty. In this connection, I would like to tell a story about my absent colleague John Reppy. For the past 25 years, he has had his wife Judith read his papers to make suggestions. Judith's own specialty is economics. She recently commented that she was amazed that anyone could have a successful career by repeating the same experiment for 25 years. How could people still be interested in how helium flows through something or other? Even scientists in fields closely related to ours think of low temperature physics in the same way. As far as they are concerned, we have been looking at the same flow of helium since the days of Kamerlingh Onnes. What could possibly be new? We need to worry about answering that question. After you return from this meeting, each of you should corner an astrophysicist and say, "Hey! Did you know that in rotating superfluid ^{3}He-B, some sort of spontaneous magnetization appears when it is cold enough? The result probably has something to do with the magnetic field generated by the rotation of superfluid neutron stars." The measurements in liquid ^{3}He will probably give the only real experimental data for the sort of calculations they are making. One of the themes of this summary is that helium research is important because liquid and solid helium are ideal model substances for testing whatever ideas might be in current fashion.

As another example of the difficulties we have in communicating the interest we feel for our research, I would like to tell you about a visitor we had at Cornell a few weeks ago. I am sure that most of you have had similar experiences. The visitor was an advisor at the highest level of the U.S. Government. He had come to Cornell to visit the Director of our Materials Science Center who has an office in the same building as our laboratory. At the end of his day of discussions on matters related to science policy he had a few minutes to kill before the cocktail hour so he was brought to our lab for a quick tour. At the time, we had only one apparatus which was not in a dewar. It had a trouble common to all low temperature labs, a leak. By coincidence, the apparatus is sited in the same place as the one used by Doug Osheroff in his thesis research. I began by describing our fortuitous discovery of the triplet-paired superfluid states of liquid ^{3}He and long before I could get to any mention of the elegance of the B phase textures in gradient fields, he interrupted. "Excuse me, but what are those things for?", he asked while pointing to the top of the cryostat. I explained that the discs of blue styrofoam and aluminum he was pointing at were used to reduce the consumption of liquid helium by blocking the radiation and controlling the convective flow of helium gas. Then he said, "It's absolutely amazing. The only other place in the world where there is something which looks like that is in a NUKE. Where do you get your styrofoam?". After that it was hard to get back to superfluid ^{3}He and it was time for him to go to dinner. I obviously failed in my goal of interesting him in low temperature physics. You have to evaluate your audience quickly and then get your message across as soon as possible.

Of course, scientific interactions are also very important within a research field and this meeting is a good example. I think that in the retrospect of 20 or 30 years from now this conference will be recognized as part of the golden era of low temperature physics. When Dwight Adams first mentioned about one and a half years ago that he might be having a meeting here, I frankly was not very enthusiastic about it. I thought, "Oh, no.

Not another meeting. What could possibly be going on that will be new? We will all have to go just to keep Dwight from being offended." Actually, a great deal has happened. This has been a very exciting meeting. I look forward to the rapid and inexpensive publication of the proceedings by the AIP. Some of the papers presented here will be real classics.

There is another important level of interaction I would also like to discuss, the communication we have with our funding agencies. The majority of us from the United States are supported by the National Science Foundation and the essential notion about the transmittal of scientific results is illustrated on the viewgraph. It's a "nugget". I am certain that analogous things exist in every other country. In almost any successful endeavor, competition is used in order to promote development. In the U.S. funding agencies, such as the NSF, the competition takes place in sort of a game in which the tokens are called "nuggets". There are periodic meetings in which the program directors in related fields get together and compare the progress which has been made by the people they have supported. There is a bit of "one-upsmanship" involved in the competition. One program director will bring out a few transparencies of recent results and say, "Here's what my guys did. Isn't that neat?" The transparencies are based upon "nuggets" sent to the program director by his "investigators". Each of the directors presents whatever nuggets he either has received or can glean from annual progress reports. In the end, the research areas which thrive are the ones with the most interesting and largest number of nuggets. The relative fertility of research fields is evaluated through such meetings. If low temperature physics is to keep pace with other fields, we must help

Fig. 1. Nuggets are not only small lumps of gold ore. They are concise statements of significant new results used by program officers at the NSF.

our program director by sending him summaries of interesting results. It won't hurt to make the transparencies for him and write a suitable brief discussion that can be used in his presentation.

Now to return to the question: "What are low temperatures good for?". The answer can be divided into three categories. The first is the aspect of the research which probably excites us the most about our work. We explore qualitatively new areas of science and try to answer fundamental questions about statistical mechanics, quantum mechanics, and the collective behavior of matter. The second class of important and interesting things that can be done at low temperatures is the study of ideal or model systems. This is primarily because of the large range of temperatures that can be controlled very carefully at temperatures less than 4 K. A large number of ideas that might originate in other branches of condensed matter or solid state physics can be tested most easily at low temperatures. This class of experiment will exist as long as people are thinking of new ideas. We have heard several excellent examples of this type of investigation this week. The third category of use of low temperature research is the advancement of technology. It is not widely recognized that it is this community which does most of the technology development for performing experiments at lower temperatures. At the banquet last night John Wheatley told us about the work in the 1960's and the development of the dilution refrigerator. The work which followed Heinz London's original suggestion and the experiments of David Edwards, and John Wheatley, and Henry Hall was very important to a small group of people who recognized that the likely outcome would be a great simplification of the technique for obtaining low temperatures. The general community was unaware of this research and does not even know the history of the development of the dilution refrigerator. Today you can visit a laboratory such as Bell Labs and walk down the hall and see commercial dilution machines in every room on each side. Those barbarians can just take their silicon thingies and screw them to the apparatus and cool them down. It is not even low temperature physics anymore. [While saying this, the speaker was standing in front of Henry Hall who had made repeated references to "silicon thingies" in another context during his talk earlier in the week.]

At this point, I want to give special recognition to some important technology advancements and achievements at this conference. Two of them were from Helsinki. One was the studies of the copper nuclear magnetic transition which Olli Lounasmaa just told us about and the other was the beautiful rotating superfluid work. The other two achievements which deserve special mention are the new milestones in cooling liquid helium. Bob Mueller described the progress of the Jülich group in cooling the mixtures and then George Pickett told us about the remarkable apparatus and results of the Lancaster group in advancing the use of copper to cool liquid ^{3}He.

CONFERENCE HIGHLIGHTS

Next we come to a partial summary of the low temperature science we have discussed this week. There is no pretense that the summary is comprehensive. There will be no discussion of the talks in this last session or most of the poster presentations. The format of the discussion will be a grouping of the talks around a central topic; then, in the spirit of my remarks the place of low temperature physics in the general scheme of modern research, I will indicate why the work is interesting. Following that, I will indicate the current status of the area and the future prospects for research in related areas.

Solid ^{3}He

The topic is interesting because solid ^{3}He has larger quantum effects than any other crystal. As a consequence, the nuclear magnetic transition occurs at a factor 10^5 higher in temperature than in most other substances. The puzzle about the long standing "discrepancy" between measurements of the magnetic susceptibility and changes in pressure of the solid in large magnetic fields seems to have been resolved. The recent pressure measurements at NBS and the new susceptibility measurements at Texas A & M are in good agreement with each other and the earlier pressure measurements of the Florida group. There is an irony in how the problem has been resolved. There never was a real discrepancy, only a misinterpretation of the data. The old values of the Weiss θ were obtained by plotting inverse nuclear susceptibility versus temperature and defining θ as the temperature intercept of the approximately straight line through the data. That quantity is not the same as the leading coefficient in the high temperature expansion of the susceptibility. Of course, it took the repetition of the earlier experiments and the more precise susceptibility measurements to make this convincing. Apparently the parameters of the three and four spin exchange model are not very sensitive to this new result. This is a pity, because I had hoped for an excuse to overthrow what still appears to be a successful theory.

With regard to work in the magnetically ordered regime of solid ^{3}He, we heard an analysis of the theoretical limits on the multiple spin parameter from Professor Masuda, a presentation by Professor Shigi of the Osaka magnetization measurements, and a progress report by Jacques Flouquet on the Grenoble neutron diffraction experiments. The reduced magnetization plot of the Osaka group is compelling. The fact that all of the data from various molar volumes seems to scale in such a simple way suggests that the theory might have too many parameters in it still. The simplest model might require only one and a half parameters instead of three or so. There are several obviously important questions remaining in this area of research. We still do not know what the microscopic form of the high field ordered phase is. The stability of the high field phase in large magnetic fields remains to be investigated. Finally, we are anxious to see the results of the polarized neutron experiments with possibly new information about the microscopic arrangement of the spins in the ordered phases.

Turbulence and Chaos

The talks which we heard from Victor Steinberg and Yoshi Maeno about studies of turbulent flow in helium mixtures gave excellent examples of the use of liquid helium as a model substance for testing general ideas. The question of the non-linear response in driven systems is being investigated in practically every subfield of condensed matter physics and chemistry. Helium is a particularly useful liquid for turbulence studies because of the large range of Prandtl numbers which can be obtained. The experiments which Maeno talked about illustrated the period doubling route to chaos which has been found in a wide variety of systems in the past few years. Because the thermal environment of liquid helium can be controlled so precisely and because the thermal conductivity of the liquid can be tuned by changing the concentration of ^{3}He we can expect liquid helium to be the standard test substance in this field as it develops. This line of research is probably only in its infancy and it will flourish only with parallel advancements in theory.

Thermometry

The maximum uncertainty in the temperature scales we are using around 1 mK is probably less than 15%. This is just not good enough if we are to successfully interpret the quantitative thermal data about liquid ^{3}He. Michael Mayberry presented the case that the difference between the heat capacity results of Greywall, Wheatley, Halperin, and most recently the Berkeley measurements on one hand and the results of the Helsinki and Cornell groups on the other hand can be traced to the temperature scales used. Richard Webb described the scheme he plans to use for measurements of noise temperature. The method looks very promising because of the advances in DC SQUIDS. Richard is right. We need to have a reliable and different thermometer used to measure the ^{3}He melting pressure and the T_c versus pressure curve of liquid ^{3}He. The possible systematic errors in platinum nmr thermometry conceivably lead to even more serious problems at temperatures much less than 1 mK.

Solid-Liquid Interfaces

The work here is another example of the use of helium as a model substance. Helium crystals are probably the easiest of all solids to grow because of the large atomic diffusion rates in the crystal and the thermal conductivity of the liquid. We heard reports by Daniel Thoulouze and Sebastian Balibar on the experiments in Grenoble and Paris in which the roughening transition has been observed for a number of faces of solid ^{4}He. At the lowest temperatures, facets are seen in every case to date. Will Saam summarized the theory of the transition between a

thermally roughened surface and the surface with facets. I am glad
that Andreev was apparently wrong in his suggestion that solid
helium would have no facets at T=0 because of the quantum motion of
the atoms.   Otherwise, we would not have been able to see the
beautiful pictures of helium crystals which the Paris group has
taken. For the future, the more difficult problem of the interface
in ^{3}He remains. The negative slope of the melting curve presents a
formidable obstacle to the experiments.

Solid Hydrogen

Solid hydrogen is the sister quantum solid to helium. The
quantum motion is not as large as in helium, but hydrogen is the
simplest of the molecular solids. The question of most interest is
whether or not any state of order in the molecular orientation of
the perishable J=1 molecules exists. I think most of the audience
was somewhat mystified after the panel discussion of Jim Gaines,
Dave Haas, Marge Klenin, Neil Sullivan and Bob Pound. My summary
of the situation is shown in Fig. 2.   There is evidently a dispute
among those in the field about whether or not a glassy state of
order exists in solid H_2 in the appropriate region of temperature
and J=1 (or ortho) concentration. The region of "frustration" is
indicated with a loop in the figure. The experiments are made
difficult by the large background arising from the molecular
conversion of the H_2 to the J=0 state and by the crystalline phase
changes in the same vicinity. The problem is interesting because
hydrogen is the model system for quadrupolar interactions in
molecular solids. We will be interested in seeing how the
controversy is resolved.

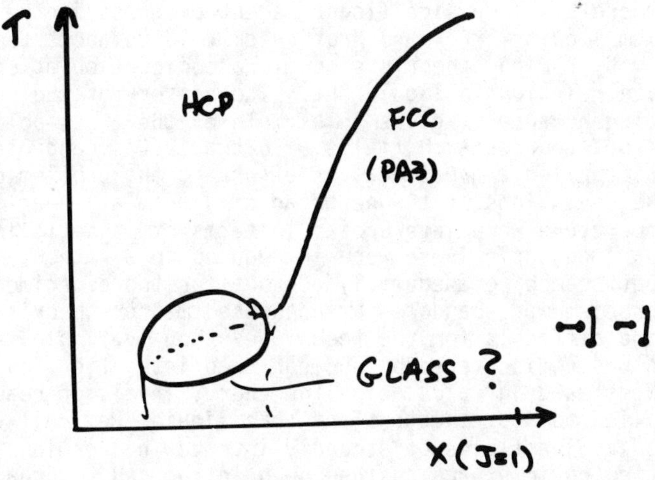

Fig. 2. The region of frustration in the temperature versus
ortho-fraction phase diagram of solid H_2.

480

Novel Fermi Systems for Study

In this session Bob Mueller told us about the Jülich experiments on the dilute mixtures of ^{3}He; and Michelle LeDuc told us about the experiments of her group in Paris in which optically polarized ^{3}He gas is condensed. These are conditions of ^{3}He which are in new limits for studies of statistical mechanics. The Jülich group has succeeded in cooling the dilute mixtures down to a record low temperature, 215 microKelvin. Unfortunately, there are no signs yet of the BCS pairing transition. Their results for the heat capacity are in good agreement with the earlier measurements down to 3 mK by Wheatley's group at Illinois. Therefore it is quite unlikely that there are any serious difficulties with thermometry. At the lowest temperatures of this work the ^{3}He quasi-particles have a very long lifetime. The transport properties in this regime are interesting even without the pairing transition. Henry Glyde pointed out that the weak interaction between polarized atoms in liquid ^{3}He makes a rigorous theory easier to calculate than that for the normal liquid in low magnetic fields. The Paris group is actually beginning to produce such a liquid through the use of techniques adapted from atomic physics. All of this work is in its infancy. It is safe to predict that a recurrent theme in future meetings will be the discussion of the properties of these and other unusual quantum fluids (such as spin aligned atomic hydrogen and deuterium). Despite the experimental difficulties, these are probably the most fertile areas for future research.

Theory of Helium Liquids

After a long pause in which the subject has apparently lain fallow there has been significant recent progress in the theory of the helium liquids. Eckhard Krotscheck told us about the work done primarily by nuclear theorists on microscopic calculations of the dispersion relation in liquid ^{3}He. The history of the subject is interesting because it gives an example of the cross-pollination of ideas between research fields. In the 1960's condensed matter theorists applied a great deal of effort to the problem of liquid ^{3}He. The deviations of the heat capacity from a linear increase with temperature were interpreted in terms of magnetic interactions called paramagnons. These were introduced in a rather ad hoc manner but seemed to adequately account for the experiments. The formal problem was abandoned by condensed matter theorists because it was too difficult for the techniques then available. The question was rediscovered by nuclear theorists after ten years. The methods used in calculating the energy levels of heavy nuclei were applied to neutron stars and then liquid ^{3}He. All three problems involve fluids of strongly interacting Fermions. The hypernetted chain expansions had been introduced to organize the calculations. Using the same formalism which had been successful in nuclear theory, Brown and coworkers produced a dispersion relation, illustrated in Fig. 3, which successfully predicted the temperature dependence of the effective mass in liquid ^{3}He. A new breath of life came to the subject at the same time that Dennis Greywall made his careful measurements of the heat capacity over a wide temperature range. The expertise in the calculations had shifted to a different area of physics, but liquid ^{3}He was available as the model for a nucleus with 10^{23} nucleons.

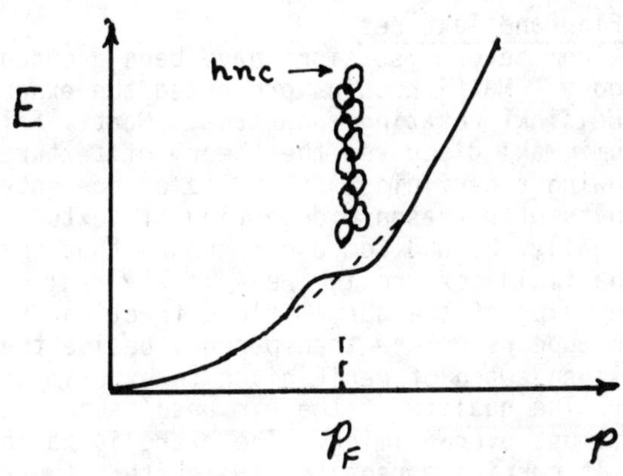

Fig. 3. The dispersion relation of liquid ^{3}He.

The physical interpretation of the temperature dependence of the effective mass is apparent from the dispersion curve. At low temperatures only small momentum excursions are excited around p_F and thus the effective mass is larger because of the flat slope of the dispersion curve near p_F. (Evidently the ^{3}He quasi-particles get trapped in the hypernetted chain?)

In the same session Peter Wölfe described the progress in the development of the phenomenological polarization potential for testing the relations between the experimentally determined Landau coefficients and for predicting various properties of the fluid. The larger values of m*/m, as measured by Greywall, lead to the most consistent set of parameters. We experimentalists owe quite a debt, particularly to the Pines group at Illinois, for the efforts made distilling the results of so many phenomena into one model.

The theory of liquid ^{4}He might also be due for a renaissance. Recent measurements of the liquid structure factor near the lambda point have been interpreted to yield values of the number of particles occupying the ground state, the Bose condensate. The standard result found in introductory statistical mechanics courses, which predicts macroscopic occupation of the ground state at a temperature of 3 K, has probably fascinated all of us. Just how this result is related to superfluidity has remained an elusive question. The interpretation of the structure factor measurements is in dispute so Larry Campbell turned his attention to the surface tension measurements of David Edwards and co-workers at Ohio State. He obtains the same result, that the condensate fraction is around 10 or 12% as the temperature goes to zero. The only parameter necessary beyond the surface tension data is the coherence length which he obtained from vortex experiments.

Superfluid ^{3}He Flow and Textures

The papers from several sessions have been grouped together under this category. Matti Krusius presented the experimental results of the Helsinki rotating apparatus. Martti Salomaa, Sandy Fetter, and Kazumi Maki discussed the theory of textures in rotating and flowing superfluids. Hans Bozler presented some preliminary results of ultrasonic detection of texture orientation. Finally, Roland Combescot showed that the inevitable distortion of the fluid texture in ^{3}He-A in any real container will lead to a finite value of the normal fluid fraction at T=0.

An asterisk appears on the transparency beside the name of Krusius. This is an award of merit given to experimentalists by the competitors. The quality of the nmr results in the rotating experiment was almost overwhelming. The award is called "The Mark of The Demon." It carries a penalty. Hereafter, the recipient will have a leak in his apparatus. However, the leak will be small. It will only appear at temperatures less than 4 K. Another such award will be given later to George Pickett and the Lancaster group.

The work presented by Krusius and Salomaa is the fruit of a very successful collaboration between Finland and the Soviet Union. Matti Krusius made it clear that he could only show a small portion of the experimental results and Martti Salomaa intrigued us with enough theory for three talks. Despite the polished quality of the results, we were left with the feeling that we had only seen the tip of the iceberg. Nuclear resonance has been a surprisingly useful probe of the vortex structure of ^{3}He-B. What is the change in state which occurs in rotating ^{3}He-B to produce an apparent spontaneous magnetization when the temperature is less than $0.6T_c$? This has been one of the most significant new things reported at this meeting.

What are these studies good for and what are the future prospects in this area? For one thing, these are the types of experiments which will be used as textbook illustrations 30 years from now. The properties of the anisotropic superfluids can be probed usefully for another century. We are only beginning to learn how to do the experiments to test the theories which Kazumi Maki and Sandy Fetter have been working on for the past half dozen years. Olli Lounasmaa has frequently said that one of his ambitions is to do an experiment which would be included in some edition of Kittel's book. The rotating experiment would qualify if Kittel revised his book to include superfluid ^{3}He.

Quasi-particle Kinematics

Henry Hall and George Pickett told us about their experiments with vibrating wires. Henry persuaded us that the wire geometry can be exploited not only as a probe of the fluid hydrodynamics but also as a dynamic "pressure gauge" of the quasiparticle excitation gas and as a local thermometer. George Pickett won a demon's mark with his talk. The Lancaster group broke the previous record for cooling pure ^{3}He and he showed data taken at a temperature of 125 microkelvin! The results were pretty remarkable too. What is responsible for the nonlinear response of their wire at even the lowest force level when the normal fraction is 10^{-6}? Are there B-phase texture phenomena like those discussed by Combescot? Tony Leggett, as usual, has given us a lot of new things to think about. Andreev reflections seem to have come of age in superfluid ^{3}He. It will be most interesting to see what experiments are devised to look for the boomerang reflections of the quasiparticles from texture lines. Peter Wölfe and Dierk Rainer also discussed Andreev reflections in the poster papers. I still have to think about why all of the quasiparticles do not collect at one end of the container. It has been ironical that the quantity apparently measured as viscosity continues to decrease with temperature in a regime in which theory says it should remain constant or even increase because of the long mean free path. The unusual momentum reflection of the quasiparticles at a surface suggested by Peter Wölfe might explain the results.

What are these studies good for? The hydrodynamics and flow properties are central topics in the study of a superfluid. The viscosity of liquid ^{3}He at T_c makes it a rather peculiar superfluid even without the anisotropy effects.

Ultrasound Spectroscopy

The sound absorption studies in superfluid ^{3}He have revealed a great abundance of collective phenomena. I tried to borrow John Ketterson's viewgraph of the table of the 18 collective modes. He left before I could get it, so I have done my best to reproduce it here [in Fig. 4]. [John's typewritten table contained so many details that even he had to stand one meter from the screen to read it.]

The physical picture which is emerging from the B phase ultrasound experiments is good "nugget" material. Most of us run into difficulties in trying to describe our work in superfluid ^{3}He to others because of the complexities of the system. People need a great deal of motivation before they are willing to think about the consequences of a tensor order parameter. It is rather pleasing to have results which can be easily related to familiar ideas in other areas of physics. John Ketterson and John Saunders described recent experiments; and Nils Schopohl and Joe Serene reviewed the

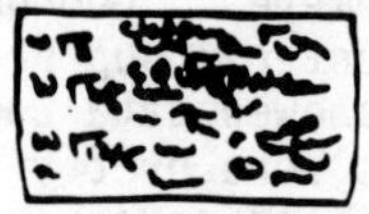

Fig. 4. A representation of John Ketterson's transparency which
listed the 18 collective modes that result from the real and
imaginary components of a 3x3 tensor order parameter.

development of the theory and the physical interpretation of the
results. The undistorted wave function of ^{3}He-B is almost a J=0
state. The "new" absorption modes are like molecular spectroscopy
for the excited J states of the wavefunction. Sound of the
appropriate energy produces J=2 and possibly J=4 distortions of the
wave function so that the radiation is absorbed by the fluid. The
interaction is weak so that the absorption lines are quite narrow.
Examples have been found of analogues to all of the line splitting
phenomena which were the essential contributions of atomic physics
to the evolution of quantum mechanics: The Zeeman effect; the
Stark effect; and the Paschen-Back effect. The elegant difference
between ^{3}He-B and the hydrogen molecule is that an Avogadro's
number of ^{3}He atoms act all at once in a collective response to the
radiation field. I was relieved that Joe Serene's only caveat

about thinking in terms of molecules was to be careful in applying the simple picture to nonlinear absorption phenomena.

Joe pointed out that the workers in superconductivity, after much effort, eventually abandoned their attempts at ultrasonic spectroscopy. In ^{3}He the effects are easy to see and the work has probably just begun. Careful studies of the A phase should be very productive. Interest in many of the fundamental questions about pairing has shifted from the superconductivity community to the liquid helium community. The situation is analagous to that which existed between condensed matter and nuclear theory. The techniques from one field were transplanted into another for nurture and growth. What's ^{3}He good for? Well apart from all of its intrinsic mysteries, it can give us new ideas about superconductivity.

CONCLUSION

At this meeting we have seen that low temperature physics remains a very fertile research field. A great deal is happening. The cooling technology continues to improve and we still have a lot of new things to discover about the quantum fluids and solids.

OTHER CONFERENCE PRESENTATIONS (TITLE ONLY)

Effects of the Gap Distortion on the Field Splitting of
Collective Modes in ^{3}He-B

Nils Schopohl

Abteilung fuer Theoretische Festkoerperphysik
Institut fuer Angwandte Physik
Universitaet Hamburg, Jungiusstr. 9
D-2000 Hamburg 36, West Germany

NMR Experiments on Spin Polarized Hydrogen

John Denker

Department of Physics, Cornell University
Ithaca, New York 14853

Neutron Scattering from Thermally-Excited
Quasi-particles in Superfluid ^{4}He

Allan Griffin

Department of Physics, University of Toronto
Toronto, Ontario M58-1A7, Canada

Surface Ferromagnetism in ^{3}He

Hans Bozler

Department of Physics, University of Southern California
Los Angeles, California 90089-0484

LIST OF PARTICIPANTS

Dr. Bernard Abraham
Argonne National Laboratory
9700 S. Cass Avenue, Bldg 223
Argonne, Illinois 60439

Dr. E. Dwight Adams
Department of Physics
University of Florida
Gainesville, Florida 32611

Mr. Glenn Agnolet
Bell Laboratories
600 Mountain Avenue
Murray Hill, New Jersey 07974

Dr. Sebastian Balibar
Laboratoire de Physique
E.N.S., 24 Rue Lhomond
75231 Paris Cedex 05, France

Mr. Robert F. Berg
Department of Physics
University of Florida
Gainesville, Florida 32611

Dr. Michel Bernier
DPHG/PSRM, CEA ORME des
MERISIERS, 91191 GIF/YVETTE,
CEDEX - FRANCE

Dr. Peter Berglund
Francis Bitter Natn'l Magnet Lab.
MIT
Building NW 14-1102a
Cambridge, Massachusetts 02139

Dr. D.S. Betts
University of Sussex, MAPS
Brighton, Sussex, BN1 9QH
Great Britain

Dr. Hans Bozler
Department of Physics
University of Southern California
Los Angeles, CA 90089-0484

Dr. Douglas F. Brewer
Physics Department
University of Sussex
Brighton BN1 9QH, England

Dr. Wolfgang Buck
Physik.-Techn. Bundesanstalt
Institut Berlin, Abbestrabel 2-12
D - 1000 Berlin, 10 W. Germany

Dr. Laurence J. Campbell
Los Alamos National Laboratory
B-262
Los Alamos, New Mexico 87545

Dr. Maurice Chapellier
Bat 510 Physique des Solides
Fac des Sciences d'ORSAY
91405 ORSAY, France

Mr. Harald Chocholass
KFA Juelich, IFF, Postfach 1913
D-5170 Juelich 1, West Germany

Dr. Roland Combescot
GPSENS 24 Rue Lhomond
75005 Paris, France

Mr. Benjamin C. Crooker
Cornell University
Physics Department
Ithaca, NY 14853

Dr. Michael Cross
Bell Laboratories
600 Mountain Avenue
Murray Hill, New Jersey 07974

Dr. Arnold Dahm
Physics Department
Case Western Reserve University
Cleveland, Ohio 44106

Mr. John Denker
G-3 Clark Hall
Cornell University
Ithaca, New York 14853

Dr. E. Roland Dobbs
Bedford College
Regent's Park, London, NW1 4NS
England

Mr. Hugh Doherty
EG&G Princeton Applied Research
P. O. Box 2565
Princeton, New Jersey 08540

Dr. David O. Edwards
Ohio State University
Physics Department
Columbus, Ohio 43210

Dr. Jim Eisenstein
Williams College
Department of Physics
Williamstown, MA 01267

Mr. Brad N. Engel
University of Florida
Department of Physics
Gainesville, Florida 32611

Dr. Georg Eska
Walther Meissner Institut fur
Tieftemperaturforschung
D-8046 Garching, W. Germany

Mr. Bob Fagaly
S.H.E. Corporation
4174 Sorrento Valley Blvd.
San Diego, California 92121

Mr. Qung Feng
University of Florida
Deparmtent of Physics
Gainesville, Florida 32611

Dr. Alexander L. Fetter
Physics Department
Stanford University
Stanford, California 94305

Dr. Jacques Flouquet
CNRS, BP 166, Centre de Tri
38042 Grenoble, Cedex, France

Dr. Juergen Franck
Department of Physics
University of Alberta
Edmonton T6G 2G1, Canada

Dr. Giorgio Frossati
Galgewater 21
2311 VZ Leiden
The Netherlands

Dr. James R. Gaines
2520 Wimbledon Road
Columbus, Ohio 43220

Dr. Henry R. Glyde
University of Delaware
Physics Department
Newark, Delaware 19711

Dr. Christopher M. Gould
Department of Physics - SSC303
University of Southern California
Los Angeles, CA 90089-0484

Dr. Dennis Greywall
Bell Laboratories
Room 1D152
Murray Hill, New Jersey 07974

Dr. Allan Griffin
Department of Physics
University of Toronto
Toronto, Ontario M58-1A7, Canada

Mr. Francis Guillon
Department of Physics and
     Astronomy
Northwestern University
Evanston, Illinois 60201

Dr. R.A. Guyer
Department of Physics
University of Massachusetts
Amherst, Massachusetts 01003

Mr. Gregory E. Haas
Department of Physics
University of Florida
Gainesville, Florida 32611

Dr. David G. Haase
Physics Department
North Carolina State University
Raleigh, North Carolina 27650

Dr. Taisto Haavasoja
Bell Laboratories
600 Mountain Avenue
Murray Hill, New Jersey 07974

Mr. Pertti Hakonen
Low Temperature Laboratory
Helsinki University of Technology
02150 Espoo 15, Finland

Dr. Henry E. Hall
LASSP, Clark Hall
Cornell University
Ithaca, New York 14853

Mr. P. Chris Hammel
Physics Department
Cornell University
Ithaca, New York 14853

Mr. Stephen I. Hernadi
Physics Department
University of Ottawa
Ottawa, Ontario, Canada K1N 6N5

Dr. J.P. Harrison
Physics Department
Queen's University
Kingston, Ontario, K7L 3N6
Canada

Dr. J.H. Hetherington
Physics Department
Michigan State University
East Lansing, Michigan

Dr. John Hook
Physics Department
Manchester University
Manchester M13 9PL, England

Dr. J. Hutchins
Physics Department
University of Sussex
Brighton, Sussex, England

Dr. Gary G. Ihas
Department of Physics
University of Florida
Gainesville, Florida 32611

Dr. John B. Ketterson
Department of Physics
Tech. Institute
Northwestern University
Evanston, Illinois 60201

Dr. Wiley Kirk
Texas A&M University
Physics Department
College Station, Texas 77843

Dr. Marjorie Klenin
North Carolina State University
Department of Physics
Box 5367
Raleigh, North Carolina 27650

Dr. Eckhard Krotscheck
ITP - Ellison Hall
University of California
Santa Barbara, California 83106

Dr. Matti Krusius
Low Temperature Laboratory
Helsinki University of Technology
02150 Espoo 15, Finland

Dr. Pradeep Kumar
Department of Physics
University of Florida
Gainesville, Florida 32611

Dr. Michele LeDuc
Laboratoire de Spectroscopie
Hertzienne de I'E.N.S.
24 Rue Lhomond
75231 Paris Cedex 05, France

Dr. David M. Lee
Department of Physics
Cornell University
Ithaca, New York 14853

Dr. Anthony J. Leggett
LASSP, Clark Hall
Cornell University
Ithaca, New York 14853

Dr. Donald Liebenberg
National Science Foundation
Low Temperature Physics Program
Washington, D.C. 20550

Ms. Ying Lin
University of Florida
Department of Physics
Gainesville, Florida 32611

Dr. Olli V. Lounasmaa
Low Temperature Laboratory
Helsinki University of Technology
SF-02150 Espoo 15, Finland

Dr. Per-Olov Lowdin
Quantum Theory Project
University of Florida
Gainesville, Florida 32611

Mr. Yoshiteru Maeno
Los Alamos National Laboratory
P-10 M764
Los Alamos, New Mexico 87545

Dr. Kazumi Maki
Department of Physics
University of Southern California
Los Angeles, CA 90089-0484

Dr. Takayoshi Mamiya
LASSP, Clark Hall
Cornell University
Ithaca, New York 14853

Dr. David B. Mast
Physics Department
Case Western Reserve University
Cleveland, Ohio 44106

Dr. Yoshika Masuda
Department of Physics
Nagoya University
Chikusa-ku, Nagoya 464, Japan

Naota Masuhara
Department of Physics
Ohio State University
Columbus, Ohio 43210

Mr. Mike Mayberry
555 Pierce #103
Albany, California 94706

Mr. Mark W. Meisel
Dept of Physics and Astronomy
Northwestern University
Evanston, Illinois 60201

Mr. Roger Mitchell
Schuster Laboratory
Brunswick Street
Manchester, England

Dr. Yukio Morii
Solid State Lab. III
Jap Atom Energy Res Inst
Tokai-mura, Ibaraki-ken 319-11
JAPAN

Dr. Robert M. Mueller
KFA/IFF, Postfach 1913
D5170 Juelich, West Germany

Siddhartha Mukherjee
Department of Physics
Ohio State University
Columbus, Ohio 43210

Dr. Yasukage Oda
Department of Physics
Ohio State University
Columbus, Ohio 43210

Mr. Zbigniew Olejniczak
Department of Physics
Texas A&M University
College Station, Texas 77843

Dr. John R. Owers-Bradley
IFF/KFA Julich
5170 - Julich, W. Germany

Mr. Jukka Pekola
Low Temperature Laboratory
Helsinki University of Technology
SF-02150 Espoo 15, Finland

Dr. Norman E. Phillips
College of Chemistry
University of California
Berkeley, California 94720

Mr. Simon Phillpot
Department of Physics
University of Florida
Gainesville, Florida 32611

Dr. George R. Pickett
Department of Physics
University of Lancaster
Lancaster LA1 4YB, England

Dr. Frank Pobell
Kernforschungsaulage
Postfach 1913, D-5170 Juelich
West Germany

Dr. Robert V. Pound
Lyman Laboratory of Physics
Harvard University
Cambridge, Massachusetts 02138

Dr. Dierk Rainer
Universitat Bayreuth
Theorectische Physik III
8580 Bayreuth, West Germany

Dr. John Reppy
Cornell University
Department of Physics
Ithaca, New York 14853

Dr. Robert C. Richardson
Department of Physics
Cornell University
Ithaca, New York 14853

Dr. Pat Roach
Argonne National Laboratory
9700 S. Cass Avenue, Bldg. 223
Argonne, Illinois 60439

Dr. Ralph Rosenbaum
Tel Aviv-University
Dept. Phys. Astron.
Ramat-Aviv, ISRAEL

Dr. William F. Saam
Physics Department
Ohio State University
Columbus, Ohio 43210

Dr. A. Sachrajda
Physics Department
Queen's University
Kingston, Ontario, K7L 3N6
Canada

Dr. Martti Salomma
Low Temperature Laboratory
Helsinki University of Technology
SF-02150 Espoo 15, Finland

Mr. Vijay Samalam
Department of Physics
University of Florida
Gainesville Florida 32611

Dr. Bimal K. Sarma
Argonne National Laboratory
9700 S. Cass Avenue, Bldg. 223
Argonne, Illinois 60439

Dr. James A. Sauls
Jadwin Hall
Princeton University
Princeton, New Jersey 08540

Dr. John Saunders
Physics Department
University of Sussex
Brighton BN1 9QH, England

Dr. Nils Schopohl
Abt. f. Theor. Festk. Physik
Inst. f. Ang. Physik
University, Hamburg, Jungiusstr 9
D-2000 Hamburg 36, W. Germany

Dr. Joseph Serene
Applied Physics Becton Center
2157 Yale Station
New Haven, Connecticut 06520

Dr. T. Shigi
Faculty of Science
Osaka City University
Sugimoto, Sumiuoshi-ku
Osaka 558, Japan

Mr. B.S. Shivaram
Department of Physics
Northwestern University
Evanston, Illinois 60201

Dr. Ralph O. Simmons
Physics Department
University of Illinois
Urbana, Illinois 61801

Dr. Kurt Skold
Argonne National Laboratory
9700 S. Cass Avenue, Bldg. 223
Argonne, Illinois 50439

Dr. Eric Smith
Physics Department
Cornell University
Ithaca, New York 14853

Dr. Paul Sokol
University of Illinois
Department of Physics
Urbana, Illinois 61808

Mr. Greg Spencer
Department of Physics
University of Florida
Gainesville, Florida 32611

Mr. Deepak Srivastava
University of Florida
Department of Physics
Gainesville, Florida 32611

Mr. Mark A. Stan
6636 Aintree Park Dr. #203
Mayfield Village, Ohio 44143

Dr. Victor Steinberg
Department of Physics
University of California
Santa Barbara, California 93106

Dr. Neil Sullivan
Department of Physics
University of Florida
Gainesville, Florida 32611

Dr. David Tanner
Department of Physics
University of Florida
Gainesville, Florida 32611

Dr. Daniel Thoulouze
C.R.T.B.T, Avenue des Martyrs
CNRS, BP 166 X
38042 Grenoble Cedex, France

Dr. Sam B. Trickey
Department of Physics
University of Florida
Gainesville, Florida 32611

Dr. Kurt Uhlig
Department of Physics
University of Florida
Gainesville, Florida 32611

Dr. Craig T. Van Degrift
National Bureau of Standards
Building 221 Room B128
Washington, D.C. 20234

Mr. Geraro Vermeullen
Rijnsburgerweg 19
2334 BR Leiden, The Netherlands

Ms. Mary Louise Vrtis
Argonne National Laboratory
9700 S. Cass Avenue, Bldg. 223
Argonne, Illinois 60439

Dr. Richard Webb
IBM Thomas J. Watson
    Research Center
P. O. Box 218
Yorktown Heights, NY 10598

Dr. John Wheatley
Group P-10, Mail Stop M764
Los Alamos National Laboratory
Los Alamos, New Mexico 87545

Mr. Douglas Wildes
Department of Physics
Cornell University
Ithaca, New York 14853

Dr. Peter Wolfle
Physik-Department
Technische Universitat Munchen
D-8046 Garching, W. Germany

Mr. Dawei Zhou
University of Florida
Department of Physics
Gainesville, Florida 32611

AIP Conference Proceedings

		L.C. Number	ISBN
No.1	Feedback and Dynamic Control of Plasmas	70-141596	0-88318-100-2
No.2	Particles and Fields - 1971 (Rochester)	71-184662	0-88318-101-0
No.3	Thermal Expansion - 1971 (Corning)	72-76970	0-88318-102-9
No.4	Superconductivity in d-and f-Band Metals (Rochester, 1971)	74-18879	0-88318-103-7
No.5	Magnetism and Magnetic Materials - 1971 (2 parts) (Chicago)	59-2468	0-88318-104-5
No.6	Particle Physics (Irvine, 1971)	72-81239	0-88318-105-3
No.7	Exploring the History of Nuclear Physics	72-81883	0-88318-106-1
No.8	Experimental Meson Spectroscopy - 1972	72-88226	0-88318-107-X
No.9	Cyclotrons - 1972 (Vancouver)	72-92798	0-88318-108-8
No.10	Magnetism and Magnetic Materials - 1972	72-623469	0-88318-109-6
No.11	Transport Phenomena - 1973 (Brown University Conference)	73-80682	0-88318-110-X
No.12	Experiments on High Energy Particle Collisions - 1973 (Vanderbilt Conference)	73-81705	0-88318-111-8
No.13	π-π Scattering - 1973 (Tallahassee Conference)	73-81704	0-88318-112-6
No.14	Particles and Fields - 1973 (APS/DPF Berkeley)	73-91923	0-88318-113-4
No.15	High Energy Collisions - 1973 (Stony Brook)	73-92324	0-88318-114-2
No.16	Causality and Physical Theories (Wayne State University, 1973)	73-93420	0-88318-115-0
No.17	Thermal Expansion - 1973 (lake of the Ozarks)	73-94415	0-88318-116-9
No.18	Magnetism and Magnetic Materials - 1973 (2 parts) (Boston)	59-2468	0-88318-117-7
No.19	Physics and the Energy Problem - 1974 (APS Chicago)	73-94416	0-88318-118-5
No.20	Tetrahedrally Bonded Amorphous Semiconductors (Yorktown Heights, 1974)	74-80145	0-88318-119-3
No.21	Experimental Meson Spectroscopy - 1974 (Boston)	74-82628	0-88318-120-7
No.22	Neutrinos - 1974 (Philadelphia)	74-82413	0-88318-121-5
No.23	Particles and Fields - 1974 (APS/DPF Williamsburg)	74-27575	0-88318-122-3
No.24	Magnetism and Magnetic Materials - 1974 (20th Annual Conference, San Francisco)	75-2647	0-88318-123-1
No.25	Efficient Use of Energy (The APS Studies on the Technical Aspects of the More Efficient Use of Energy)	75-18227	0-88318-124-X

AIP Conference Proceedings

No.50	Laser-Solid Interactions and Laser Processing - 1978 (Boston)	79-51564	0-88318-149-5
No.51	High Energy Physics with Polarized Beams and Polarized Targets (Argonne, 1978)	79-64565	0-88318-150-9
No.52	Long-Distance Neutrino Detection - 1978 (C.L. Cowan Memorial Symposium)	79-52078	0-88318-151-7
No.53	Modulated Structures - 1979 (Kailua Kona, Hawaii)	79-53846	0-88318-152-5
No.54	Meson-Nuclear Physics - 1979 (Houston)	79-53978	0-88318-153-3
No.55	Quantum Chromodynamics (La Jolla, 1978)	79-54969	0-88318-154-1
No.56	Particle Acceleration Mechanisms in Astrophysics (La Jolla, 1979)	79-55844	0-88318-155-X
No. 57	Nonlinear Dynamics and the Beam-Beam Interaction (Brookhaven, 1979)	79-57341	0-88318-156-8
No. 58	Inhomogeneous Superconductors - 1979 (Berkeley Springs, W.V.)	79-57620	0-88318-157-6
No. 59	Particles and Fields - 1979 (APS/DPF Montreal)	80-66631	0-88318-158-4
No. 60	History of the ZGS (Argonne, 1979)	80-67694	0-88318-159-2
No. 61	Aspects of the Kinetics and Dynamics of Surface Reactions (La Jolla Institute, 1979)	80-68004	0-88318-160-6
No. 62	High Energy e^+e^- Interactions (Vanderbilt , 1980)	80-53377	0-88318-161-4
No. 63	Supernovae Spectra (La Jolla, 1980)	80-70019	0-88318-162-2
No. 64	Laboratory EXAFS Facilities - 1980 (Univ. of Washington)	80-70579	0-88318-163-0
No. 65	Optics in Four Dimensions - 1980 (ICO, Ensenada)	80-70771	0-88318-164-9
No. 66	Physics in the Automotive Industry - 1980 (APS/AAPT Topical Conference)	80-70987	0-88318-165-7
No. 67	Experimental Meson Spectroscopy - 1980 (Sixth International Conference , Brookhaven)	80-71123	0-88318-166-5
No. 68	High Energy Physics - 1980 (XX International Conference, Madison)	81-65032	0-88318-167-3
No. 69	Polarization Phenomena in Nuclear Physics - 1980 (Fifth International Symposium, Santa Fe)	81-65107	0-88318-168-1
No. 70	Chemistry and Physics of Coal Utilization - 1980 (APS, Morgantown)	81-65106	0-88318-169-X
No. 71	Group Theory and its Applications in Physics - 1980 (Latin American School of Physics, Mexico City)	81-66132	0-88318-170-3
No. 72	Weak Interactions as a Probe of Unification (Virginia Polytechnic Institute - 1980)	81-67184	0-88318-171-1
No. 73	Tetrahedrally Bonded Amorphous Semiconductors (Carefree, Arizona, 1981)	81-67419	0-88318-172-X
No.74	Perturbative Quantum Chromodynamics (Tallahassee, 1981)	81-70372	0-88318-173-8

No. 75	Low Energy X-ray Diagnostics—1981 (Monterey)	81-69841	0-88318-174-6
No. 76	Nonlinear Properties of Internal Waves (La Jolla Institute, 1981)	81-71062	0-88318-175-4
No. 77	Gamma Ray Transients and Related Astrophysical Phenomena (La Jolla Institute, 1981)	81-71543	0-88318-176-2
No. 78	Shock Waves in Condensed Matter – 1981 (Menlo Park)	82-70014	0-88318-177-0
No. 79	Pion Production and Absorption in Nuclei – 1981 (Indiana University Cyclotron Facility)	82-70678	0-88318-178-9
No. 80	Polarized Proton Ion Sources (Ann Arbor, 1981)	82-71025	0-88318-179-7
No. 81	Particles and Fields – 1981: Testing the Standard Model (APS/DPF, Santa Cruz)	82-71156	0-88318-180-0
No. 82	Interpretation of Climate and Photochemical Models, Ozone and Temperature Measurements (La Jolla Institute, 1981)	82-071345	0-88318-181-9
No. 83	The Galactic Center (Cal. Inst. of Tech., 1982)	82-071635	0-88318-182-7
No. 84	Physics in the Steel Industry (APS.AISI, Lehigh University, 1981)	82-072033	0-88318-183-5
No. 85	Proton-Antiproton Collider Physics – 1981 (Madison, Wisconsin)	82-072141	0-88318-184-3
No. 86	Momentum Wave Functions – 1982 (Adelaide, Australia)	82-072375	0-88318-185-1
No. 87	Physics of High Energy Particle Accelerators (Fermilab Summer School, 1981)	82-072421	0-88318-186-X
No. 88	Mathematical Methods in Hydrodynamics and Integrability in Dynamical Systems (La Jolla Institute, 1981)	82-072462	0-88318-187-8
No. 89	Neutron Scattering – 1981 (Argonne National Laboratory)	82-073094	0-88318-188-6
No. 90	Laser Techniques for Extreme Ultraviolt Spectroscopy (Boulder, 1982)	82-073205	0-88318-189-4
No. 91	Laser Acceleration of Particles (Los Alamos, 1982)	82-073361	0-88318-190-8
No. 92	The State of Particle Accelerators and High Energy Physics (Fermilab, 1981)	82-073861	0-88318-191-6
No. 93	Novel Results in Particle Physics (Vanderbilt, 1982)	82-73954	0-88318-192-4
No. 94	X-Ray and Atomic Inner-Shell Physics-1982 (International Conference, U. of Oregon)	82-74075	0-88318-193-2
No. 95	High Energy Spin Physics – 1982 (Brookhaven National Laboratory)	83-70154	0-88318-194-0
No. 96	Science Underground (Los Alamos, 1982)	83-70377	0-88318-195-9

No. 97	The Interaction Between Medium Energy Nucleons in Nuclei-1982 (Indiana University)	83-70649	0-88318-196-7
No. 98	Particles and Fields - 1982 (APS/DPF University of Maryland)	83-70807	0-88318-197-5
No. 99	Neutrino Mass and Gauge Structure of Weak Interactions (Telemark, 1982)	83-71072	0-88318-198-3
No. 100	Excimer Lasers - 1983 (OSA, Lake Tahoe, Nevada)	83-71437	0-88318-199-1
No. 101	Positron-Electron Pairs in Astrophysics (Goddard Space Flight Center, 1983)	83-71926	0-88318-200-9
No. 102	Intense Medium Energy Sources of Strangeness (UC-Santa Cruz, 1983)	83-72261	0-88318-201-7
No. 103	Quantum Fluids and Solids - 1983 (Sanibel Island, Florida)	83-72440	0-88318-202-5